WHAT WENT WRONG?

FIFTH EDITION

WHAT WENT WRONG?

Case Histories of Process Plant Disasters and How They Could Have Been Avoided

FIFTH EDITION

TREVOR KLETZ

AMSTERDAM • BOSTON • HEIDELBERG • LONDON
NEW YORK • OXFORD • PARIS • SAN DIEGO
SAN FRANCISCO • SINGAPORE • SYDNEY • TOKYO
Gulf Professional Publishing is an imprint of Elsevier

Gulf Professional Publishing is an imprint of Elsevier
30 Corporate Drive, Suite 400, Burlington, MA 01803, USA
Linacre House, Jordan Hill, Oxford OX2 82P

Library of Congress Cataloging-in-Publication Data
Kletz, Trevor A.
    What went wrong? : case histories of process plant
disasters and how they could have been avoided /
Trevor Kletz. —5th ed.
        p. cm.
    Includes bibliographical references and index.
    ISBN 978-1-85617-531-9 (hardcover : alk. paper) 1.
Chemical plants—Accidents.   I. Title.
    TP155.5.K54 2009
    363.11'966—dc22

                                                                2009011194

British Library Cataloguing in Publication Data
A catalogue record for this book is available from the British Library

ISBN 13: 978-1-85617-531-9

For all information on all Elsevier Gulf Professional Publishing
publications visit our Web site at www.elsevierdirect.com

Printed in the United States of America

Transferred to Digital Printing in 2013

Working together to grow
libraries in developing countries

www.elsevier.com | www.bookaid.org | www.sabre.org

ELSEVIER   BOOK AID
            International   Sabre Foundation

To Denise
Who waited while I
"scorned delights and lived laborious days"
but never saw the results.

Only that shall happen
Which has happened,
Only that occur
Which has occurred;
There is nothing new
Beneath the sun. . .
For as wisdom grows, vexation grows;
To increase learning is to increase heartache.

—Ecclesiastes 1: 9 and 18

Contents

PART A
WHAT WENT WRONG?

PART B

STILL GOING WRONG

Acknowledgments

Thanks are due to the staffs of the companies where the incidents occurred for allowing me to describe their mistakes; to many colleagues, past and present, especially to Professor F. P. Lees for his ideas and advice; and to the U.K. Science and Engineering Research Council and the Leverhulme Trust for their financial support of the first edition.

Preface

In 1968, after many years' experience in plant operations, I was appointed safety adviser to the heavy organic chemicals division (later the petrochemicals division) of Imperial Chemical Industries. My appointment followed a number of serious fires in the 1960s, and therefore I was mainly concerned with process hazards rather than those of a mechanical nature. Today I would be called a process safety adviser.

One of my tasks was to pass on to design and operating staff details of accidents that had occurred and the lessons that should be learned. This book contains a selection of the reports I collected from many different companies, as well as many later reports. Although most have been published before, they were scattered among many different publications, some with small circulations.

The purpose here is to show what has gone wrong in the past and to suggest how similar incidents might be prevented in the future. Unfortunately, the history of the process industries shows that many incidents are repeated after a lapse of a few years. People move on, and the lessons are forgotten. This book will help keep the memories alive.

The advice is given in good faith but without warranty. Readers should satisfy themselves that it applies to their circumstances. In fact, you may feel that some of my recommendations are not appropriate for your company. Fair enough, but if the incidents could occur in your company, and you do not wish to adopt my advice, then please do something else instead. But do not ignore the incidents.

To quote the advice of John Bunyan, written more than 300 years ago,

What of my dross thou findest there, be bold
To throw away, but yet preserve the gold.
What if my gold be wrapped up in ore?
None throws away the apple for the core:
But if thou shalt cast all away as vain ...

You have been warned what will happen.

You may believe that the accidents could not happen at your plant because you have systems to prevent them. Are you are sure that they are always followed, everywhere, all the time? Perhaps they are followed most of the time but someone turns a blind eye when a job is urgent. Also remember that systems have limitations. All they can do is make the most of people's knowledge and experience by applying them in a systematic way. If people lack knowledge and experience, the systems are empty shells.

Many of the accidents I describe occurred in plants that had such systems, but the systems were not always followed. The accidents happened because of various management failures: failure to convince people that they should follow the systems, failure to detect previous violations (by audits, spot checks, or just keeping an open eye), or deliberately turning a blind eye to avoid conflict or to get a job done quickly. The first step down the road to many a serious accident occurred when someone turned a blind eye to a missing blind (see Chapter 1).

The incidents described could occur in many different types of plants and are therefore of widespread interest. Some of them illustrate the hazards involved in activities such as preparing equipment for maintenance and modifying plants. Others illustrate the hazards associated with widely used equipment, such as storage tanks and hoses, and with that universal component of all plants and processes: people. Other incidents illustrate the need for techniques, such as hazard and operability studies, and protective devices, such as emergency isolation valves.

You will notice that most of the incidents are very simple. No esoteric knowledge or detailed study was required to prevent them—only a knowledge of what had happened before, which this book provides.

Only a few incidents started with the sudden failure of a major component. Most started with a flaw in a minor component, an instrument that was out of order or not believed, a poor procedure, or a failure to follow procedures or good engineering practice. *For want of a nail, a kingdom was lost.*

Many of the incidents described could be discussed under more than one heading. Therefore, cross-references have been included.

If an incident that happened in your plant is described, you may notice that one or two details have been changed. Sometimes this has been done to make it harder for people to tell where the incident occurred. Sometimes this has been done to make a complicated story simpler but without affecting the essential message. Sometimes—and this is the most likely reason—the incident did not happen in your plant at all. Another plant had a similar incident.

Many of the incidents did not actually result in death, serious injury, or serious damage—they were so-called near misses, although they were really near accidents. But they could have had much more serious consequences. We should learn from these near misses, as well as from incidents that had serious results.

Most of the incidents described occurred at so-called major hazard plants or storage installations—that is, those containing large quantities of flammable, explosive, or toxic chemicals. The lessons learned apply particularly to such plants. However, most of the incidents could have occurred at plants handling smaller quantities of materials or less hazardous materials, and the consequences, though less serious, would be serious enough. At a major-hazard plant, opening up a pump that is not isolated could cause (and has caused) a major fire or explosion. At other plants, this would cause a smaller fire or a release of corrosive chemicals—still enough to kill or injure the employee on the job. Even if the contents of the plant are harmless, there is still a waste of materials. The lessons to be learned therefore apply throughout the process industries.

For the second edition of this book, I added more incidents, extended the sections on Bhopal and Mexico City, and added chapters on some little-known but quite common hazards and on accidents in computer-controlled plants.

For the third edition, I added sections or chapters on heat exchangers, furnaces, inherently safer design, and runaway reactions, and extended many other chapters. Although I have read many accident reports since the first edition appeared, most have merely reinforced the messages of the book, and I added only those incidents that tell us something new.

For the fourth edition, I added further incidents to every chapter.

For the fifth edition, Part A of this book, changes have been minor. A supplement to *What Went Wrong?* called *Still Going Wrong* was published in 2003. It is reprinted as Part B, and many reports on incidents that have occurred since then or become available since then have been added.

There is, however, one difference between Parts A and B. In Part A I emphasized the immediate technical causes of the accidents and the changes in design and methods of working needed to prevent them from happening again. In Part B I have, whenever possible, discussed also the underlying weaknesses in the management systems. It is not possible to do this in every case, as the information is not always available. Too many reports still describe only the immediate technical causes. I do not blame their authors for this. Most of them are close to the "coal-face." They want to solve the immediate technical problems and get the plant back on line in a safe manner as soon as they can, so they concentrate on the immediate technical causes. More senior people, before approving the reports, should look for the underlying weaknesses that result in poor designs, poor methods of working, failures to learn from the past, tendencies to blame people who make occasional but inevitable errors, and so on. They should also see that changes that cannot be made on the existing plants are fed back to the design organizations, both in-house and contractors, for use in the future. Because of this difference in approach, I have not merged the contents of the two original books but left them as they were. There are therefore chapters in Parts A and B with the same or similar titles.

In Part A, some of the chapters covered different types of equipment, whereas others covered procedures such as maintenance or modifications. In Part B, most of the chapters cover procedures, but a number of reports on explosions and leaks are collected under these headings. This part also emphasizes the multiple causes of accidents. As a result, the accidents described in the chapter on the management of change, for example, also have other causes, whereas some incidents in other chapters also involve the management of change. Similarly, several scattered reports show that some accidents cannot be prevented by more detailed instructions but only by giving people a better understanding of the process. This makes the allocation of incidents to chapters rather arbitrary, so I have included many cross-references and a comprehensive index.

Missing from this book is a chapter on human error. This is because all accidents are due to human error. Someone, usually a manager or supervisor, has to decide what to do; someone, often a designer, has to decide how to do it; someone, usually an operator, has to do it. All may make errors, of different sorts, for different reasons. Human errors are too diverse to be treated as a single group, and I find it useful to classify them as follows:

Mistakes. They occur because someone does not know what to do. To prevent them, we need better training or instructions or changes to the plant design or method of working so that the task is easier.

Violations or noncompliance. They occur because someone decides not to follow instructions or recognized good practice. They are the only sort of error for which blame might be justified. It is not justified if the supervisors had turned a blind eye or if the violations had been going on for a long period and, unknown to the supervisors, had become custom and practice. Many violations are made with good intentions, and if the instructions are wrong a violation may prevent an accident. To prevent violations, we should explain the reason for instructions, carry out occasional checks to make sure they are being followed, and never turn a blind eye when they are not. We should also, when possible, simplify procedures that are difficult to follow.

Mismatches. The job is beyond the mental or physical ability of the person asked to do it, perhaps beyond anyone's ability. We should change the design or method of working.

Slips and lapses of attention. These are inevitable from time to time, so we should change designs or methods of working so as to remove opportunities for human error.

This classification is discussed in more detail in Chapter 38. An underlying principle behind this book is that whenever possible we should remove situations that are error-prone rather than expect people to never make errors. There is more about human error in my book *An Engineer's*

View of Human Error, 3rd edition (Institution of Chemical Engineers, Rugby, United Kingdom, 2001).

Another feature illustrated in almost every chapter is the way that the lessons of past accidents are soon forgotten (or never learned) and the accidents are allowed to happen again (see Section 16.10).

As in my monthly safety newsletters, I have preserved the anonymity of the companies where the accidents occurred, except when this is mentioned in the title of a published report. When no reference is cited, the information came from a private communication or my own experience.

The book is intended for all who work in industry (especially the chemical, oil, and other process industries) and are involved in production, maintenance, or design at any level. It is not just a book for safety professionals; it is intended for all who design, operate, or maintain plants, especially process plants, including people at the highest levels. Many of them are not chemical engineers but mechanical or other engineers, chemists, or other scientists. Often the only safety information senior managers or directors get is a periodic summary. They should sample the detail.

If you decide to recommend a course of action, try not to yield to pressure, obstacles, complacency, or example, but do yield to sound technical arguments. I used to tell my safety colleagues in industry that a job wasn't finished when they gave their advice. It was not finished until their advice was followed or they were convinced by technical arguments that it should be changed.

In science it is permissible to say that we do not know the answer to a problem, but this is not possible in plant design and operation. We have to make a decision even though the evidence is conflicting. Also, to quote David Pye, "It is quite impossible for any design to be the logical outcome of the requirements simply because the requirements being in conflict, their logical outcome is an impossibility." Information on what has gone wrong in the past can help us find the best balance between these conflicting requirements.

Many of the incidents I describe did not have serious results. By good fortune no one was killed or injured and damage was slight. For example, a leak of flammable liquid did not ignite or corrosion was spotted in time. Do not ignore these incidents. Next time you may not be so lucky.

In the following pages, I criticize the performance of some organizations. However, I am not suggesting that they neglected safety to save money. A few may do so, but the vast majority did not. Most accidents occur because the people in charge did not see the hazards (what could occur) or underestimated the risk (the probability that it will occur), because they did not know what more could be done to remove the hazards or reduce the risk, or because they allowed standards of performance to slip, all common human failings. Cock-ups are far more common than conspiracies to cut costs.

Since the first edition of Part A was published in 1985, the press and public have become more likely to look for someone to blame when something goes wrong (see Section 34.9). The old legal principle of "no liability without fault" is being replaced with "those in charge should pay compensation whether or not they were negligent." This increases the pressure for better safety, but it also makes some companies reluctant to publish all the facts, even internally, so that others can learn from them. There may be no net gain. In describing well-known accidents such as Bhopal (see Section 21.1), I have emphasized features that most other writers have minimized or ignored.

Most of the incidents described were the result of not following good engineering practice. Some violated the law, and many more would if they occurred today. In the United States, they would violate OSHA 1910.147 (1990) on the Control of Hazardous Energy (Lock Out/Tag Out) and the Process Safety Management (PSM) Law (OSHA 1910.119, in force since 1992), which applies to listed chemicals above a threshold quantity. The PSM law requires companies to follow good engineering practice, codes, industry consensus standards, and even the company's own standards. OSHA could view failure to follow any of these as violations.

In the United Kingdom, the Health and Safety at Work Act (1974) and regulations made under it require "occupiers" to provide a safe plant and system of work and adequate instruction, training, and supervision. In the European Community, occupiers of major hazard sites are required to produce a "safety case," which describes how hazards have been assessed and are kept under control. Many other countries have similar legislation, though standards of enforcement vary.

As a result of OSHA 1910.119 and similar legislation, there has been a growth of interest in process safety management systems and publications on them. This is welcome, but we must not forget their limitations. Some managers seem to think a good system is all that is needed to ensure safety. However, all a system can do is harness the knowledge and experience of people. If knowledge and experience have been downsized away, the system is an empty shell. Knowledge and experience without a system will achieve less than their full potential. Without knowledge and experience, a system will achieve nothing. We are not going to prevent downsizing, but we can ensure that the lessons of the past are not forgotten. The book tries to contribute to the achievement of that aim.

How to Use This Book

1. Read it right through. As you do so, ask yourself if the incidents could occur in *your* plant, and, if so, *write down* what you have done or intend to do to prevent them from occurring.

2. Use it as a deskside book on safety. Dip into it at odd moments or pick a subject for the staff meeting, the safety committee or bulletin, or the plant inspection.
3. Refer to it when you become interested in something new as the result of an incident, a change in responsibility, or a new problem in design. However, this book does not claim to comprehensively review process safety and loss prevention. For that, refer to *Lees' Loss Prevention in the Process Industries*, 3rd edition, edited by M. S. Mannan (Elsevier, 2005, three volumes).
4. Use the incidents to train new staff, managers, foremen, and operators so they know what will happen if they do not follow recognized procedures and good operating practice.
5. If you are a teacher, use the incidents to tell your students why accidents occur and to illustrate scientific principles.

In the training of both plant staff and students, the material can be used as lecture material or, better, as discussion material (those present discuss and agree among themselves what they think should be done to prevent similar incidents from happening again). The use of case histories in this way is discussed in my book, *Lessons from Disaster: How Organizations Have No Memory and Accidents Recur* (Institution of Chemical Engineers/Gulf Publishing Co., 1993, Chapter 10).
6. At a staff meeting, give each person an extract from the book and ask him or her to describe at the next meeting what has been or should be done to prevent a similar accident occurring in the plant or equipment that the individual designs, operates, or maintains.
7. If you want to be critical, send a copy of the book, open at the appropriate page, to people who have allowed one of the accidents described to happen again. They may read the book and avoid further unnecessary accidents.

A high price has been paid for the information in this book: many persons killed and billions of dollars worth of equipment destroyed. You get this information for the price of the book. It will be the best bargain you have ever received if you use the information to prevent similar incidents at your plant.

Trevor Kletz

This book will make a traveller of thee,
If by its counsel thou wilt ruled be.
It will direct thee to a safer land
If thou wilt its directions understand.

—Adapted from R. Vaughan Williams, libretto for
The Pilgrim's Progress

Units and Nomenclature

I have used units likely to be most familiar to the majority of my readers. Although I welcome the increasing use of SI units, many people still use imperial units—they are more familiar with a 1-in. pipe than a 25-mm pipe.

Short lengths are therefore quoted in inches but longer lengths in meters and feet:

$$1 \text{ in.} = 25.4 \text{ mm}$$

$$1 \text{m} = 3.28 \text{ ft or } 1.09 \text{ yd}$$

Volumes are quoted in cubic meters (m^3) and also in U.S. gallons:

$$1 \text{ m}^3 = 264 \text{ U.S. gallons}$$
$$= 220 \text{ imperial gallons}$$
$$= 35.3 \text{ ft}^3$$

A tank 30 ft tall by 40 ft diameter has a volume of 1,068 m^3 (280,000 U.S. gallons); a tank 15 ft tall by 20 ft diameter has a volume of 133 m^3 (35,250 U.S. gallons).

Masses are quoted in kilograms (kg) or tons:

$$1 \text{kg} = 2.20 \text{ lb}$$
$$1,000 \text{ kg} = 1 \text{ metric tonne} = 1.10 \text{ short (U.S.) tons}$$
$$= 0.98 \text{ long (U.K.) ton}$$

Temperatures are quoted in °C and °F.

Pressures are quoted in pounds force per square inch (psi) and also in bars. As it is not usual to refer to bar gauge, I have, for example, referred to "a gauge pressure of 90 psi (6 bar)," rather than "a pressure of 90 psig":

$$1 \text{ bar} = 14.50 \text{ psi}$$
$$= 1 \text{ atmosphere (atm)}$$
$$= 1 \text{ kg/cm}^2$$
$$= 100 \text{ kilopascals (kPa)}$$

Very small gauge pressures are quoted in inches water gauge, as this gives a picture:

$$1 \text{ in. water gauge} = 0.036 \text{ psi}$$
$$= 2.5 \times 10^{-3} \text{ bar}$$
$$= 0.2 \text{ kPa}$$

A NOTE ABOUT NOMENCLATURE

Different words are used, in different countries, to describe the same job or piece of equipment. Some of the principal differences between the United States and the United Kingdom are listed here. Within each country, however, there are differences between companies. More Britons understand U.S. terms than Americans understand British ones.

Management Terms

Job	United States	United Kingdom
Operator of plant	Operator	Process worker
Operator in charge of others	Lead operator	Chargehand, or assistant foreman or junior supervisor
Highest level normally reached by promotion from operator	Foreman	Foreman or supervisor
First level of professional management (usually in charge of a single unit)	Supervisor	Plant manager
Second level of professional management	Superintendent	Section or area manager
Senior manager in charge of site containing many units	Plant manager	Works or factory manager
Machine worker	Craftsman or mechanic	Fitter, electrician, and the like

The different meanings of the terms *supervisor* and *plant manager* in the United States and the United Kingdom should be noted.

In this book I have used the term *foreman* as it is understood in both countries, though its use in the United Kingdom is becoming outdated. *Manager* is used to describe any professionally qualified person in charge of a unit or group of units. That is, it includes people who, in many U.S. companies, would be described as supervisors or superintendents.

Certain items of plant equipment have different names in the two countries. Some common examples are as follows:

Chemical Engineering Terms

United States	United Kingdom
Accumulator	Reflux drum
Agitator	Mixer or stirrer
Air masks	Breathing apparatus (BA)
Blind	Slip-plate
Carrier	Refrigeration plant
Cascading effects	Knock-on (or domino effects)
Check valve	Nonreturn valve
Clogged (of filter)	Blinded
Consensus standard	Code of practice
Conservation vent	Pressure/vacuum valve
Dike, berm	Bund
Discharge valve	Delivery valve
Division (in electrical area classification)	Zone
Downspout	Downcomer
Expansion joint	Bellows
Explosion proof	Flameproof
Faucet	Tap
Fiberglass-reinforced plastic (FRP)	Glass-reinforced plastic (GRP)
Figure-8 plate	Spectacle plate
Flame arrestor	Flame trap
Flashlight	Torch
Fractionation	Distillation
Gasoline	Petrol
Gauging (of tanks)	Dipping
Generator	Dynamo or alternator
Ground	Earth
Horizontal cylindrical tank	Bullet
Hydro (Canada)	Electricity
Install	Fit

(Continued)

United States	United Kingdom
Insulation	Lagging
Interlock[*]	Trip[*]
Inventory	Stock
Lift-truck	Forklift truck
Loading rack	Gantry
Manway	Manhole
Mill water	Cooling water
Nozzle	Branch
OSHA (Occupational Safety and Health Administration)	Health and Safety Executive
Pedestal, pier	Plinth
Pipe diameter (internal)	Pipe bore
Pipe rack	Pipebridge
Plugged	Choked
Rent	Hire
Rupture disc or frangible	Bursting disc
Scrutinize	Vet
Seized (of a valve)	Stuck shut
Shutdown	Permanent shutdown
Sieve tray	Perforated plate
Siphon tube	Dip tube
Spade	Slip-plate
Sparger or sparge pump	Spray nozzle
Spigot	Tap
Spool piece	Bobbin piece
Stack	Chimney
Stator	Armature
Takeaway (from a meeting)	Outcome or conclusion[†]
Tank car	Rail tanker or rail tank wagon
Tank truck	Road tanker or road tank wagon
Torch	Cutting or welding torch
Tower	Column

United States	United Kingdom
Tow motor	Forklift truck
Tray	Plate
Turnaround	Shutdown
Utility hole	Manhole
Valve cheater	Wheel dog
Water seal	Lute
Wrench	Spanner
C-wrench	Adjustable spanner
Written note	Chit
$M	Thousand dollars
SMM	$M or million dollars
STP	60°F, 1 atmosphere
32°F, 1 atmosphere	STP
NTP	32°F, 1 atmosphere

*In the United Kingdom, *interlock* is used to describe a device that prevents someone opening one valve while another is open (or closed). *Trip* describes an automatic device that closes (or opens) a valve when a temperature, pressure, flow, and so on reach a preset value.
†In the United Kingdom, a takeaway is an outlet selling meals for consumption off the premises.

Firefighting Terms

United States	United Kingdom
Dry chemical	Dry powder
Dry powder	Dry powder for metal fires
Egress	Escape
Evolutions	Drills
Excelsior (for fire tests)	Wood wool
Fire classification:	
Class A: Solids	Class A: Solids
Class B: Liquids and gases	Class B: Liquids
Class C: Electrical	Class C: Gases
Class D: Metals	Class D: Metals
Fire stream	Jet

(Continued)

United States	United Kingdom
Nozzle	Branchpipe
Open butt	Hose without branchpipe
Rate density	Application rate
Siamese connection	Collecting breeching
Sprinkler systems:	
Branch pipe	Range pipe
Cross main	Distribution pipe
Feed main	Main distribution pipe
Standpipe	Dry riser
Tip	Nozzle
Wye connection	Dividing breeching

WHAT WENT WRONG?

1

Preparation for Maintenance

Mr. Randall (factory inspector) said he was surprised at the system of work, as he knew the company's safety documents were very impressive. Unfortunately they were not acted upon.

—Health and Safety at Work, April 1996

The following pages describe accidents that occurred because equipment was not adequately prepared for maintenance. Sometimes the equipment was not isolated from hazardous materials; sometimes it was not identified correctly and so the wrong equipment was opened up; sometimes hazardous materials were not removed [1, 2].*

Entry to vessels is discussed in Chapters 11 and 24.

1.1 ISOLATION

1.1.1 Failure to Isolate

A pump was being dismantled for repair. When the cover was removed, hot oil, above its auto-ignition temperature, came out and caught fire. Three men were killed, and the plant was destroyed. Examination of the

---

*End-of-chapter references are indicated by a number inside brackets. Items not referenced are private communications or based on the author's experience.

doi:10.1016/B978-1-85617-531-9.00001-9

wreckage after the fire showed that the pump suction valve was open and the drain valve shut [3].

The pump had been awaiting repair for several days when a permit-to-work was issued at 8 a.m. on the day of the fire. The foreman who issued the permit should have checked ahead of time that the pump suction and delivery valves were shut and the drain valve open. He claimed that he did so. Either his recollection was incorrect or, after he inspected the valves and before work started, someone closed the drain valve and opened the suction valve. When the valves were closed, there was no indication—on them—of *why* they were closed. An operator who was not aware that the pump was to be maintained might have opened the suction valve and shut the drain valve so that the pump could be put on line quickly if required.

A complicating factor was that the maintenance team originally intended to work only on the pump bearings. When team members found that they had to open up the pump, they told the process team, but no further checks of the isolations were carried out.

It was not customary in the company concerned to isolate equipment under repair by slip-plates, only by closed valves. But after the fire, the company introduced the following rules:

(a) Equipment under repair must be isolated by slip-plates (blinds or spades) or physical disconnection unless the job to be done will be so quick that fitting slip-plates (or disconnecting pipework) would take as long as the main job and be as hazardous. If hot work is to be carried out or a vessel is to be entered, then slip-plating or physical disconnection must always take place.

(b) Valves isolating equipment under maintenance, including valves that have to be closed while slip-plates are fitted (or pipework disconnected), must be locked shut with a padlock and chain or similar device. A notice fixed to the valve is not sufficient.

(c) For fluids at gauge pressures above 600 psi (40 bar) or at a temperature near or above the auto-ignition point, double block and bleed valves should be installed—not for use as main isolations but so that slip-plates can be inserted safely (Figure 1-1).

(d) If there is any change in the work to be done, the permit-to-work must be withdrawn and a new one issued.

A similar but more serious incident occurred in a polyethylene plant in 1989. A take-off branch was dismantled to clear a choke. The 8-in. valve isolating it from the reactor loop (the Demco valve in Figure 1-2) was open, and hot ethylene under pressure came out and exploded, killing 23 people, injuring more than 130, and causing extensive damage. Debris was thrown 10 km (6 miles), and the subsequent fire caused two liquefied petroleum gas tanks to burst.

TYPE A. FOR LOW-RISK FLUIDS

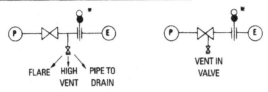

| SPADE POSITION FOR FLEXIBLE LINES | RING FOR RIGID LINES | SPECTACLE SPADE FOR LINES IN FREQUENT USE |

TYPE B. FOR HAZARDOUS FLUIDS WITH VENT TO CHECK ISOLATION

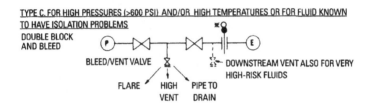

FLARE   HIGH   PIPE TO
        VENT   DRAIN

VENT IN VALVE

ALTERNATIVE DESTINATIONS ACCORDING TO HAZARD

TYPE C. FOR HIGH PRESSURES (>600 PSI) AND/OR HIGH TEMPERATURES OR FOR FLUID KNOWN
TO HAVE ISOLATION PROBLEMS

DOUBLE BLOCK
AND BLEED

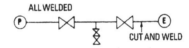

BLEED/VENT VALVE

DOWNSTREAM VENT ALSO FOR VERY
HIGH-RISK FLUIDS

FLARE   HIGH   PIPE TO
        VENT   DRAIN

TYPE D. FOR STEAM ABOVE 600 PSI

ALL WELDED

CUT AND WELD

E = EQUIPMENT UNDER MAINTENANCE
P = PLANT UP TO PRESSURE
* = OR SPADE OR RING AS REQUIRED

FIGURE 1-1   Summary of isolation methods.

The valve was operated by compressed air, and the two air hoses, one to open the valve and one to close it, were connected up the wrong way around. The two connectors should have been different in size or design so that this could not occur. In addition, they were not disconnected, and a lockout device on the valve—a mechanical stop—had been removed. It is also bad practice to carry out work on equipment isolated from hot flammable gas under pressure by a single isolation valve. The take-off branch should have been slip-plated, and double block and bleed valves should have been provided so the slip-plate could be inserted safely (Figure 1-1) [16, 17].

There was another similarity to the first incident. In this case, the equipment also had been prepared for repair and then had to wait for

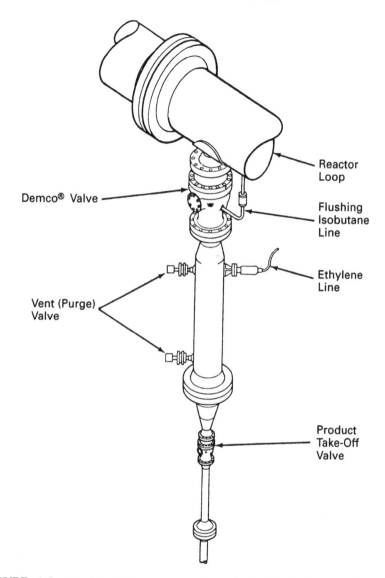

Reactor
Loop

Demco® Valve

Flushing
Isobutane
Line

Ethylene
Line

Vent (Purge)
Valve

Product
Take-Off
Valve

FIGURE 1-2 The take-off branch was dismantled with the Demco valve open. *(Illustration courtesy of the U.S. Department of Labor.)*

a couple of days until the maintenance team was able to work on it. During this period, the air lines were reconnected, the lockout removed, and the isolation valve opened.

In both incidents, the procedures were poor and were not followed. It is unlikely that the accidents occurred the first time this happened. If the managers had kept their eyes open, they might have seen that the procedures were not being followed.

The 1988 explosion and fire on the Piper Alpha oil platform in the North Sea, which killed 163 people, was also caused by poor isolation. A pump relief valve was removed for overhaul and the open end blanked. Another shift, not knowing that the relief valve was missing, started up the pump. The blank was probably not tight, and light oil leaked past it and exploded in the confined processing area. The official report [18] concluded "that the operating staff had no commitment to working to the written procedure; and that the procedure was knowingly and flagrantly disregarded." The loss of life was greater on Piper Alpha than on the other two incidents because oil platforms are very congested and escape is difficult.

Section 18.1 describes similar incidents.

1.1.2 Isolations Removed Too Soon

An ethylene compressor was shut down for maintenance and correctly isolated by slip-plates. When repairs were complete, the slip-plates were removed before the machine was tried out. During the tryout, some ethylene leaked through the closed isolation valves into the machine. The ethylene/air mixture was ignited, either by a hot spot in the machine or by copper acetylide on the copper valve gaskets. The compressor was severely damaged.

Isolations should not be removed until maintenance is complete. It is good practice to issue three work permits—one for inserting slip-plates (or disconnecting pipework), one for the main job, and one for removing slip-plates (or restoring disconnections).

A similar incident occurred on a solids drier. Before maintenance started, the end cover was removed, and the inlet line was disconnected. When maintenance was complete, the end cover was replaced, and at the same time the inlet pipe was reconnected. The final job was to cut off the guide pins on the cover with a cutting disc. The atmosphere outside (but not inside) the drier was tested, and no flammable gas was detected. While cutting was in progress, an explosion occurred in the drier. Some solvent had leaked into the inlet pipe and then drained into the drier [19]. The inlet line should not have been reconnected before the guide pins were cut off.

1.1.3 Inadequate Isolation

A reactor was prepared for maintenance and washed out. No welding needed to be done, and no entry was required, so it was decided not to slip-plate off the reactor but to rely on valve isolations. Some flammable vapor leaked through the closed valves into the reactor and was ignited by a high-speed abrasive wheel, which was being used to cut through one of the pipelines attached to the vessel. The reactor head was blown off and killed two men. It was estimated that 7 kg of hydrocarbon vapor could have caused the explosion.

After the accident, demonstration cuts were made in the workshop. It was found that as the abrasive wheel broke through the pipe wall, a small flame occurred, and the pipe itself glowed dull red.

The explosion could have been prevented by isolating the reactor by slip-plates or physical disconnection. This incident and the others described show that valves are not good enough.

1.1.4   Isolation of Service Lines

A mechanic was affected by fumes while working on a steam drum. One of the steam lines from the drum was used for stripping a process column operating at a gauge pressure of 30 psi (2 bar). A valve on the line to the column was closed, but the line was not slip-plated. When the steam pressure was blown off, vapors from the column came back through the leaking valve into the steam lines (Figure 1-3).

The company concerned normally used slip-plates to isolate equipment under repair. On this occasion, no slip-plate was fitted because it was "only" a steam line. However, steam and other service lines in plant areas are easily contaminated by process materials, especially when there is a direct connection to process equipment. In these cases, the equipment under repair should be positively isolated by slip-plating or disconnection before maintenance.

When a plant was taken out of use, the cooling water lines were left full of water. Dismantling started nearly 20 years later. When a mechanic cut a cooling water line open with a torch, there was a small fire. Bacteria had degraded impurities in the water, forming hydrogen and methane [20].

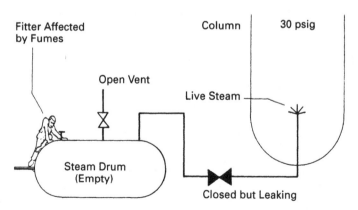

FIGURE 1-3   Contamination of a steam drum by process materials.

Plants should be emptied before they are mothballed or left for dismantling. Apart from the hazard just described, water can freeze and rupture lines (see Section 9.1.1).

Many years ago, river water was used for the water layer in a large kerosene storage tank. Bacterial decomposition of impurities formed methane, which exploded. As so often happens, the source of ignition was never found [21].

1.1.5 Isolations Not Removed

While a plant was on line, an operator noticed a slip-plate on a tank vent. The slip-plate had been fitted to isolate the tank from the blowdown system while the tank was under maintenance. When the maintenance was complete, the slip-plate was overlooked. Fortunately, the tank, an old one, was stronger than it needed to be for the duty, or it would have burst.

If a vessel has to be isolated from the vent or blowdown line, do not slip-plate it off, but whenever possible, disconnect it and leave the vessel vented to atmosphere (as shown in Figure 1-4).

If the vent line forms part of a blowdown system, it will have to be blanked to prevent air being sucked in. Make sure the blank is put on the flare side of the disconnection, not on the tank side (Figure 1-4). Note that if the tank is to be entered, the joint nearest the tank should be broken.

If a vent line has to be slip-plated because the line is too rigid to be moved, then the vents should be slip-plated last and de-slip-plated first. If all slip-plates inserted are listed on a register, they are less likely to be overlooked.

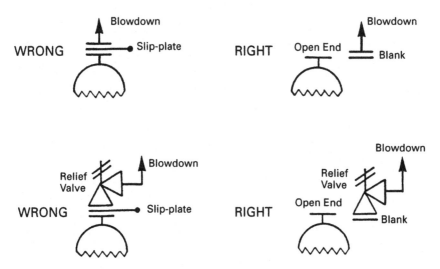

FIGURE 1-4   The right and wrong ways to isolate a vent line.

1.1.6 Some Miscellaneous Incidents Involving Isolation for Maintenance

(a) A slip-plate that had been in position for many months, perhaps years, was relied on to isolate equipment. It had corroded right through (Figure 1-5). Slip-plates in position for a long time should be removed and inspected before being used as maintenance isolations. (Such slip-plates should be registered for inspection every few years.)

(b) A slip-plate with a short tag was overlooked and left in position when maintenance was complete. Tags should be at least 130 mm long on lines up to and including 6 in. diameter and at least 150 mm long on larger lines. Figure-8 plates are better than slip-plates, as their position can be seen at a glance; figure-8 plates should be used on lines that have to be slip-plated regularly. Although the initial cost is higher, they are always available on the job, while slip-plates tend to disappear and have to be replaced.

(c) On several occasions, small bore branches have been covered by insulation, overlooked, and not isolated.

(d) On several occasions, thin slip-plates have been used and have become bowed; they are then difficult to remove. Figure 1-6 shows a thin slip-plate that has been subjected to a gauge pressure of 470 psi (32 bar).

Slip-plates should normally be designed to withstand the same pressure as the piping. However, in some older plants that have not been designed to take full-thickness slip-plates, it may be impossible to insert them. A compromise will be necessary.

(e) A butane pump was isolated for repair by valves only. When it was opened up, the pump and adjoining lines were found to be full of hydrate, a compound of water and butane that stays solid at a higher temperature than ice. A steam hose was used to clear the choke. Soon afterward there was a leak of butane, which was ignited by a furnace 40 m away, and exploded. The suction valve was also blocked by ice and was one turn open [22].

If you are not convinced that all isolation valves should be backed up by slip-plates before maintenance takes place, at least back up valves on lines containing materials that might turn solid and then melt.

1.1.7 Electrical Isolation

When an electrical supply has been isolated, it is normal practice to check that the right switches have been locked or fuses removed by trying to start the equipment that has been isolated. However, this system is not foolproof, as the following incidents illustrate.

FIGURE 1-5   A slip-plate left in position for many months had corroded right through.

FIGURE 1-6   A slip-plate bowed by a gauge pressure of 470 psi (32 bar).

In one case, the wrong circuit was isolated, but the circuit that should have been isolated was dead because the power supply had failed. It was restored while work was being carried out. In another case, the circuit that should have been isolated fed outside lighting. The circuit was dead because it was controlled by a photo-eye control [41].

On several occasions, maintenance teams have not realized that by isolating a circuit they have also isolated equipment that was still needed. In one case, they isolated heat tracing tape and, without realizing it, also isolated a ventilation fan. The wiring was not in accordance with the drawings [42]. In another case, maintenance team members isolated a power supply without realizing that they were also isolating the power to nitrogen blanketing equipment and an oxygen analyzer and alarm. Air leaked into the unit and was not detected, and an explosion occurred [43].

An unusual case of inadvertent reconnection occurred when a contract electrician pulled a cable, and it came out of the junction box. He thought he had pulled it loose, so he replaced it, but it had been deliberately disconnected [41].

1.2 IDENTIFICATION

1.2.1 The Need for Tagging

On many occasions, the wrong pipeline or piece of equipment has been broken into. Consider these examples:

(a) A joint that had to be broken was marked with chalk. The mechanic broke another joint that had an old chalk mark on it. He was splashed with a corrosive chemical.

(b) An out-of-service pipeline was marked with chalk at the point where it was to be cut. Before the mechanic could start work, a heavy rain washed off the chalk mark. The mechanic "remembered" where the chalk mark had been. He was found cutting his way with a hacksaw through a line containing a hazardous chemical.

(c) Water was dripping from a joint on a line on a pipebridge. Scaffolding was erected to provide access for repair. But to avoid having to climb up onto the scaffold, the process foreman pointed out the leaking joint from the ground and asked a mechanic to remake the joint in the "water line." The joint was actually in a carbon monoxide line. So when the mechanic broke the joint, he was overcome and, because of the poor access, was rescued only with difficulty.

If the process foreman had gone up to the joint on the pipebridge to fit an identifying tag, he would have realized that the water was dripping out of the carbon monoxide line.

(d) The bonnet had to be removed from a steam valve. It was pointed out to the mechanic from the floor above. He went down a flight of stairs, approached the valve from the side, and removed the bonnet from a compressed air valve. It flew off, grazing his face.

(e) Six slip-plates were inserted to isolate a tank for entry. When the work inside the tank was complete, six slip-plates were removed. Unfortunately, one of those removed was a permanent slip-plate left in position to prevent contamination. One of the temporary slip-plates was left behind.

(f) A mechanic was asked to repair autoclave No. 3. He removed the top manhole cover and then went down to the floor below to remove a manhole cover there. Instead of removing the cover from the manhole on autoclave No. 3, he removed the cover from No. 4, which contained vinyl chloride and nitrogen at a gauge pressure of 70 psi (5 bar). Polymer had formed around the inside of the manhole, so when he removed the bolts, there was no immediate evidence of pressure inside the vessel. Almost immediately afterward, the pressure blew off the cover. The mechanic and two other men were blown to the ground and killed, and the vinyl chloride was ignited [23].

(g) When a man tried to start the building ventilation fans, he found that the control and power panels had been removed. Contractors were removing surplus equipment and thought that these panels were supposed to be removed. The surplus equipment should have been clearly marked [44].

(h) A section of a chlorine gas line had been renewed and had to be heat treated. The operator who was asked to prepare the line and issue the permit-to-work misunderstood his instructions and thought a vent line had to be treated. There would be no need to gas-free this line, and he allowed the work to go ahead. It went ahead, on the correct line; the chlorine reacted with the iron, a 0.5-m length burned away, and 350 kg of chlorine escaped. To quote from the report, "at no stage on the day of the incident was the job thoroughly inspected by the issuer [of the permit-to-work] or the plant manager [supervisor in most U.S. companies]." The plant manager had inspected the permit and the heat treatment equipment but did not visit the site. He saw no reason to doubt the operator's belief that the line to be treated was the vent line [45]. Tagging would have prevented heat treatment of a line full of chlorine.

Incidents like these and many more could be prevented by fitting a numbered tag to the joint or valve and putting that number on the work permit. In incident (c), the foreman would have had to go up onto the scaffold to fix the tag. Accidents have occurred, however, despite tagging systems.

In one plant, a mechanic did not check the tag number and broke a joint that had been tagged for an earlier job; the tag had been left in position. Tags should be removed when jobs are complete.

FIGURE 1-7   Numbering pumps like this leads to error.

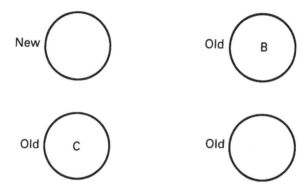

FIGURE 1-8   Which is crystallizer A?

In another plant, the foreman allowed a planner to fix the tags for him and did not check that they were fixed to the right equipment. The foreman prepared one line for maintenance, but the tags were on another.

1.2.2   The Need for Clear, Unambiguous Labeling

(a) A row of pumps was labeled as shown in Figure 1-7. A mechanic was asked to repair No. 7. Not unreasonably, he assumed that No. 7 was the end one. He did not check the numbers. Hot oil came out of the pump when he dismantled it.

(b) There were four crystallizers in a plant, three old ones and one just installed. A man was asked to repair A. When he went onto the structure, he saw that two were labeled B and C but the other two were not labeled. He assumed that A was the old unlabeled crystallizer and started work on it. Actually, A was the new crystallizer. The original three were called B, C, and D. Crystallizer A was reserved for a possible future addition for which space was left (Figure 1-8).

(c) The labels on two air coolers were arranged as shown in Figure 1-9. The B label was on the side of the B cooler farthest away from the B fan and near the A fan. Not unreasonably, workers who were asked to overhaul the B fan assumed it was the one next to the B label and overhauled it. The power had not been isolated. But fortunately, the overhaul was nearly complete before someone started the fan.

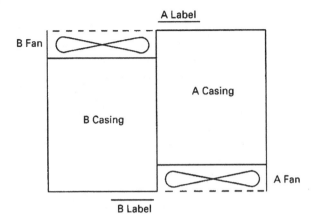

FIGURE 1-9 Which is the A fan?

(d) Some pump numbers were painted on the coupling *guards*. Before long, repairs were carried out on the couplings of two adjacent pumps. You can guess what happened. Now, the pump numbers are painted on the pump bodies. It would be even better to paint the numbers on the plinths.

(e) On one unit the pumps and compressors were numbered J1001 onward. When the unit's allocation of numbers was used up, numbers from JA1001 onward were used. J1001 and JA1001 sound alike (say them aloud). An operator was asked to prepare JA1001—a small pump—for repair. He thought the foreman said J1001 and went to it. J1001 was a 40,000 HP compressor. Fortunately, the size of the machine made him hesitate. He asked the foreman if he really wanted the compressor shut down.

1.2.3 The Need for Clear Instructions

(a) A permit was issued for modifications to the walls of a room. The maintenance workers started work on the ceilings as well and cut through live electric cables.

(b) A permit was issued for welding on the top only of a tank, which had been removed from the plant. When the job was complete, the welders rolled the tank over so that another part became the top. Some residue, which had been covered by water, caught fire.

(c) Because a lead operator on a chlorine storage unit was rather busy, he asked the second operator to issue a permit for heat treatment of a line. The second operator misunderstood his instructions and issued a permit for the wrong line. The lead operator's supervisor checked the permit and inspected the heat treatment equipment but did not look at the line. The line actually heat treated contained chlorine, and the heat was sufficient for the iron and the chlorine to react and "burn" a hole in the line; 350 kg of chlorine escaped. Afterward, the lead operator said he thought it was obvious that the line to be heat treated was the one that had been renewed the day before [24].

(d) An electrician was asked, in writing, to remove a fuse labeled FU-5. He did so. Unfortunately, he removed a fuse labeled FU-5 from the fuseboard that supplied the control room, not from the fuseboard that supplied the equipment room [25]. Not only were his instructions ambiguous, but the labeling system was poor.

(e) An operator asked an electrician to disconnect the cable leading to a piece of equipment that was to be modified. The operator checked the disconnection and signed the permit-to-work for the modification. A second operator certified that the preparation had been carried out correctly.

The construction worker who was to carry out the modification checked the cable with a current detector and found that the wrong one had been disconnected. It was then found that the cable was incorrectly described on the written instructions given to the operators. The description of the cable was not entirely clear, but instead of querying it, the first operator decided what he thought was the correct cable and asked the electrician to disconnect it. The second operator, or checker, had not been trained to check cables [32].

This incident shows the weakness of checking procedures. The first operator may assume that if anything is wrong the checker will pick it up; the checker may become casual because he has never known the first operator to make an error (see Sections 3.2.7b and 14.5c).

1.2.4   Identification of Relief Valves

Two relief valves, identical in appearance, were removed from a plant during a shutdown and sent to the workshops for overhaul. One relief valve was set to operate at a gauge pressure of 15 psi (1 bar) and the other at 30 psi (2 bar). The set pressures were stamped on the flanges, but this did not prevent the valves from being interchanged.

A number of similar incidents have occurred in other plants.

Such incidents can be prevented, or at least made much less likely, by tying a numbered tag to the relief valve when it is removed and tying another tag with the same number to the flange.

1.2.5   Make Sure You Find the Right Line

There was a leak on the line supplying steam to a plant. To avoid a shutdown, a hot tap and stopple was carried out—that is, the line was bypassed and the leaking section plugged off (stoppled) while in use. The job went well mechanically, but the leak continued. It was then found that the leak was not coming from the steam line but from a hot condensate line next to it. The condensate flashed as it leaked, and the leak looked like a steam leak [26].

1.3 REMOVAL OF HAZARDS

Many accidents have occurred because equipment, though isolated correctly, was not completely freed from hazardous materials or because the pressure inside it was not completely blown off and the workers carrying out the repair were not made aware of this.

1.3.1 Equipment Not Gas Freed

It is usual to test for the presence of flammable gas or vapor with a combustible gas detector before maintenance, especially welding or other hot work, is allowed to start. The following incidents show what can happen if these tests are not carried out or not carried out thoroughly. Large pieces of equipment or those of complex shape should be tested in several places, using detector heads at the ends of long leads if necessary (see Section 5.4.2d).

(a) An explosion occurred in a 4,000-m^3 underground storage tank at Sheffield Gas Works, England, in October 1973. Six people were killed, 29 injured, and the tank was destroyed. The tank top was thrown into the air, turned over, and deposited upside down on the bottom of the tank.

The tank had contained a light naphtha and had not been thoroughly cleaned before repairs started. It had been filled with water and then emptied, but some naphtha remained in various nooks and crannies. (It might, for example, have got into the hollow roof supports through pinholes or cracks and then drained out when the tank was emptied.) No tests were carried out with combustible gas detectors.

It is believed that the vapor was ignited by welding near an open vent. The body of the welder was found 30 m (100 ft) up on the top of a neighboring gasholder, still holding a welding torch.

According to the incident report, there was no clear division of responsibilities between the Gas Board and the contractor who was carrying out the repairs. "Where, as in this case, a special risk is likely to arise due to the nature of the work performed (and the owner of the premises has special knowledge of it), the owner must retain sufficient control of the operation to ensure that contractors' employees are properly protected against the risk" [4].

(b) A bottom manhole was removed from an empty tank still full of gasoline vapor. Vapor came out of the manhole and caught fire. As the vapor burned, air was sucked into the tank through the vent until the contents became explosive. The tank then blew up [5].

(c) Welding had to be carried out—during a shutdown—on a relief valve tailpipe. It was disconnected at both ends. Four hours later the atmosphere at the end farthest from the relief valve was tested with a combustible gas detector. The head of the detector was pushed as far down the tailpipe as it would go; no gas was detected, and a work permit was issued. While the relief valve discharge flange was being ground, a flash and bang occurred at the other end of the tailpipe. Fortunately, no one was hurt. Gas in the tailpipe—20 m long and containing a number of bends—had not dispersed and had not been detected by a test at the other end of the pipe.

Before allowing welding or similar operations on a pipeline that has or could have contained flammable gas or liquid, (1) sweep out the line with steam or nitrogen from end to end, and (2) test at the point at which welding will be carried out. If necessary, a hole may have to be drilled in the pipeline.

(d) Solids in a vessel can "hold" gas that is released only slowly. A reactor, which contained propylene and a layer of polypropylene granules 1 to 1.5 m thick, had to be prepared for maintenance. It was purged with nitrogen six times. A test near the manhole showed that only a trace of propylene was present, less than 5% of the lower explosive limit (LEL). However, when the reactor was filled with water, gas was emitted, and gas detectors in the surrounding area registered 60% of the LEL.

The vessel had been prepared for maintenance in a similar way on three previous occasions, but there were then far fewer granules in the reactor [14] (see Sections 11.1a and b).

(e) A label had to be welded onto an empty drum. As the drum was brand new, no precautions were taken, and no tests were carried out. The drum exploded, breaking the welder's leg. The manufacturer had cleaned the drum with a flammable solvent, had not gas freed it, and had not warned the customer [15].

(f) In 1992, the catwalks and ladders were being removed with oxyacetylene torches from a group of tanks so the tanks could be moved. An empty tank that had contained ethanol exploded, killing three men. The ethanol vapor had leaked out of a faulty seal on the gauge hatch; it was ignited by a torch, and the flame traveled back into the tank. The men who were killed had taken combustible gas detectors onto the job, but no one knew whether they had used them correctly or had used them at all. Gas testing should be carried out by the operating team before it issues a permit-to-work; because the tanks would have had to be gas freed before they were moved, this should have been done before hot work started [27].

(g) In fluidized bed catalytic cracker units, air is blown into large vessels called regenerators to burn carbon off the catalyst. The regenerators are vented to the air, so there should be no need to test or inert them before maintenance. However, on one occasion when a manway cover was being removed, 50 hours after the unit had shut down, an explosion occurred inside the vessel, and flames appeared at various openings in the ducts connected to it.

Carbon is usually burned off before a shutdown. On this occasion, the air blower failed, and the unit had to shut down at once. Steam was blown into the regenerator, and most of the catalyst was removed. However, the steam reacted with the carbon on the remaining catalyst, forming hydrogen and carbon monoxide. When the manway cover was removed, air entered the regenerator, and an explosion occurred. The source of ignition was the hot catalyst, which was still at about 600°C (1,100°F) [33]. Older regenerators are fitted with a spare blower. Some plants connect up mobile blowers if their single blower fails.

This incident shows the importance, during hazard and operability studies (see Chapter 18), of considering abnormal conditions, such as failure of utilities, as well as normal operation.

1.3.2 Conditions Can Change after Testing

As already stated, it is usual to test for the presence of flammable gas or vapor with a combustible gas detector before maintenance, especially welding or other hot work, starts. Several incidents have occurred because tests were carried out several hours beforehand and conditions changed.

(a) An old propylene line that had been out of use for 12 years had to be modified for reuse. For the past two years it had been open at one end and blanked at the other. The first job was welding a flange onto the open end. This was done without incident. The second job was to fit a 1-in. branch 60 m from the open end. A hole was drilled in the pipe and the inside of the line tested. No gas was detected. Fortunately, a few hours later, just before welding was about to start, the inside of the pipe was tested again, and flammable gas was detected. It is believed that some gas had remained in the line for 12 years and a slight rise in temperature had caused it to move along the pipeline. Some people might have decided that a line out of use for 12 years did not need testing at all. Fortunately, the men concerned did not take this view. They tested the inside of the line and tested again immediately before welding started.

(b) A test for benzene in the atmosphere was carried out eight hours before a job started. During this time the concentration of benzene rose.

(c) An acid tank was prepared for welding and a permit issued. The maintenance team was not able to start for 40 days. During this time, a small amount of acid that had been left in the tank attacked the metal, producing hydrogen. No further tests were carried out. When welding started, an explosion occurred [6].

(d) A branch had to be welded onto a pipeline that was close to the ground. A small excavation, between ½ and 1 m deep (20 and 40 in.), was made to provide access to the bottom of the pipeline. The atmosphere in the excavation was tested with a combustible gas detector, and because no gas was detected, a welding permit was issued. Half an hour later, after the welder had started work, a small fire occurred in the excavation. Some hydrocarbons had leaked out of the ground. This incident shows that it may not be sufficient to test just before welding starts. It may be necessary to carry out continuous tests using a portable combustible gas detector alarm.

(e) The sewer from a chemical plant discharged into a river. The river wall was lined with steel plates, and a welder was burning holes in one of them, just downstream of the outlet, so that a crane could remove it. The atmosphere was tested for flammable gas before work started. After a break the welder started again. There was a flash fire, which did not last long but killed the welder. An underground pipeline was leaking, and it seems that the liquid had collected in a sump and then overflowed into the sewer.

Section 11.5 describes another fatality caused by hazardous materials in drains.

1.3.3   Hazards Can Come Out of Drains, Vents, and Other Openings

A number of incidents have occurred because gas or vapor came out of drains or vents while work was in progress:

(a) Welding had to be carried out on a pipeline 6 m (20 ft) above the ground. Tests inside and near the pipeline were negative, so a work permit was issued. A piece of hot welding slag bounced off a pipeline and fell onto a sump 6 m below and 2.5 m (8 ft) to the side. The cover on the sump was loose, and some oil inside caught fire. Welding jobs should be boxed in with fire-resistant sheets. Nevertheless, some sparks or pieces of slag may reach the ground. So drains and sumps should be covered.

(b) While an electrician was installing a new light on the outside wall of a building, he was affected by fumes coming out of a ventilation duct 0.6 m (2 ft) away. When the job was planned, the electrical hazards were considered and also the hazards of working on ladders. But it did not occur to anyone that harmful or unpleasant fumes might come out of the duct. Yet ventilation systems are installed to get rid of fumes.

(c) Radioactive material was transferred into transport casks by remote handling in a shielded cell. Checks showed that the radiation level outside the cell was low, but no one thought about the roof. Several years later, a technician walked across the flat roof while a transfer was taking place below. Fortunately, she was carrying a radiation detector, and when it alarmed, she left at once. The radiation stream to the roof was greater than 50 mSv/hr, and the technician received a dose of about 1 mSv. (The International Committee on Radiological Protection recommends that no one be exposed to more than 50 mSv in a single year or more than 20 mSv/yr [2 rem/yr] averaged over five years. In practice, most radiation workers receive far smaller doses.) Several similar incidents have been reported [34].

Not many readers will handle radioactive materials, but this incident and the previous one do show how easy it is to overlook some of the routes by which hazardous materials or effects can escape from containment.

1.3.4 Liquid Can Be Left in Lines

When a line is drained or blown clear, liquid may be left in low-lying sections and run out when the line is broken. This is particularly hazardous if overhead lines have to be broken. Liquid splashes down onto the ground. Funnels and hoses should be used to catch spillages.

When possible, drain points in a pipeline should be fitted at low points, and slip-plates should be fitted at high points.

1.3.5 Service Lines May Contain Hazardous Materials

Section 1.1.4 described how fumes got into a steam drum because it was not properly isolated. Even when service lines are not directly connected to process materials, they should always be tested before maintenance, particularly if hot work is permitted on them, as the following incidents show:

(a) A steam line was blown down and cold cut. Then a plug was hammered into one of the open ends. A welder struck an arc ready to weld in the plug. An explosion occurred, and the plug was blown out of the pipeline, fortunately missing the welder. Acid had leaked into the pipeline through a corroded heating coil in an acid tank and had reacted with the iron of the steam pipe, producing hydrogen.

(b) While a welder was working on the water line leading to a waste heat boiler, gas came out of a broken joint and caught fire. The welder was burned but not seriously. There was a leaking tube in the waste heat boiler. Normally, water leaked into the process stream. However, on shutting down the plant, pressure was taken off the water side before it was taken off the process side, thus reversing the leak direction. The water side should have been kept up to pressure until the process side was depressured. In addition, the inside of the water lines should have been tested with a combustible gas detector.

See also Section 5.4.2b.

1.3.6   Trapped Pressure

Even though equipment is isolated by slip-plates and the pressure has been blown off through valves or by cracking a joint, pressure may still be trapped elsewhere in the equipment, as the following incidents show:

(a) This incident occurred on an all-welded line. The valves were welded in. To clear a choke, a fitter removed the bonnet and inside of a valve. He saw that the seat was choked with solid and started to chip it away. As he did so, a jet of corrosive chemical came out under pressure from behind the solid, hit him in the face, pushed his goggles aside, and entered his eye.

(b) An old acid line was being dismantled. The first joint was opened without trouble. But when the second joint was opened, acid came out under pressure and splashed the fitter and his assistant in their faces. Acid had attacked the pipe, building up gas pressure in some parts and blocking it with sludge in others.

(c) A joint on an acid line, known to be choked, was carefully broken, but only a trickle of acid came out. More bolts were removed, and the joint pulled apart, but no more acid came. When the last bolt was removed and the joint pulled wide apart, a sudden burst of pressure blew acid into the fitter's face.

In all three cases the lines were correctly isolated from operating equipment. Work permits specified that goggles should be worn and stated, "Beware of trapped pressure."

To avoid injuries of this sort, we should use protective hoods or helmets when breaking joints on lines that might contain corrosive liquids trapped under pressure, either because the pressure cannot be blown off through a valve or because lines may contain solid deposits.

Other incidents due to trapped pressure and clearing chokes are described in Sections 17.1 and 17.2.

1.3.7 Equipment Sent Outside the Plant

When a piece of equipment is sent to a workshop or to another company for repair or modification we should, whenever possible, make sure that it is spotlessly clean before it leaves the plant. Contractors are usually not familiar with chemicals and do not know how to handle them.

Occasionally, however, it may be impossible to be certain that a piece of equipment is spotlessly clean, especially if it has contained a residual oil or a material that polymerizes. If this is the case, or if there is some doubt about its cleanliness, then the hazards and the necessary precautions should be made known to the workshop or the other company. This can be done by attaching a certificate to the equipment. This certificate is not a work permit. It does not authorize any work but describes the state of the equipment and gives the other company sufficient information to enable it to carry out the repair or modification safely. Before issuing the certificate, the engineer in charge should discuss with the other company the methods it proposes to use. If the problems are complex, a member of the plant staff may have to visit the other company. The following incidents show the need for these precautions:

(a) A large heat exchanger, 2.4 m long by 2.6 m (8 ft by 8 ft 6 in.) in diameter, was sent to another company for retubing. It contained about 800 tubes of 2½-in. (64 mm) diameter, and about 80 of these tubes had been plugged. The tubes had contained a process material that tends to form chokes, and the shell had contained steam.

Before the exchanger left the plant, the free tubes were cleaned with high-pressure water jets. The plugged tubes were opened up by drilling ⅜-in. (9.5 mm) holes through the plugs to relieve any trapped pressure. But these holes were not big enough to allow the tubes to be cleaned.

A certificate was attached to the exchanger stating that welding and burning were allowed but only to the shell. The contractor, having removed most of the tubes, decided to put workers into the shell to grind out the plugged tubes. He telephoned the plant and asked if it would be safe to let workers enter the shell. He did not say why he wanted them to do so.

The plant engineer who took the telephone call said that the shell side was clean and therefore entering it would be safe. He was not told that the workers were going into it to grind out some of the tubes.

Two men went into the shell and started grinding. Fumes affected them, and the job was left until the next day. Another three workers then restarted the job and were affected so badly that they were hospitalized. Fortunately, they soon recovered.

The certificate attached to the exchanger when it left the plant should have contained much more information. It should have said that the plugged tubes had not been cleaned and that they contained a chemical that gave off fumes when heated. Better still, the plugged tubes should have been opened up and cleaned. The contractor would have to remove the plugs, so why not remove them before they left the plant?

(b) At least two serious titanium fires have occurred when scrap metal dealers used torches to cut up heat exchangers containing titanium tubes [28]. Once titanium (melting point about 1,660°C [3,020°F]) is molten, it burns readily in air. Titanium sent for scrap should be clearly labeled with a warning note.

Do your instructions cover the points mentioned in this section?

1.4   PROCEDURES NOT FOLLOWED

It is usual, before a piece of equipment is maintained, to give the maintenance team a permit-to-work that sets out the following:

1. What is to be done
2. How the equipment is isolated and identified
3. What hazards, if any, remain
4. What precautions should be taken

This section describes incidents that occurred because of loopholes in the procedure for issuing work permits or because the procedure was not followed. There is no clear distinction between these two categories. Often the procedure does not cover, or seem to cover, all circumstances. Those concerned use this as the reason, or excuse, for a shortcut, as in the following two incidents.

1.4.1   Equipment Used after a Permit Has Been Issued

(a) A plumber foreman was given a work permit to modify a pipeline. At 4 p.m. the plumbers went home, intending to complete the job on the following day. During the evening, the process foreman wanted to use the line the plumbers were working on. He checked that the line was safe to use, and he asked the shift maintenance man to sign off the permit. The next morning, the plumbers, not knowing that their permit had been withdrawn, started work on the line while it was in use.

To prevent similar incidents from happening, (1) it should be made clear that permits can only be signed off by the person who has accepted them (or a person who has taken over that person's responsibilities), and (2) there should be two copies of every permit, one kept by the maintenance team and one left in the book in the process team's possession.

(b) A manhole cover was removed from a reactor so some extra catalyst could be put in. After the cover had been removed, it was found that the necessary manpower would not be available until the next day. So it was decided to replace the manhole cover and regenerate the catalyst overnight. By this time it was evening, and the maintenance foreman had gone home and left the work permit in his office, which was locked. The reactor was therefore boxed up and catalyst regeneration carried out with the permit still in force. The next day a fitter, armed with the work permit, proceeded to remove the manhole cover again and, while doing so, was drenched with process liquid. Fortunately, the liquid was mostly water, and he was not injured.

The reactor should not have been boxed up and put on line until the original permit had been handed back. If it was locked up, then the maintenance supervisor should have been called in. Except in an emergency, plant operations should never be carried out while a work permit is in force on the equipment concerned.

1.4.2 Protective Clothing Not Worn

The following incidents are typical of many:

(a) A permit issued for work to be carried out on an acid line stated that goggles must be worn. Although the line had been drained, there might have been some trapped pressure (see Section 1.3.6). The man doing the job did not wear goggles and was splashed in the eye.

At first, it seemed the injury was entirely the fault of the injured man and no one else could have done anything to prevent it. However, further investigation showed that all permits issued asked for goggles to be worn, even for repairs to water lines. The maintenance workers therefore frequently ignored this instruction, and the managers turned a blind eye. No one told the fitter that on this job, goggles were really necessary.

It is bad management for those issuing work permits to cover themselves by asking for more protective clothing than is really necessary. They should ask only for what is necessary and then *insist* that it be worn.

Why did they ask for more than was necessary in this case? Perhaps someone was reprimanded because he asked for less protective clothing than his supervisor considered necessary. That person and his colleagues then decided to cover themselves by asking for every-thing every time. If we give people the discretion to decide what is necessary, then inevitably they will at times come to a different deci-sion than we would. We may discuss this with them but should not reprimand them.

(b) Two men were told to wear air masks while repairing a compressor, which handled gas containing hydrogen sulfide. The compressor had been swept out, but traces of gas might have been left in it. One of the men had difficulty handling a heavy valve that was close to the floor and removed his mask. He was overcome by gas—hydro-gen sulfide or possibly nitrogen.

Again, it is easy to blame the man. But he had been asked to do a job that was difficult to perform while wearing an air mask. The plant staff members resisted the temptation to blame him—the easy way out. Instead, they looked for suitable lifting aids [7].

Section 3.2 discusses similar incidents. Rather than blame workers who make mistakes or disobey instructions, we should try to remove the opportunities for error by changing the work situation—that is, the design or method of operation.

(c) Work permits asked for goggles to be worn. They were not always worn and, inevitably, someone was injured. This incident differs from (a) in that goggles were always necessary on this unit.

Investigation showed that the foreman and manager knew that goggles were not always worn. But they turned a blind eye to avoid dispute and to avoid delaying the job. The workers knew this and said to themselves, "Wearing goggles cannot be important." The foreman and manager were therefore responsible for the inevitable injury. People doing routine tasks become careless. Foremen and managers cannot be expected to stand over them all the time, but they can make occasional checks to see that the correct precautions are taken. And they can comment when they see rules being flouted. A friendly word *before* an accident is better than punitive action afterwards.

1.4.3   Jobs Near Plant Boundaries

Before a permit to weld or carry out other hot work is issued, it is nor-mal practice to make sure there are no leaks of flammable gas or liquid *nearby* and no abnormal conditions that make a leak likely. The mean-ing of *nearby* depends on the nature of the material that might leak, the slope of the ground, and so on. For highly flammable liquids, 15 m (50 ft) is often used.

Fires have occurred because a leak in one unit was set alight by weld-ing in the unit next door. Before welding or other hot work is permitted within 15 m (50 ft), say, of a unit boundary, the foreman of the unit next door should countersign it.

Similar hazards arise when a pipeline belonging to one unit passes through another unit.

Suppose a pipeline belonging to area A passes through area B and that this pipeline has to be broken in area B (Figure 1-10).

The person doing the job is exposed to two distinct hazards: those due to the contents of the pipeline (these are understood by the area A foreman) and those due to work going on in area B (these are understood by the area B foreman). If the work permit for the pipeline is issued by the area A foreman, then the area B foreman should countersign it. If B issues it, then A should countersign it. The system should be covered by local instructions and clearly understood.

An incident occurred because the area A foreman issued a permit for work to be done on a flow transmitter in a pipeline in area B. The area B foreman issued a permit for grinding in area B. He checked that no flammable gas was present and had the drains covered. He did not know about the work on the flowmeter. A spark set fire to a drain line on the flowmeter, which had been left open.

What would happen in your plant?

1.4.4 Maintenance Work Over Water

A welder was constructing a new pipeline in a pipe trench, while 20 m (65 ft) away a slip-plate was being removed from another pipe, which had contained light oil. Although the pipe had been blown with nitrogen, it was realized that a small amount of the oil would probably spill when the joint was broken. But it was believed that the vapor would not spread to the welders. Unfortunately, the pipe trench was flooded after heavy rain, and the oil spread across the water surface and was ignited by the welder's torch. One of the men working on the slip-plate 20 m away was badly burned and later died.

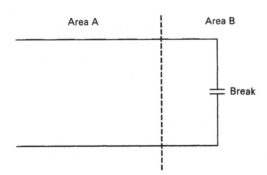

FIGURE 1-10   Who should authorize the pipeline break?

The first lesson from the incident is that *welding should not be allowed over large pools of water.* Spillages some distance away might be ignited. In 1970, 35 tons of gasoline were spilled on the Manchester Ship Canal, England; 1 km (0.6 mile) away, 2½ hours later, the gasoline caught fire, killing six men [8].

The second lesson is that *when large joints have to be broken regularly, a proper means of draining the line should be provided.* The contents should not be allowed to spill onto the ground when the joint is broken.

Why was a permit issued to remove a slip-plate 20 m away from a welding job? Although vapor should not normally spread this far, the two jobs were rather close together.

The foremen who issued the two permits were primarily responsible for operating a unit some distance away. As they were busy with the running plant, they did not visit the pipe trench as often as they might. Had they visited it immediately before allowing the de-slip-plating job to start, they would have realized that the two jobs were close together. They might have realized that oil would spread across the water in the trench.

After the incident, special day foremen were appointed to supervise construction jobs and interface with the construction teams. The construction teams like this system because they deal with only one process foreman instead of four shift foremen.

For another incident involving a construction team, see Section 5.4.2b.

1.4.5 Misunderstandings

Many incidents have occurred because of misunderstandings of the meanings of words and phrases. The following incidents are typical:

(a) A permit was issued to remove a pump for overhaul. The pump was defused, removed, and the open ends blanked. The next morning, the maintenance foreman signed the permit to show that the job—removing the pump—was complete. The morning shift lead operator glanced at the permit. Seeing that the job was complete, he asked the electrician to replace the fuses. The electrician replaced them and signed the permit to show that he had done so. By this time the afternoon shift lead operator had come on duty. He went out to check the pump and found that it was not there.

The job on the permit was to remove the pump for overhaul. Permits are sometimes issued to remove a pump, overhaul it, and replace it. But in this case, the permit was just for removal (see Section 1.1.2). When the maintenance foreman signed the permit to show that the job was complete, he meant that the job of *removal* was complete. The lead operator, however, did not read the permit thoroughly. He assumed that the *overhaul* was complete.

The main message is clear: read permits carefully; don't just glance at them.

When a maintenance worker signs a permit to show that the job is complete, he means he has completed the job *he thought he had to do*. This may not be the same as the job he was expected to do. The process team should therefore always inspect the job to make sure that the one completed is the one wanted.

When handing over or handing back a permit, the maintenance and process people should speak to each other. It is not good practice to leave a permit on the table for someone to sign when he or she comes in.

(b) When a work permit is issued to excavate the ground, it is normal practice for an electrician to certify that there are no buried cables. What, however, is an excavation? A contractor asked for and received a work permit to "level and scrape the ground." No excavation was requested, so the process foreman did not consult the electricians. The contractor used a mechanical shovel, removed several feet of dirt from the ground, and cut through a live electric cable. The word *excavation* needs careful definition.

(c) A construction worker was wearing a plastic protective suit, supplied by breathing air, when the air supply suddenly stopped. Fortunately, he was rescued without injury. A mechanic had isolated the breathing air supply to change a filter.

The plant had what everyone involved thought was a good system: before anyone used breathing air or did any work on the air system, that person was supposed to tell the control room. Unfortunately, the supervisor and the standby operator both thought that the other was going to do so. The mechanic did contact the control room before starting work, but the control room staff told him that no one was using breathing air. To make sure, both the mechanic and someone from the control room had a look around, but the check was rather casual as neither of them expected to find anyone. The air was in use in an out-of-the-way part of the site, and neither of them noticed the job [29].

This system of working was not really very good. No work should be allowed on the breathing air system (or any other system) without a permit-to-work, as people will say okay with less thought than they will sign a form. Users of breathing air should sign a book in the control room or collect a tag, not just tell someone they are going to use the air.

(d) Electricity supplies to a boiler and a water treatment unit were isolated for replacement of electrical equipment. The supervisor in charge of the work kept in touch with the two units by radio. After the replacement work had been carried out, he received a message that testing was complete. The message actually said that the testing on the water treatment unit was complete, but the supervisor took it to mean that testing on both units was complete. He therefore announced over his radio that power would be restored to both units. The crew working on the boiler plant did not hear this message and continued testing. When power was restored, arcing occurred at a test point on the boiler plant. Fortunately, the electrician carrying out the test was wearing high-voltage gloves and safety glasses, or he might have been killed [35].

The report says that when communicating by radio, no action should be taken until the message has been acknowledged, but this is nowhere near sufficient. Power should not have been restored until the crews working on both units had signed off written permits-to-work. Verbal communication alone is never adequate.

(e) While someone was writing out a permit-to-work in duplicate, the lower copy moved under the carbon paper, and a wrong line was crossed out on the lower copy. It was given to the man who was to repair the equipment, and as a result of the error he thought the plant was free from acid. While he was breaking a joint, sulfuric acid came out and burned him on the face and neck [46].

1.4.6   Excavations

A report from the U.S. Department of Energy (DOE) says that events involving the unexpected discovery of underground utilities during excavations or trenching operations have occurred at its facilities [36]. Several of the events resulted in electric shock; one caused serious injury, and many others were near misses. The serious injury occurred when a sump 1 m (3.3 ft) deep was being constructed in the basement of a building, and a compressed air hammer hit a 13,200-volt power line. Before any excavation (or leveling of the ground; see Section 1.4.5b) is authorized, the electrical department should certify that no electric cables are present or that any present have been isolated. If any underground pipelines are present, they should be identified, from drawings or with metal detectors, and excavations nearby (say, within 1 m [3.3 ft]) should be carried out by hand.

In another incident, a backhoe ruptured a 3-in. polyethylene natural gas pipeline; fortunately, the gas did not ignite. The drawings were complex and cluttered, and the contractor overlooked the pipeline. A metal detector was not used. This would have detected the pipe as a metal wire was fixed to it, a good practice. In a third incident, a worker was hand-digging a trench, as an electric conduit was believed to be present. It was actually an old transfer line for radioactive waste, and he received a small dose of radioactivity. The planner had misread the drawing.

1.4.7   A Permit to Work Dangerously?

A permit system is necessary for the safe conduct of maintenance operations. But issuing a permit in itself does not make the job safe. It merely provides an opportunity to check what has been done to make the equipment
safe, to review the precautions necessary, and to inform those who will have
to carry out the job. The necessity of saying this is shown by the following
quotation from an official report:

> [T]hey found themselves in difficulty with the adjustment of some scrapers on
> heavy rollers. The firm's solution was to issue a work permit, but it was in fact a per-
> mit to live dangerously rather than a permit to work in safety. It permitted the fitter
> to work on the moving machinery with the guards removed. A second permit was
> issued to the first aid man to enable him to stand close to the jaws of death ready to
> extricate, or die in the attempt to extricate, the poor fitter after he was dragged into
> the machinery. In fact, there was a simple solution. It was quite possible to extend
> the adjustment controls outside the guard so that the machinery could be adjusted,
> while still in motion, from a place of safety. [9]

1.5   QUALITY OF MAINTENANCE

Many accidents have occurred because maintenance work was not
carried out in accordance with the (often unwritten) rules of good engineering practice, as the following incidents and Section 10.4.5 show.

1.5.1   The Right and Wrong Ways to Break a Joint

(a) One of the causes of the fire described in Section 1.1.1 was the fact that
the joint was broken incorrectly. The fitter removed all the nuts from
the pump cover and then used a wedge to release the cover, which was
held tightly on the studs. It came off suddenly, followed by a stream of
hot oil.

   The correct way to break a joint is to slacken the nuts farthest
away from you and then spring the joint faces apart, using a wedge
if necessary. If any liquid or gas is present under pressure, then
the pressure can be allowed to blow off slowly or the joint can be
retightened.

(b) Another incident was the result of poor preparation and poor workmanship. A valve had to be changed. The valves on both sides were
closed and a drain valve in between opened. A flow through the drain
valve showed that one of the isolating valves was leaking, so the drain
valve was closed and a message left for the employees working the next
shift, telling them to open the drain valve before work started. Nothing
was written on the permit-to-work. The message was not passed on;
the drain valve was not opened, and the fitter broke the joint the wrong

way, removing all the bolts. The joint blew apart, and the fitter received head injuries from which he will never fully recover [37].

(c) It is not only flammable oils that cause accidents. In another incident, two workers were badly scalded when removing the cover from a large valve on a hot water line, although the gauge pressure was only 9 in. of water (0.33 psi or 0.023 bar). They removed all the nuts, attached the cover to a chain block, and tried to lift it. To release the cover they tried to rock it. The cover suddenly released itself, and hot water flowed out onto the workers' legs.

1.5.2   Use of Excessive Force

A joint on an 8-in. line containing a hot solvent had to be remade. The two sides were ¾-in. out of line. There was a crane in the plant at the time, so it was decided to use it to lift one of the lines slightly. The lifting strap pulled against a ¾-in. branch and broke it off (Figure 1-11).

It was not a good idea to use a crane for a job like this on a line full of process material. Fortunately, the leaking vapor did not ignite, although nearby water was being pumped out of an excavation. At one time a diesel pump would have been used, but the use of diesel pumps had been banned only a few months before the incident.

Section 2.11.1 describes an explosion caused by the failure of nuts that had been tightened with excessive force.

1.5.3   Ignorance of Material Strength

(a) When a plant came back on line after a long shutdown, some of the flanges had been secured with stud bolts and nuts instead of ordinary bolts and nuts. And some of the stud bolts were located so that

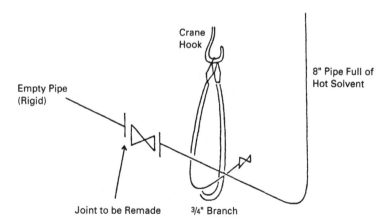

FIGURE 1-11   A branch broke when a crane was used to move a live line.

more protruded on one side than on the other. On some flanges, one of the nuts was secured by only two or three threads (Figure 1-12).

Nobody knows why this had been done. Probably one nut was tighter than the other, and in attempting to tighten this nut, the whole stud was screwed through the second nut. Whatever the reason, it produced a dangerous situation because the pressure on different parts of the flange was not the same.

In addition, stud bolts should not be indiscriminately mixed with ordinary bolts or used in their place. They are often made of different grades of steel and produce a different tension.

In the plant concerned, for the eight-bolt joints the bolts were changed one bolt at a time. Four-bolt joints were secured with clamps until the next shutdown.

(b) There was a leak on a large fuel-gas system operating at gasholder pressure. To avoid a shutdown, a wooden box was built around the leak and filled with concrete. It was intended as a temporary job but was so successful that it lasted for many years.

On other occasions, leaks have been successfully boxed in or encased in concrete. But the operation can only be done at low pressures, and expert advice is needed, as shown by the following incident.

There was a bad steam leak from the bonnet gasket of a 3-in. steam valve at a gauge pressure of 300 psi (20 bar). An attempt to clamp the bonnet was unsuccessful, so the shift crew decided to encase the valve in a box. Crew members made one 36 in. long, 24 in. wide, and 14 in. deep out of ¼-in. steel plate. Plate of this thickness is strong, but the shape of the box was unsuitable for pressure and could hardly have held a gauge pressure of more than 50 psi (3 bar), even if the welds had been full penetration, which they were not (Figure 1-13).

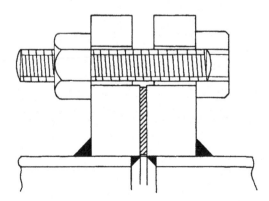

FIGURE 1-12 Nuts fitted incorrectly to studs.

FIGURE 1-13 This steel box was quite incapable of containing a leak of steam at a gauge pressure of 300 psi (20 bar).

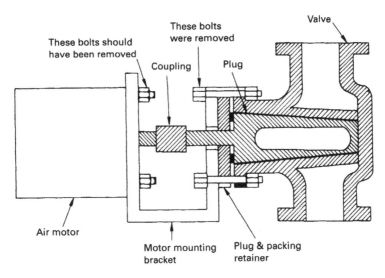

FIGURE 1-14 Wrong nuts undone to remove the valve actuator.

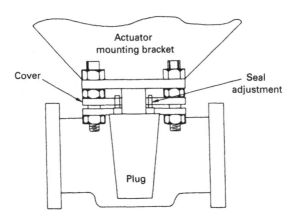

FIGURE 1-15 Another example of wrong nuts undone to remove the valve actuator.

The box was fitted with a vent and valve. When the valve was closed, the box started to swell, and the valve was quickly opened.

A piece of 2-in. by 2-in. angle iron was then welded around the box to strengthen it. The vent valve was closed. A few minutes later the box exploded. Fortunately, the mechanic—if he deserves the title—had moved away.

This did not happen in a back-street firm but in a major international company.

These incidents show the need for continual vigilance. We cannot assume that because we employ qualified craftspeople and graduate engineers they will never carry out repairs in a foolish or unsafe manner.

1.5.4   Failure to Understand How Things Work or How They Are Constructed

(a) Several spillages have occurred from power-operated valves while the actuators were being removed because the bolts holding the valve bonnets in position were removed in error. Figures 1-14 and 1-15 show how two such incidents occurred. The second system is particularly vulnerable because in trying to unscrew the nuts that hold the actuator mounting bracket in place, the stud may unscrew out of the lower nuts. This incident could be classified as due to poor design [10].

The first incident resulted in the release of 70 to 100 tons of vinyl chloride. There was little wind, and the cloud of vapor and mist drifted slowly backward and forward. After an hour, when the cloud was about 240 m across and 1.5 m deep, it ignited. Some of the vinyl chloride had entered buildings, and it exploded, destroying the buildings. The rest burned outside and caused several vinyl chloride tanks to burst, adding further fuel to the fire. Remarkably, only one man was killed. The injured included spectators who arrived to watch the fire [30].

(b) A similar accident occurred on a common type of ball valve. Two workers were asked to fit a drain line below the valve. There was not much room, so they decided to remove what they thought was a distance piece or adaptor below the valve but that was in fact the lower part of the valve body (Figure 1-16). When they had removed three bolts and loosened the fourth, it got dark, and they left the job to the next day.

The valve was the drain valve on a small tank containing liquefied petroleum gas (LPG). The 5 tons of LPG that were in the tank escaped over two to three hours but fortunately did not catch fire. However, 2,000 people who lived near the plant were evacuated from their homes [11].

(c) In canned pumps, the moving part of the electric motor—the rotor—is immersed in the process liquid; there is no gland, and gland leaks cannot occur.

The fixed part of the electric motor—the stator—is not immersed in the process liquid and is separated from the rotor by a stainless steel can (Figure 1-17).

If there is a hole in the can, process liquid can get into the stator compartment. A pressure relief plug is therefore fitted to the compartment

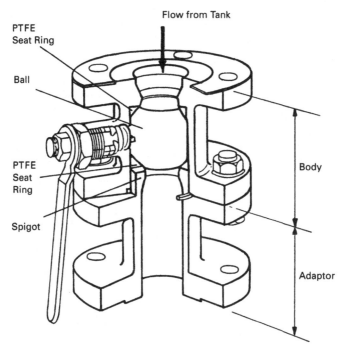

FIGURE 1-16 Valve dismantled in error.

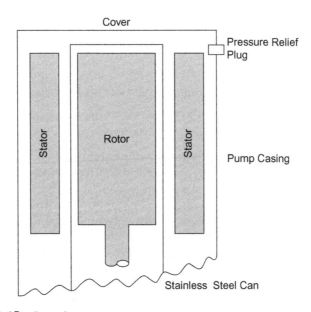

FIGURE 1-17 Canned pumps.

and should be used before the compartment is opened for work on the stator. Warning plates, reminding us to do this, are often fitted to the pumps.

The stator compartment of a pump was opened up without the pressure relief plug being used. There was a hole in the can. This had caused a pressure buildup in the stator compartment. When the cover was unbolted, it was blown off and hit a scaffold pole 2 m (6.5 ft) above. On the way up, it hit a man on the knee, and the escaping process vapor caused eye irritation. Persons working on the pump did not know the purpose of the plug, and the warning notice was missing.

For a more detailed diagram and description of a canned pump, see reference 12.

(d) On several occasions, fitters have removed thermowells without realizing that this would result in a leak. They did not realize that the thermowell—the pocket into which a thermocouple or other temperature-measuring device sits—is in direct contact with the process fluid. A serious fire that started this way is described in reference 13.

(e) A high-pressure reciprocating ammonia pump (known as an injector) had run for 23 years without serious problems when the crankshaft suddenly fractured, due to fatigue, and the plungers came out of the cylinders. The ammonia killed two men. No one realized that a failure of the motion work would produce a massive release of ammonia. If people had realized this could happen, they would have installed remotely operated emergency isolation valves (see Section 7.2.1). These would have greatly reduced the size of the leak but would not have acted quickly enough to prevent the fatalities [31].

1.5.5   Treating the Symptoms Instead of the Disease

The following incidents and Section 10.5.3 show what can happen if we continue to repair faults but never ask why so many faults occur:

(a) A cylinder lining on a high-pressure compressor was changed 27 times in nine years. On 11 occasions it was found to be cracked, and on the other 16 occasions it showed signs of wear. No one asked why it had to be changed so often. Everyone just went on changing it. Finally, a bit of the lining got caught between the piston and the cylinder head and split the cylinder.

(b) While a man was unbolting some ¾-in. bolts, one of them sheared. The sudden jerk caused a back strain and absence from work. During the investigation of the accident, seven bolts that had been similarly sheared on previous occasions were found nearby. It was clear that the bolts sheared frequently. If, instead of simply replacing them and carrying on, the workers had reported the failures, then a more suitable bolt material could have been found.

Why did they not report the failures? If they had reported them, would anything have been done? The accident would not have occurred if the foreman or the engineer, on their plant tours, had noticed the broken bolts and asked why there were so many.

(c) A line frequently choked. As a result of attempts to clear the chokes, the line was hammered almost flat in several places. It would have been better to have replaced the line with a larger one or with a line that had a greater fall, more gentle bends, or rodding points.

1.5.6   Flameproof Electrical Equipment

On many occasions, detailed inspections of flameproof electrical equipment have shown that many items were faulty. For example, at one plant, a first look around indicated that nothing much was wrong. A more thorough inspection, paying particular attention to equipment not readily accessible and that could be examined only from a ladder, showed that out of 121 items examined, 33 needed repair. The faults included missing and loose screws, gaps too large, broken glasses, and incorrect glands. Not all the faults would have made the equipment a source of ignition, but many would have done so.

Why were there so many faults? Before this inspection, there had been no regular inspections. Many electricians did not understand why flameproof equipment was used and what would happen if it were badly maintained. Spare screws and screwdrivers of the special types used were not in stock, so there was no way of replacing those lost.

Regular inspections were set up. Electricians were trained in the reasons why flameproof equipment is used, and spares were stocked. In addition, it was found that in many cases flameproof equipment was not really necessary. Division (Zone) 2 equipment—cheaper to buy and easier to maintain—could be used instead.

1.5.7   Botching

Section 1.5.3a described a botched job. Here are two more:

(a) A pressure vessel was fitted with a quick-opening lid, with a 10-in. diameter, secured by four eyebolts (Figure 1-18). They had to be replaced, as the threads were corroded. Instead of replacing the whole eyebolt, a well-meaning person decided to save time by simply cutting the eyes off the bolts and welding new studs onto them. As soon as the vessel was pressurized (with compressed air), the new studs, which had been made brittle by the welding, failed, and the lid flew off. Fortunately, a short length of chain restrained it, and it did not fly very far [38]. (See Sections 13.5 and 17.1 for the hazards of quick-opening lids.)

(b) A screwdriver was left in the steering column of a truck after the truck was serviced. The truck and semitrailer crashed, and the servicing company had to pay $250,000 in damages. To quote from the report, "Workplaces need to be as rigorous as the aviation and medical industries in ensuring that all tools are accounted for when servicing is completed" [39].

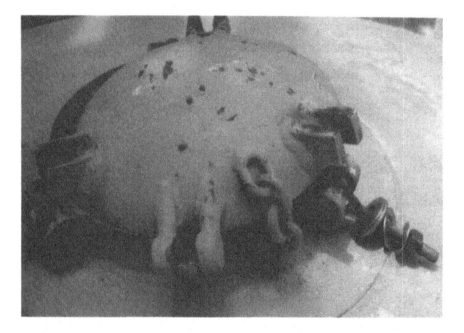

FIGURE 1-18   Instead of replacing the eyebolts, new studs were welded in place of the threaded portions. They were made brittle by the heat and failed in use. Fortunately, the chain prevented the lid from going into orbit.

1.5.8   Who Should Decide How to Carry Out a Repair?

The following report raises two interesting questions:

- Who decides how a maintenance job should be carried out?
- How should we clear chokes in small bore lines?

A sample point on the suction line of two water pumps became choked, and a maintenance worker was asked to clear it. He was not told how to do so—craftspeople dislike people from other departments telling them how to do their jobs—but the operators assumed he would use water under pressure or a rod. Instead he used compressed air at a gauge pressure of 115 psi (8 bar), and a pocket of air caused the pumps to lose suction (Figure 1-19).

The results were not serious. The pumps supplied water to cool the hot gases leaving an incinerator; when the water flow stopped, a high-temperature trip shut down the burner. The incinerator was new, it was still undergoing tests, and the job had not been done before. The water was recycled, and ash in it probably caused the choke [40].

According to the report, the maintenance worker should have been given more detailed instructions. But, as it also points out, some skills

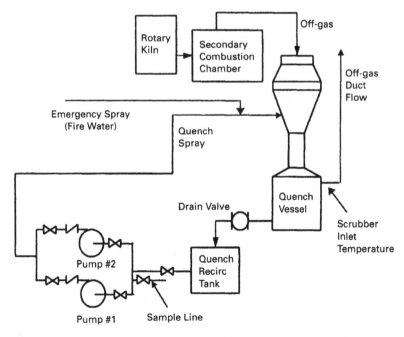

FIGURE 1-19 Simplified drawing of incinerator quench recirculation system. *(Illustration courtesy of the U.S. Department of Energy.)*

are skills of the craft; we should be able to assume that craftspeople are aware of them and should not need to give them detailed instructions on each and every occasion. We should not need to tell them, for example, how to break a joint every time they are asked to do so (but see Section 1.5.1). Where do we draw the line?

Craftspeople (and operators) ought to be taught, as part of their safety training, that compressed gases should not be used to clear chokes. There is a lot of energy in a compressed gas, and it can accelerate a plug to great speed, putting it into orbit if there is an open end or breaking a pipeline if the plug hits a bend (see Section 17.2).

This incident shows how much we can learn from a simple event if we treat it as a learning experience and do not say, "No one was hurt, and there was no damage, so let's forget about it."

1.6 A PERSONAL NOTE

The recommendations described in this chapter go further than some companies consider necessary. For example, companies may put Do Not Operate notices on valves instead of locks, or to save time they may turn a blind eye to occasional shortcuts. Nevertheless, bitter experience has convinced me that the recommendations are necessary.

In 1968, after 16 of years' experience in production, I was transferred to a new position in safety. It was an unusual move at the time for someone with my background, but five deaths from three serious fires in three years, two of them the results of poor preparation for maintenance, convinced senior management that more resources should be devoted to safety and that it could no longer be left to nontechnical people and elderly foremen. Since then I have read scores of reports about other accidents that happened for this reason. Some were serious; others were near misses.

When I retired from industry, one of my first tasks was to sort the many accident reports I had collected. The thickest folder by far was one labeled "Preparation for Maintenance." Some of the incidents from that folder, together with more recent ones, are described in this chapter.

If you decide my recommendations are not right for your organization, please do not ignore the accidents I have described. Check that your procedures will prevent them, or they will happen again.

There are more reports on maintenance accidents in Chapter 23.

References

1. T. A. Kletz, in D. A. Crowl and S. S. Grossel (editors), *Handbook of Toxic Materials Handling and Management*, Marcel-Dekker, New York, 1994, Chapter 11.
2. M. S. Mannan (editor), *Lees' Loss Prevention in the Process Industries*, 3rd edition, Elsevier, Boston, MA, 2005, Chapter 21.

3. T. A. Kletz, *Lessons from Disaster: How Organizations Have No Memory and Accidents Recur*, co-published by Institution of Chemical Engineers, Rugby, UK and Gulf Publishing Co., Houston, TX, 1993.

4. *Annual Report of the Chief Inspector of Factories for 1974*, Her Majesty's Stationery Office, London, 1975, p. 19.

5. *The Bulletin, The Journal of the Society for Petroleum Acts Administration*, Oct. 1970, p. 68.

6. *Chemical Safety Summary*, Chemical Industries Association, London, July–Sept. 1980, p. 15.

7. *Petroleum Review*, Apr. 1982, p. 34.

8. *Annual Report of Her Majesty's Inspectors of Explosives for 1970*, Her Majesty's Stationery Office, London, 1971, p. 19.

9. *Health and Safety—Manufacturing and Service Industries 1979*, Her Majesty's Stationery Office, London, 1981, p. 62.

10. T. A. Kletz, *Hydrocarbon Processing*, Vol. 61, No. 3, Mar. 1982, p. 207.

11. Health and Safety Executive, *Leakage of Propane at Whitefriars Glass Limited, Wealdstone, Middlesex, 20 November 1980*, Her Majesty's Stationery Office, London, 1981.

12. G. R. Webster, *The Chemical Engineer*, Feb. 1979, p. 91.

13. *Petroleum Review*, Oct. 1981, p. 21.

14. J. H. Christiansen and L. E. Jørgensen, *Proceedings of the Fourth International Symposium on Loss Prevention and Safety Promotion in the Process Industries* (held in Harrogate, UK, Sept. 1983), Institution of Chemical Engineers, Rugby, UK, p. L9.

15. L. G. Britton and J. A. Smith, *Plant/Operations Progress*, Vol. 7, No. 1, Jan. 1988, p. 53.

16. *The Phillips 66 Company Houston Chemical Complex Explosion and Fire*, U.S. Dept. of Labor, Washington, DC, Apr. 1990.

17. *Oil and Gas Journal*, May 28, 1990, p. 36.

18. W. D. Cullen, *The Public Inquiry into the Piper Alpha Disaster*, Her Majesty's Stationery Office, London, 1990, especially paragraphs 6.109, 6.187, 11.3, and 18.29.

19. *Loss Prevention Bulletin*, No. 091, Feb. 1990, p. 17.

20. *Potential Fire Hazard from Anaerobic Decomposition in Cooling Water System*, Safety Note No. DOE/EH-0109, U.S. Dept. of Energy, Washington, DC, Oct. 1989.

21. H. E. Watts, *Report on Explosion in Kerosene Tank at Killingholme, Lincs.*, Her Majesty's Stationery Office, London, 1938.

22. *Loss Prevention Bulletin*, No. 099, Apr. 1991, p. 9.

23. C. H. Vervalin, *Hydrocarbon Processing*, Vol. 51, No. 12, Dec. 1972, p. 52.

24. *Loss Prevention Bulletin*, No. 098, Apr. 1991, p. 25.

25. *Labeling Errors Cause Accidents*, Safety Note No. DOE/EH-0328, U.S. Dept. of Energy, Washington, DC, July 1993.

26. C. W. Ramsay, *Plant/Operations Progress*, Vol. 9, No. 2, Apr. 1990, p. 117.

27. *Occupational Health and Safety Observer*, Vol. 2, No. 6, U.S. Dept. of Energy, Washington, DC, June 1993, p. 6.

28. B. A. Prine, "Analysis of Titanium/Carbon Steel Heat Exchanger Fire," *Paper presented at AIChE Loss Prevention Symposium*, Aug. 1991.

29. *Occupational Health and Safety Observer*, Vol. 2, No. 1, U.S. Dept of Energy, Washington, DC, Sept. 1993, p. 3.

30. *Loss Prevention Bulletin*, No. 100, Aug. 1991, p. 35.

31. P. J. Nightingale, "Major Incident Following the Failure of an Ammonia Injector on a Urea Plant," *Paper presented at AIChE Ammonia Symposium*, Aug. 1990.

32. *Operating Experience Weekly Summary*, No. 96-50, Office of Nuclear and Facility Safety, U.S. Dept. of Energy, Washington, DC, 1996, p. 9.

33. *Loss Prevention Bulletin*, No. 107, Oct. 1992, p. 17.

34. *Operating Experience Weekly Summary*, No. 96-44, Office of Nuclear and Facility Safety, U.S. Dept. of Energy, Washington, DC, 1996, p. 2.

35. *Occupational Safety and Health Observer*, Vol. 3, Nos. 7 and 8, U.S. Dept. of Energy, Washington, DC, July–Aug. 1994, p. 1.

36. *Underground Utilities Detection and Excavation,* Safety Notice No. DOE/EH-0541, Office of Nuclear and Facility Safety, U.S. Dept. of Energy, Washington, DC, 1996.

37. *Effective Shift Handover—A Literature Review,* Report No. OTO 96 003, Health and Safety Executive, Sheffield, UK, 1996, p. 7.

38. R. E. Sanders and W. L. Spier, *Process Safety Progress,* Vol. 15, No. 4, Winter 1996, p. 189.

39. *Safety Management,* South Africa, Sept. 1995, p. 2. Though reported in a South African magazine, the incident occurred in Australia.

40. *Operating Experience Weekly Summary,* Nos. 96-46 and 96-49, Office of Nuclear and Safety Facility, U.S. Dept. of Energy, Washington, DC, 1996, p. 9 (of 96-46) and 2 (of 96-49).

41. *Operating Experience Weekly Summary,* No. 97-20, Office of Nuclear and Facility Safety, U.S. Dept. of Energy, Washington, DC, 1997, p. 2.

42. *Operating Experience Weekly Summary,* No. 97-15, Office of Nuclear and Facility Safety, U.S. Dept. of Energy, Washington, DC, 1997, p. 7.

43. J. A. Senecal, *Journal of Loss Prevention in the Process Industries,* Vol. 4, No. 5, 1991, p. 332.

44. *Operating Experience Weekly Summary,* No. 97-22, Office of Nuclear and Facility Safety, U.S. Dept. of Energy, Washington, DC, 1997, p. 3.

45. *Loss Prevention Bulletin,* No. 098, Apr. 1991, p. 25.

46. Health and Safety Executive, *Dangerous Maintenance,* Her Majesty's Stationery Office, London, 1987, p. 8.

Modifications

I consider it right that every talented man should be at liberty to make improvements, but that the supposed improvements should be duly considered by proper judges.

—George Stephenson, 1841

Many accidents have occurred because changes were made in plants or processes and these changes had unforeseen side effects. This chapter describes a number of such incidents. How to prevent similar changes in the future is discussed. Some of the incidents are taken from references 1 and 2, where others are described as well.

2.1   STARTUP MODIFICATIONS

Startup is a time when many modifications may have to be made. It is always a time of intense pressure. It is therefore not surprising that some modifications introduced during startup have had serious unforeseen consequences.

At one plant, a repeat relief and blowdown review was carried out one year after startup. The startup team had been well aware of the need to look for the consequences of modifications and had tried to do so as modifications were made. Nevertheless, the repeat relief and blowdown review brought to light 12 instances in which the assumptions of the original review were no longer true and additional or larger relief valves, or changes in the position of a relief valve, were necessary. Figure 2-1 shows some examples.

doi:10.1016/B978-1-85617-531-9.00002-0
45

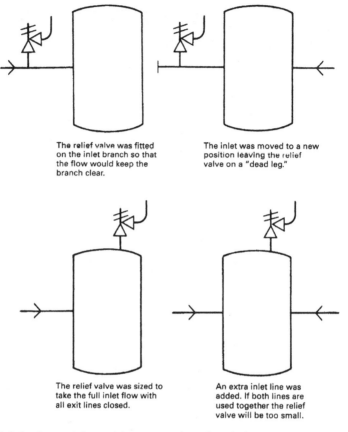

The relief valve was fitted on the inlet branch so that the flow would keep the branch clear.

The inlet was moved to a new position leaving the relief valve on a "dead leg."

The relief valve was sized to take the full inlet flow with all exit lines closed.

An extra inlet line was added. If both lines are used together the relief valve will be too small.

FIGURE 2-1 Some of the modifications made to the relief system in a plant during its first year on line.

The line diagrams had been kept up to date despite the pressures on the plant staff during startup. This made it easier to repeat the relief and blowdown review. The plant staff members were so impressed by the results that they decided to have another look at the relief and blowdown after another year.

Section 5.5.2c describes a late change in design that had unforeseen results.

2.2 MINOR MODIFICATIONS

This term is used to describe modifications so inexpensive that either they do not require formal financial sanction or the sanction is easily

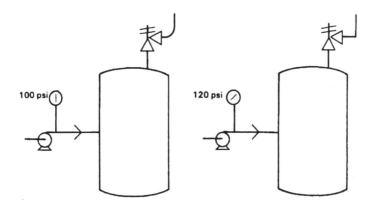

The vessel was designed to withstand the maximum pressure the pump could deliver. The relief valve was not designed to take the maximum flow from the pump.

The pump actually proved capable of producing 20 psi more than design. If the exit from the vessel is isolated when the pump is running the vessel will be overpressured.

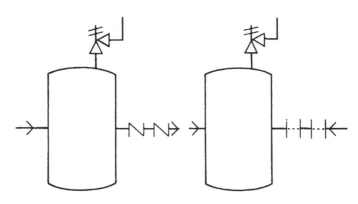

The relief valve was sized on the assumption that two non-return valves in series would prevent backflow into the vessel.

Both nonreturn valves corroded, allowing back-flow to take place.

FIGURE 2-1 Continued

obtained. They therefore may not receive the same detailed consideration as a more expensive modification.

(a) A modification so simple that it required only a work permit resulted in the end blowing off a tank and fatal injuries to two men working in the area.

The tank was used for storing a liquid product that melts at 97°C (206°F). It was therefore heated by a steam coil using steam at a

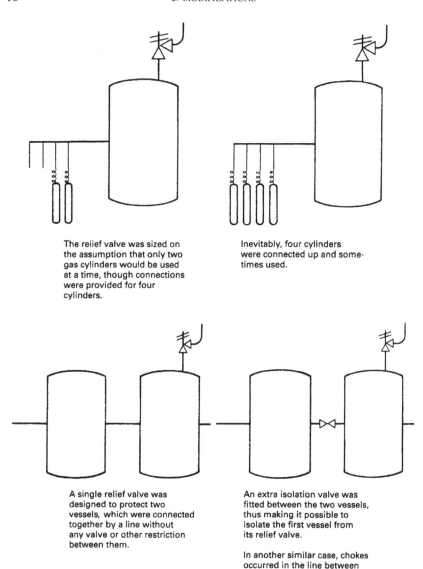

The relief valve was sized on the assumption that only two gas cylinders would be used at a time, though connections were provided for four cylinders.

Inevitably, four cylinders were connected up and sometimes used.

A single relief valve was designed to protect two vessels, which were connected together by a line without any valve or other restriction between them.

An extra isolation valve was fitted between the two vessels, thus making it possible to isolate the first vessel from its relief valve.

In another similar case, chokes occurred in the line between the two vessels.

FIGURE 2-1   Continued

gauge pressure of 100 psi (7 bar). At the time of the incident, the tank was almost empty and was being prepared to receive some product. The inlet line was being blown with compressed air to prove that it was clear—the normal procedure before filling the tank. The air was not getting through, and the operator suspected a choke in the pipeline.

In fact, the vent on the tank was choked. The gauge air pressure (75 psi or 5 bar) was sufficient to burst the tank (design gauge pressure 5 psi or 0.3 bar). Originally the tank had a 6-in.-diameter vent. But at some time this was blanked off, and a 3-in.-diameter dip branch was used instead as the vent.

Several other things were wrong. The vent was not heated; its location made it difficult to inspect. Most important of all, neither manager, supervisors, nor operators recognized that if the vent choked, the air pressure was sufficient to burst the tank. Nevertheless, if the 6-in. vent had not been blanked, the incident would not have occurred (see also Section 12.1). Everyone should know the maximum and minimum temperatures and pressures that their equipment can withstand.

(b) A reactor was fitted with a bypass (Figure 2-2a). The remotely operated valves A, B, and C were interlocked so that C had to be open before A or B could be closed. It was found that the valves leaked, so hand-operated isolation valves (a, b, and c) were installed in series with them (Figure 2-2b).

After closing A and B, the operators were instructed to go outside and close the corresponding hand valves a and b. This destroyed the interlocking. One day, an operator could not get A and B to close. He had forgotten to open C. He decided that A and B were faulty and closed a and b. Flow stopped. The tubes in the furnace were overheated. One of them burst, and the lives of the rest were shortened.

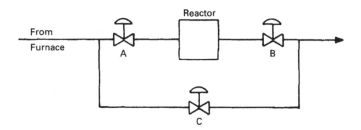

FIGURE 2-2a   Original reactor bypass.

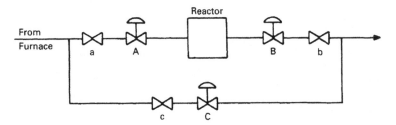

FIGURE 2-2b   Modified reactor bypass.

(c) A let-down valve was a bottleneck, so a second let-down valve was added in parallel (Figure 2-3). During installation, the check valve was hidden beneath insulation and was not noticed, and the parallel line was joined to the original line downstream of the check valve where there was a convenient branch. The upstream equipment was thus connected directly to the downstream equipment, bypassing the relief valve.

A blockage occurred downstream. The new let-down valve was leaking, and the downstream equipment was overpressured and burst.

When the modification was designed, the designers assumed that the new line would join the original line immediately after the relief valve. If they realized the importance of this, they did not draw attention to it, and they did not check that the modification had been made correctly [13].

Modifications should always be marked on a line diagram before they are approved, and the person who authorizes them should always inspect the finished modification to make sure that his or her intentions have been followed.

(d) A group of three rooms in a control center was pressurized to prevent the entry of hazardous vapors. A fan blew air into room 1, and it passed through louvers into room 2 and from there into room 3. A pressure controller measured the pressure in room 1. For an unknown reason, the louvers between rooms 1 and 2 were blocked. Six years passed before anyone realized that the pressures in the other rooms were not under control [20]. The company had a procedure for the control of modifications, but those concerned did not use it when the louvers were blocked, perhaps because they thought it applied only

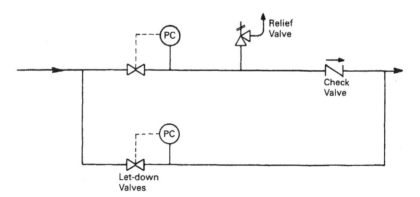

FIGURE 2-3 The second *(lower)* let-down valve was joined to the original line downstream of the check valve. When this let-down valve leaked, the downstream equipment was overpressured.

to plant equipment and not to buildings. What would happen in your plant?

In another building, gusting winds caused transient increases in pressure that tripped the ventilation fans. The problem was overcome by increasing the fan speeds. Unfortunately, the increased flow of cold air through a doorway cooled a fire water main so much that it froze and cracked [21].

(e) Other minor modifications that have had serious effects on plant safety are as follows:

1. Removing a restriction plate that limits the flow into a vessel and that has been taken into account when sizing the vessel's relief valve. A length of narrow bore pipe is safer than a restriction plate, as it is less easily removed.
2. Fitting a larger trim into a control valve when the size of the trim limits the flow into a vessel and has been taken into account when sizing the vessel's relief valve.
3. Fitting a substandard drain valve (see Section 8.2a).
4. Replacing a metal duct or pipe by a hose (see Section 15.3).
5. Solid was scraped off a flaker—a rotating steel drum—by a steel knife. After the knife was replaced by a plastic one, an explosion occurred, probably because more dust was produced [14].
6. Without consulting the manufacturer, the owner of a set of hot tapping equipment made a small modification: he installed a larger vent valve to speed up its use. As a result, the equipment could no longer withstand the pressure and was violently ejected from a pipeline operating at a gauge pressure of 40 bar (600 psi) [22].
7. Making a small change in the size of a valve spindle and thus changing its natural frequency of vibration (see Section 9.1.2a).
8. Changing the level in a vessel (see Sections 2.6i, 22.2d, and 22.2e).

2.3 MODIFICATIONS MADE DURING MAINTENANCE

Even when systems for controlling modifications have been set up, modifications often slip in unchecked during maintenance. (Someone decides, for what he or she thinks is a good reason, to make a slight change.)

Many years ago, a special network of air lines was installed for use with air masks only. A special branch was taken off the top of the compressed air main as it entered the works (Figure 2-4).

For 30 years this system was used without any complaint. Then one day a man got a face full of water while wearing an air mask inside a vessel. Fortunately, he was able to signal to the standby man that something was wrong, and he was rescued before he suffered any harm.

Investigators found that the compressed air main had been renewed and that the branch to the breathing apparatus network had been moved to the *bottom* of the main. When a slug of water got into the main, it all went into the catchpot, which filled up more quickly than it could empty. Unfortunately, everyone had forgotten why the branch came off the top of the main, and nobody realized that this was important.

A similar incident occurred on a fuel-gas system. When a corroded main was renewed, a branch to a furnace was taken off the bottom of the main instead of the top. A slug of liquid filled up the catchpot and extinguished the burners.

Some hot (370°C [700°F]) pipework was supported by spring hangers to minimize stress as it was heated and cooled. The atmosphere was corrosive, and the spring hangers became impaired. They were removed, and the pipework was left solidly supported. It could not withstand the stress, and a condenser fractured; hot heat-transfer oil was released and caught fire.

No one, it seems, realized the importance of the supports. Unlike the pipework, they were not protected against corrosion and were removed with little or no thought about the consequences [23].

As the result of some problems with a compressor, changes were made to the design of a shaft labyrinth. A new one was ordered and installed, but a spare of the old design was left in the store. Eight years later, after the staff had changed, the part had to be replaced. You've guessed what happened: the old spare was withdrawn from the store and installed.

A similar incident occurred on a boiler. After the No. 1 roof tube had failed several times, it was replaced by a thicker tube, and the change

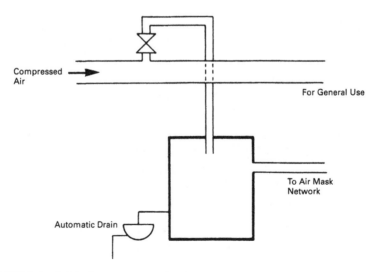

FIGURE 2-4   Original arrangement of air lines.

marked on the drawings. Some time later, a small leak developed in this tube, and a length had to be replaced. No one looked at the drawings, and a 0.5 m length of standard tube was welded in. The discontinuity caused turbulence, local overheating, and rapid failure [24].

Moving people is a modification of which the consequences are rarely considered. (See Chapter 26.)

2.4   TEMPORARY MODIFICATIONS

(a) The most famous of all temporary modifications is the temporary pipe installed in the Nypro Factory at Flixborough, in the United Kingdom, in 1974. It failed two months later, causing the release of about 50 tons of hot cyclohexane. The cyclohexane mixed with the air and exploded, killing 28 people and destroying the plant [3, 25, 26].

At the Flixborough plant, there were six reactors in series. Each reactor was slightly lower than the one before so that the liquid in them flowed by gravity from No. 1 down to No. 6 through short 28-in.-diameter connecting pipes (Figure 2-5). To allow for expansion, each 28-in. pipe contained a bellows (expansion joint).

One of the reactors developed a crack and had to be removed. (The crack was the result of a process modification; see Section 2.6b.) It was replaced by a temporary 20-in. pipe, which had two bends in it, to allow for the difference in height. The existing bellows were left in position at both ends of the temporary pipe (Figure 2-5).

The design of the pipe and support left much to be desired. The pipe was not properly supported; it merely rested on scaffolding. Because there was a bellows at each end, it was free to rotate or "squirm" and did so when the pressure rose a little above the normal level. This caused the bellows to fail.

No professionally qualified engineer was in the plant at the time the temporary pipe was built. The men who designed and built it

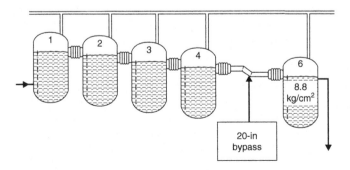

FIGURE 2-5   Arrangement of reactors and temporary pipe at Flixborough.

(*design* is hardly the word because the only drawing was a full-scale sketch in chalk on the workshop floor) did not know how to design large pipes required to operate at high temperatures (150°C [300°F]) and gauge pressures (150 psi or 10 bar). Few engineers have the specialized knowledge to design highly stressed piping. But in addition, the engineers at Flixborough did not know that design by experts was necessary.

They did not know what they did not know [25, 26]. As a result of the Flixborough explosion, many companies introduced procedures to manage change.

(b) A reactor was cooled by a supply of brine to the jacket. The brine system had to be shut down for repair, so town water was used instead. The pressure of the town water (a gauge pressure of 130 psi or 9 bar) was higher than that of the brine, and the reactor collapsed.

A modification approval form, containing 20 questions, had been completed before the modification was made, but this was treated as a formality, and the questions were answered in a perfunctory manner [6].

For another temporary modification, see Section 5.5.1.

2.5   SANCTIONED MODIFICATIONS

This term is used to describe modifications for which the money has to be authorized by a senior manager or a committee. These modifications cannot, therefore, be done in a hurry. Justifications have to be written out and people persuaded. Although the systems are (or have been in the past) designed primarily to control cost rather than safety, they usually result in careful consideration of the proposal by technical personnel. Unforeseen consequences may come to mind, though not always. Sometimes sanction is obtained before detailed design has been carried out, and the design may then escape detailed considerations. Nevertheless, it is harder to find examples of serious incidents caused by sanctioned modifications. The following might almost rank as a startup modification. Though the change was agreed upon more than a year before startup, it occurred after the initial design had been studied and approved.

(a) A low-pressure refrigerated ethylene tank was provided with a relief valve set at a gauge pressure of about 1.5 psi (0.1 bar), which discharged to a vent stack. After the design had been completed, it was realized that cold gas coming out of the stack would, when the wind speed was low, drift down to ground level, where it might be ignited. The stack was too low to be used as a flarestack—the radiation at ground level would be too high—and was not strong enough to be extended. What could be done?

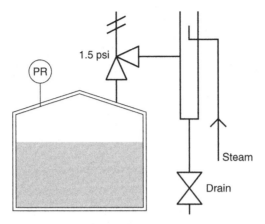

FIGURE 2-6 Liquid ethylene tank venting arrangements.

Someone suggested putting steam up the stack to disperse the cold vapor. This seemed a good idea, and the suggestion was adopted (Figure 2-6).

As the cold vapor flowed up the stack, it met condensate flowing down. The condensate froze and completely blocked the 8-in.-diameter stack. The tank was overpressured and ruptured. Fortunately, the rupture was a small one and the escaping ethylene did not ignite. It was dispersed with steam while the tank was emptied.

Should the design team have foreseen that the condensate might freeze? A hazard and operability study (see Chapter 18) would probably have drawn attention to the hazard.

After the tank was repaired, the vent stack was replaced by a flarestack.

(See also Sections 3.3.1b, 6.2b, 8.1.6, and 9.2.1 g.)

(b) A new loading gantry was built for filling tank trucks with liquefied petroleum gas. The ground was sloped so that any spillages would run away from the tanker and would not heat it if they caught fire. As a result of this change in design, it was found that the level indicator would not read correctly when the tank truck was located on sloping ground. The design had to be modified again so that the wheels stood on level ground, but the ground in between and around them was sloped.

(c) A "carbon copy" plant was built with the floors 3 m (10 ft) apart instead of 2.4 m (8 ft) apart, as 3 m was the company standard. The increased height was too much for convective flow, and efficiency was lost.

(d) When construction of a nuclear power station was well under way, an advisory committee suggested that a zirconium liner should be added to part of the cooling circuit. The operating company did not think the liner was necessary, but it was cheap to install; proving that it was unnecessary would have cost more than installing the liner, so the company installed it. Bits of it came loose and blocked the cooling circuit; the reactor overheated and was damaged, but there was no release of radioactivity. The liner was not replaced when the reactor was repaired [15].

The operator of a plant, not an advisory committee, however eminent, is responsible for the safety of the plant and should not follow advice that it believes to be wrong just to save costs or avoid arguments.

Another sanctioned modification is described in Section 12.4.6.

2.6   PROCESS MODIFICATIONS

So far we have discussed modifications to the plant equipment. Accidents can also occur because changes to process materials or conditions had unforeseen results, as the following cases and Section 19.5 show:

(a) A hydrogenation reactor developed a pressure drop. Various causes were considered—catalyst quality, size, distribution, activation, reactant quality, distribution, and degradation—before the true cause was found.

The hydrogen came from another plant and was passed through a charcoal filter to remove traces of oil before it left the supplying plant. Changes of charcoal were infrequent, and the initial stock lasted several years. Reordering resulted in a finer charcoal being supplied and charged without the significance of the change being recognized. Over a long period, the new charcoal passed through its support into the line to the other plant. Small amounts of the charcoal partially clogged the ⅜-in. (10 mm) distribution holes in the catalyst retaining plate (Figure 2-7). There was a big loss of production.

The cause of the pressure drop was difficult to find because it was due to a change in another plant.

(b) At one time, it was common to pour water over equipment that was too hot or that was leaking fumes. The water was taken from the nearest convenient supply. At Flixborough, cyclohexane vapor was leaking from the stirrer gland on one of the reactors. To condense the leaking vapor, water was poured over the top of the reactor. Plant cooling water was used because it was conveniently available.

Unfortunately, the water contained nitrates, which caused stress corrosion cracking of the mild steel reactor. The reactor was removed for repair, and the temporary pipe that replaced it later failed and caused the explosion (see Section 2.4a).

Nitrate-induced cracking is well known to metallurgists but was not well known to other engineers at the time. Before you poured

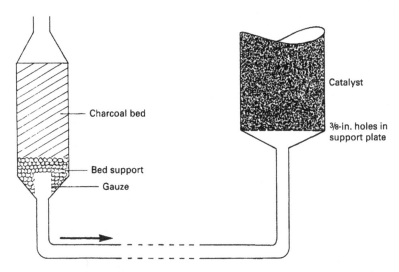

FIGURE 2-7   Hydrogen purification system.

water over equipment—emergencies apart—would you ask what the water contained and what its effect would be on the equipment?

Pouring water over equipment is a change from normal operating practice. It should therefore be treated as a modification. For a description of nitrate cracking of mild steel, see reference 4.

(c) The following incident shows how difficult it is to foresee all the results of a change and how effects can be produced a long way downstream of the place where the change is made:

Some radioactive bromine (half-life 36 hours), in the form of ammonium bromide, was put into a brine stream as a radioactive tracer. At another plant 30 km (20 miles) away, the brine stream was electrolyzed to produce chlorine. Radioactive bromine entered the chlorine stream and subsequently concentrated in the base of a distillation column, which removed heavy ends. This column was fitted with a radioactive-level controller. The radioactive bromine affected the level controller, which registered a low level and closed the bottom valve on the column. The column became flooded. There was no injury, but production was interrupted.

(d) A minor change in operating conditions can sometimes have devastating results. A nitration reaction was carried out at low temperature, and then the reactor was heated to 90°C and kept at this temperature for 30 minutes; it was then cooled. After a year's operation, someone decided to let the batch cool by heat loss to the surroundings, with no one in attendance, as soon as the temperature reached 90°C. An explosion occurred; the building was wrecked, and parts of the reactor were found 75 m away [7].

(e) The heating in a building had to be shut down over a weekend for repairs. There were fears that the water in the sprinkler system might freeze, so it was replaced by alcohol. A fire occurred and was fed by the sprinklers!

(f) Aqueous ammonia was added to a plant to reduce corrosion. The corrosion stopped, but the liquid droplets caused erosion, a pipe failure, and a substantial fire [8].

(g) Three vacuum stills were fitted with steam ejectors and direct contact condensers. The water was cooled in a small cooling tower and recycled, and the small amount of vapor carried over from the stills dispersed in the tower. The tower was fitted with a fan, but to save electricity, the operators switched off the fan and found that they could get adequate cooling without it. However, the flammable vapors from the stills were no longer dispersed so effectively. This was discovered when an operator, about to light a furnace a few meters away, tested the atmosphere inside the furnace in the usual way—with a combustible gas detector—and found that gas was present.

No one realized, when a site for the furnace was decided, that flammable vapors could come out of the cooling tower. Direct contact condensers are not common, but flammable vapors can appear in many cooling towers if there are leaks on water-cooled heat exchangers. After the incident, a combustible gas detector was mounted permanently between the furnace and the tower (Figure 2-8).

(h) To meet new Environmental Protection Agency requirements, the fuel oil used for two emergency diesel generators was changed to a low-sulfur grade (from 0.3% maximum to 0.05% maximum). The lubricating oil used contained an additive that neutralized the sulfuric acid formed during combustion. With less acid produced, the excess additive formed carbon deposits, which built up behind the piston rings, causing them to damage the cylinder walls. Fortunately, the problem was found after test runs and solved by changing the type of lubricating oil, before the generators were needed in an emergency [27].

(i) A sparger was removed for inspection and found to be corroded, though it had been in use for 30 years and had never been known to corrode before. The problem was traced to a change in the level in the vessel so that the sparger was repeatedly wetted and then dried [28].

(j) The storage tank on a small detergent bottling plant was washed out every week. A small amount of dilute washings was allowed to flow into the dike and from there to drain. The operators carrying out the washing had to work in the dike and got their feet wet, so they connected a hose to the dike drain valve, put the other end into the sewer, and left it there. You've guessed right again. After a few months, someone left the drain valve open. When the tank was filled,

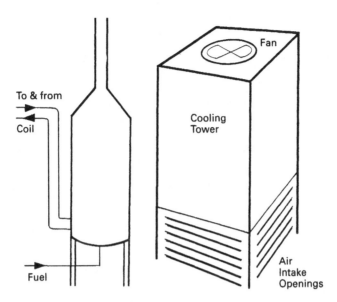

FIGURE 2-8   Gas entered the furnace when the cooling tower fan was switched off.

20 m^3 of detergent went down the drain. It overloaded the sewage plant, and a 3-m-high wall of foam moved down the local river [29].

(k) The duck pond at a company guesthouse was full of weeds, so the company water chemist was asked for advice. He added an herbicide to the pond. It was also a detergent; it wetted the ducks' feathers, and the ducks sank.

2.7 NEW TOOLS

The introduction of new tools can have unforeseen side effects:

(a) On several occasions, radioactive level indicators have been affected by radiography being carried out on welds up to 70 m away.

(b) This incident did not occur in the process industries, but nevertheless is a good example of the way a new tool can introduce unforeseen hazards:

A natural gas company employed a contractor to install a 2-in. plastic natural gas main to operate at a gauge pressure of 60 psi (4 bar) along a street. The contractor used a pneumatic boring technique. In doing so, he bored right through a 6-in. sewer pipe serving one of the houses on the street.

The occupant of the house, finding that his sewer was obstructed, engaged another contractor to clear it. The contractor used an auger and ruptured the plastic gas pipe. Within three minutes, the natural gas had traveled 12 m up the sewer pipe into the house and exploded. Two people were killed and four injured. The house was destroyed, and the houses on both sides were damaged.

After the explosion, it was found that the gas main had passed through a number of other sewer pipes [5].

2.8 ORGANIZATIONAL CHANGES

These changes can also have unforeseen side effects, as shown by the following incidents:

(a) A plant used sulfuric acid and caustic soda in small quantities, so the two substances were supplied in similar plastic containers called polycrates (Figure 2-9). While an operator was on his day off, someone decided it would be more convenient to have a polycrate of acid and a polycrate of alkali on each side (Figure 2-10). When the operator

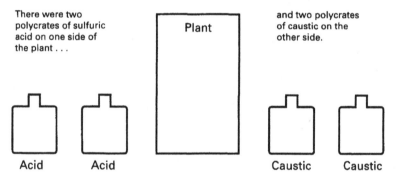

FIGURE 2-9   Original layout of acid and caustic containers.

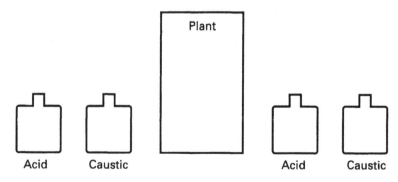

FIGURE 2-10   Modified layout of acid and caustic containers.

came back, no one told him about the change. Without checking the labels, he poured some excess acid into a caustic crate. There was a violent reaction, and the operator was sprayed in the face. Fortunately, he was wearing goggles.

We should tell people about changes made while they were away. In addition, if incompatible chemicals are handled at the same plant, then, whenever possible, the containers should differ in size, shape, or color, and the labels should be large and easily seen from eye level.

(b) The staff of a plant decided to exhibit work permits so that workers on the job could more readily see them—a good idea.

The permits were usually put in plastic bags and tied to the equipment. But sometimes they were rolled up and inserted into the open ends of scaffold poles.

One day a man put a permit into the open end of a pipe. He probably thought that it was a scaffold pole or defunct pipe. Unfortunately, it was the air bleed into a vacuum system. A motor valve controlled the air rate. The permit got sucked into the valve and blocked it. The vacuum could not be broken, product was sucked into the vacuum system, and the plant had to be shut down for cleaning for two days.

(c) Section 2.3 described some of the results of moving people. See Chapter 26 for more on organizational change.

2.9 GRADUAL CHANGES

These are the most difficult to control. Often, we do not realize that a change is taking place until it is too late. For example, over the years, steam consumption at a plant had gradually fallen. Flows through the mains became too low to prevent condensate from accumulating. On one of the mains, an inaccessible steam trap had been isolated, and the other main had settled slightly. Neither of these mattered when the steam flow was large, but it gradually fell. Condensate accumulated, and finally water hammer fractured the mains.

Oil fields that produce sweet (that is, hydrogen-sulfide-free) oil and gas can gradually become sour. If this is not detected in time, there can be risks to life and unexpected corrosion.

In ammonia plants, the furnace tubes end in pigtails—flexible pipes that allow expansion to take place. On one plant, over the years, many small changes were made to the pigtails' design. The net effect was to shorten the bending length and thus increase the stress. Ultimately, 54 tubes failed, producing a spectacular fire [9].

In the United Kingdom, cars are usually about 1.35 m (53 in.) high. During the 1990s, a number of taller models were introduced with heights of 1.6 to 1.8 m (62 to 70 in.). They gave better visibility, but the center of gravity rose, and the cars became less stable when cornering. An expensive model had to be withdrawn for modification [38].

Most incidents have occurred before. In 1906, in the United Kingdom, there was a sharp curve in the railway line outside Salisbury rail station. The speed limit was 30 mph, but drivers of trains that did not stop at the station often went faster. A new engine design was introduced, similar to those already in use but with a larger boiler and thus a higher center of gravity. When the train was driven around the curve at excessive speed, it came off the rails, killing 28 people. Afterward, all trains were required to stop at the station [39].

2.10 MODIFICATION CHAINS

We make a small change to a plant or new design. A few weeks or months later, we realize that the change had or will have a consequence we did not foresee and a further change is required; later still, further changes are required, and in the end we may wish we had never made the original change, but it may be too late to go back.

For example, small leaks through relief valves may cause pollution, so rupture discs were fitted below the relief valves (Figure 2-11a). (On other occasions they have been fitted to prevent corrosion of the relief valves.) We soon realized that if there is a pinhole in a rupture disc, the pressure in the space between the disc and the relief valve will rise until it is the same as the pressure below the disc. The disc will then not rupture until the pressure below it rises to about twice the design rupture pressure. Therefore, to prevent the interspace pressure rising, small vents to atmosphere were fitted between the discs and the relief valves (Figure 2-11b).

This is okay if the disc is there to prevent corrosion, but if the disc is intended to prevent pollution, it defeats the object of the disc. Pressure gauges were therefore fitted to the vents and the operators asked to read them every few hours (Figure 2-11c).

Many of the relief valves were on the tops of distillation columns and other high points, so the operators were reluctant to read the pressure gauges. They were therefore brought down to ground level and connected to the vents by long lengths of narrow pipe (Figure 2-11d).

These long lengths of pipe got broken or kinked or liquid collected in them. Sometimes operators disconnected them so the pressure always read zero. The gauges and long lengths of pipe were therefore replaced by excess flow valves, which vent small leaks from pinholes but close if the rupture disc ruptures (Figure 2-11e).

Unfortunately, the excess flow valves were fitted with female threads, and many operators are trained to screw plugs into any open female threads they see. So some of the excess flow valves became plugged.

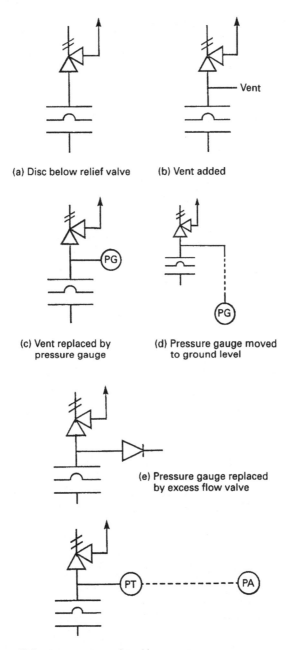

(a) Disc below relief valve   (b) Vent added

(c) Vent replaced by   (d) Pressure gauge moved
    pressure gauge   to ground level

(e) Pressure gauge replaced
    by excess flow valve

(f) Pressure gauge replaced by pressure
    transmitter alarming in control room

FIGURE 2-11   A modification chain—rupture discs below relief valves.

Pressure transmitters, alarming in the control room, were therefore fitted in place of the excess flow valves (Figure 2-11f). This was an expensive solution. Perhaps it would have been better to remove the rupture discs and prevent leaks to the atmosphere by taking more care over the machining and lapping of the relief valves.

A tank truck containing liquefied petroleum gas was fitted with a rupture disc below its relief valve, and a pressure gauge was fitted to the interspace. When it arrived at its destination, in Thailand, the customer telephoned the supplier, in Holland, to say the tank was empty, as the pressure gauge read zero [10].

For other examples of modification chains, see references 11 and 12.

2.11   MODIFICATIONS MADE TO IMPROVE THE ENVIRONMENT

Modifications made to improve the environment have sometimes produced unforeseen hazards [16]. We should, of course, try to improve the environment, but before making any changes we should try to foresee their results, as described in Section 2.12.

2.11.1   Explosions in Compressor Houses

A number of compressor houses and other buildings have been destroyed or seriously damaged, and the occupants killed, when leaks of flammable gas or vapor exploded. Indoors, a building can be destroyed by the explosion of a few tens of kilograms of flammable gas, but outdoors, several tons or tens of tons are needed. During the 1960s and 1970s, most new compressor houses and many other buildings in which flammable materials were handled were built without walls so that natural ventilation could disperse any leaks that occurred; the walls of many existing buildings were pulled down.

In recent years, many closed buildings have again been built to meet new noise regulations. The buildings are usually provided with forced ventilation, but this is much less effective than natural ventilation and is usually designed for the comfort of the operators rather than the dispersion of leaks.

The noise radiation from compressors can be reduced in other ways, for example, by surrounding the compressor with acoustic insulation. Any gap between the compressor and the insulation should be purged with air.

The leaks that lead to explosions in compressor houses are often not from a compressor but from other equipment, such as pipe joints. One such leak occurred because a spiral-wound gasket had been replaced by a compressed asbestos fiber one, probably as a temporary measure, seven years earlier. Once installed, it was replaced by a similar one during subsequent maintenance [30].

Another explosion, which killed one man and destroyed three natural gas compressors and the building housing them, started when five of the eight nuts that held a bypass cap on a suction valve failed, as the result of fatigue. They had been overtightened. The emergency shutdown system failed to operate when gas was detected and again when an attempt was made to operate it manually. It was checked only once per year. The source of ignition was believed to be the electrical equipment on the gas engine that drove the compressor [31].

In recent years there has been a rapid growth in the number of combined heat and power (CHP) and combined cycle gas turbine (CCGT) plants, driven mainly by gas turbines using natural gas, sometimes with liquid fuel available as standby. Governments have encouraged the construction of these plants, as their efficiency is high and they produce less carbon dioxide than conventional coal and oil-burning power stations. However, they present some hazards, as gas turbines are noisy and are therefore usually enclosed.

In addition, they are usually constructed without isolation valves on the fuel supply lines. As a result, the final connection in the pipework cannot be leak-tested. In practice, it is tested as far as possible at the manufacturer's works but often not leak-tested onsite. Reference 32 reviews the fuel leaks that have occurred, including a major explosion at a CCGT plant in England in 1996 due to the explosion of a leak of naphtha from a pipe joint. One man was seriously injured, and a 600-m^3 chamber was lifted off its foundations. The reference also reviews the precautions that should be taken. They include selecting a site where noise reduction is not required or can be achieved without enclosure. If enclosure is essential, then a high ventilation rate is needed; it is often designed to keep the turbine cool and is far too low to disperse gas leaks. Care must be taken to avoid stagnant pockets.

A reaction occasionally ran away and released vapor through a vent into the surrounding building. The vapor condensed to form a flammable fog. It had never been known to ignite, but nevertheless the company issued a strong but nonbinding recommendation that the walls of the building should be removed. One plant decided not to follow the recommendation. As a result, an explosion removed most of the walls. The source of ignition was never found [33].

2.11.2   Aerosols and Other Uses of CFCs

During the 1980s, it became recognized that chlorofluorocarbons (CFCs), widely used as aerosol propellants, are damaging the ozone layer, and aerosol manufacturers were asked to use other propellants. Some manufacturers already used butane, a cheaper material, and other manufacturers started to use it. The result was a series of fires and explosions.

The change was made quickly with little consideration of the hazards of handling butane. The reports on some of the fires that occurred say the hazards were not understood and that elementary safety precautions were lacking. One United Kingdom company was prosecuted for failing to train employees in the hazards of butane, in fire evacuation procedures, and in emergency shutdown procedures. These actions were, of course, not necessary or less necessary when CFCs were used. Following this fire, factory inspectors visited other aerosol factories and found much that could be improved. The manufacturers of the filling machines agreed to modify them so that they would be suitable for handling butane. This, apparently, had not been considered before.

CFCs have been widely used as cleaning solvents, as they are non-flammable and their toxicity is low. Now, flammable solvents are coming back into favor. A news item from a manufacturer described "a new ozone-friendly cleaning process for the electronics industry," which "uses a unique hydrocarbon-alcohol formulation." It did not remind readers that the mixture is flammable and that they should check that their equipment and procedures are suitable.

Bromochlorofluorocarbons (BCFs or halons) have been widely used for firefighting. They were considered wonder chemicals when first used, but their manufacture has now ceased, though existing stocks may still be used. Alternative, though less effective, materials, such as fluorinated hydrocarbons, are available. Let us hope there will not be a return to the use of carbon dioxide for the automatic protection of rooms containing electrical equipment. If the carbon dioxide is accidentally discharged while someone is in the room, they will be asphyxiated, but accidental discharge of halon will not cause serious harm. Of course, procedures require the carbon dioxide supply to be isolated before anyone enters the room, but these procedures have been known to break down.

A liquid chlorine tank was kept cool by a refrigeration system that used CFCs. In 1976, the local management decided to use ammonia instead. Management was unaware that ammonia and chlorine react to form explosive nitrogen trichloride. Some of the ammonia leaked into the chlorine, and the nitrogen trichloride that was formed exploded in a pipeline connected to the tank; six men were killed, though the report does not say whether they were killed by the explosion or by the chlorine.

2.11.3   Vent Systems

During the 1970s and 1980s, there was increasing pressure to collect the discharges from tank vents, gasoline filling, and the like for destruction or absorption, instead of discharging them into the atmosphere, particularly in areas subject to photochemical smog. A 1976 report said that when gasoline recovery systems were installed in the San Diego area, more than 20 fires occurred in four months. In time, the problems were overcome, but it seems that the recovery systems were introduced too quickly and without sufficient testing.

As vent collection systems normally contain vapor/air mixtures, they are inherently unsafe. They normally operate outside the flammable range, and precautions are taken to prevent them from entering it, but it is difficult to think of everything that might go wrong. For example, an explosion occurred in a system that collected flammable vapor and air from the vents on a number of tanks and fed the mixture into a furnace. The system was designed to run at 10% of the lower explosion limit, but when the system was isolated in error, the vapor concentration rose. When the flow was restored, a plug of rich gas was fed into the furnace, where it mixed with air and exploded [17]. Reference 34 describes 10 other incidents.

At other times the burning of waste products in furnaces to save fuel and reduce pollution has caused corrosion and tube failure.

A fire in a bulk storage facility at Coode Island, Melbourne, Australia, in August 1991 caused extensive damage and many complaints about the pollution caused by the smoke plume, but there were no injuries. The tank vents were connected together and piped to a carbon bed vapor recovery system. There were no flame arrestors in the pipework. Whatever the cause of the initial fire or explosion, the vent collection system provided a means of spreading the fire from one tank to another.

In the past it was difficult to prevent the spread of explosions through vent systems, as flame arrestors were effective only when located at the ends of pipes. Effective inline detonation arrestors are now available. Like all flame arrestors they will, of course, need regular cleaning, something that is often neglected. In other cases, when tanks have been overfilled, liquid has contaminated other tanks through common vent systems, and this has led to runaway reactions.

Carbon beds are often used for absorbing vapors in vent systems, but absorption produces heating, and the beds may catch fire, particularly if they are used to absorb ketones, aldehydes, organic acids, and organic sulfur compounds. References 35 to 37 describe some fires and ways of preventing them.

In 1984, an explosion in a water pumping station at Abbeystead in the United Kingdom killed 16 people, most of them local residents who were visiting the plant. Water was pumped from one river to another through a tunnel. When pumping was stopped, some water was allowed to drain out of the tunnel and leave a void. Methane from the rocks below accumulated in the void and, when pumping was restarted, was pushed through vent valves into a valve house, where it exploded [18].

It is surprising that the vent was routed into an underground pump house. It seems that this was done because the local authority objected to any vents that might spoil the view.

A small factory in a residential area in the United Kingdom recovered solvent by distillation. The cooling water supply to the condenser, after giving trouble for several weeks, finally failed, and hot vapors were discharged from a vent inside a building. They exploded, killing one man, injuring another, and seriously damaging the factory. Some of the surrounding houses were slightly damaged, and five drums landed outside the factory, one on a house.

There were no operating or emergency instructions and no indication of cooling water flow, and drums were stored too near buildings. But by far the most serious error was allowing the vent pipe to discharge inside the building. If it had discharged outside, the vapor would have dispersed harmlessly, or at worst, there would have been a small fire on the end of the vent pipe. Vent pipes are designed to vent, so this was not an unforeseen leak. The vent pipe may have been placed indoors to try to minimize smells that had caused some complaints [19].

Increasingly, safety, health, and the environment are becoming parts of the same SHE department in industry. This should help to avoid incidents such as those described in Section 2.11. Unfortunately, there are few signs of a similar integration in government departments.

2.12   CONTROL OF MODIFICATIONS

This chapter is a story of repeated incompetence. How can we prevent modifications from producing unforeseen and undesirable side effects? References 1 and 2 propose a three-pronged approach:

1. Before any modification, however inexpensive, temporary or permanent, is made to a plant or process or to a safety procedure, it should be authorized in writing by a process engineer and a maintenance engineer—that is, by professionally qualified staff, usually the first level of professionally qualified staff. Before authorizing the modification, they should make sure there will be no unforeseen consequences and that it is in accordance with safety and engineering standards. When the modification is complete, they should inspect it to make sure their intentions have been followed and that it "looks right." What does not look right is usually not right and should at least be checked.

2. The managers and engineers who authorize modifications cannot be expected to stare at a drawing and hope that the consequences will show up. They must be provided with an aid, such as a list of questions to be answered. Such an aid is shown in references 1 and 2.

Large or complex modifications should be subjected to a hazard and operability study (see Chapter 18).

3. It is not sufficient to issue instructions about (1) and the aid described in (2). We must convince all concerned, particularly foremen, that they should not carry out unauthorized modifications. This can be done by discussing typical incidents, such as those described here; those illustrated in the Institution of Chemical Engineers (United Kingdom) Safety Training Package No. 025, *Modifications—The Management of Change;* or better still, incidents that have occurred in your own company.

Engineers are not the only ones who fail to see the possible results of the changes they make, but at least many now make systematic attempts to do so. This is more than can be said about some other occupations. For an example, see Chapter 39.

References

1. T. A. Kletz., *Chemical Engineering Progress,* Vol. 72, No. 11, Nov. 1976, p. 48.
2. M. S. Mannan (editor), *Lees' Loss Prevention in the Process Industries,* 3rd edition, Elsevier, Boston, Mass, 2005, Chapter 21.
3. *The Flixborough Cyclohexane Disaster,* Her Majesty's Stationery Office, London, 1975.
4. *Guide Notes on the Safe Use of Stainless Steel,* Institution of Chemical Engineers, Rugby, UK, 1978.
5. A note issued by the U.S. National Transportation Safety Board on Nov. 12, 1976.
6. *Chemical Safety Summary,* Vol. 56, No. 221, Chemical Industries Association, London, 1985, p. 6.
7. L. Silver, *Loss Prevention,* Vol. 1, 1967, p. 58.
8. A. H. Searson, *Loss Prevention,* Vol. 6, 1972, p. 58.
9. C. S. McCoy, M. D. Dillenback, and D. J. Truax, *Plant/Operations Progress,* Vol. 5, No. 3, July 1986, p. 165.
10. *Hazardous Cargo Bulletin,* Jan. 1985, p. 31.
11. T. A. Kletz, *Plant/Operations Progress,* Vol. 5, No. 3, July 1986, p. 136.
12. R. E. Sanders, *Chemical Process Safety—Learning from Case Histories,* Butterworth-Heinemann, Oxford, UK, 1999.
13. *Loss Prevention Bulletin,* No. 098, Apr. 1991, p. 13.
14. S. J. Skinner, *Plant/Operations Progress,* Vol. 8, No. 4, Oct 1989, p. 211.
15. D. Mosey, *Reactor Accidents,* Butterworth Scientific, London, 1990, p. 45.
16. T. A. Kletz, *Process Safety Progress,* Vol. 12, No. 3, July 1993, p. 147.
17. S. E. Anderson, A. M. Dowell, and J. B. Mynagh, *Plant/Operations Progress,* Vol. 11, No. 2, Apr. 1992, p. 85.
18. Health and Safety Executive, *The Abbeystead Explosion,* Her Majesty's Stationery Office, London, 1985.
19. Health and Safety Executive, *The Explosion and Fire at Chemstar Ltd., 6 September 1981,* Her Majesty's Stationery Office, London, 1982.
20. *Operating Experience Weekly Summary,* No. 96-47, Office of Nuclear and Safety Facility, U.S. Dept. of Energy, Washington, DC, 1996, p. 3.
21. *Operating Experience Weekly Summary,* No. 96-52, Office of Nuclear and Safety Facility, U.S. Dept. of Energy, Washington, DC, 1996, p. 8.
22. S. J. Brown, *Plant/Operations Progress,* Vol. 5, No. 11, Jan. 1987, p. 20.
23. R. E. Sanders, *Plant/Operations Progress,* Vol. 15, No. 3, Fall 1996, p. 150.

24. *Loss Prevention Bulletin,* No. 119, Oct. 1994, p. 17.
25. T. A. Kletz, *Learning from Accidents,* 3rd edition, Butterworth-Heinemann, 2001, Chapter 8.
26. M. S. Mannan (editor), *Lees' Loss Prevention in the Process Industries,* 3rd edition, Elsevier, Boston, Mass, 2005, Appendix 2.
27. *Operating Experience Weekly Summary,* No. 97-01, Office of Nuclear and Facility Safety, U.S. Dept. of Energy, Washington, DC, 1997, p. 2.
28. W. Zacky, remarks made at 2nd biennial Canadian Conference on Process Safety and Loss Management, Toronto, Canada, June 1995.
29. C. Whetton and P-J. Bots, *Loss Prevention Bulletin,* Vol. 128, Apr. 1996, p. 7.
30. J. A. McDiarmid and G. J. T. North, *Plant/Operations Progress,* Vol. 8, No. 2, 1989, p. 96.
31. *Loss Prevention Bulletin,* No. 127, p. 6.
32. R. C. Santon, "Explosion Hazards at CHP and CCGT Plants," *Hazards XIII: Process Safety—The Future,* Symposium Series No. 141, Institution of Chemical Engineers, Rugby, UK, 1997.
33. W. B. Howard, "Case Histories of Two Incidents Following Process Safety Reviews," *Proceedings of the Thirty-first Annual Loss Prevention Symposium,* AIChE, New York, 1997.
34. F. E. Self and J. D. Hill, "Safety Considerations When Treating VOC Streams with Thermal Oxidizers," *Proceedings of the Thirty-first Annual Loss Prevention Symposium,* AIChE, New York, 1997.
35. M. J. Chapman and D. L. Field, *Loss Prevention,* Vol. 12, 1979, p. 136, including discussion.
36. R. E. Sherman, et al, *Process Safety Progress,* Vol. 15, No. 3, Fall 1996, p. 148.
37. G. R. Astbury, *Loss Prevention Bulletin,* No. 134, 1997, p. 7.
38. *Daily Telegraph* (London), Motoring Supplement, Nov. 22, 1997, p. C1.
39. S. Hall, *British Railway Disasters,* Ian Allan, Shepperton, UK, p. 178.

CHAPTER

3

Accidents Said to Be Due to Human Error

Teach us, Lord, to accept the limitations of man.

—Forms of Prayer for Jewish Worship

3.1 INTRODUCTION

In earlier editions of this book, this chapter was titled "Accidents Due to Human Error," but I have altered it for two reasons. First, because as explained in the Preface, all accidents are due to human error—by those who decide what to do, those who decide how to do it, or those who actually do it. However, when a report says that an accident was due to human error, the writer usually means an error by an operator or maintenance worker, the last person in the chain of events leading to the accident who had an opportunity to prevent it.

Sometimes accidents are said to be due to equipment failure, but equipment failure is due to errors by designers or errors by those who operate, install, or maintain the equipment. Some accidents are said to be due to organizational error, but organizations, like equipment, have no minds of their own. The term *organizational errors* is a euphemism for errors by managers who fail to recognize the weaknesses in organization or to take action to change it. Often these weaknesses are never designed or intended but just arise by custom and practice and become the common law of the organization.

Managers and designers and those who advise them can make errors of the four types listed in the Preface: namely slips and lapses of attention,

doi:10.1016/B978-1-85617-531-9.00003-2

violations, mistakes, and mismatches. Slips and lapses of attention are unlikely, as these executives usually have time to check their decisions; most errors by designers and managers are mistakes—that is, they are due to inadequate training or instructions, but some are due to violations or mind-sets.

The second reason for renaming this chapter is that it describes accidents caused by those slips and lapses of attention that even well-trained and well-motivated persons make from time to time. For example, they forget to close a valve or close the wrong valve. They know what they should do, want to do it, and are physically and mentally capable of doing it, but they forget to do it. Exhortation, punishment, or further training will have no effect. We must either accept an occasional error or change the work situation so as to remove the opportunities for error.

These errors occur, not in spite of the fact that someone is well trained but *because* he or she is well trained. Routine operations are relegated to the lower levels of the brain and are not continuously monitored by the conscious mind. We would never get through the day if everything we did required our full attention. When the normal pattern or program of actions is interrupted for any reason, errors are likely to occur. These slips are similar to those we make in everyday life. Reason and Mycielska [1] have described the psychology of such slips.

I describe some accidents that occurred because employees were not adequately trained (mistakes). Sometimes they lacked basic knowledge; sometimes they lacked sophisticated skills.

Errors also occur because people deliberately decide not to carry out instructions that they consider unnecessary or incorrect. These are called violations or noncompliance. For example, they may not wear all the protective clothing or take the other precautions specified on a permit-to-work, as discussed in Section 1.4.2. Before blaming anyone we should ask the following questions both before and after accidents of this type:

- Are the rules known and understood? Is it possible to follow them?
- Are the rules, such as wearing protective clothing, really necessary? (See Section 1.4.2a.)
- Can the job be simplified? If the correct method is difficult and an incorrect method is easy, people are likely to use the incorrect method.
- Do people understand the reasons for the rules? We do not live in a society in which people will follow the rules just because they are told to do so.
- Have breaches of the rules been ignored in the past?
- There is a narrow line between initiative and rule breaking. What would have happened if no accident had occurred?
- If the rules are wrong, violating them may prevent an accident.

3.2   ACCIDENTS THAT COULD BE PREVENTED BY CHANGING THE PLANT DESIGN OR METHOD OF WORKING

3.2.1   "There Is Nothing Wrong with the Design, but the Equipment Wasn't Assembled Correctly"

How often has a designer made this statement after a piece of equipment has failed? The designer is usually correct, but we should use designs that are impossible (or difficult) to assemble incorrectly or that are unlikely to fail if assembled incorrectly. Consider these examples:

(a) In some compressors, it is possible to interchange suction and delivery valves. Damage and leaks have developed as a result. Valves should be designed so they cannot be interchanged.

(b) With many types of screwed couplings and compression couplings, it is easy to use the wrong ring. Accidents have occurred as a result. Flanged or welded pipes should therefore be used except on small-bore lines carrying nonhazardous materials.

(c) Loose-backing flanges require more care during joint making than fixed flanges. Fixed flanges are therefore preferred.

(d) Bellows (expansion joints) should be installed with great care, because unless specially designed, they cannot withstand any sideways thrust. With hazardous materials, it is therefore good practice to avoid the need for bellows by designing expansion bends into the pipework.

(e) A runaway reaction occurred in a polymerization reactor. A rupture disc failed to burst. It had been fitted on the wrong side of the vacuum support, thus raising its bursting pressure from a gauge pressure of 150 psi (10 bar) to about 400 psi (27 bar) (Figures 3-1a and 3-1b).

The polymer escaped through some of the flanged joints, burying the reactor in a brown polymer that looked like molasses candy (treacle toffee). The reactor was fitted with class 150 flanges. If these are overpressured, the bolts will stretch, and the flanges will leak,

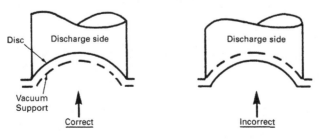

FIGURE 3-1a   Arrangements of rupture disc and vacuum support.

FIGURE 3-1b Because a rupture disc was fitted to the wrong side of a vacuum support, the flanges leaked, covering the reactor with "candy."

thus preventing the vessel from bursting (provided the pressure does not rise too rapidly). But this may not occur with flanges of a higher pressure rating.

The best way to prevent accidents such as this is to use types of rupture disc that are harder to assemble incorrectly and can be checked for correct installation after assembly. It is possible to get discs that have been permanently attached to their vacuum supports by the manufacturer and fitted with a projecting tag, which carries the words *vent side* on one side. The tag also gives the pressure rating.

A small rupture disc failed to operate; it was then found that the manufacturer had inadvertently supplied two discs that nested one on top of the other and appeared to be one. Most discs are individually boxed, but some are supplied stacked and should be carefully checked. Some small discs are supplied with gaskets already glued to them, and these are particularly likely to stick together (see Section 5.3 g and Section 9.1.3).

3.2.2   Wrong Valve Opened

The pump feeding an oil stream to the tubes of a furnace failed. The operator closed the oil valve and intended to open a steam valve to purge

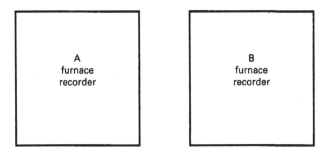

FIGURE 3-2   Layout of recorders on panel.

the furnace tubes. He opened the wrong valve, there was no flow to the furnace, and the tubes were overheated and collapsed.

This incident is typical of those that would at one time have been blamed on human failing—the operator was at fault, and there was nothing anyone else could have done. In fact, investigation showed the following:

1. The access to the steam valve was poor, and it was difficult to see which was the right valve.
2. There was no indication in the control room to show that there was no flow through the furnace coils.   .
3. There was no low-flow alarm or trip on the furnace.

3.2.3 Would You Climb over a Pipe or Walk 90 m (300 ft)?

To repair a flowmeter, a man had to walk six times from the orifice plate to the transmitter and back. To get from one to the other, he had to walk 45 m, cross a 30-in.-diameter pipe by a footbridge, and walk 45 m back—a total of 540 m (1,800 ft) for the whole job. Instead, he climbed over the pipe; while doing so, he hurt his back. Is it reasonable to expect someone to repeatedly walk 90 m (300 ft) to avoid climbing over a pipe?

3.2.4 An Error While Testing a Trip

Two furnaces were each fitted with a temperature recorder controller and high-temperature trip. The two recorders were side by side on the instrument panel in the control room, with the recorder for A furnace on the left (Figure 3-2).

An instrument mechanic was asked to test the trip on A furnace. He put the controller on manual and then went behind the panel. His next step was to take the cover off the back of the controller, disconnect one of the leads, apply a gradually increasing potential from a potentiometer, and note the reading at which the trip would operate if it was on auto control.

The mechanic, who had done the job many times before, took the cover off the back of B recorder, the one on the left behind the panel (Figure 3-3), and disconnected one of the leads. The effect was the same as if the recorder had registered a high temperature. The controller closed the fuel gas valve, shutting down the furnace and the rest of the plant.

We all know that the recorder on the left, viewed from the front of the panel, will be on the right when viewed from behind the panel, but the mechanic had his mind set on the words "the one on the left."

The backs of the two recorders should have been labeled A and B in large letters. Better still, the connections for the potentiometer should have been at the front of the panel.

3.2.5　Poor Layout of Instructions

A batch went wrong. Investigation showed that the operator had charged 104 kg of one constituent instead of 104 g (0.104 kg). The instructions to the operator were set out as shown in Table 3-1 (the names of the ingredients being changed). With instructions like these, it is easy for the operator to get confused.

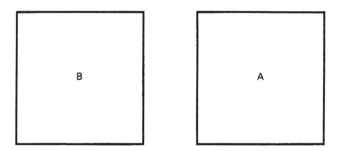

FIGURE 3-3　Layout of recorders behind panel.

TABLE 3-1　Operator Instructions

Blending Ingredients	Quantity (tons)
Marmalade	3.75
Oxtail soup	0.250
Pepper	0.104 kg
Baked beans	0.020
Raspberry jam	0.006
TOTAL	4.026

3.2.6   An Inaccurate Reading Not Noticed on an Instrument at Thigh Level

A reactor was being started up. It was filled with reaction mixture from another reactor, which was already on line, and the panel operator started to add fresh feed, gradually increasing the flow while he watched the temperature on a recorder conveniently situated at eye level. He intended to start a flow of cooling water to the reaction cooler as soon as the temperature started to rise—the usual method.

Unfortunately, there was a fault in the temperature recorder, and although the temperature actually rose, this was not indicated. The result was a runaway reaction. The rise in temperature was, however, indicated on a six-point temperature recorder at a lower level on the panel, but the operator did not notice this (Figure 3-4).

An interesting feature of this incident was that no one blamed the operator. The plant manager said he would probably have made the same mistake because the check instrument was at a low level (about 1 m above the floor) and because a change in one temperature on a six-point recorder in that position is not obvious unless you are actually looking for it. It is not the sort of thing you notice out of the corner of your eye.

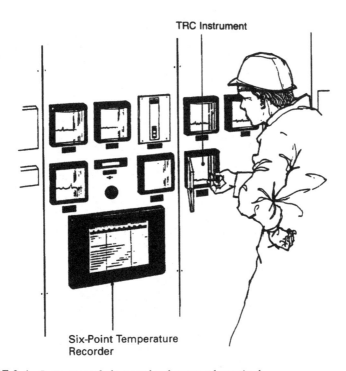

TRC Instrument

Six-Point Temperature Recorder

FIGURE 3-4   Instruments below eye level may not be noticed.

3.2.7   Closing Valves in Error

(a) Figure 3-5 shows part of a plant in which five reactors were in parallel. There were two gas-feed lines with cross connections between them. Oxygen was also fed to the reactors, but the oxygen lines are not shown. At the time of the incident, only two reactors, Nos. 1 and 4, were on line.

The operator thought valve B was open, so he shut valve A. This stopped the flow of gas to No. 1 reactor. A ratio controller managed the oxygen flow, but it had a zero error, and a small flow of oxygen continued. When the operator realized his mistake and restored the gas flow, the reactor contained excess oxygen, and an explosion occurred, not actually in the reactor but in the downstream waste heat boiler. Four men were killed.

Here we have a situation where a simple error by an operator produced serious consequences. The explosion was not, however, the operator's fault but was the result of bad design and lack of protective equipment.

We would never knowingly tolerate a situation in which accidental operation of a valve resulted in the overpressuring of a vessel. We would install a relief valve. In the same way, accidental operation of a valve should not be allowed to result in explosion or runaway reaction.

(b) The switch in the power supply to a safety interlock system was normally locked in the open position, even during shutdowns, to prevent accidental isolation. One day an operator was asked to lock it closed. He was so used to locking it open that he locked it in the wrong position.

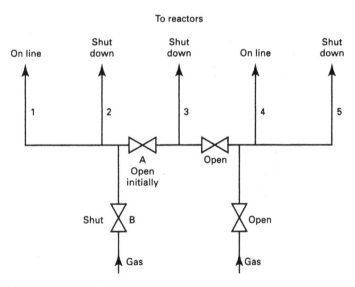

FIGURE 3-5   Accidental closing of a valve can cause an explosion.

Breaking a habit is difficult. Another operator who was asked to check did not spot the error. As Sections 1.2.3e and 14.5c show, checking is often ineffective because the checker expects to find everything in order. According to the report [7], the operators were disciplined, but this will not prevent another incident, as the errors were not deliberate. A better method of working might involve using a key that can be removed only when the switch is in one position. This incident occurred in a nuclear power station but could just as easily occur in the process industries.

3.2.8 An Explosion in a Batch Reactor

Figure 3-6 shows a batch reaction system. A batch of glycerol was placed in the reactor and circulated through a heat exchanger, which could act as both a heater and a cooler. Initially it was used as a heater, and when the temperature reached 115°C, ethylene oxide was added. The reaction was exothermic, and the exchanger was now used as a cooler.

The ethylene oxide pump could not be started unless the following occurred:

1. The circulation pump was running.
2. The temperature was above 115°C (240°F), as otherwise the ethylene oxide would not react.
3. The temperature was below 125°C (257°F), as otherwise the reaction was too fast.

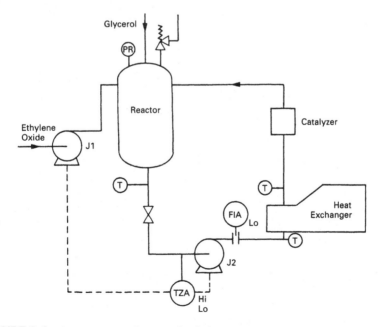

FIGURE 3-6   Arrangements of reactor circulating system.

Despite these precautions, an explosion occurred. One day, when the ethylene oxide addition was started, the pressure in the reactor rose. This showed that the ethylene oxide was not reacting. The operator decided that perhaps the temperature point was reading low or perhaps a bit more heat was required to start the reaction, so he adjusted the trip setting and allowed the indicated temperature to rise to 200°C (390°F). Still the pressure did not fall.

He then suspected that his theory might be wrong. Could he have forgotten to open the valve at the base of the reactor? He found it shut and opened it. Three tons of unreacted ethylene oxide, together with the glycerol, passed through the heater and catalyzer, and a violent, uncontrolled reaction occurred. The reactor burst, and the escaping gases exploded. Two men were injured. One, 160 m (525 ft) away, was hit by flying debris, and the other was blown off the top of a tank truck.

Although the indicated temperature had risen, the temperature of the reactor's contents had not. Pump J2, running with a closed suction valve, got hot, and the heat affected the temperature point, which was close to the pump.

Why Did It Happen?

1. The immediate cause of the explosion was an operator forgetting to open a valve. It was not due to lack of knowledge, training, or instructions but was another of those slips that even well-trained, well-motivated, capable people make from time to time.
2. If the operator had not opened the valve when he found it shut, the explosion could have been avoided. However, it is hard to blame him. His action was instinctive. What would you do if you found something undone that you should have done some time ago?
3. The explosion was due to a failure to heed warning signs. The high pressure in the reactor was an early warning, but the operator had another theory to explain it. He stuck to this theory until the evidence against it was overwhelming. This is known as a mind-set or tunnel vision.

   Once we form an incorrect belief about something, we tend to stick to it. If new evidence arrives, we tend to think of other explanations for it and focus on the evidence that supports our mind-set.

   The other temperature points would have helped the operator diagnose the trouble. But he did not look at them. He probably thought there was no point in doing so. All the temperature points were bound to read the same. The need for checking one reading by another should have been covered in the operator's training.
4. The explosion was due to a failure to directly measure the property that we wish to know. The temperature point was not measuring the temperature in the reactor but the temperature near the pump. This got hot because the pump was running with a closed suction valve. Similarly, the trip initiator on J2 showed that its motor was energized. It did not prove that there was a flow.

5. The explosion occurred because key instruments were not kept in working order. The flow indicator and low flow alarm (FIA) were out of order. They often were, and the operators had found that the plant could be operated without them. If there is no flow, they thought, J2 will have stopped, and this will stop J1.
6. The operator should not have raised the interlock setting, though doing so did not in itself cause the explosion. (However, he did try to use his intelligence and think why the reaction was not occurring. Unfortunately, he was wrong.)

What Should We Do?

It is no use telling the operator to be more careful. We have to recognize that the possibility of an error—forgetting to open the valve—is inherent in the work situation. If we want to prevent an error, we must change the work situation. That is, change the design and/or the method of operation—the hardware and/or the software.

The original report blamed the operator for the explosion. But his failure to open the valve might have been foreseen:

1. The temperature should be measured in the reactor or as close to it as possible. We should always try to measure the property we wish to know directly, rather than measuring another property from which the property we wish to know can be inferred. The designers assumed that the temperature near the pump would be the same as that in the reactor. It will not be if there is no circulation. The designers also assumed that if the pump is energized, then liquid is circulating, but this is not always the case.
2. Operators should not be allowed to change trip settings at will. Different temperatures are needed for different batches. But even so, only someone who is given written permission to do so should make the adjustment.
3. More effort might have been made to keep the flow indicator alarm in working order.
4. A high-pressure trip should be installed on the reactor.
5. Operators should be trained to "look before they leap" when they find valves wrongly set. See also Section 3.3.5a. Other accidents that occurred because operators failed to carry out simple tasks are described in Sections 13.5 and 17.1.

3.3 ACCIDENTS THAT COULD BE PREVENTED BY BETTER TRAINING

As we shall see, often it is not a lack of sophisticated training that results in accidents but ignorance of the basic requirements of the job or the basic properties of the materials and equipment handled.

3.3.1 Readings Ignored

Many accidents have occurred because operators apparently thought their job was just to write down readings and not to respond to them.

(a) The temperature controller on the base of a distillation column went out of order at 5 a.m. and drew a straight line. This was not noticed. During the next seven hours, the following readings were abnormal:
1. Six tray temperatures (one rose from 145°C [293°F] to 255°C [490°F])
2. Level in base of still (low)
3. Level in reflux drum (high)
4. Take-off rate from reflux drum (high)

Most of these parameters were recorded on the panel. The operator wrote them all down on the record sheet.

Finally, at 12 noon, the reflux drum overflowed, and there was a spillage of flammable oil. From 7 a.m. onward, a trainee served as operator, but a lead operator was present, and the foreman visited the control room from time to time.

(b) Section 2.5a describes how a low-pressure liquefied ethylene storage tank split when the vent pipe became plugged with ice. For 11 hours before the split occurred, the gauge pressure in the tank was reading 2 psi (0.13 bar). This pressure was above the set point of the relief valve (a gauge pressure of 1.5 psi, or 0.1 bar) and was the full-scale reading on the pressure gauge. The operators entered this reading on the record sheet but took no other action and did not even draw it to the attention of the foremen or managers when they visited the control room [2]. Section 8.1.6 describes a similar incident.

(c) The governor assembly and guard on a steam engine disintegrated with a loud bang, scattering bits over the floor. Fortunately, no one was injured. It was then found that the lubricating oil gauge pressure had been only 8 psi (0.5 bar) instead of 25 psi (1.7 bar) for at least "several months." In this case, the pressure was not written down on the record sheet.

(d) The level measuring instrument and alarm on a feed tank were out of order, so the tank was hand-dipped every shift. When the plant was shut down, the operators stopped dipping the tank. The plant that supplied the feed was not shut down. It continued to supply feed into the feed tank until it overflowed. In this case, readings were not ignored but simply not taken. There were some errors in the stock sheets, and the tank contained more than expected. However, if the operators had continued to dip the tank every shift, the error would have been detected before the tank overflowed.

How can we prevent similar incidents from happening again?

1. Emphasize in operator training that operators should take action on unusual readings, not just write them down. Make sure they know the action to take.

2. Mark control limits in red on record sheets. If readings are outside these limits, some action is required.
3. Continue to take certain readings, such as tank levels, even when the plant is shut down. Tank levels are particularly liable to rise or fall when they should be steady.

3.3.2   Warnings Ignored

When they receive a warning, many operators are too ready to assume the alarm is out of order. They thus ignore it or send for the instrument mechanic. By the time the mechanic confirms that the alarm is correct, it is too late. Consider these examples:

(a) During the morning shift, an operator noticed that a tank level was falling faster than usual. He reported that the level gauge was out of order and asked an instrument mechanic to check it. It was afternoon before the mechanic could do so. He reported that it was correct. The operator then looked around and found a leaking drain valve. Ten tons of material had been lost.

(b) After making some modifications to a pump, the pump was used to transfer some liquid. When the transfer was complete, the operator pressed the stop button on the control panel and saw that the pump running light went out. He also closed a remotely operated valve in the pump delivery line.

Several hours later, the high-temperature alarm on the pump sounded. Because the operator had stopped the pump and seen the pump running light go out, he assumed the alarm was faulty and ignored it. Soon afterward, an explosion occurred in the pump.

When the pump was modified, an error was introduced into the circuit. As a result, pressing the stop button did not stop the pump but merely switched off the pump running light. The pump continued running against a closed delivery valve, overheated, and the material in it decomposed explosively.

Operator training should emphasize the importance of responding to alarms. *They might be correct!* If operators ignore alarms, it may be because experience has taught them that alarms are unreliable. Are your alarms adequately maintained (see also Section 17.10)?

(c) A car on a pleasure ride in a leisure park developed a fault. The system was shut down automatically. The operator could not see the stranded car, assumed the trip was spurious, and restarted the ride with an override key. There were several collisions, and six people were injured. The company was fined, as the operator's training was described as "woefully inadequate" [8].

(d) An electron beam accelerator, used to irradiate cancer patients, broke down. After repair, the energy-level indicator showed 36 MeV when the energy-level selection keys were set for lower levels. The operators assumed that the needle was stuck at 36 MeV and carried on.

The needle was not stuck. The machine delivered 36 MeV no matter what level was selected, and some patients got three to seven times more radiation than their doctors had prescribed. The beam was narrower than it should have been, and the radiation went deeper.

What went wrong? As well as the operators ignoring the warning reading, several other errors were made:

- The repairs had been botched, though it is not clear whether the contract person who did the repair did not know what to do or simply carried out a quick fix.
- The hospital physics service staff members were supposed to check, after repairs, that the energy level selected and the energy level indicated agreed. They did not check, as no one told them there had been a repair.
- The physics service was also supposed to carry out routine checks every day, but because few, if any, faults were found, the test interval was increased to a month. I doubt if anyone calculated the fractional dead time or hazard rate; the report does not say.
- A discrepancy between the energy level selected and the energy level indicated should trip the machine. However, the interlock had been easily bypassed by changing from automatic to manual control [9].

The incident was not simply the result of errors by the operating, repair, or physics staff members. They had been doing the wrong things for some time, but no one had noticed (or if they had noticed, they did nothing). This is typical of human error accidents. Many people fail, many things are wrong, and it is unfair to put all the blame on the person who adds the last straw.

3.3.3   Ignorance of Hazards

This section presents a number of incidents that occurred because of ignorance of the most elementary properties of materials and equipment.

(a) A man who wanted some gasoline for cleaning decided to siphon it out of the tank of a company vehicle. He inserted a length of rubber tubing into the gasoline tank. Then, to fill the tubing and start the siphon, he held the hose against the suction nozzle of an industrial vacuum cleaner. The gasoline caught fire. Two vehicles were destroyed and eleven damaged. This occurred in a branch of a large organization, not a small company.

(b) A new cooler was being pressure-tested using a water pump driven by compressed air. A plug blew out, injuring the two men on the job. It was then found that the pressure gauge had been fitted to the air supply instead of the cooler. The pressure had been taken far above the test pressure.

(c) An operator had to empty some tank trucks by gravity. He had been instructed to do the following:

1. Open the valve on top of the tank.
2. Open the drain valve.
3. When the tank was empty, close the valve on top of the tank.

He had to climb onto the top of the tank twice. He therefore decided to close the vent before emptying the tank. To his surprise, the tank was sucked in.

(d) At one plant it was discovered that contractors' employees were using welding cylinders to inflate pneumatic tires. The welders' torches made a good fit on the tire valves.

3.3.4 Ignorance of Scientific Principles

The following incidents differ from those just described in that the operators, though generally competent, did not fully understand the scientific principles involved:

(a) A waste product had to be dissolved in methanol. The correct procedure was to put the waste in an empty vessel, box it up, evacuate it, break the vacuum with nitrogen, and add methanol. When the waste had dissolved, the solution was moved to another vessel, the dissolving vessel evacuated again, and the vacuum broken with nitrogen.

If this procedure is followed, a fire or explosion is impossible because air and methanol are never in the vessel together. However, to reduce the amount of work, the operators added the methanol as soon as the waste was in the vessel, without bothering to evacuate or add nitrogen. Inevitably, a fire occurred, and a man was injured. As often happens, the source of ignition was never identified.

It is easy to say that the fire occurred because the operators did not follow the rules. But why did they not follow the rules? Perhaps because they did not understand that if air and a flammable vapor are mixed, an explosion may occur and that we cannot rely on removing all sources of ignition. To quote from an official report on a similar incident, "we do feel that operators' level of awareness about hazards to which they may be exposing themselves has not increased at the same rate as has the level of personal responsibility which has been delegated to them" [3]. Also, the managers should have checked from time to time that the correct procedure was being followed.

(b) Welding had to take place near the roof of a storage tank that contained a volatile flammable liquid. There was a vent pipe on the roof of the tank, protected by a flame arrestor. The welding might have ignited vapor coming out of this vent. The foreman therefore fitted a hose to the end of the vent pipe. The other end of the flex was placed on the ground so that the vapor now came out at ground level.

The liquid in the tank was soluble in water. As an additional precaution, the foreman therefore put the end of the flex in a drum of water. When the tank was emptied, the water first rose up the hose, and then the tank was sucked in. The tank, like most such tanks, was designed for a vacuum of only 2½ in. water gauge (0.1 psi or 0.6 kPa) and would collapse at a vacuum of about 6 in. water gauge (0.2 psi or 1.5 kPa).

If the tank had been filled instead of emptied, it might have burst, because it was designed to withstand a pressure of only 8 in. water gauge (0.3 psi or 2 kPa) and would burst at about three times this pressure. Whether it burst or not would have depended on the depth of water above the end of the flex.

This incident occurred because the foreman, though a man of great experience, did not understand how a lute works. He did not realize how fragile storage tanks usually are (see also Section 5.3).

(c) The emergency blowdown valves in a plant were hydraulically operated and were kept shut by oil under pressure. One day the valves opened, and the pressure in the plant blew off. It was then discovered that (unknown to the manager) the foremen, contrary to the instructions, were closing the oil supply valve "in case the pressure in the oil system failed"—a most unlikely occurrence and much less likely than the oil pressure leaking away from an isolated system.

Accidents that occurred because maintenance workers did not understand how things work or how they were constructed were described in Section 1.5.4.

3.3.5   Errors in Diagnosis

(a) The incident described in Section 3.2.8 is a good example of an error in diagnosis.

The operator correctly diagnosed that the rise in pressure in the reactor was due to a failure of the ethylene oxide to react. He decided that the temperature indicator might be reading high and that the temperature was therefore too low for reaction to start or that the reaction for some reason was sluggish to start and required a little more heat. He therefore raised the setting on the temperature interlock and allowed the temperature to rise.

His diagnosis, though wrong, was not absurd. However, having made a diagnosis, he developed a mind-set. That is, he stuck to it even though further evidence did not support it. The temperature rose, but the pressure did not fall. Instead of looking for another explanation or stopping the addition of ethylene oxide, he raised the temperature further and continued to do so until it reached 200°C

(390°F) instead of the usual 120°C (250°F). Only then did he realize that his diagnosis might be incorrect.

In developing a mind-set the operator was behaving like most of us. If we think we have found the solution to a problem, we become so committed to our theory that we close our eyes to evidence that does not support it. Specific training and practice in diagnostic skills may make it less likely that operators will make errors in diagnosis.

Duncan and co-workers [4] have described one method. Abnormal readings are marked on a drawing of the control panel (or a simulated screen). The operator is asked to diagnose the reasons for them and say what action he or she would take. The problems gradually get more difficult.

Attempts to change mind-sets that have been in existence for a long time, perhaps for all a person's life or career, give rise to what is called cognitive dissonance, literally an unpleasant noise in the mind, and are particularly difficult to overcome. General examples are the beliefs that most accidents are due to human error, that managers can do little to prevent them and that the lost-time accident rate is a good measure of process safety.

(b) The accident at Three Mile Island in 1979 provided another example of an error in diagnosis [5]. There were several indications that the level in the primary water circuit was low, but two instruments indicated a high level. The operators believed these two readings and ignored the others. Their training had emphasized the hazard of too much water and the action to take but had not told them what to do if there was too little water in the system.

For more examples of accidents caused by human error and a discussion of responsibility, see reference 6. Section 38.10 discusses psychological and sociological aspects of human errors.

Chapter 7 of reference 6 describes methods for estimating the probability of errors. Such methods can be valuable aids to management judgment, but we should not forget that "Human beings sometimes have an ability to simply recognize the right thing to do. Judgment is rarely a calculated weighing of all options, which we are not good at anyway, but instead an unconscious form of pattern recognition" [10]. An example is driving a car. We do have time to review every possible option, but we act without thinking in a way that experience has taught us. Young drivers have high accident rates because they have still to acquire that experience.

References

1. J. Reason and K. Mycielska, *Absent Minded? The Psychology of Mental Lapses and Everyday Errors*, Prentice-Hall, Englewood Cliffs, NJ, 1982.
2. T. A. Kletz, *Chemical Engineering Progress*, Vol. 70, No. 7, Apr. 1974, p. 80.

3. *Annual Report of Her Majesty's Inspectors of Explosives for 1970*, Her Majesty's Stationery Office, London, 1971.

4. E. E. Marshall, et al, *The Chemical Engineer*, No. 365, Feb. 1981, p. 66.

5. T. A. Kletz, *Learning from Accidents*, 3rd edition, Butterworth-Heinemann, Oxford, UK, 2001, Chapter 11.

6. T. A. Kletz, *An Engineer's View of Human Error*, 3rd edition, Institution of Chemical Engineers, Rugby, UK, 2001.

7. *Lockout/Tagout Programs*, Safety Notice No. DOE/EH-0540, Office of Nuclear and Facility Safety, U.S. Dept. of Energy, Washington, DC, 1996.

8. *Health and Safety at Work*, Nov. 1991, p. 10.

9. *Report on the Accident with the Linear Accelerator at the University Clinical Hospital of Zaragoza in December 1990*, Translation No. 91-11401 (8498e/813e), International Atomic Energy Agency, 1991.

10. A. Gwande, *A Surgeon's Notes on an Imperfect Science*, Profile Books, London, 2002, p. 248.

Labeling

In my exploratory wanderings I would often ask what this or that pipe was conveying and at what pressure. Often enough there was no answer to my query, and a hole would have to be drilled to discover what the pipe contained.

—A U.K. gas works in 1916, described by Norman Swindin,
Engineering without Wheels

Many incidents have occurred because equipment was not clearly labeled. Some of these incidents have already been described in the section on the identification of equipment under maintenance (Section 1.2).

Seeing that equipment is clearly and adequately labeled and checking from time to time to make sure that the labels are still there is a dull job, providing no opportunity to exercise our technical or intellectual skills. Nevertheless, it is as important as more demanding tasks are. One of the signs of good managers, foremen, operators, and designers is that they see to the dull jobs as well as those that are interesting. If you want to judge a team, look at its labels as well as the technical problems it has solved.

4.1 LABELING OF EQUIPMENT

(a) Small leaks of carbon monoxide from the glands of a compressor were collected by a fan and discharged outside the building. A man working near the compressor was affected by carbon monoxide. It was then found that a damper in the fan delivery line was shut.

There was no label or other indication to show when the damper was closed and when it was open.

In a similar incident, a furnace damper was closed in error. It was operated pneumatically. There was no indication on the control knob to show which was the open position and which was the closed position.

(b) On several occasions it has been found that the labels on fuses or switchgear and the labels on the equipment they supply do not agree. The wrong fuses have then been withdrawn. Regular surveys should be made to confirm that such labels are correct. Labels are a sort of protective equipment and, like all protective equipment, should be checked from time to time.

(c) Sample points are often unlabeled. As a result, the wrong material has often been sampled. This usually comes to light when the analysis results are received, but sometimes a hazard develops. For example, a new employee took a sample of butane instead of a higher boiling liquid. The sample was placed in a refrigerator, which became filled with vapor. Fortunately, it did not ignite.

(d) Service lines are often not labeled. A fitter was asked to connect a steam supply at a gauge pressure of 200 psi (13 bar) to a process line to clear a choke. By mistake, he connected up a steam supply at a gauge pressure of 40 psi (3 bar). Neither supply was labeled, and the 40 psi supply was not fitted with a check valve. The process material came back into the steam supply line. Later, the steam supply was used to disperse a small leak. Suddenly the steam caught fire. It is good practice to use a different type of connector on each type of service point.

(e) Two tank trucks were parked near each other in a filling bay. They were labeled as shown in Figure 4-1. The filler said to the drivers, "Number eight is ready." He meant that the No. 8 tank was ready, but the driver assumed that the tank attached to the No. 8 tractor was ready. He got into the No. 8 tractor and drove away. Tank No. 4 was still filling. Fortunately, the tank truck was fitted with a device to prevent it from departing when the filling hose was connected [1], and the driver was able to drive only a few yards. If possible, tanks and tractors should be given entirely different sets of numbers.

(f) Nitrogen was supplied in tank cars that were also used for oxygen. Before filling the tank cars with oxygen, the filling connections were

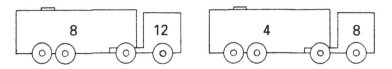

FIGURE 4-1   Arrangement of tank trailers and tractors.

changed, and hinged boards on both sides of the tanker were folded down so that they read Oxygen instead of Nitrogen.

A tank car was fitted with nitrogen connections and labeled Nitrogen. Probably because of vibration, one of the hinged boards fell down so that it read Oxygen. The filling station staff therefore changed the connections and put oxygen in the tank car. Later, some nitrogen tank trucks were filled from the tank car—which was labeled Nitrogen on the other side—and supplied to a customer who wanted nitrogen. The customer off-loaded the oxygen into his plant, thinking it was nitrogen (Figure 4-2).

The mistake was found when the customer looked at his weigh-bridge figures and noticed that on arrival the tanker had weighed 3 tons more than usual. A check then showed that the plant nitrogen system contained 30% oxygen.

Analyze all nitrogen tankers before off-loading (see Section 12.3.4).

(g) A British Airways 747 plane had to make an emergency landing after sparks were seen coming out of an air conditioning vent. A motor bearing in a humidifier had failed, causing a short circuit, and the miniature circuit breakers (MCBs), which should have protected the circuit, had not done so. The reason: 25 amp circuit breakers had been installed instead of 2.5 amp ones. The fault current, estimated at 14 to 23 amps, was high enough to melt parts of the copper wire.

MCBs have been confused before. Different ratings look alike, and the part numbers are hard to read and are usually of the forms 123456-2.5 and 123456-25 [8].

(h) A lifting device had a design capacity of 15 tons, but in error it was fitted with a label showing 20 tons. As a result, it was tested every year, for eight years, with a load of 1.5 times the indicated load—that is, with a load of 30 tons. This stressed the lifting device beyond its yield point, though there was no visible effect. The ultimate load, at which the device would fail, was much higher, but it is bad practice to take equipment above its yield point [9].

(i) Notices should be visible. On more than one occasion, someone has entered a section of a plant without the required protective clothing because a door normally propped open shielded the warning notice [10].

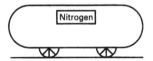 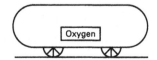

FIGURE 4-2 Arrangement of labels on tank cars. The Nitrogen label folds down to read Oxygen.

(j) A powder was conveyed in large plastic bags in a container fitted with a door. When someone started to open the door, the weight of the powder caused the bags to burst open, and he escaped injury only by leaping aside. The doors were intended to carry labels saying that it is dangerous to open them, but the one on this container was missing. However, a label is not sufficient; the door should have been locked.

4.2   LABELING OF INSTRUMENTS

(a) Plant pressures are usually transmitted from the plant to the control room by a pneumatic signal. This pneumatic signal, which is generated within the pressure-sensing element, usually has a gauge pressure in the range of 3 to 15 psi, covering the plant pressure from zero to maximum. For example, 3 to 15 psi (0.2 to 1 bar) might correspond to 0 to 1,200 psi plant pressure (0 to 80 bar).

The receiving gauge in the control room works on the transmitted pneumatic pressure, 15 psi giving full scale, but has its dial calibrated in terms of the plant pressure that it is indicating. The Bourdon tube of such a gauge is capable of withstanding only a limited amount of overpressure above 15 psi before it will burst. Furthermore, the material of the Bourdon tube is chosen for air and may be unsuitable for direct measurement of the process fluid pressure.

A pressure gauge of this sort with a scale reading up to 1,200 psi was installed directly in the plant. The plant gauge pressure was 800 psi, and the gauge was damaged. Gauges of this type should have the maximum safe working pressure clearly marked in red letters on the face.

(b) A worker, who was pressure-testing some pipework with a hand-operated hydraulic pump, told his foreman that he could not get the gauge reading above 200 psi. The foreman told him to pump harder. He did and burst the pipeline.

The gauge he was using was calibrated in atmospheres and not psi. The word *ats* was in small letters, and in any case the worker did not know what it meant.

If more than one sort of unit is used in your plant for measuring pressure or any other property, then the units used should be marked on instruments in large, clear letters. It is a good practice to use different colors for different units. Everyone should be aware of the differences between the units. However, it is better to avoid the use of different units.

(c) An extraordinary case of confusion between units occurred on a piece of equipment manufactured in Europe for a customer in England. The manufacturers were asked to measure all temperatures in °F and were told how to convert °C to °F.

A damper on the equipment was operated by a lever, whose position was indicated by a scale, calibrated in degrees of arc. These were converted to °F!

A medical journal reported that patients suffering from paracetamol poisoning should be nursed at 30 to 40 degrees. In the next issue, it said that this referred to the angle in bed, not the temperature [7].

(d) An operator was told to control the temperature of a reactor at 60°C (140°F). He set the set point of the temperature controller at 60. The scale actually indicated 0% to 100% of a temperature range of 0 to 200 degrees C, so the set point was really 120°C. This caused a runaway reaction, which overpressured the vessel. Liquid was discharged and injured the operator [2].

(e) An error in testing made more probable by poor labeling is described in Section 3.2.4.

(f) Although digital instruments have many advantages, there are times when analog readings are better. One of the raw materials for a batch reaction had to be weighed. The project team intended to install a weighing machine with a digital display, but an experienced operator asked for an analog scale instead because, he said, he was more likely to misread a figure than a position on a scale.

(g) A catalyst arrived in cylinders and was egged into the plant with nitrogen at a gauge pressure of 30 psi (2 bar). The gauge on the pressure regulator had two scales. The inner one, which was normally used, indicated 0 to 200 psig in divisions of 10 psi, so it was normally set at three divisions.

The regulator developed a fault and had to be changed. The gauge on the new one also had two scales. The inner one indicated 0 to 280 kg/cm^2 gauge (a kg/cm^2 is almost the same as a bar) in intervals of 10 kg/cm^2; the outer one indicated psig. The inner one thus looked like the inner scale on the old gauge, so the operators set the pointer at three divisions on it. Long before the pressure reached two divisions, corresponding to a gauge pressure of 20 kg/cm^2 or 300 psi, the cylinder burst. Figure 4-3 shows the results. The estimated bursting pressure was 215 psig (15 kg/cm^2 gauge) [11].

4.3 LABELING OF CHEMICALS

4.3.1 Poor or Missing Labels

One incident is described in Section 2.8a. Several incidents have occurred because drums or bottles were unlabeled and people assumed that they contained the material usually handled at the plant. In one case, six drums of hypo (sodium hypochlorite) had to be added to a tank of

FIGURE 4-3   The result of pressurizing a cylinder to "two divisions" on a scale graduated in kg/cm² instead of psi. *(Photo courtesy of the Institution of Chemical Engineers.)*

water. Some of the drums were not labeled. One, which contained sulfuric acid, was added after some of the genuine hypo and chlorine was given off. The fumes affected the workers who were adding the material in the drums.

In another case, an unlabeled drum smelled like methylethylketone (MEK), so it was assumed to be MEK and was fed to the plant. Actually, it contained ethanol and a bit of MEK. Fortunately, the only result was a ruined batch.

Mononitro-o-xylene was manufactured by the nitration of o-xylene. An operator required some o-xylene to complete a series of batches. He found a tank labeled Xylene in another part of the plant and ran some of it into drums. It was then charged to the reactor. There was a violent reaction, a rupture disc blew, and about 600 gal of acid were discharged into the air through a vent pipe. Passersby and schoolchildren were affected and needed first aid. The tank actually contained methanol and had contained it for eight months, but the label had not been changed though the engineering department had been asked to change it (note that if the vent pipe had discharged into a catchpot instead of the open air, the results of the runaway would have been trivial) [4].

Some nitric acid had to be flown from the United States to the United Kingdom. Several U.S. regulations were broken: the acid was packed in glass bottles instead of metal ones and was surrounded by sawdust instead of nonflammable material, and the boxes containing the bottles were not labeled as hazardous or marked This Side Up. The boxes were therefore loaded into the cargo aircraft on their sides, and the bottles leaked. Smoke entered the flight deck, and the crew decided to land, but while doing so the plane crashed, probably as the result of poor visibility on the flight deck, and the crew was killed. It is not clear why a common material of commerce had to be flown across the Atlantic [5].

Inspections showed that two cooling towers contained asbestos. Sticky warning labels were affixed to them. No maintenance work was carried out on the towers until three years later. By this time, the labels had been washed away. Nine members of the maintenance team removed filters from the towers without wearing protective equipment and may have been exposed to asbestos dust. Fortunately, the asbestos was of a nonfriable type [12].

4.3.2 Similar Names Confused

Several incidents have occurred because similar names were confused. The famous case involving Nutrimaster (a food additive for animals) and Firemaster (a fire retardant) is well known. The two materials were supplied in similar bags. A bag of Firemaster, delivered instead of Nutrimaster, was mixed into animal feeding stuffs, causing an epidemic of illness among the farm animals. Farmers and their families were also affected [3].

In another case, a manufacturer of animal feedstuffs bought a starch additive from a Dutch company for incorporation in a milk substitute for calves. The Dutch company was out of stock, so it asked its U.K. affiliate company to supply the additive; the Dutch company quoted the product number. Unfortunately, the U.K. affiliate used this number to describe a different additive, which was highly toxic. As a result, 68,000 calves were affected, and 4,600 died. Chemicals (and equipment) should be ordered by name and not just by a catalog number [6].

A unit used small amounts of sodium sulfite and potassium sulfate. It was custom and practice to call these two chemicals simply sulfite and sulfate. During a busy period, someone from another unit was asked to help and was told to prepare a batch of sulfate. The only sulfate he knew was aluminum sulfate, so he prepared a batch of it. Fortunately, the error was spotted before the sulfate was used [13].

Other chemicals have been confused, with resultant accident or injury:

1. Washing soda (sodium carbonate) and caustic soda (sodium hydroxide)
2. Sodium nitrite and sodium nitrate
3. Sodium hydrosulfide and sodium sulfide
4. Ice and dry ice (solid carbon dioxide)
5. Photographers' hypo (sodium thiosulfate solution) and ordinary hypo (sodium hypochlorite solution)

In the last case, a load of photographers' hypo was added to a tank containing the other sort of hypo. The two sorts of hypo reacted together, giving off fumes.

4.4 LABELS NOT UNDERSTOOD

Finally, even the best labels are of no use if they are not understood:

(a) The word *slops* means different things to different people. A tank truck collected a load of slops from a refinery. The driver did not realize that the slops were flammable. He took insufficient care, and they caught fire. He thought slops were dirty water.

(b) A demolition contractor was required to use air masks while demolishing an old tank. He obtained several cylinders of compressed air, painted gray. Finding that they would be insufficient, he sent a truck for another cylinder. The driver returned with a black cylinder. None of the workers, including the man in charge of the air masks, noticed the change or, if they did, attached any importance to it. When the new cylinder was brought into use, a welder's face piece caught fire. Fortunately, he pulled it off at once and was not injured.

The black cylinder had contained oxygen. All persons responsible for handling cylinders, particularly persons in charge of air masks, should be familiar with the color codes for cylinders.

References

1. T. A. Kletz, *Loss Prevention*, Vol. 10, 1976, p. 151.
2. R. Fritz, *Safety Management* (South Africa), Jan. 1982, p. 27.
3. J. Egginton, *Bitter Harvest*, Seeker and Warburg, London, 1980.
4. *Health and Safety at Work*, Vol. 8, No. 12, Dec. 1986, p. 8; and Vol. 9, No. 4, Apr. 1987, p. 37.
5. J. D. Lewis, *Hazardous Cargo Bulletin*, Feb. 1985, p. 44.
6. *Risk and Loss Management*, Vol. 2, No. 1, Jan. 1985, p. 21.
7. *Atom*, No. 400, Feb. 1990, p. 38.
8. *Bulletin 3/96*, Air Accident Investigation Branch, Defence Research Establishment, Farnborough, UK.
9. *Operating Experience Weekly Summary*, No. 97-13, Office of Nuclear and Facility Safety, U.S. Dept. of Energy, Washington, DC, 1997, p. 5.
10. *Operating Experience Weekly Summary*, No. 97-20, Office of Nuclear and Facility Safety, U.S. Dept. of Energy, Washington, DC, 1997, p. 7.
11. *Loss Prevention Bulletin*, No. 135, June 1997, p. 12.
12. *Operating Experience Weekly Summary*, No. 96-43, Office of Nuclear and Facility Safety, U.S. Dept. of Energy, Washington, DC, 1996, p. 2.
13. C. Whetton, *Chemical Technology Europe*, Vol. 3, No. 4, July/Aug. 1996, p. 17.

5

Storage Tanks

Once a consensus has developed about how to treat a particular disease, there is huge urge in medicine to follow the herd. Yet blindly playing follow the leader can doom us to going the wrong way if the leaders are poorly informed.

—Ian Ayres [22]

Are engineers any different?

No item of equipment is involved in more accidents than the storage tank, probably because storage tanks are fragile and easily damaged by slight overpressure or vacuum. Fortunately, the majority of accidents involving tanks do not cause injury, but they do cause damage, loss of material, and interruption of production. (However, Section 30.13.1 describes an overflow that resulted in a devastating explosion.)

5.1 OVERFILLING

Most cases of overfilling are the result of lack of attention, valves placed at the wrong setting, errors in level indicators, and so on (see Section 3.3.1d). For this reason, many companies fit high-level alarms to storage tanks. However, overfilling has still occurred because the alarms were not tested regularly or the warnings were ignored (see Section 3.3.2a).

Whether a high-level alarm is needed depends on the rate of filling and on the size of the batches being transferred into the receiving tank. If these are big enough to cause overfilling, a high-level alarm is desirable.

Spillages resulting from overfilling should be retained in tank dikes (bunds). But very often the drain valves on the dikes—installed so that rainwater can be removed—have been left open, and the spillage is lost to drain (see Section 5.5.2c).

Drain valves should normally be locked shut. In addition, they should be inspected weekly to make sure they are closed and locked.

5.1.1   Alarms and Trips Can Make Overfilling More Likely

A high-level trip or alarm may actually *increase* the frequency of over-filling incidents if its limitations are not understood.

At one plant, a tank was filled every evening with enough raw material for the following day. The operator watched the level. When the tank was full, he shut down the filling pump and closed the inlet valve. After several years, inevitably, one day he allowed his attention to wander, and the tank overflowed. It was then fitted with a high-level trip, which shut down the filling pump automatically.

To everyone's surprise the tank overflowed again a year later.

It had been assumed that the operator would continue to watch the level and that the trip would take over on the odd occasion when the operator failed to do so. Coincident failure of the trip was most unlikely. However, the operator no longer watched the level now that he was supplied with a trip. The manager knew that he was not doing so, but he decided that the trip was giving the operator more time for his other duties. The trip had the normal failure rate for such equipment, about once in two years, so another spillage after about two years was inevitable. A reliable operator had been replaced by a less reliable trip.

If a spillage about once in five years (or however often we think the operator will fail) is unacceptable, then it is necessary to have two protective devices, one trip (or alarm) to act as a process controller and another to take over when the controller fails. It is unrealistic to expect an operator to watch a level when a trip (or alarm) is provided (see Section 14.7a).

5.1.2   Overfilling Due to Change of Duty

On more than one occasion, tanks have overflowed because the contents were replaced by a liquid of lower specific gravity. The operators did not realize that the level indicator measured weight, not volume. For example, at one plant a tank that had contained gasoline (specific gravity 0.81) was used for storing pentane (specific gravity 0.69). The tank overflowed when the level indicator said it was only 85% full. The level indicator was a DP cell, which measures weight. Another incident is described in Section 8.2b. If the level indicator measures weight, it is good practice to fit a high-level alarm, which measures volume.

5.1.3 Overfilling by Gravity

Liquid is sometimes transferred from one tank to another by gravity. Overfilling occurs when liquid flows from a tall tank to a shorter one. On one occasion, an overflow occurred when liquid was transferred from one tank to another of the same height several hundred meters away. The operators did not realize that a slight slope in the ground was sufficient to cause the lower tank to overflow.

5.2 OVERPRESSURING

Most storage tanks are designed to withstand a gauge pressure of only 8 in. of water (0.3 psi or 2 kPa) and will burst at about three times this pressure. They are thus easily damaged. Most storage tanks are designed so they will burst at the roof/wall weld, thus avoiding any spillage, but older tanks may not be designed this way.

Tanks designed to fail at the roof/wall weld have failed at the base/wall weld because this weld was corroded or fatigued or because holding-down bolts were missing (Figure 5-1). Corrosion is most likely to occur in tanks containing a water layer or when spill absorbents have been placed

FIGURE 5-1 Corrosion and missing holding-down bolts caused this tank to fail at the base instead of the top.

around the base. Frequent emptying of a tank can cause fatigue failure of the base/wall weld. This can be prevented by leaving about 1 m depth of liquid in the tank when it is emptied [12].

5.2.1 Overpressuring with Liquid

Suppose a tank is designed to be filled at a rate of x m^3/hr. Many tanks, particularly those built some years ago, are provided with a vent big enough to pass x m^3/hr of air but not x m^3/hr of liquid. If the tank is overfilled, the delivery pump pressure will almost certainly be large enough to cause the tank to fail.

If the tank vent is not large enough to pass the liquid inlet rate, then the tank should be fitted with a hinged manhole cover or similar overflow device. Proprietary devices are available.

This overflow device should be fitted to the roof near the wall. If it is fitted near the center of the roof, the height of liquid above the top of the walls may exceed 8 in., and the tank may be overpressured (see Figure 5-2a).

Similarly, if the vent is designed to pass liquid, it should be fitted near the edge of the roof, and its top should not be more than 8 in. above the tops of the walls. Vessels have been overpressured because their vent pipes were too long (see Figure 5-2b). Tanks in which hydrogen may be evolved should be fitted with a vent at the highest point as well as an overflow (see Section 16.2).

An 80-m^3 fiberglass-reinforced plastic tank containing acid was blown apart at the base as the result of overpressure. The vent had been slip-plated so the tank could be entered for inspection. The steel slip-plate was covered with a corrosion-resistant sheet of polytetrafluoroethylene. Afterward, when the slip-plate was removed, the sheet was left behind. This did not matter at the time, as the tank was also vented through an overflow line, which discharged into a sewer. A year later the sewer had to be maintained, so the overflow line was slip-plated to prevent acid from entering it during the overhaul. The operators were told to fill the tank

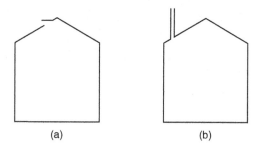

(a)   (b)

FIGURE 5-2 A tank may be overpressured if the vent or overflow is more than 8 in. above the tops of the walls.

slowly and watch the level. When they started to fill the tank, the reading on the level indicator rose rapidly, and the tank ruptured at the base. The level indicator was actually measuring the increasing pressure of the air in the tank as the liquid level rose and compressed the air in the tank [16].

5.2.2 Overpressuring with Gas or Vapor

This has usually occurred because those concerned did not realize that tanks are incapable of withstanding the pressure of the compressed air supply and that the vent may be too small to pass the inlet gas rate, as in the following incidents:

(a) There was a choke on the exit line from a small tank. To try to clear the choke, the operator held a compressed air hose against the open end at the top of the level glass. The gauge pressure of the compressed air was 100 psi (7 bar), and the top of the tank was blown off (Figure 5-3).

(b) An old vessel, intended for use as a low-pressure storage tank, had been installed in a new position by a contractor who decided to pressure-test it. He could not find a water hose to match the hose connection on the vessel, so he decided to use compressed air. The vessel ruptured. Another incident in which a storage vessel was ruptured by compressed air is described in Section 2.2a.

(c) On other occasions, tanks have been ruptured because the failure of a level controller (e.g., on a high-pressure distillation column) allowed a gas stream to enter the tank via the bottoms transfer line (Figure 5-4). Pressure vessels have also been ruptured in this way (see Section 9.2.2d). The precautions necessary to prevent this from occurring are analyzed in detail in reference 1.

(d) A storage tank for refrigerated butane was being brought back into service after maintenance. The tank was swept out with carbon dioxide to remove the air, and the refrigerated butane was then added. As the tank cooled down, some of the butane vaporized, and a 2-in. vent

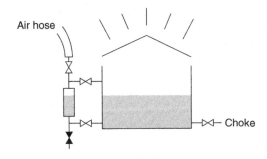

FIGURE 5-3 Tank top blown off by compressed air.

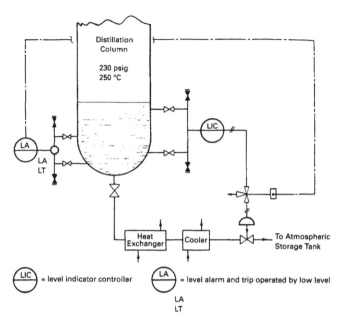

FIGURE 5-4   How failure of a level controller can overpressure a tank.

was left open to prevent the pressure from rising. This was not large enough, so the operator opened a 6-in. vent. The pressure continued to rise. Both relief valves on the tank had been set at too high a pressure, and the butane addition rate was rather high. The tank floor became convex, and the holding-down fittings around the base were pulled out of the ground, but fortunately, the tank did not leak. The relief valves should have been set at a gauge pressure of 1.0 psi (0.07 bar)—the pressure in the tank probably reached 1.5 to 2 psi (0.1 to 0.14 bar) [13].

5.3   SUCKING IN

This is by far the most common way in which tanks are damaged. The ways in which it occurs are legion. Some are listed here. Sometimes it seems that operators show great ingenuity in devising new ways of sucking in tanks!

Many of the incidents occurred because operators did not realize how fragile tanks are. They can be overpressured easily but sucked in much more easily. Whereas most tanks are designed to withstand a gauge pressure of 8 in. of water (0.3 psi or 2 kPa), they are designed to withstand a vacuum of only 2½ in. of water (0.1 psi or 0.6 kPa). This is the hydrostatic pressure at the bottom of a cup of tea. Some incidents occurred because operators did not understand how a vacuum works. See, for example, the incidents already described in Sections 3.3.3c and 3.3.4b.

The following are some of the ways by which tanks have been sucked in. In some cases, the vent was made ineffective. In others the vent was too small.

(a) Three vents were fitted with flame arrestors, which were not cleaned. After two years they choked. The flame arrestors were scheduled for regular cleaning (every six months), but this had been neglected due to the pressure of work. If you have flame arrestors on your tanks, are you sure they are necessary (see Section 6.2g)?

(b) A loose blank was put on top of the vent to prevent fumes from coming out near a walkway.

(c) After a tank had been cleaned, a plastic bag was tied over the vent to keep dirt from getting in. It was a hot day. When a sudden shower cooled the tank, it collapsed.

(d) A tank was boxed up with some water inside. Rust formation used up some of the oxygen in the air (see Section 11.1d).

(e) While a tank was being steamed, a sudden thunderstorm cooled it so quickly that air could not be drawn in fast enough. When steaming out a tank, a manhole should be opened. Estimates of the vent area required a range from 10 in. diameter to 20 in. diameter. On other occasions, vent lines have been isolated too soon after steaming stopped. Tanks that have been steamed may require several hours to cool.

(f) Cold liquid was added to a tank containing hot liquid.

(g) A pressure/vacuum valve (conservation vent) was assembled incorrectly—the pressure and vacuum pallets were interchanged. Valves should be designed so that this cannot occur (see Section 3.2.1a).

(h) A pressure/vacuum valve was corroded by the contents of the tank.

(i) A larger pump was connected to the tank, and it was emptied more quickly than the air could get in through the vent.

(j) Before emptying a tank truck, the driver propped the manhole lid open. It fell shut.

(k) A tank was fitted with an overflow, which came down to ground level. There was no other vent. When the tank was overfilled, the contents siphoned out (Figure 5-5). The tank should have been fitted with a vent on its roof, as well as the liquid overflow.

(l) A vent was almost blocked by polymer (Figure 5-6). The liquid in the tank was inhibited to prevent polymerization, but the vapor that condensed on the roof was not inhibited. The vent was inspected regularly, but the polymer was not noticed. Now a wooden rod is pushed through the vent to prove it is clear. (The other end of the rod should be enlarged so it cannot fall into the tank.)

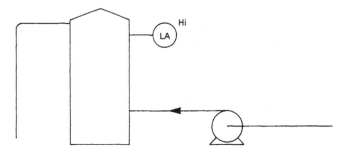

FIGURE 5-5   Overflow to ground level can cause a tank to collapse if there is no other vent.

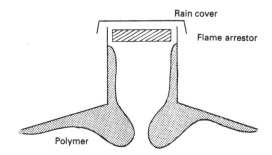

FIGURE 5-6   Vent almost blocked by polymer.

(m) Water was added too quickly to a tank that had contained a solution of ammonia in water. To prevent the tank collapsing, the vent would have had to be 30 in. in diameter! This is impractical, so the water should therefore be added slowly through a restriction orifice or, better, a narrow bore pipe.

It is clear from these descriptions that we cannot prevent tanks from being sucked in by writing lists of dos and don'ts or by altering plant designs, except in a few cases (see items g and h). We can prevent these incidents only by increasing people's knowledge and understanding of the strength of storage tanks and of the way they work, particularly the way a vacuum works.

The need for such training is shown by the action taken following one of the incidents. Only the roof had been sucked in, and it was concave instead of convex. The engineer in charge decided to blow the tank back to the correct shape by water pressure. He gave instructions for this to be done. A few hours later, he went to see how the job was progressing. He found that the tank had been filled with water and that a hand-operated hydraulic

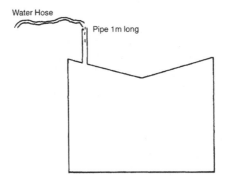

FIGURE 5-7 Method of restoring a tank with a concave roof to its original shape.

pump, normally used for pressure-testing pipework, was being connected to the tank. He had it removed, and he replaced the vent with a vertical pipe, 1 m long. He dribbled water into the pipe from a hose, and as he did so, the tank was restored to its original shape (Figure 5-7) to the amazement of onlookers. The static pressure of the water in the pipe was sufficient.

5.4 EXPLOSIONS

Explosions in the vapor spaces of fixed-roof storage tanks have been numerous. One estimate puts the probability of an explosion at about once in 1,000 years per tank, based on historical records. According to a 1997 report, 25 to 30 storage tank explosions occur per year in Canada alone [17]. The reason for the large number of explosions is that explosive mixtures are present in the vapor spaces of many storage tanks. It is almost impossible to be certain that a source of ignition will never turn up, particularly if the liquid in the tank has a low conductivity so that static charges can accumulate on the liquid. For this reason, many companies do not allow explosive mixtures to form. They insist that fixed-roof storage tanks containing hydrocarbons above their flash points are blanketed with nitrogen (see Section 5.6.3). Other companies insist that such hydrocarbons are stored only in floating-roof tanks.

Nonhydrocarbons usually have a higher conductivity than hydrocarbons. (Nonhydrocarbons with a symmetrical molecule, such as diethyl ether and carbon disulfide, have a low conductivity.) Charges of static electricity can rapidly drain away to earth (provided the equipment is grounded), and the risk of ignition is much lower. Many companies therefore store these materials in fixed-roof tanks without nitrogen blanketing [2].

External sources of ignition, such as lightning (Figure 5-8) or welding near an open vent, can also trigger a tank explosion. Sample and dip

This tank contains explosive vapor

LIGHTNING could blow it up

KEEP ALL SAMPLE AND DIP HOLES

closed

FIGURE 5-8

holes and other openings should be kept closed or protected by flame arrestors. These are liable to choke and need regular inspection (see Sections 5.3a, 6.2g, and 14.2.4).

5.4.1 A Typical Tank Explosion

A large tank blew up 40 minutes after the start of a blending operation in which one grade of naphtha was being added to another. The fire was soon put out, and the naphtha was moved to another tank. The next day, blending was resumed; 40 minutes later, another explosion occurred.

The tanks were not nitrogen-blanketed, and there was an explosive mixture of naphtha vapor and air above the liquid in the tanks. The source of ignition was static electricity. The pumping rate was rather high, so that the naphtha flowing through the pump and lines acquired a charge. A spark passed between the liquid in the tank and the roof or walls of the tank, igniting the vapor-air mixture.

These explosions led to an extensive series of investigations into the formation of static electricity [3]. There are several ways of preventing similar explosions:

1. Use nitrogen blanketing or floating-roof tanks.
2. Use antistatic additives; they increase the conductivity of the liquid so that charges can drain away rapidly to earth (provided equipment is

grounded). However, make sure that the additives do not deposit on catalysts or interfere with chemical operations in other ways.

3. Minimize the formation of static electricity by keeping pumping rates low (less than 3 m/s for pure liquids but less than 1 m/s if water is present) and avoiding splash filling. Filters and other restrictions should be followed by a long length of straight line to allow charges to decay.

It is difficult to feel confident that suggestion 3 can always be achieved; therefore, suggestions 1 and 2 are recommended.

For more information on static electricity, see Chapter 15.

5.4.2   Some Unusual Tank Explosions

(a) A new tank was being filled with water for hydrostatic testing when an explosion occurred. Two welders who were working on the roof, finishing the handrails, were injured, fortunately not seriously.

The tank had been filled with water through a pipeline that had previously contained gasoline. A few liters left in the line were flushed into the tank by the water and floated on top of it. The welders ignited the vapor.

No one should be allowed to go onto the roof of a tank while it is being filled with water for testing. One of the reasons for filling it with water is to make sure that the tank and its foundations are strong enough. If we were sure they were, we would not need to test. People should be kept out of the way, in case these structures are not strong.

(b) During the construction of a new tank, the contractors decided to connect the nitrogen line to the tank. They knew better, they said, than to connect the process lines without authority. But nitrogen was inert and therefore safe.

The new tank and an existing one were designed to be on balance with each other to save nitrogen (Figure 5-9), but the contractors did not understand this. The valve to the new tank was closed but leaking. Nitrogen and methanol vapor entered the tank, and a welder who was completing the inlet line to the tank ignited the vapor. The roof was blown right off. By great good fortune, it landed on a patch of empty ground just big enough to contain it (Figure 5-10).

(c) The roof of an old gasoline tank had to be repaired. The tank was steamed out and cleaned, and tests with a combustible gas detector showed that no flammable gas or vapor was present. A welder was therefore allowed to start work. Soon afterward, a small flash of flame singed his hair.

The roof was made from plates, which overlapped each other by about 4 in. and which were welded together on the top side only—an old method of construction that is not now used (Figure 5-11). It is

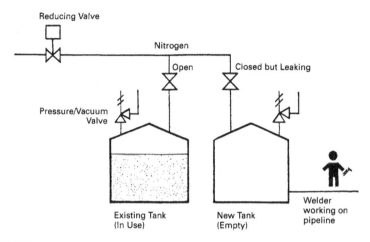

FIGURE 5-9   If tanks are on balance, the nitrogen entering one tank is inevitably mixed with vapor.

FIGURE 5-10   When an explosion occurred in a tank, the roof landed on an area just big enough to contain it.

FIGURE 5-11   An old method of tank construction allows liquid to enter the gap between the plates.

believed that some gasoline entered the space between the plates and became trapped by rust and scale. The heat from the welding vaporized the gasoline, and it blew out of the molten weld. At the time, the suggestion was made that the tank should be filled with water, but this cannot be done without the risk of overpressuring the tank (see Section 5.2.1).

(d) Some welding had to take place at the top of a 38-m^3 tank containing about 10 m^3 of hydrocarbons and water. Instead of emptying the tank and sweeping out the remaining vapor with steam or nitrogen—the usual procedure—the people in charge tested the inside of the tank, just below a roof opening, with a combustible gas detector. As they got a zero reading, they decided to go ahead with the welding. The roof was blown off the tank, landing 30 m (100 ft) away, killing one man and injuring another.

It is believed that a hot speck, loosened by the welding, fell off the inside of the roof and ignited a flammable mixture that was present near the surface of the liquid [18]. Tests have shown that the atmosphere in a tank can take a long time to reach equilibrium when the liquid level is low. As stated in Section 1.3.1, large vessels should be tested in several places.

(e) During the manufacture of zinc, metallic impurities are removed by addition of zinc slurry or powder to an acidic solution of zinc salts in a number of tanks fitted with vents, overflows, and extract fans. Hydrogen is produced and has to be removed. In a new plant, there was no inerting and the so-called "basis of safety" (really a basis of danger) was to operate with the hydrogen concentration either below the lower flammable limit (4%) or above the upper flammable limit (75%) and to pass rapidly between the two, as follows (Figure 5-12):

- When zinc was added to a tank, the extract fans were operated at full rate with the vent closed. No air could be sucked in, and the concentration of hydrogen rose rapidly above the upper flammable limit (75%).
- The fan speed was then lowered.
- When the rate of production of hydrogen fell, the fans were again switched to full rate, this time with the vent open. Air was sucked in, and the concentration of hydrogen fell rapidly below the lower flammable limit.

After three months of operation, an explosion occurred in a 400-m^3 tank, which fortunately was fitted with explosion relief. Three weeks later, another explosion blew the roof off another tank, and the Australian Department of Mines ordered the closure of the plant.

The source of ignition was never found, but a report [19] on the explosion lists six possible causes, thus confirming the view—well known to everyone except those who designed and operated the

Stage	Vent	Fan speed	Objective
Start	Shut	High	Remove air and get above UEL quickly
Run	Shut	Low	Stay above UEL
Near end	Open	High	Suck in air and get below LEL quickly

Results:
After three months of operation: mild explosion; vent panel lifted.
Three weeks later: another explosion; roof blown off tank.

Modifications:

Tanks inerted except for a few, which were sparged with air to keep below the lower explosive limit.

UEL = upper explosive limit
LEL = lower explosive limit

FIGURE 5-12   An attempt to avoid explosions by passing quickly through the explosive range was not successful. *(Reproduced with permission of the American Institute of Chemical Engineers. Copyright © 1995 AIChE. All rights reserved.)*

plant—that sources of ignition are so numerous that we can never be sure they will not turn up even though we do what we can to remove known sources. Flammable mixtures should not be deliberately allowed to form except under rigidly defined circumstances where the chance of an occasional ignition is accepted. This is particularly true where hydrogen is handled, as it is more easily ignited than most other gases or vapors.

The plant was restarted after 23 days. Most of the tanks were now blanketed with nitrogen, but a few, which were difficult to blanket, were fitted with an air sparge system designed to keep the hydrogen concentration well below 25% of the lower flammable limit.

(f) Paper mills use large quantities of water, and the water is usually recycled. Buffer storage is needed, and at one paper mill, it took the form of a 740-m^3 tank. Experience showed that this was insufficient, and another tank of the same size was installed alongside. To simplify installation, it was not connected in parallel with the original tank but on balance with it, as shown in Figure 5-13. A week after the new tank was brought into use, welders were completing the handrails on the roof when an explosion occurred in the tank. Two welders were killed, and the tank was blown 20 m into the air, landing on a nearby building.

Investigation showed that the explosion was due to hydrogen formed by anaerobic bacteria. In the original tank, the splashing of the inlet liquor aerated the water and prevented anaerobic conditions. This did not apply in the new tank [20].

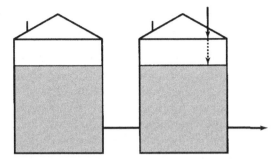

FIGURE 5-13 Extra buffer storage for water was provided by installing a second tank on balance with the first one. Lack of aeration allowed hydrogen-forming bacteria to grow, and an explosion occurred. *(Reproduced with permission of the American Institute of Chemical Engineers. Copyright © 1995 AIChE. All rights reserved.)*

The incident shows once again how a simple modification, in this case adding liquid to the bottom of a tank instead of the top, can produce an unforeseen hazard. In the oil and chemical industries we are taught to add liquid to the bottom of a tank, not the top, to prevent splashing, the production of mist, and the generation of static electricity (see Section 5.4.1). No rule is universal.

Hydrogen produced by corrosion has also turned up in some unexpected places (see Section 16.2). As mentioned in Section 1.1.4, bacterial action on river water can also produce methane.

Fires and explosions that occurred while repairing or demolishing storage tanks containing traces of heavy oil are described in Section 12.4.1, and an explosion of a different type is described at the end of Section 1.1.4.

5.4.3 An Explosion in an Old Pressure Vessel Used as a Storage Tank

Sometimes old pressure vessels are used as storage tanks. It would seem that by using a stronger vessel than is necessary, we achieve greater safety. But this may not be the case, as if the vessel fails, it will do so more spectacularly (see Section 2.2a).

A tank truck hit a pipeline leading to a group of tanks. The pipeline went over the top of the dike wall, and it broke off inside the dike. The engine of the truck ignited the spillage, starting a dike fire, which damaged or destroyed 21 tanks and five tank trucks.

An old 100-m³ pressure vessel, a vertical cylinder, designed for a gauge pressure of 5 psi (0.3 bar), was being used to store, at atmospheric pressure, a liquid of flash point 40°C. The fire heated the vessel to above 40°C and ignited the vapor coming out of the vent; the fire flashed back into the tank, where an explosion occurred. The vessel burst at the bottom seam, and the entire vessel, except for the base, and contents went into orbit like a rocket [4].

If the liquid had been stored in an ordinary low-pressure storage tank with a weak seam roof, then the roof would have come off, and the burning liquid would have been retained in the rest of the tank.

The incident also shows the importance of cooling, with water, all tanks or vessels exposed to fire. It is particularly important to cool vessels. They fail more catastrophically, either by internal explosion or because the rise in temperature weakens the metal (see Sections 2.6b and 8.1).

Another tank explosion is described in Section 16.2a.

5.5   FLOATING-ROOF TANKS

This section describes some incidents that could have occurred only on floating-roof tanks.

5.5.1   How to Sink the Roof

A choke occurred in the flexible pipe that drained the roof of a floating-roof tank. It was decided to drain rainwater off the roof with a hose. To prime the hose and establish a siphon, the hose was connected to the water supply. It was intended to open the valve on the water supply for just long enough to fill the hose. This valve would then be closed and the drain valve opened (Figure 5-14). However, the water valve was opened in error and left open, with the drain valve shut. Water flowed onto the floating roof, and it sank in 30 minutes (see also Section 18.8).

Temporary modifications should be examined with the same thoroughness as permanent ones (see Section 2.4).

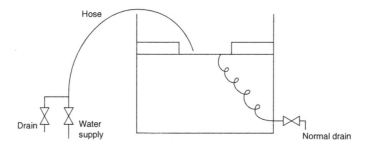

FIGURE 5-14   How to sink the roof of a floating-roof tank.

5.5.2 Fires and Explosions

(a) Most fires on floating-roof tanks are small rim fires caused by vapor leaking through the seals. The source of ignition is often atmospheric electricity. It can be eliminated as a source of ignition by fitting shunts—strips of metal—about every meter or so around the rim to ground the roof to the tank walls.

Many rim fires have been extinguished by a worker using a hand-held fire extinguisher. However, in 1979, a rim fire had just been extinguished when a pontoon compartment exploded, killing a fire-fighter. It is believed that there was a hole in the pontoon and some of the liquid in the tank leaked into it.

Workers should not go onto floating-roof tanks to extinguish rim fires [5]. If fixed firefighting equipment is not provided, foam should be supplied from a monitor.

(b) The roof of a floating-roof tank had to be replaced. The tank was emptied, purged with nitrogen, and steamed for six days. Each of the float chambers was steamed for four hours. Rust and sludge were removed from the tank. Demolition of the roof was then started.

Fourteen days later, a small fire occurred. About a gallon of gasoline came out of one of the hollow legs that support the roof when it is off-float and was ignited by a spark. The fire was put out with dry powder. It is believed that the bottom of the hollow leg was blocked with sludge and that, as cutting took place near the leg, the leg moved and disturbed the sludge (Figure 5-15).

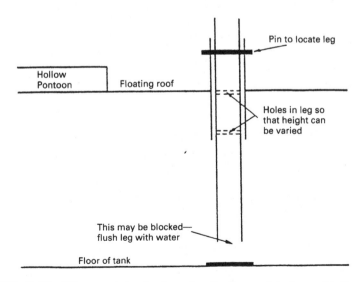

FIGURE 5-15 Oil trapped in the leg of a floating-roof tank caught fire during demolition.

Before welding or burning is permitted on floating-roof tanks, the legs should be flushed with water from the top. On some tanks, the bottoms of the legs are sealed. Holes should be drilled in them so they can be flushed through.

(c) Sometimes a floating roof is inside a fixed-roof tank. In many cases, this reduces the concentration of vapor in the vapor space below the explosive limit. But in other cases, it can increase the hazard, because vapor that was previously too rich to explode is brought into the explosive range.

A serious fire that started in a tank filled with an internal floating roof is described in reference 6.

As a result of a late change in design, the level at which a floating roof came off-float had been raised, but this was not marked on the drawings that were given to the operators. As a result, without intending to, they took the roof off-float. The pressure/vacuum valve (conservation vent) opened, allowing air to be sucked into the space beneath the floating roof.

When the tank was refilled with warm crude oil at 37°C (100°F), vapor was pushed out into the space above the floating roof and then out into the atmosphere through vents on the fixed-roof tank (Figure 5-16). This vapor was ignited at a boiler house some distance away.

The fire flashed back to the storage tank, and the vapor burned as it came out of the vents. Pumping was therefore stopped. Vapor no longer came out of the vents, air got in, and a mild explosion occurred inside the fixed-roof tank. This forced the floating roof down like a piston, and some of the crude oil came up through the seal past the side of the floating roof and out of the vents on the fixed-roof tank. This oil caught fire, causing a number of pipeline joints to fail, and this caused further oil leakages. One small tank burst; fortunately, it had a weak seam roof. More than 50 fire appliances and 200 firefighters attended, and the fire was under control in a few hours.

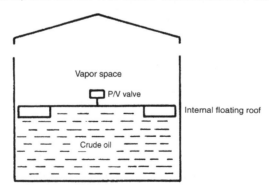

FIGURE 5-16   Tank with internal floating roof.

The water level outside the dike rose because the dike drain valve had been left open, and the dike wall was damaged by the firefighting operations. The firefighters pumped some of the water into another dike, but it ran out because the drain valve on this dike had also been left open.

An overhead power cable was damaged by the fire and fell down, giving someone an electric shock. The refinery staff members therefore isolated the power to all the cables in the area. Unfortunately, they did not tell the firefighters what they were going to do. Some electricity-driven pumps that were pumping away some of the excess water stopped, and the water level rose even further. Despite a foam cover, oil floating on top of the water was ignited by a fire engine that was parked in the water. The fire spread rapidly for 150 m. Eight firefighters were killed and two were seriously injured. A naphtha tank ruptured, causing a further spread of the fire, and it took 15 hours to bring it under control.

The main lessons from this incident are as follows:

1. Keep plant modifications under control and keep drawings up to date (see Chapter 2).
2. Do not take floating-roof tanks off-float except when they are being emptied for repair.
3. Keep dike drain valves locked shut. Check regularly to make sure they are shut.
4. Plan now how to get rid of firefighting water. If the drains will not take it, it will have to be pumped away.
5. During a fire, keep in close touch with the firefighters and tell them what you propose to do.

(d) Roof cracks led to an extensive fire on a large (94,000-m^3) tank containing crude oil. The cracking was due to fatigue, the result of movement of the roof in high winds, and a repair program was in hand. A few days before the fire, oil was seen seeping from several cracks, up to 11 in. long, on the single-skin section of the floating roof, but the tank was kept in use, and no attempt was made to remove the oil. The oil was ignited, it is believed, by hot particles of carbon dislodged from a flarestack 108 m (350 ft) away and 76 m (170 ft) high, the same height as the tank. The fire caused the leaks to increase, and the tank was severely damaged. Six firefighters were injured when a release of oil into the dike caused the fire to escalate. The fire lasted 36 hours, 25,000 tons of oil were burned, and neighboring tanks, 60 m away, were damaged. The insulation on one of these tanks caught fire, and the tank was sucked in, but the precise mechanism was not clear [9, 10].

The release of oil into the dike was due to boilover—that is, production of steam from the firefighting foam by the hot oil. As the steam leaves the tank, it brings oil with it. Boilover usually occurs when the heat from the burning oil reaches the water layer at the bottom of the tank, but in this case it occurred earlier than usual when the heat reached pockets of water trapped on the sunken roof [14].

Most large floating roofs are made from a single layer of steel, except around the edges, where there are hollow pontoons to give the roof its buoyancy. The single layer of steel is liable to crack, and any spillage should be covered with foam and then removed as soon as possible. Double-deck roofs are obviously safer but much more expensive [14].

For more information on tank fires, see reference 23.

5.6   MISCELLANEOUS INCIDENTS

5.6.1   A Tank Rises Out of the Ground

A tank was installed in a concrete-lined pit. The pit was then filled with sand, and a layer of concrete 6 in. thick was put over the top. Water accumulated in the pit, and the buoyancy of the tank was sufficient to break the holding-down bolts and push it through the concrete covering.

A sump and pump had been provided for the removal of water. But either the pump-out line had become blocked or pumping had not been carried out regularly [7].

Underground tanks are not recommended for plant areas. As the ground is often contaminated by corrosive chemical the tanks may be corroded and this is difficult to detect by inspection as the soil has first to be removed.

5.6.2   Foundation Problems

Part of the sand foundation beneath a 12-year-old tank subsided. Water collected in the space that was left and caused corrosion. This was not detected because the insulation on the tank came right down to the ground.

When the corrosion had reduced the wall thickness from 6 mm to 2 mm, the floor of the tank collapsed along a length of 2.5 m (8 ft), and 30,000 m^3 of hot fuel oil came out. Most of it was collected in the dike. However, some leaked into other dikes through rabbit holes in the earth walls.

All storage tanks should be scheduled for inspection every few years. And on insulated tanks, the insulation should finish 200 mm (8 in.) above the base so that checks can be made for corrosion.

Tanks containing liquefied gases that are kept liquid by refrigeration sometimes have electric heaters beneath their bases to prevent freezing of the ground. When such a heater on a liquefied propylene tank failed, the tank became distorted and leaked—but fortunately, the leak did not ignite. Failure of the heater should activate an alarm. As stated in Section 5.2, frequent complete emptying of a tank can weaken the base/wall weld.

5.6.3   Nitrogen Blanketing

Section 5.4.1 discussed the need for nitrogen blanketing. However, if it is to be effective, it must be designed and operated correctly.

Incorrect Design

On one group of tanks, the reducing valve on the nitrogen supply was installed at ground level (Figure 5-17). Hydrocarbon vapor condensed in the vertical section of the line and effectively isolated the tank from the nitrogen blanketing.

The reducing valve should have been installed at roof height. Check your tanks—there may be more like this one.

Incorrect Operation

An explosion and fire occurred on a fixed-roof tank that was supposed to be blanketed with nitrogen. After the explosion, it was found that the nitrogen supply had been isolated. Six months before the explosion, the manager had personally checked that the nitrogen blanketing was in operation. But no later check had been carried out [8].

All safety equipment and systems should be scheduled for regular inspection and test. Nitrogen blanketing systems should be inspected at least weekly. It is not sufficient to check that the nitrogen is open to the tank. The atmosphere in the tank should be tested with a portable oxygen analyzer to make sure that the oxygen concentration is below 5%.

Large tanks (say, over $1,000 \, m^3$) blanketed with nitrogen should be fitted with low-pressure alarms to give immediate warning of the loss of nitrogen blanketing.

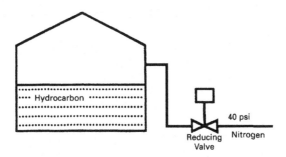

FIGURE 5-17   Incorrect installation of nitrogen blanketing.

5.6.4   Brittle Failure

On several occasions, a tank has split open rapidly from top to bottom, as if it were fitted with a zipper and someone pulled it. An official report [15] describes one incident in detail:

The tank, which was nearly full, contained 15,000 m³ of diesel oil, which surged out of the failed tank like a tsunami, washing over the dike walls. About 3,000 m³ escaped from the site into a river that supplied drinking water for neighboring towns, disrupting supplies for a week. Fortunately, no one was killed.

The collapse was due to a brittle failure that started at a flaw in the shell about 2.4 m (8 ft) above the base. The fault had been there since the tank was built more than 40 years earlier, and the combination of a full tank and a low temperature triggered the collapse. For most of the 40 years, the tank had been used for the storage of a fuel oil that had to be kept warm; the high temperature prevented a brittle failure. However, two years before the collapse, the tank had been dismantled, re-erected on a new site, and used for the storage of diesel oil at ambient temperature.

The flaw was close to the edge of a plate, and if the contractor that moved the tank had cut it up along the welds—the usual practice—some or all of the flaw might have been removed. However, the tank was cut up close to the welds but away from them. The flaw was obscured by rust and residue and could not be seen.

The owner and contractor are strongly criticized in the report for not complying with the relevant American Petroleum Institute codes. They did not radiograph all T-joints (the flaw was close to a T-joint and would have been detected), and they did not realize that the grade of steel used and the quality of the original welding were not up to modern standards. The comments about the engineers in charge are similar to those made in the Flixborough report (see Section 2.4a): their lack of qualifications "does not necessarily affect their ability to perform many aspects of a project engineer's job. However, when tough technical issues arise, such as whether to accept defective welds, a stronger technical background is required. If help on such matters was available . . . , there is no evidence that . . . utilized it" (p. 69 of the report).

The summing up of the report reminds us of similar comments made about many serious accidents in other industries: the company (a large, independent oil refiner)

> failed to take any active or effective role in controlling its contractors or establish any procedures which might lead to a quality job. It was a passive consumer of the worst kind—apathetic as to potential problems, ignorant of actual events, unwilling to take any engaged role. Its employees were both institutionally and often personally unable to respond in any other way. Both the details and the big picture equally escaped [the company's] attention. Compared against the applicable standards,

its industry peers, or even common sense [the company's] conduct and procedures can only be considered grossly negligent. The structural collapse . . . can be directly traced to the supervisory bankruptcy at [the company]. (p. 79 of the report)

The report also includes a list of other similar tank collapses: six in the U.S. in the period 1978–1986 (p. 102). A similar incident involving a liquefied propane tank occurred in Qatar in 1977 (see Section 8.1.5).

5.7 FIBERGLASS-REINFORCED (FRP) TANKS

Tanks made from fiberglass-reinforced (FRP) plastic are being increasingly used, but a number of failures have occurred. In the United Kingdom, 30 catastrophic failures are known to have occurred during the period 1973–1980, and a 1996 report shows that they seem to have been continuing at a similar rate [21]. The following typify the catastrophic failures that have occurred [11]:

(a) A 50-m^3 tank made from bolted sections failed because the bolts holding the steel reinforcements together were overstressed. The contents—liquid clay—pushed over a wall and ran into the street.

(b) Ninety cubic meters of sulfuric acid was spilled when a tank failed as the result of stress corrosion cracking. It had not been inspected regularly, and the company was not aware that acid can affect FRP tanks. The failure was so sudden that part of the dike wall was washed away.

(c) Another tank, used to store a hot, acidic liquid, failed because it was heated above its design temperature and damaged when digging out residues. Again, it had not been inspected regularly, and the company was not aware of the effects of acid.

(d) Forty-five cubic meters of 10% caustic soda solution was spilled when the end came off a horizontal cylindrical tank. The polypropylene lining was leaking, and the caustic soda attacked the FRP.

(e) Three hundred fifty cubic meters of hot water was spilled and knocked over a wall when a tank failed at a brewery. The grade of FRP used was unsuitable, and the tank had never been inspected during the three years it had been in use. Another failure of a plastic hot water tank is described in Section 12.2.

(f) Thirty tons of acid were spilled when a tank failed. A weld was below standard, and stress corrosion cracking occurred. There had been no regular inspections.

(g) An internal lining failed as the result of bending stresses, and the acidic contents attacked the FRP. Cracks in the tank had been noticed and repaired, but no one investigated why they had occurred. Finally, the tank failed catastrophically, and the contents knocked over a wall.

(h) An FRP tank leaked near a manway after only 18 months in service. The wall thickness was too low, the welding was substandard, and this poor construction was not detected during inspection. The tank failed the first time it was filled to 85% capacity, suggesting that it was never tested properly after installation [21].

These incidents show that, to prevent failures of FRP tanks, we should take the following precautions:

1. Use equipment designed for the conditions of use.
2. Know the limitations of the equipment.
3. Inspect regularly.
4. Do not repair faults and carry on until their cause is known.

These rules, of course, apply generally, but they are particularly applicable to FRP tanks.

References

1. T. A. Kletz and H. G. Lawley, in A. E. Green (editor), *High Risk Safety Technology*, Wiley, Chichester, UK, 1982, Chapter 2.1.
2. T. A. Kletz, "Hard Analysis—A Quantitative Approach to Safety," Symposium Series No. 34, Institution of Chemical Engineers, 1971, p. 75.
3. A. Klinkenberg and J. L. van der Minne, *Electrostatics in the Petroleum Industry*, Elsevier, Amsterdam, 1958.
4. *Loss Prevention*, Vol. 7, 1972, p. 119; and Manufacturing Chemists Association, *Case History No. 1887*, Washington, DC, 1972.
5. D. K. McKibben, "Safe Design of Atmospheric Pressure Vessels," Paper presented at Seminar on Prevention of Fires and Explosions in the Hydrocarbon Industries, Institute of Gas Technology, Chicago, June 21–26, 1982.
6. Press release issued by the City of Philadelphia Office of the City Representatives, Dec. 12, 1975.
7. *Petroleum Review*, Oct. 1974, p. 683.
8. T. A. Kletz, *Learning from Accidents*, 3rd edition, Butterworth-Heinemann, Oxford, UK, 2001, Chapter 6.
9. *Report of the Investigation into the Fire at Amoco Refinery, 30 August 1983*, Dyfed County Fire Brigade, UK.
10. *Hazardous Cargo Bulletin*, Sept. 1983, p. 32; and Dec. 1983, p. 32.
11. T. E. Maddison, *Loss Prevention Bulletin*, No. 076, Aug. 1987, p. 31.
12. D. Nevill and G. C. White, "Research into the Structural Integrity of LNG Tanks," *New Directions in Process Safety*, Symposium Series No. 124, Institution of Chemical Engineers, 1991, p. 425 (discussion).
13. *Loss Prevention Bulletin*, No. 106, Aug. 1992, p. 5.
14. L. Streinbrecher, *Loss Prevention Bulletin*, No. 088, Aug. 1989, p. 25.
15. *Report of the Investigation into the Collapse of Tank 1338*, Commonwealth of Pennsylvania Department of Environmental Resources, Harrisburg, PA., June 1988.
16. R. E. Sanders, *Chemical Process Safety—Learning from Case Histories*, Butterworth-Heinemann, Boston, MA, 1999, p. 99.
17. F. E. Self, J. D. Hill, "Safety Considerations When Treating VOC Streams with Thermal Oxidizers," *Proceedings of the Thirty-first Annual Loss Prevention Symposium*, AIChE, New York, 1997.

18. *Loss Prevention Bulletin*, No. 131, Oct. 1996, p. 8.
19. N. Maddison, "Explosion Hazards in Large Scale Purification by Metal Dust," *Hazards XII—European Advances in Process Safety*, Symposium Series No. 134, Institution of Chemical Engineers, Rugby, UK, 1994.
20. R. S. Rowbottom, *Pulp and Paper Canada*, Vol. 90, No. 4, 1989, p. T138.
21. A. Trevitt, *The Chemical Engineer*, No. 27, 1996, p. 27.
22. I. Ayres, *Super Crunchers*, Bantam Dell, New York, 2007, p. 89.
23. *Liquid Hydrocarbon Tank Fires*, Institution of Chemical Engineers, Rugby, UK, 2005.

Stacks

The [U.S.] Institute of Medicine estimates that it took "an average of 17 years for new knowledge ... to be incorporated into practice, and even then application [was] highly uneven." Progress in medical science occurred one funeral at a time. If doctors didn't learn something in medical school or in their residency, there was a good chance they never would.

—Ian Ayres [10]

Are engineers any different?

Stacks, like storage tanks, have been the sites of numerous explosions. They have also been known to choke.

6.1 STACK EXPLOSIONS

(a) Figure 6-1 shows the results of an explosion in a large flarestack. The stack was supposed to be purged with nitrogen. However, the flow was not measured and had been cut back almost to zero to save nitrogen. Air leaked in through the large bolted joint between unmachined surfaces. The flare had not been lit for some time. Shortly after it was relit, the explosion occurred—the next time some gas was put into the stack. The mixture of gas and air moved up the stack and the pilot flame ignited it.

The following precautions should prevent similar incidents from happening again:

1. Stacks should be welded. They should not contain bolted joints between unmachined surfaces.

FIGURE 6-1   Base of flarestack.

2. There should be a continuous flow of gas up every stack to prevent air diffusing down and to sweep away small leaks of air into the stack. The continuous flow of gas does not have to be nitrogen—a waste-gas stream is effective. But if gas is not being flared continuously, it is usual to keep nitrogen flowing at a linear velocity of 0.03 to 0.06 m/s (1.2 to 1.4 in/s). The flow of gas should be measured. A higher rate is required if hydrogen or hot condensable gases are being flared. If possible, hydrogen should be discharged through a separate vent stack and not mixed with other gases in a flarestack.

3. The atmosphere inside every stack should be monitored regularly, say daily, for oxygen content. Large stacks should be fitted with oxygen

analyzers that alarm at 5% (2% if hydrogen is present). Small stacks should be checked with a portable analyzer.

These recommendations apply to vent stacks as well as flarestacks.

(b) Despite the publicity given to the incident just described, another stack explosion occurred nine months later in the same plant.

To prevent leaks of carbon monoxide and hydrogen from the glands of a number of compressors getting into the atmosphere of the compressor house, they were sucked away by a fan and discharged through a small vent stack. Air leaked into the duct because there was a poor seal between the duct and the compressor. The mixture of air and gas was ignited by lightning.

The explosion would not have occurred if the recommendations made after the first explosion had been followed—if there had been a flow of inert gas into the vent collection system and if the atmosphere inside had been tested regularly for oxygen.

Why were they not followed? Perhaps because it was not obvious that recommendations made after an explosion on a large flarestack applied to a small vent stack.

(c) Vent stacks have been ignited by lightning or in other ways on many occasions. On several occasions, a group of 10 or more stacks have been ignited simultaneously. This is not dangerous provided that the following conditions apply:

1. The gas mixture in the stack is not flammable so that the flame cannot travel down the stack.
2. The flame does not impinge on overhead equipment. (Remember that in a wind, it may bend at an angle of 45 degrees.)
3. The flame can be extinguished by isolating the supply of gas or by injecting steam or an increased quantity of nitrogen. (The gas passing up the stack will have to contain more than 90% nitrogen to prevent it from forming a flammable mixture with air.)

(d) A flarestack and the associated blowdown lines were prepared for maintenance by steaming for 16 hours. The next job was to isolate the system from the plant by turning a figure-8 plate in the 0.9 m (35 in.) blowdown line. As it was difficult to turn the figure-8 plate while steam escaped from the joint, the steam purge was replaced by a nitrogen purge two hours beforehand.

When the plate had been removed for turning, leaving a gap about 50 mm (2 in.) across, there was an explosion. A man was blown off the platform and killed.

The steam flow was 0.55 ton/hr, but the nitrogen flow was only 0.4 ton/hr, the most that could be made available. As the system cooled, air was drawn in. Some liquid hydrocarbon had been left in a blowdown vessel, and the air and hydrocarbon vapor formed a flammable mixture. According to the report, this moved up the stack and was ignited by the pilot burner, which was still lit. It is possible, however, that it was ignited by the maintenance operations.

As the steam was hot and the nitrogen was cold, much more nitrogen than steam was needed to prevent air from being drawn into the stack. After the explosion, calculations showed that 1.6 tons/hr were necessary, four times as much as the amount supplied. After the explosion, the company decided to use only nitrogen in the future, not steam [5].

Should the staff have foreseen that steam in the system would cool and that the nitrogen flow would be too small to replace it? Probably the method used seemed so simple and obvious that no one stopped to ask if there were any hazards.

(e) Three explosions occurred in a flarestack fitted, near the tip, with a water seal, which was intended to act as a flame arrestor and prevent flames from passing down the stack. The problems started when, as a result of incorrect valve settings, hot air was added to the stack that was burning methane. The methane/air mixture was in the explosive range, and as the gas was hot (300°C [570°F]), the flashback speed from the flare (12 m/s) was above the linear speed of the gas (10 m/s in the tip, 5 m/s in the stack). An explosion occurred, which probably damaged the water seal, though no one realized this at the time. Steam was automatically injected into the stack, and the flow of methane was tripped. This extinguished the flame. When flow was restarted, a second explosion occurred, and as the water seal was damaged, this one traveled right down the stack into the knockout drum at the bottom. Flow was again restarted, and this time the explosion was louder. The operating team then decided to shut down the plant [6]. We should not restart a plant after an explosion (or other hazardous event) until we know why it occurred.

(f) Another explosion, reported in 1997, occurred, like that described in (a), because the nitrogen flow to a stack was too low. It was cut back by an inexperienced operator; there was no low-flow alarm or high-oxygen alarm [7]. The author shows commendable frankness in describing the incident so that others may learn from it, but nowhere in the report (or editorial comment) is there any indication that the lessons learned were familiar ones, described in published reports decades before.

For other stack explosions, see Section 7.13c and references 1 and 11.

6.2   BLOCKED STACKS

(a) Section 2.5a described how an 8-in.-diameter vent stack became blocked by ice because cold vapor at −100°C (−150°F) and steam were passed up the stack together. The cold gas met the condensate running down the walls and caused it to freeze. A liquefied gas

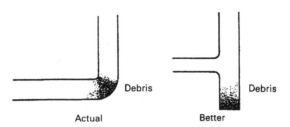

Actual   Better

FIGURE 6-2 Flarestack after fall of debris.

tank was overpressured, and a small split resulted. The stack was designed to operate without steam. But the steam was then introduced to make sure that the cold gas dispersed and did not drift down to ground level.

(b) The vent stack was replaced by a 14-in.-diameter flarestack with a supply of steam to a ring around the top of the stack. A few years later, this stack choked again, this time because of a deposit of refractory debris from the tip, cemented together by ice (as some condensate from the steam had found its way down the stack). Fortunately, in this case someone noticed the high pressure in the tank before any damage occurred. There was no boot at the bottom of the stack to collect debris (Figure 6-2). A boot was fitted [2].

(c) On other occasions, blowdown lines or stacks have become blocked in cold weather because benzene or cyclohexane, both of which have freezing points of 5°C (40°F), was discharged through them. Steam tracing of the lines or stacks may be necessary.

(d) Blowdown lines should never be designed with a dip in them, or liquid may accumulate in the dip and exert a back pressure. This has caused vessels to be overpressured [3].

(e) A blowdown line that was not adequately supported sagged when exposed to fire and caused a vessel to be overpressured.

(f) Water seals have frozen in cold weather. They should not be used except in locales where freezing cannot occur. Flare and vent systems should be simple. It is better to avoid water seals than install steam heating systems and low-temperature alarms, which might fail.

(g) Vent stacks are sometimes fitted with flame arrestors to prevent a flame on the end of the stack from traveling back down the stack. The arrestors are liable to choke unless regularly cleaned. They are also unnecessary, because unless the gas mixture in the stack is flammable, the flame cannot travel down the stack. If the gas mixture in the stack is flammable, then it may be ignited in some other way. Stacks should therefore be swept by a continuous flow of gas to prevent a flammable mixture from forming, as discussed in Section 6.1.

There are, however, two cases in which flame arrestors in vent stacks are justified:

1. If the gas being vented can decompose without the addition of air; an example is ethylene oxide. Whenever possible, such gases should be diluted with nitrogen. If this is not always possible, a flame arrestor may be used.

2. In the vent pipes of storage tanks containing a flammable mixture of vapor and air (Section 5.4.1). Such flame traps should be inspected regularly and cleaned if necessary. Section 5.3a described how a tank was sucked in because the flame arrestors on all three vents had not been cleaned for two years.

A type of flame arrestor that can be easily removed for inspection without using tools is described in reference 4.

(h) Molecular seals have been choked by carbon from incompletely burned gas, and water seals could be choked in the same way. For this reason, many companies prefer not to use them. If they are partly choked, burning liquid or particles of hot carbon may be expelled when flaring rates are high [9] (see Section 5.5.2d).

(i) The relief valve on a liquid hydrogen tank discharged to atmosphere through a short stack. The escaping hydrogen caught fire. The fire service poured water down the stack; the water froze, and the tank was overpressured and split. The fire should have been extinguished by injecting nitrogen up the stack, as discussed in Section 6.1c.

The common theme of many of these items is that blowdown lines and flare- and vent stacks should be kept simple because they are part of the pressure relief system. Avoid flame arrestors, molecular seals, water seals, and U-bends. Avoid steam, which brings with it rust and scale and may freeze.

6.3   HEAT RADIATION

The maximum heat radiation that people are exposed to from a flarestack should not exceed 4.7 kW/m^2 ($1,500 \text{ Btu/ft}^2/\text{hr}$), about three times the peak solar radiation in the tropics. Even this amount of radiation can be withstood without injury for only a minute or two. The maximum to which people may be exposed continuously is about 1.7 kW/m^2 ($500 \text{ Btu/ft}^2/\text{hr}$). In the neighborhood of flarestacks (say, wherever the radiation could exceed 1.7 kW/m^2), the temperatures reached by cables, roofing materials, and plastic equipment should all be reviewed to make sure they cannot be damaged [8, 9].

There are more reports of explosions in stacks in Sections 36.3 and 36.4.

References

1. J. L. Kilby, *Chemical Engineering Progress*, June 1968, p. 419.
2. T. A. Kletz, *Chemical Engineering Progress*, Vol. 70, No. 4, Apr. 1974, p. 80.
3. T. J. Laney, in C. H. Vervalin (editor), *Fire Protection Manual for Hydrocarbon Processing Plants*, 3rd edition, Vol. 1, Gulf Publishing Co., Houston, TX, 1985, p. 101.
4. T. A. Kletz, *Learning from Accidents*, 3rd edition, Butterworth-Heinemann, Boston, MA, 2001, Section 7.6.
5. *Loss Prevention Bulletin*, No. 107, Oct. 1992, p. 23.
6. V. M. Desai, *Process Safety Progress*, Vol. 15, No. 3, Fall 1996, p. 166.
7. T. Fishwick, *Loss Prevention Bulletin*, No. 135, June 1997, p. 18.
8. M. S. Mannan, *Lees' Loss Prevention in the Process Industries*, 3rd edition, Butterworth-Heinemann, Boston, 2005, Chapter 16.
9. D. Shore, *Journal of Loss Prevention in the Process Industries*, Vol. 9, No. 6, Nov. 1996, p. 363.
10. I. Ayres, *Super Crunchers*, Bantam Dell, New York, 2007, p. 89.
11. I. M. Shaluf, *Disaster Prevention and Management*, Vol. 17, No. 1, 2008, p. 6.

Leaks

A small leak will sink a great ship.

—Thomas Fuller, 1732

Leaks of process materials are the process industries' biggest hazard. Most of the materials handled will not burn or explode unless mixed with air in certain proportions. To prevent fires and explosions, we must therefore keep the fuel in the plant and the air out of the plant. The latter is relatively easy because most plants operate at pressure. Nitrogen is widely used to keep air out of low-pressure equipment, such as storage tanks (Section 5.4), stacks (Section 6.1), centrifuges (Section 10.1), and equipment that is depressured for maintenance (Section 1.3).

The main problem in preventing fires and explosions is thus preventing the process material from leaking out of the plant—that is, maintaining plant integrity. Similarly, if toxic or corrosive materials are handled, they are hazardous only when they leak.

Many leaks have been discussed under other headings, including leaks that occurred during maintenance (Chapter 1), as the result of errors (Chapter 3), or as the result of overfilling storage tanks (Section 5.1). Other leaks have occurred as the result of pipe or vessel failures (Chapter 9), whereas leaks of liquefied flammable gas are discussed in Chapter 8 and leaks from pumps and relief valves in Chapter 10.

Here, we discuss some other sources of leaks and the isolation and control of the leaking material.

doi:10.1016/B978-1-85617-531-9.00007-X

7.1   SOME COMMON SOURCES OF LEAKS

7.1.1   Small Cocks

Small cocks have often been knocked open or have vibrated open, as shown in Figure 7-1a. To prevent this from happening the valves should be installed so that when they are open the valve handle points upward, as shown in Figure 7-1b. Cocks should never be used as the sole isolation valve (and preferably not at all) on lines carrying hazardous materials, particularly flammable or toxic liquids, at pressures above their atmospheric boiling points (for example, liquefied flammable gases and most heat transfer oils when hot). These liquids turn to vapor and spray when they leak and can spread long distances. It is good practice to use other types of valves for the first isolation valve, as shown in Figure 7-1c.

7.1.2   Drain Valves and Vents

Many leaks have occurred because workers left drain valves open while draining water from storage tanks or process equipment and then returned to find that oil was running out instead of water.

In one incident, a man was draining water, through a 2-in.-diameter line, from a small distillation column rundown tank containing benzene. He left the water running for a few minutes to attend to other jobs. Either there was less water than usual or he was away longer than expected. He returned to find benzene running out of the drain line. Before he could close it, the benzene was ignited by the furnace, which heated the distillation column. The operator was badly burned and died from his injuries.

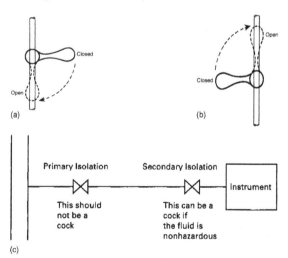

FIGURE 7-1   (a) Vibration can cause the cock to open. (b) A better arrangement. (c) Small cocks should not be used as primary isolation values.

The furnace was too near the drain point (it was about 10 m away), and the slope of the ground allowed the benzene to spread toward the furnace. Nevertheless, the fire would not have occurred if the drain valve had not been left unattended.

Spring-loaded ball valves should be used for drain valves. They have to be held open, and they close automatically if released. The size of drain valves should be kept as small as practicable. With liquefied flammable gases and other flashing liquids, ¾ in. should be the maximum allowed.

Drain valves that are used only occasionally to empty equipment for maintenance should be blanked when not in use. Regular surveys should be made to see that the blanks are in position. On one plant, a survey after a turnaround showed that 50 blanks were loose, each hanging on one bolt.

If water has to be drained regularly from liquefied flammable gases or other flashing liquids, and if a spring-loaded valve cannot be used, then a remotely operated emergency isolation valve (see Section 7.2.1) should be installed in the drain line.

When flammable materials are used, drain valves should not be located above hot pipework or equipment. A fire on an ethylene plant started when a mixture of water and naphtha was drained through a ¾-in. drain valve onto pipework at 315°C (600°F). It took a long time to replace damaged control and electric cables [21].

Drain valves should not be located above places where pools of water are liable to form, as leaks may then spread a long way (see Section 1.4.4).

While drain valves are installed to get rid of unwanted liquid, vent lines get rid of unwanted gas or vapor. They should be located so that the vapor is unlikely to ignite, so that damage is minimal if it does ignite, and so that people are not affected by the gas or vapor discharged. One fire destroyed a small plant. It started because the vent on a distillation column condenser discharged into the control room, possibly to prevent pollution of the surroundings, which had given rise to complaints about the smell [1] (see Section 2.11.3).

An electrician went up a ladder to repair a light fitting and was affected by fumes coming out of a vent about a meter away. The electrical hazards and the hazards of working from a ladder were considered, but no one thought about the hazards introduced by the vent—yet vents are designed to vent.

While contractors were working in a building, they inadvertently burned some insulation material. The ventilation system spread the fumes around the building. The fumes affected two people, and an expensive experiment taking place in a laboratory was ruined [15]. Before authorizing hot work in a building, consider the effects of any fumes that might be produced and, if necessary, switch off or isolate the ventilation system.

7.1.3   Open Containers

Buckets and other open-topped containers should never be used for collecting drips of flammable, toxic, or corrosive liquids or for carrying small quantities about the plant. Drips, reject samples, and so on should be collected in closed metal cans, and the caps should be fitted before the cans are moved.

One man was badly burned when he was carrying gasoline in a bucket and it caught fire. The source of ignition was never found. Another man was carrying phenol in a bucket when he slipped and fell. The phenol spilled onto his legs. One-half hour later he was dead. A third man was moving a small open-topped drum containing hot cleaning fluid. He slipped; liquid splashed onto him and scalded him.

A worker was draining hot tar from a portable kettle into a bucket when it caught fire. As he stepped back his glove stuck to the handle of the bucket, tipping it up and spilling the burning tar over the ground. The drain valve on the kettle was leaking, and this allowed the fire to spread. Two small liquefied-petroleum-gas containers (about 100 L), a trailer, and the kettle were destroyed. The end of one of the tanks was thrown 40 m [22].

Other incidents are described in Sections 12.2c and 15.1.

These incidents may seem trivial compared with those described in other pages. But for those concerned, they were their Flixborough.

Similarly, glass sample bottles should never be carried by hand. Workers have been injured when bottles they were carrying knocked against projections and broke. Bottles should be carried in baskets or other containers, such as those used for soft drinks. Bottles containing particularly hazardous chemicals, such as phenol, should be carried in closed containers.

Flammable liquids should, of course, never be used for cleaning floors or for cleaning up spillages of dirty oil. Use nonflammable solvents or water plus detergents.

7.1.4   Level and Sight Glasses

Failures of level glasses and sight glasses have caused many serious incidents. A leak of ethylene and an explosion that destroyed a plant may have been due to a level glass failure [2].

Level glasses and sight glasses (except magnetic types) should not be used on vessels containing flashing flammable or toxic liquids—that is, liquids under pressure above their normal boiling points. When level glasses are used, they should be fitted with ball check cocks, which prevent a massive leak if the glass breaks. Unfortunately, people who did not understand their purpose have sometimes removed the balls. The hand valves must be fully opened or the balls cannot operate (Figure 7-2).

These cocks contain
a ball, which will
isolate the sight glass
if the glass breaks

The cocks must be
fully open to allow
the ball to seat
if the glass breaks

Never remove the
balls. Check that
they are there

FIGURE 7-2   Ball check cocks.

A batch reactor was fitted with a rupture disc. A sight glass was fitted in a branch off the vent line so the disc could be inspected. When a runaway reaction occurred, the sudden rise in temperature and pressure broke the sight glass. Large amounts of flammable mist and vapor were discharged into the building, where they exploded, killing 11 people who had left the building but were standing outside [23]. The same reference describes other sight glass failures.

7.1.5   Plugs

On many occasions, screwed plugs have blown out of equipment:

(a) A ½-in. plug was fitted in a bellows (expansion joint) so that after pressure testing, in a horizontal position, water could be completely drained out. Soon after the bellows was installed, the plug blew out, followed by a jet of hot oil 30 m (100 ft) long.

Plugs installed to facilitate pressure testing should be welded in position. However, it is bad practice to seal weld over an ordinary screwed plug. If the thread corrodes, the full pressure is applied to the seal. A specially designed plug with a full-strength weld should be used.

(b) A 1-in. plug blew out of a pump body, followed by a stream of oil at 370°C (700°F) and a gauge pressure of 250 psi (17 bar). The oil caught fire and caused extensive damage. The plug had been held by only one or two threads and had been in use for 18 years.

Following this incident, surveys at other plants brought to light many other screwed plugs, some held by only a few threads and some made from the wrong grade of steel. At one plant, which did not allow the use of screwed plugs, several 2-in. plugs were found, held by only one thread. They had been in use for 10 years and were supplied as part of a compressor package.

A survey of all plugs is recommended.

(c) A similar incident is described in Section 9.1.6e. A screwed nipple and valve, installed for pressure testing, blew out of an oil line.

(d) The hinge-pin retaining plug on a standard swing check valve worked loose and blew out. Gas leaked out at a rate of 2 tons/hr until the plant could be shut down.

This incident emphasizes the point made in Section 7.2.1b. Check valves have a bad name among many plant operators, but no item of equipment can be expected to function correctly if it is never maintained.

(e) A valve was being overhauled in a workshop. A screwed plug was stuck in the outlet. To loosen the plug, the valve was heated with a welding torch. It shattered. The valve was in the closed position, and some water was trapped between the valve and the plug. Valves should normally be opened before they are maintained.

7.1.6   Hoses

Hoses are a frequent source of leaks. The most common reasons have been the following:

1. The hose was made of the wrong material.
2. The hose was damaged.
3. The connections were not made correctly. In particular, screwed joints were secured by only a few threads, different threads were combined, or gaskets were missing.
4. The hose was fixed to the connector or to the plant by a screwed clip of the type used for automobile hoses (Jubilee clips). These are unsuitable for industrial use. Bolted clamps should be used.
5. The hose was disconnected before the pressure had been blown off, sometimes because there was no vent valve through which it could be blown off.
6. The hose was used for a service such as steam or nitrogen, and the service valve was closed before the process valve. As a result, process materials entered the hose.

These points are illustrated by the following incidents:

(a) It was decided to inject live steam at a gauge pressure of 100 psi (7 bar) into a distillation column to see if this improved its performance. An operator was standing in the position shown in Figure 7-3 and was about to close the inlet valve to the column when the hose burst. He was showered with hot, corrosive liquid. He was standing

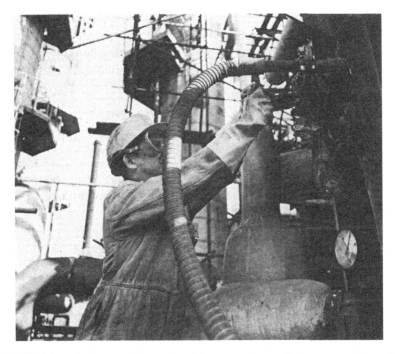

FIGURE 7-3 This hose burst, injuring the operator. It was the wrong type, was damaged, and was badly located.

on an access platform. The leak prevented him from reaching the access ladder. He had to wait until someone fetched a portable ladder.

Investigation of an incident showed the following:

1. The hose was made of reinforced rubber, the wrong material. A stainless steel hose should have been used.
2. The hose was damaged.
3. The steam valve at the other end of the hose was closed just before the column inlet valve, thus allowing process material to enter the hose. The operators knew this was not normal practice. But they closed the steam valve first because they knew the hose was damaged and wanted to avoid subjecting it to the full steam pressure.

The right type of hose should have been used, it should have been in good condition, and the process valve should have been closed first. In addition, a valve on a hose should not be in a position to which access is so poor. If no other valve was available, a steel pipe should have been fitted to the valve so that the end of the hose was in a safer place.

All hoses should be inspected and tested regularly and marked to show that they have been approved for use. A good practice is to change the color of the label every 6 or 12 months. This incident is a good illustration of the way both operators and managers become so used to the hazards of process materials that they fail to establish and maintain proper precautions. How often had the wrong hose or a damaged hose been used before? Why had the foremen or the supervisors not noticed them?

(b) A tank truck containing 60% oleum arrived at a plant. The truck's hose was damaged, so the operators found a hose intended for use with 20% oleum. After 45 minutes it leaked, and there was a large spillage. The operators assumed the hose must have been damaged. They replaced it with a similar one, and after 15 minutes another spillage occurred.

This incident illustrates the mind-sets described in Section 3.3.5. Having assumed that a hose used for 20% oleum would be suitable for any sort of oleum, the operators stuck to their opinion even though the hose leaked. They thus had an "action replay." Sections 5.4.1 and 6.1c describe other action replays.

Do your operators know which hoses are suitable for which materials? Mistakes are less likely if the number of different types used is kept to a minimum.

(c) A radioactive sludge was being pumped from a tank through a 1.5-in.-diameter hose into a moveable container. The pump stopped working, and a mechanic, asked to investigate, disconnected the hose using a quick-release coupling. Sludge was sprayed over three people standing up to 3 m away. There was no means of venting the pressure in the hose before uncoupling it. A choke in the hose prevented it from venting into the container, and it could not be vented through the positive displacement pump. The quick-release coupling could not be cracked open in the same way as a flange [24].

(d) To keep railway brake hoses clean before use, soft plastic "top hat" plugs were fitted into the ends. Each plug consisted of a closed cylinder, 7 mm long, which fitted into the end of a brake pipe, and a narrow lip, which was supposed to prevent it from going in too far (Figure 7-4, *left*). The plugs were colored red, the same color as the end of the brake pipe, so not surprisingly, a hose was fitted to a coach with the plug still in place. When the brakes were tested, the soft plug distorted and allowed compressed air to pass. Ultimately the plug moved into a position where it obstructed the pipe; a train failed to stop when required and overran by several miles. Fortunately, the line was clear.

A more rigid plug with a larger lip, as fitted to the other end of each pipe (Figure 7-4, *right*), would have caused the brake test to fail. The larger lip, and a different color, would have made the plug more visible. However, plastic bags tied over the ends would be a better way of keeping the hoses clean [25].

Other hose failures are described in Section 13.2.

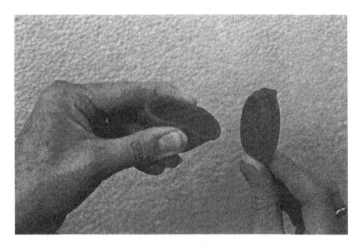

FIGURE 7-4 A soft plastic "top hat" plug *(left)* was fitted to the end of a railway carriage brake pipe to keep it clean. It was the same color as the end of the pipe and was not noticed and removed before the pipe was installed. The brakes failed, and the train overran. A more rigid plug with a larger lip *(right)* would have failed the brake test and would have been more visible. *(Photo courtesy of Roger Ford.)*

7.1.7 Cooling Coils

The cooling coil in a storage tank developed a small leak. To prevent the liquid in the tank from leaking into the cooling water, the coil was isolated but kept up to pressure by closing and slip-plating the exit water valve but leaving the inlet valve open. The tank contained an aqueous solution of a toxic acid, so a small leak of water into the tank contents did not matter and was far preferable to a leak of the acid into the cooling water. Another coil provided all the cooling necessary.

Ten years later, there was a pressure surge on the cooling water lines when the cooling water pumps were changed over; this caused a sample valve on the inlet water line to the coil to leak inside the building. The leaking water was contaminated with acid, which had been lying in the coil for 10 years since the leak first occurred. There were no instructions for the changeover of the cooling water pumps, and on the occasion of the incident the valves were operated in an unusual order.

7.2 CONTROL OF LEAKS

7.2.1 Emergency Isolation Valves (EIVs)

Many fires have been prevented or quickly extinguished by remotely operated emergency isolation valves. We cannot install them in the lines leading to all equipment that *might* leak. However, we can install them in the lines leading to equipment that, experience shows, is particularly liable to leak (for example, very hot or cold pumps or drain lines, as described in Section 7.1.2) or in lines from which, if a leak did occur, a large quantity of material, say 50 tons or more, would be spilled (for example, the bottoms pumps or reflux pumps on large distillation columns).

In all these cases, once the leak starts, particularly if it ignites, it is usually impossible to approach the normal hand-isolation valves to close them. Emergency isolation valves are discussed in detail in reference 3, and the following incidents show how useful they can be. They can be operated electrically, pneumatically, or in some cases, hydraulically:

(a) A leak of light oil from a pump caught fire. The flames were 10 m (33 ft) high. From the control room, the operator closed a remotely operated valve in the pump suction line. The flames soon died down, and the fire burned itself out in 20 minutes. It would have been impossible to close a hand-operated valve in the same position. And if the emergency valve had not been provided, the fire would have burned for many hours. The emergency valve had been tested regularly. It could not be fully closed during testing but was closed part way.

Backflow from the delivery side of the pump was prevented by a check (nonreturn) valve. In addition, a control valve and a hand valve well away from the fire were closed (Figure 7-5).

(b) The bearing on the feed pump to a furnace failed, causing a gland failure and a leak of hot oil. The oil caught fire, but an emergency isolation valve in the pump suction line was closed immediately, and the fire soon died out (Figure 7-6).

The control valve in the delivery line to the furnace was also closed. Unfortunately, the line through the heat exchanger bypassed this valve. In the heat of the moment, no one remembered to close the valve in the bypass line. In addition, the check (nonreturn) valve did not hold. Closing a hand valve next to the furnace, which was about 30 m (100 ft) from the fire, stopped the return flow of oil from the furnace. Afterward, another EIV was installed in the pump delivery line.

After the fire, the check valves on all three pumps were found to be out of order. On one, the seat had become unscrewed. On another, the fulcrum pin was badly worn. On the third, the pin was worn right through and the flap was loose. The valves had not been inspected since the plant was built.

Check valves have a bad name among many plant operators. However, this is because many of these valves are never inspected or tested. No equipment, especially that containing moving parts, can be expected to work correctly forever without inspection and repair. When check valves are relied on for emergency isolation, they should be scheduled for regular inspection.

Figure 7-7 shows a fluidic check valve that contains no moving parts. There is a low resistance to flow out of the tangential opening

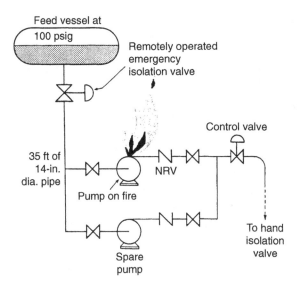

FIGURE 7-5 An emergency isolation valve stopped a fire.

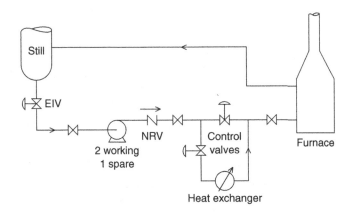

FIGURE 7-6 Another emergency isolation valve stopped a fire.

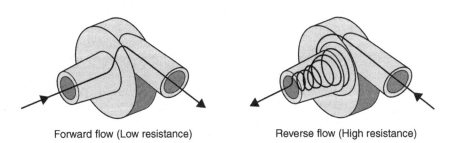

FIGURE 7-7 Fluidic check valve. *(Illustration courtesy of AEA Technology.)*

but high resistance to flow, 200 times higher, in the other direction. There is thus at times a small flow in the wrong direction, but if this can be tolerated the valves are very reliable and good at stopping pressure surges that might damage upstream equipment [26].

The fire did not affect the EIV. But it was close to it, and the incident drew attention to the need to either place EIVs where they are unlikely to be affected by fire or to provide them with fire protection. Fire-resistant sacks and boxes are available [9, 10]. The impulse lines—electrical or pneumatic—leading to EIVs should also be fire-protected [11].

If control valves are used for emergency isolation, a special switch may be necessary, out on the plant, to close them in an emergency so that operators do not have to go to the control room to alter the set points on the controllers.

Note that the operation of an emergency isolation valve should automatically shut down any pump in the line and trip the fuel supply to any furnace.

(c) In contrast, on other occasions, EIVs failed to control fires because the installation was not up to standard. In one case, a fire burned for six hours because the button controlling the EIV was too close to the leaking pump for anyone to operate it safely [4]. It should have been at least 10 m (33 ft) away. In another case, an EIV failed to work because it had not been tested regularly. All EIVs should be tested regularly, say monthly. If they cannot be closed without upsetting production, they should be closed part way and tested fully at shutdowns.

(d) Emergency isolation valves are, of course, of no value if they are not used when required. Sometimes when there has been a leak of a hazardous material, the operators have been tempted to try to isolate the leak without shutting down the plant. In doing so, they have taken unnecessary risks. For example, there was a bad leak of propylene on a pump inside a building. Four workers were badly injured. Afterwards, a lot of money was spent on moving the pumps into the open air, surrounding them with a steam curtain [5], and fitting remotely operated isolation valves and blowdown valves. If another leak should occur, then it would be possible to stop the leak by closing the pump suction valve, opening the blowdown valve, and switching off the pump motor without any need for anyone to go near the pumps [16] (see Section 8.1.3).

Eight years went by before another bad leak occurred. When it did occur, the area around the pumps was filled with a visible cloud of propylene vapor 1 m (3 ft) deep. Instead of using the emergency equipment, which would have stopped the flow of propylene and shut down the plant, two experienced foremen went into the compound, shut down the leaking pump, and started the spare up in its place. Fortunately the leak did not fire.

Afterward, when one of the foremen was back in his office, he realized the risk he had been taking. He complained that he should not be expected to take such risks. He had forgotten, in his eagerness to maintain production, that emergency equipment had been provided to avoid the need for such risk taking.

Other incidents that might have been controlled by EIVs are described in Sections 1.5.4e and 16.1g.

EIVs should close quickly but not too quickly, or they may produce hammer pressures in the pipework, especially if the valves are located in long lines. An extra 30 seconds closing time is unlikely to be serious. Similarly, EIVs should not open too quickly. If there is a control valve in the pipework, it should also be closed to back up the EIV; afterwards, it should be opened last, as it will open slowly [17].

The actuators fitted to EIVs should be somewhat more powerful than those recommended by manufacturers, especially if the liquid in the line is viscous. Some manufacturers do not allow for valve-packing friction forces [18]. EIVs, like all safety equipment, should be tested regularly (see Section 14.2.2).

7.2.2 Other Methods of Controlling Leaks

The following methods have been used successfully:

1. Injecting water so that it leaks out instead of oil. This method can, of course, be used only when the water pressure is higher than the oil pressure.
2. Reducing the plant pressure, thus reducing the size of the leak.
3. Closing an isolation valve some distance away.
4. Freezing a pipeline. This method requires time to organize the necessary equipment and can only be used with materials of relatively high freezing points, such as water or benzene.
5. Injecting a sealing fluid into a leaking flange or valve gland using a proprietary process such as Furmaniting. Caution: Accidents have occurred because correct procedures were not followed. Take care that bolts are not overstressed [12].
6. Confining the spread of the leak by water spray [6, 8] or steam curtains [5]. The latter have to be permanently installed, but the former can be temporary or permanent.
7. Controlling the evaporation from liquid pools by covering with foam. This method can be used for chlorine and ammonia spillages as well as hydrocarbon spillages if suitable foams are used.

8. Adding a less volatile liquid to a spillage to reduce its volatility. When some liquefied petroleum gas (LPG) got into the drains, some gas oil was poured down them to absorb the LPG and reduce the chance of an explosion.

7.2.3   How Not to Control a Leak

On many occasions employees have entered a cloud of flammable gas or vapor to isolate a leak. In the incident described in Section 7.2.1d, this was done to avoid shutting down the plant. More often, it has been done because there was no other way of stopping the leak. The persons concerned would have been badly burned if the leak had ignited while they were inside the cloud.

It would be going too far to say that no one should ever enter a cloud of flammable vapor to isolate a leak. There have been occasions when, by taking a risk for a minute, a worker has isolated a leak that would otherwise have spread a long way and probably ignited, perhaps exploded. However, we should try to avoid putting people in such situations by providing remotely operated emergency isolation valves to isolate likely sources of leaks.

It may be possible to isolate a leak by hand by forcing back the vapor with water spray and protecting the worker who closes the valve in the same way. The National Fire Protection Association can provide a set of slides or a film showing how this is done.

It is possible to measure the extent of a leak of flammable gas or vapor with a combustible gas detector. If the leak is small, a person may be allowed (but not expected) to put hands, suitably protected, inside the flammable cloud. But only in the most exceptional circumstances should a person be allowed to put more of his or her body into the cloud.

7.3   LEAKS ONTO WATER, WET GROUND, OR INSULATION

7.3.1   Leaks onto Water or Wet Ground

Section 1.4.4 describes two leaks onto pools of water that spread much farther than anyone expected. One was ignited by a welder 20 m (65 ft) away, and the other spillage, onto a canal, caught fire 1 km (0.6 mile) away.

In other cases, spillages of oil have soaked into the ground and have then come to the surface after heavy rain. A spillage of gasoline in Essex, England, in 1966, came back to the surface two years later. The vapor accumulated on the ground floor of a house, ignited, and blew a hole in the stairs, injuring two people. A trench 7 m (23 ft) deep was dug to recover the rest of the gasoline [7]. Spillages of oil have leaked into sewers and from there into houses.

If a substantial quantity of oil is spilled into the ground, attempts should be made to recover it by digging a well or trench.

7.3.2 Leaks onto Insulation

When organic compounds come into contact with many hot insulation materials, they can degrade, and the auto-ignition temperature can fall by 100 to 200 degrees C (180 to 360 degrees F). Many fires have started in this way (see Section 12.4.4). Most of them have been small, but some have been serious. For example, in a plant in Belgium in 1989, ethylene oxide (EO) leaked through a hairline crack in a weld on a distillation column and contaminated the rock wool insulation on a level indicator. The EO then reacted with moisture to form nonvolatile polyethylene glycols. The metal covering of the insulation was removed so that the level indicator could be repaired. Air leaked in, and later the same day the polyethylene glycols ignited. This heated the wall of the piping system, in which there was no flow. The heat caused the EO to decompose explosively—a well-known reaction—and the decomposition traveled into the distillation column, which exploded. Figure 7-8 shows the result.

The source of ignition of the polyethylene glycol was probably autoignition of the degraded material. The report recommends the use of nonabsorbent insulation for equipment containing heat-sensitive materials such as EO [19, 20].

In another incident, a long-chain alcohol leaked into the insulation of a pipeline. When the covering over the insulation was opened, allowing air to enter, the temperature (60°C [140°F]) was sufficient for ignition [19].

7.4 DETECTION OF LEAKS

On many occasions, combustible gas detectors have detected a leak soon after it started, and action to control it has been taken promptly. Installation of these detectors is strongly recommended whenever liquefied flammable gases or other flashing liquids are handled or when experience shows there is a significant chance of a leak [3]. Detectors are also available for common toxic gases. However, these detectors do not do away with the need for regular patrols of the plant by operators. Several plants that have invested heavily in gas detectors report that, nevertheless, half the leaks that have occurred have been detected by people.

On one plant, liquid leaks drained into a sump that was fitted with a level detector. When a leak actually occurred, it dripped onto a hot pipe and evaporated and was not detected for many hours (see Section 20.2.4).

FIGURE 7-8 The ethylene oxide plant after the fire and explosion. (*Photo courtesy of BP Chemicals Limited.*)

A similar incident occurred in another plant. The liquid in the plant was cold, so a low-temperature alarm was installed in the sump. It was tested with cold water and worked well. When a leak occurred, the leaking liquid, which was acidic, reacted with the steelwork on its way to the sump and warmed up; the temperature element could not, of course, tell the difference between warm air and warm liquid, and it failed to detect the leak.

Whenever possible we should measure directly what we need to know and not some other property from which it can be deduced (see Section 14.6).

Electric cables that detect liquid leaks are available. They can be run through areas where leaks are possible, and they indicate the presence and location of a leak.

Large leaks can be detected by comparing flow rates in different parts of a plant. This can be done automatically on plants that are computer controlled.

7.5   FUGITIVE EMISSIONS

There is increasing interest in fugitive emissions, small continuous leaks from flanges, valve and pump glands, relief valves, and the like, which produce small but ever-present concentrations of chemicals in the workplace environment and some of which may produce long-term toxic effects. Reference 13 summarizes published data on the leak rates from various items of equipment and ways of reducing them. According to reference 14, more than half the total emission from a refinery comes from valve glands. Actual figures are as follows:

Source	Percentage of Total Emission
Flanges	8
Valves	57
Compressors	3
Pumps	12
Relief valves	1
Separators	4
Cooling towers	7
Drains	8

Some other leaks are described in Chapter 28.

References

1. Health and Safety Executive, *The Explosion and Fire at Chemstar Ltd.*, *6 September 1981*, Her Majesty's Stationery Office, London, 1982.
2. C. T. Adcock and J. D. Weldon, *Chemical Engineering Progress*, Vol. 63, No. 8, Aug. 1967, p. 54.
3. T. A. Kletz, *Chemical Engineering Progress*, Vol. 71, No. 9, Sept. 1975, p. 63.
4. T. A. Kletz, *Hydrocarbon Processing*, Vol. 58, No. 1, Jan. 1979, p. 243.
5. H. G. Simpson, *Power and Works Engineering*, May 8, 1974, p. 8.
6. J. McQuaid and R. D. Fitzpatrick, *J. of Hazardous Materials*, Vol. 5, 1983, p. 121; and J. McQuaid, *J. of Hazardous Materials*, Vol. 5, 1983, p. 135.
7. *Petroleum Times*, Apr. 11, 1969.
8. K. Moodie, *Plant/Operations Progress*, Vol. 4, No. 4, Oct. 1985, p. 234.
9. M. Symalla, *Hydrocarbon Processing*, Vol. 63, No. 7, July 1984, p. 83.
10. A. E. Choquette, *Hydrocarbon Processing*, Vol. 63, No. 7, July 1984, p. 85.
11. G. K. Castle, *Hydrocarbon Processing*, Vol. 63, No. 3, Mar. 1984, p. 113.
12. H. J. Maushagen, *Loss Prevention Bulletin*, No. 055, Feb. 1984, p. 1.
13. British Occupational Hygiene Society, *Fugitive Emissions of Vapours from Process Equipment, Report in Science Reviews*, Northwood, UK, 1984.
14. C. R. Freeberg and C. W. Arni, *Chemical Engineering Progress*, Vol. 78, No. 6, June 1982, p. 35.
15. *Coordinating Construction/Maintenance Plans with Facility Manager May Deter Unexpected Problems and Accidents*, Safety Note DOE/EH-0127, U.S. Dept. of Energy, Washington, DC, Apr. 1990.
16. T. A. Kletz, *Lessons from Disaster: How Organizations Have No Memory and Accidents Recur*, co-published by Institution of Chemical Engineers, Rugby, UK, and Gulf Publishing Co., Houston, TX, 1993, Sections 2.2 and 8.10.
17. B. T. Matusz and D. L. Sadler, "A Comprehensive Program for Preventing Cyclohexane Oxidation Process Piping Failures," *Paper presented at AIChE Loss Prevention Symposium*, Houston, TX, Mar./Apr. 1993.
18. *Undersized Valve Actuators*, Safety Note DOE/EH-0113, U.S. Dept. of Energy, Washington, DC, Oct. 1989.
19. *Loss Prevention Bulletin*, No. 100, Aug. 1991, p. 1.
20. L. Britton, *Plant/Operations Progress*, Vol. 9, No. 2, Apr. 1990, p. 75, and Vol. 10, No. 1, Jan. 1991, p. 27.
21. *Loss Prevention Bulletin*, No. 131, Oct. 1996, p. 13.
22. *Occupational Health and Safety Observer*, Vol. 3, No. 12, U.S. Dept. of Energy, Washington, DC, Dec. 1994, p. 4.
23. *Loss Prevention Bulletin*, No. 134, Apr. 1997, pp. 3, 5, and 6.
24. *Operating Experience Weekly Summary*, No. 96-45, Office of Nuclear and Facility Safety, U.S. Dept. of Energy, Washington, DC, 1996, p. 10.
25. R. Ford, *Modern Railways*, Vol. 53, No. 579, Dec. 1996, p. 752.
26. Literature can be obtained from manufacturers. Enter "fluid check valves" in Google. There are other designs besides the one shown in Figure 7-7.

8

Liquefied Flammable Gases

When a design engineer started learning to fly he was struck by the readiness of pilots to accept and follow procedures. He asked his flight instructor what accounted for the difference. The instructor said, "It is very simple. Unlike pilots, design engineers don't go down with their planes."

—Based on a quotation from Ian Ayres [24]

This chapter discusses recommendations made many years ago that are not always followed. It describes a number of incidents involving liquefied flammable gases (LFG) that could have occurred only with these materials (or other flashing flammable liquids).

The property of LFG that makes it so hazardous is that it is usually stored and handled under pressure at temperatures above normal boiling points. Any leak thus flashes, much of it turning to vapor and spray. This can spread for hundreds of meters before it reaches a source of ignition.

The amount of vapor and spray produced can far exceed the theoretical amount of vapor produced, estimated by heat balance [1]. The vapor carries some of the liquid with it as spray. It may evaporate on contact with the air. Whether it does or not, it is just as likely to burn or explode.

Any flammable liquid under pressure above its normal boiling point will behave like LFG. Liquefied flammable gases are merely the most common example of a flashing liquid. Most unconfined vapor cloud explosions, including the one at Flixborough (Section 2.4), have been due to leaks of such flashing liquids [2].

The term *liquefied petroleum gas* (LPG) is often used to describe those liquefied flammable gases that are derived from petroleum. The term *LFG* is preferred when the discussion applies to all liquefied flammable

doi:10.1016/B978-1-85617-531-9.00008-1

gases. It includes materials such as ethylene oxide, vinyl chloride, and methylamines, which behave similarly so far as their flashing and flammable properties are concerned.

LFGs stored at atmospheric pressure and low temperature behave differently from those stored under pressure at atmospheric temperature. Incidents involving these materials are described in Sections 2.5, 5.2.2d, 5.6.2, and 8.1.5.

8.1   MAJOR LEAKS

8.1.1. Feyzin

The bursting of a large pressure vessel at Feyzin, France, in 1966 was at the time one of the worst incidents involving LFG that had ever occurred but has since been overshadowed by the events at Mexico City (see Section 8.1.4). It caused many companies to revise their standards for the storing and handling of these materials. Because no detailed account has been published, it is described here. The information is based on references 3 through 6 and on a discussion with someone who visited the site soon after the fire.

An operator had to drain water from a 1,200-m^3 spherical storage vessel nearly full of propane (Figure 8-1). He opened valves A and B. When traces of oil showed that the draining was nearly complete, he shut A and then cracked it to complete the draining. No flow came. He opened A fully. The choke—presumably hydrate, a compound of water and a light hydrocarbon with a melting point above 0°C (32°F)—cleared suddenly, and the operator and two others were splashed with liquid. The handle came off valve A, and they could not get it back on. Valve B was frozen and could not be moved. Access was poor because the drain valves were immediately below the tank, which was only 1.4 m (4.5 ft) above the ground.

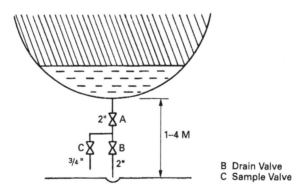

FIGURE 8-1   Drain valves underneath propane tank at Feyzin.

A visible cloud of vapor, 1 m (3 ft) deep, spread for 150 m (1,500 ft) and was ignited by a car that had stopped on a nearby road 25 minutes after the leak started. The police had closed the road, but the driver approached from a side road. The fire flashed back to the sphere, which was surrounded by flames. There was no explosion. The sphere was fitted with water sprays. But the system was designed to deliver only half the quantity of water normally recommended (0.2 U.S. gal/ft^2 min or 8 L/m^2 min), and the supply was inadequate. When the fire brigade started to use its hoses, the supply to the spheres ran dry. The firefighters seemed to have used most of the available water for cooling neighboring spheres to stop the fire from spreading, in the belief that the relief valve would protect the vessel on fire.

The ground under the sphere was level so that any propane that did not evaporate or burn immediately collected under the sphere and burned later.

Ninety minutes after the fire started, the sphere burst. Ten out of 12 firefighters within 50 m (160 ft) were killed. Men 140 m (460 ft) away were badly burned by a wave of propane that came over the compound wall. Altogether, 15 to 18 men were killed (reports differ), and about 80 were injured. The area was abandoned. Flying debris broke the legs of an adjacent sphere, which fell over. Its relief valve discharged liquid, which added to the fire, and 45 minutes later this sphere burst. Altogether, five spheres and two other pressure vessels burst, and three were damaged. The fire spread to gasoline and fuel oil tanks.

At first it was thought that the spheres burst because their relief valves were too small. But later it was realized that the metal in the upper portions of the spheres was softened by the heat and lost its strength. Below the liquid level, the boiling liquid kept the metal cool. Incidents such as this one in which a vessel bursts because the metal gets too hot are known as boiling liquid expanding vapor explosions, or BLEVEs.

To prevent such incidents from occurring, many companies—after Feyzin—adopted recommendations similar to the following.

Recommendations to Prevent a Fire from Starting [7]

- Restrict the size of the second drain valve to ¾ in., and place it at least 1 m from the first valve. The drain line should be robust and firmly supported. Its end should be located outside the shadow of the tank.
- Fit a remotely controlled emergency isolation valve (see Section 7.2.1) in the drain line.
- Ensure that new installations are provided with only one connection below the liquid level, fully welded up to a first remotely operated fire-safe isolation valve located clear of the tank area.
- Install combustible gas detectors to provide early warning of a leak.

Recommendations to Prevent a Fire from Escalating [7, 8]

- Insulate vessels with a fire-resistant insulation, such as vermiculite concrete. This is available as an immediate barrier to heat input. Unlike water spray, it does not have to be commissioned. In some countries, mounding is used instead of conventional insulation. The tank is completely covered with clean sand or other clean material. Portions of the covering must be removed from time to time so that the outside of the tank can be inspected.
- Provide water spray or deluge (unless the vessel is mounded). If insulation is provided, then water deluge at a rate of 0.06 U.S. gal/ft^2 min (2.4 L/m^2 min) is sufficient. If insulation is not provided, then water spray at a rate of 0.2 U.S. gal/ft^2 min (8 L/m^2 min) is necessary. (Deluge water is poured on the top of a vessel; spray is directed at the entire surface.)
- Slope the ground so that any spillage runs off to a collection pit.
- Fit an emergency depressuring valve so that the pressure in the vessel can be reduced to one-fifth of design in 10 minutes to reduce the strain on the metal [13]. The time can be increased to 30 minutes if the vessel is insulated and to one hour if, in addition, the ground is sloped.

Figure 8-2 summarizes these proposals.

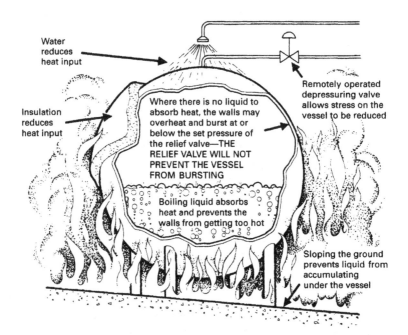

FIGURE 8-2   Methods of protecting a pressure vessel against fire.

8.1.2   Duque De Caxias

A similar incident to that at Feyzin occurred at this refinery in Brazil in 1972. According to press reports, the relief valve failed to open when the pressure in an LPG sphere rose. To try to reduce the pressure, the operators opened the drain valve. Little water came out, and the LPG that followed it caused the valve to freeze, and the flow could not be stopped. There was only one drain valve. The LPG ignited, the vessel BLEVEd, and 37 people were killed.

If the operators did, in fact, try to reduce the pressure by draining water, they did not realize that the vapor pressure above a liquid is the same whatever the quantity present.

8.1.3   United Kingdom

This fire occurred some years ago because those concerned did not fully appreciate the difference in behavior between liquid hydrocarbons, such as naphtha or gasoline, and LFGs. The vapor from a spillage of gasoline will usually spread only a short distance—about the diameter of the pool but can go farther (see Section 30.13). But the vapor from a spillage of LFG can spread for hundreds of meters.

Some equipment that had been designed and used for handling gasoline and similar liquids was adapted to handle propylene. A leak occurred from the gland of a high-pressure reciprocating pump operating at a gauge pressure of 3,625 psi (250 bar) because of the failure of the studs holding the gland in position. The pump was located in an unventilated building. But the vapor escaped through a large doorway opposite the pump and was ignited by a furnace 75 m (250 ft) away. Four men were badly burned. The vapor from a spillage of the same amount of gasoline in the same position would probably not have spread anywhere near the furnace.

After the fire, the pump (and others) was relocated in the open air, under a canopy, so that small leaks would be dispersed by natural ventilation. It was surrounded by a steam curtain to disperse larger leaks. This would not have been necessary if the pump could have been located more than 150 m (500 ft) from sources of ignition. Gas detectors were installed to give early warning of any leaks. Emergency isolation valves (Section 7.2.1) were provided so that the pumps could be isolated safely from a distance [9]. What happened when another leak occurred is described in Section 7.2.1d.

Note that a common factor in the incidents described in Sections 8.1.1 through 8.1.3 was a failure by those concerned to understand the properties of the materials and equipment.

8.1.4   Mexico City

The fire and explosion at a processing plant and distribution center for liquefied petroleum gas (LPG—actually 80% butane, 20% propane) in San Juanico, a suburb of Mexico City, in November 1984, was one of the worst incidents that has ever occurred in the oil and chemical industries, exceeded only by Bhopal. According to official figures, 542 people were killed, 4,248 injured, and about 10,000 made homeless; but unofficial estimates are higher. The disaster started when an 8-in. LPG pipeline ruptured. The reason for the failure is not known, but according to one report [14] a tank was overfilled and the inlet line overpressured. It is not clear why the relief valve did not lift. The gas cloud covered an area of 200 m (650 ft) by 150 m (500 ft) before it was ignited, probably by a ground-level flare, about 5 to 10 minutes after the leak started. The gas cloud burned and disappeared, but a flame was left burning near the broken pipe, and this flame heated an LPG sphere, which BLEVEd, causing further damage and further BLEVEs. Altogether, four spheres and 15 cylindrical tanks BLEVEd during the next 1½ hours, and some of the tanks landed up to 1,200 m (4,000 ft) from the plant [15].

Most of those killed and injured were members of the public who were living in a shantytown near the plant. When the plant was built, the nearest houses were 360 m (1,200 ft) away, but homes had been allowed to encroach on the intervening ground until the nearest houses were only 130 m (430 ft) from the plant.

Although much of the plant was only a few years old (some parts were 20 years old), most of the recommendations made in Section 8.1.1 and taken from a report published in 1970 [7] do not seem to have been followed in the design. For example, there were no gas detectors, the water deluge system was inadequate (or failed to operate), there was little or no fire insulation (even the legs of the spheres were not insulated), bunds around the vessels allowed LPG to accumulate where it could do most harm, and there were many connections to vessels below the liquid level [16]. In addition, the plant seems to have been congested and was much too near concentrations of people. A typical recommended distance for a large LPG processing area is 600 m (2,000 ft) [17], not 360 m (1,200 ft), the original distance, or 130 m (430 ft), the distance at the time. At Bhopal also (see Section 21.1), uncontrolled spread of a shantytown was responsible for the large number of casualties.

8.1.5   Qatar

The incidents described so far involved LFG stored under pressure at atmospheric temperature. LFG can also be stored at a low temperature and atmospheric pressure, and this method is often preferred for large storages as, because of the low pressure, the leak rate through a hole of a given size is smaller, and because of the low temperature, the evaporation rate is smaller. However, when deciding on the method of storage to be used, the probability of a leak from the installation as a whole, including the refrigeration and vaporization plants, should be considered. There will be no gain if we reduce the chance of a leak from the storage tanks but introduce extra equipment that is more likely to leak.

In 1977, the technical press reported that a major leak from a 20,000-m^3 liquefied propane tank in Qatar had ignited and that the resulting fire and explosion had killed seven people and caused extensive damage to the rest of the plant [18]. There had also been a leak the year before, but it had not ignited, and the tank had been repaired. The propane was stored at $-42°C$ ($-44°F$) and atmospheric pressure. No detailed report on the incident was issued, for legal reasons, but a member of the company concerned published several papers [19–21], which gave new recommendations for the construction of tanks for refrigerated LFG, and it is thus possible to read between the lines and surmise what probably happened.

The new recommendations said that refrigerated LFG tanks should be made from materials such as 9% nickel steel, which will not propagate a crack if one should start. Previously, the policy of many companies was to prevent cracks rather than rely on the crack-arresting properties of the tank material. It thus seems that at Qatar a crack started, despite the precautions taken to prevent such an occurrence, and then propagated rapidly. The reason for the initial crack is not known. It may have been connected with the repairs that were carried out following the crack the previous year. According to one report [22], this crack was due to weld attack by bacteria in the seawater used for pressure testing. However, the cause of the crack is of secondary importance compared with the fact that once it appeared, it spread rapidly.

It is now widely recognized that we cannot prevent fires and explosions by eliminating sources of ignition. We do what we can, but they are still liable to turn up. So we try to prevent the formation of explosive mixtures. Similarly, it is now argued that we cannot eliminate all causes of cracks, and so we should make sure that any that do occur do not propagate.

At Qatar, the liquid came out with such force that it spilled over the dike wall. Conventional dike walls also have the disadvantage that a large area of liquid is exposed to the atmosphere if a leak occurs. For these reasons it is now usual to surround cryogenic storage tanks with a concrete wall, about 1 m from the tank and the full height of the tank. If the tank is not made from crack-resistant material, then the concrete wall should be designed to withstand the effects of a sudden release of liquid.

8.1.6   Ethyl Chloride Plant

In 1994 a leak of impure ethyl chloride (boiling point 12°C [54°F]) caught fire, 1½ hours after it started, and damaged the plant so extensively that it had to be rebuilt. Fortunately, no one was killed or injured. The leak started at a flange assembly on the delivery of a pump (Figure 8-3), probably the result of corrosion of the flanges but possibly caused by the failure of a plastic bellows. The official report [23] made the following points:

- The split flanges were badly corroded; their thickness was reduced, and the bolt holes were much enlarged. (Split flanges are not a good feature because they expose twice as much surface to the effects of corrosion.) There was no system for identifying critical items, the failure of which could have serious results, and registering them for regular inspection. Maintenance was on a breakdown basis, and there were no formal records that could be used to identify items needing regular inspection or replacement.
- The leak could have been stopped as soon as it was detected if an emergency isolation valve (Section 7.2.1) had been fitted in the pump suction line. On the rebuilt plant, such valves were fitted on the pump suction lines, more combustible gas detectors and more extensive insulation were installed, plastic pump bodies were replaced by metal ones, and spillages were directed to collection pits. The plant

FIGURE 8-3   Split flange assembly similar to one that leaked. *(Photo courtesy of the UK Health and Safety Executive.)*

was built in 1972, when these features were not common practice; many improvements had been made since then, but they did not go far enough. Most of those made after the fire could have been made beforehand.

- The source of ignition may have been a box containing electrical equipment. It had a badly fitted or incorrect type of plug, which could have allowed water to enter and to cause arcing.
- The fire was more serious than it would normally have been because the inventory in the plant, about 70 tons, was about twice the usual amount. Some of the overheads from a reactor were collected in a slops drum and recycled. The inventory in the drum was usually small. At about 9 a.m. on the day of the fire, the recycle pump failed. As a result, the level in the drum rose, and the level in the reactor fell. The operator noticed the fall in the reactor level (but not the rise in the drum level) and recycled product to maintain the level. At 8 p.m. the supervisor noticed that the high-level alarm on the slops drum was lit; he found that the recycle pump had failed, and he changed over to the spare; it leaked 25 minutes later. Section 3.3.1 describes another occasion when operators failed to notice unusual readings for 11 hours.
- The major hazard on the site as a whole was the storage and use of chlorine. So much attention was devoted to this that other hazards received less attention than they should have.
- The official report sums up the lessons of the fire by saying that it might have been prevented or its severity greatly reduced if a more detailed assessment of the inherent hazards and risks of the plant had been carried out by the company beforehand and if adequate records had been kept to build up a history on which an inspection and replacement program could have been based.

8.2 MINOR LEAKS

(a) After Feyzin (see Section 8.1.1), one company spent a lot of money improving the standard of its LFG storage facilities—in particular, the water-draining arrangements—so as to comply with the recommendations made in Section 8.1.1.

Less than a year later, a small leak from a passing drain valve on a pipeline caught fire. It was soon extinguished by closing the valve. But an investigation disclosed the following:

1. There should have been two valves in series or a single valve and a blank.
2. The valve was made of brass and was of a type stocked for use on domestic water systems. It was not the correct pressure rating for LFG.

3. The valve was screwed onto the pipeline, though the company's codes made it clear that only flanged or welded joints were allowed.
4. It was never discovered who installed this unauthorized substandard drain point. An attempt had been made to publicize the lessons of Feyzin, the company's standards, and the reasons for them. However, this did not prevent the installation of the drain point. Note that a number of people must have been involved. Besides the man who actually fitted it and his foreman, someone must have issued a work permit and accepted it back (when he should have inspected the job), and several persons must have used the drain point. Many must have passed by. If only one of them had recognized the substandard construction and drawn it to the attention of those responsible, the fire would not have occurred [10].

Like the plants in our gardens, our plants grow unwanted branches while our backs are turned.

(b) A propane sphere was filled with water to inert it during repair work. When the repairs were complete, the water was drained from the sphere, and propane vapor was admitted to the top to replace the water. The instruction stated that draining should stop when 5 m³ of water was left in the vessel. But no one was present when this stage was reached. All the water drained out, followed by propane. Fortunately, it did not ignite. The job had been left because the operators did not realize that the level indicator, which measured weight, would indicate a level of water almost twice the actual level. Other similar incidents are described in Section 5.1.2. If nitrogen is available, it should be used instead of water for inerting vessels. Or if water is used, it should be replaced by nitrogen when it is drained. Before filling any equipment with water, always check that it is strong enough to take the weight of the water [11].

8.3   OTHER LEAKS

Numerous leaks of LFG, mainly minor but occasionally more serious, have occurred from the following items of equipment.

8.3.1   Flanged Joints

The size and frequency of leaks can be reduced by using spiral-wound gaskets in place of compressed asbestos fiber ones. Screwed joints should not be used.

8.3.2 Pump Seals

The leak size can be reduced by using double mechanical seals or a mechanical seal and a throttle bush, the space between the two being vented to a safe place. Major leaks may still occur, however, as a result of the collapse of the bearing or seal. LFG pumps should therefore be fitted with emergency isolation valves (see Section 7.2.1), particularly if the temperature is low or the inventory that can leak out is high.

8.3.3 Level Glasses

These should not be used with flashing flammable liquids (see Section 7.1.4).

8.3.4 Sample Points

These should not exceed ¼-in. diameter.

8.3.5 Small Branches

These should be physically robust and well supported so they cannot be knocked off accidentally or vibrate until they fail by fatigue.

8.3.6 Equipment Made from Grades of Steel Unsuitable for Use at Low Temperatures

Materials of construction should be chosen so the equipment will withstand the lowest temperature that can be reached during abnormal operation. In the past, materials have been used that will withstand normal operating temperatures but may become brittle at lower temperatures reached during plant upsets or abnormal operation, for example, when the pressure in a vessel containing liquefied petroleum gas is reduced and the vessel is cooled by evaporation of the liquid (see Section 10.5.2). Some spectacular failures have resulted [12] (see Section 9.2.1g).

Wholesale replacement of such materials in existing plants is impractical, and there is no universal solution. Some lines can be replaced in different grades of steel. Sometimes low-temperature trips or alarms can be used. Sometimes the need to watch the temperature closely during startup has to be impressed on operators.

Other leaks of LFG are described in Sections 1.1.6e, 1.5.4a, 5.2.2d, 9.1.6d, 9.1.6f, and 13.4.

8.4   SAFETY IN THE DESIGN OF PLANTS HANDLING LIQUEFIED LIGHT HYDROCARBONS

The following is a summary of a report issued in 1971. Most of the advice is still applicable, though it is not always followed. Most of the recommendations also apply to other liquefied flammable gases.

Experience shows that the cloud of vapor generated from any major escape of liquefied hydrocarbon is likely to find a source of ignition, often with disastrous results. The basic approach to safety in processing, storage, and handling of liquefied hydrocarbons must therefore be one of prevention, aiming to eliminate accidental escape wherever possible and to ensure that whatever spillage does occur is restricted to a manageable quantity that can be dispersed safely:

1. Particular care should be taken in the selection of materials for the construction of equipment, which may at any time contain liquefied hydrocarbons at subzero temperatures.
2. The capacity of fire relief valves should be determined as described in American Petroleum Institute standards.
3. Particular care should be taken with the arrangement of vent lines from relief valves on low-pressure refrigerated storage tanks to avoid causing excessive back pressure.
4. Liquid relief valves should be provided on pipelines or other equipment that may be endangered by thermal expansion of locked-in liquid.
5. Small branches on pressure vessels and major pipelines should be supported mechanically to prevent them from breaking off.
6. Particular care should be taken to protect equipment against fire exposure by a suitable combination of water cooling and fireproof insulation to ensure that metal temperatures cannot rise sufficiently to cause failure at or below relief valve set pressures.
7. No attempt should be made to extinguish a liquefied hydrocarbon fire except by cutting off the supply of hydrocarbon, nor should a cloud of vapor from an escape that is not on fire be deliberately ignited.
8. Pumps should in general be fitted with mechanical seals instead of packed glands to reduce leakage.
9. Process draining and sampling facilities should be designed to withstand mechanical breakage, to minimize the risk of blockage by ice or hydrate, and to restrict the quantity of any spillage. There should be a robust connection and first isolation on the plant or storage vessel and a second valve, of not more than ¾ in. size for draining or ¼ in. for sampling, separated from the first valve by at least 1 m (3 ft) of piping. The discharge pipe from the drain of not more than ¾ in. bore should deliver clear of the vessel and be supported to prevent breakage by jet forces. Both valves should have means of actuation that

cannot be readily removed. Samples should be taken only into a bomb through a closed ¼ in. bore piping system.

10. Pressure storage vessels should preferably be designed with only one connection below the liquid level, fully welded up to a first remotely operated fire-safe isolation valve located clear of the area of the tank and behind a diversion wall.

11. Valve connections should be provided on process vessels for disposal of residues of liquefied hydrocarbons, preferably to a closed flare system. No bleed direct to atmosphere should be of more than 1 in. bore.

12. Remotely controlled isolation valves should be provided on items of equipment that are liable to leak significantly in service.

13. Discharge of heavy vapor from relief valves and blowdowns should be vented to a closed system, preferably with a flarestack, except when it is possible to discharge to atmosphere at sufficient velocity to ensure safe dilution by jet mixing with air (but this may be impossible on environmental grounds).

14. Remotely operated isolation valves should be installed in liquid and vapor connections that are regularly broken to atmosphere, particularly the flexible hose connections used in tank wagon operation. They are better than excess flow valves.

15. Whenever possible, equipment should be located at the safe distances from sources of ignition. The horizontal extent of Division (Zone) 2 areas in electrical classification should be taken the safe distance.

16. The ground under pressure storage vessels should be impervious and should slope so that any liquid spillage will flow away from the vessels to a catchment area where it can be safely disposed of or can burn if it ignites without causing further hazard. Suitable diversion and retaining walls should be provided to prevent uncontrolled spread of the spillage. The height of the walls should be suitably limited in relation to their distance apart to allow minor leakage to be dispersed by natural air movement. The retention capacity for liquid should be decided in relation to the amount likely to escape allowing for flash-off and boil-off from the ground.

17. Low-pressure refrigerated storage tanks should be fully bunded (diked), and the floor of the bund should be sloped so that spillage flows preferentially away from the tank.

18. The principle of diverting liquid spillage away from equipment should be applied in process areas wherever possible.

19. In plant or storage areas where safe distances from sources of ignition cannot be met or in areas near a factory perimeter adjacent to public roads or property, the installation of a steam curtain should be considered.

20. Flammable gas detectors should be installed in areas where experience shows there is a significant chance of a leak or a large amount will escape if there is a leak.

References

1. J. D. Reed, "Containment of Leaks from Vessels Containing Liquefied Gases with Particular Reference to Ammonia," *Proceedings of the First International Symposium on Loss Prevention and Safety Promotion in the Process Industries*, Elsevier, Amsterdam, 1977, p. 191.
2. J. A. Davenport, *Chemical Engineering Progress*, Vol. 73, No. 9, Sept. 1977, p. 54.
3. *The Engineer*, Mar. 25, 1966, p. 475.
4. *Paris Match*, No. 875, Jan. 15, 1966.
5. *Fire*, Special Supplement, Feb. 1966.
6. *Petroleum Times*, Jan. 21, 1966, p. 132.
7. Imperial Chemical Industries Ltd., *Liquefied Flammable Gases—Storage and Handling*, Royal Society for the Prevention of Accidents, Birmingham, UK, 1970.
8. T. A. Kletz, *Hydrocarbon Processing*, Vol. 56, No. 8, Aug. 1977, p. 98.
9. T. A. Kletz, *Lessons from Disaster: How Organizations Have No Memory and Accidents Recur*, co-published by Institution of Chemical Engineers, Rugby, UK, 1993, Section 2.2.
10. T. A. Kletz, *Chemical Engineering Progress*, Vol. 72, No. 11, Nov. 1976, p. 48.
11. *Petroleum Review*, Apr. 1982, p. 35.
12. A. L. M. van Eijnatten, *Chemical Engineering Progress*, Vol. 73, Sept. 1977.
13. R. S. Sonti, *Chemical Engineering*, Jan. 23, 1984, p. 66.
14. J. A. Davenport, "Hazards and Protection of Pressure Storage of Liquefied Petroleum Gases," *Proceedings of the Fifth International Symposium on Loss Prevention and Safety Promotion in the Process Industries*, Société de Chimie Industrielle, Paris, 1986, p. 22-1.
15. *BLEVE! The Tragedy of San Juanico*, Skandia International Insurance Corporation, Stockholm, Sweden, 1985.
16. *The Chemical Engineer*, No. 418, Oct. 1985, p. 16 (contains a series of articles on Mexico City).
17. T. A. Kletz, *Loss Prevention*, Vol. 13, 1980, p. 147.
18. *Middle East Econ. Digest*, Vol. 21, Apr. 15, 1977, and July 1, 1977.
19. N. J. Cupurus, "Cryogenic Storage Facilities for LNG and NGL," *Proceedings of the Tenth World Petroleum Congress*, Heyden, London, 1979, p. 119 (Panel Discussion 17, Paper 3).
20. N. J. Cupurus, "Developments in Cryogenic Storage Tanks," Paper presented at 6th International Conference on Liquefied Natural Gas, Kyoto, Japan, 1980.
21. N. J. Cupurus, "Storage of LNG and NGL," Paper presented at Seminar on LNG and NGL in Western Europe in the 1980s, Oslo, Norway, Apr. 2, 1981, Session II.
22. *One Hundred Largest Losses—A Thirty Year Review of Property Damage Losses in the Hydrocarbon-Chemical Industries*, 9th edition, M&M Protection Consultants, Chicago, 1986.
23. Health and Safety Executive, *The Chemical Release and Fire at the Associated Octel Company Limited, HSE Books*, Sudbury, UK, 1996.
24. I. Ayres, *Super Crunchers*, Bantam Dell, New York, 2007, p. 89.

Pipe and Vessel Failures

It happens, like as not,
There's an explosion and good-bye the pot!
These vessels are so violent when they split
Our very walls can scarce stand up to it.

—Geoffrey Chaucer, "The Canon Yeoman's Tale," c. 1386

9.1 PIPE FAILURES

Davenport [1] has listed more than 60 major leaks of flammable materials, most of which resulted in serious fires or unconfined vapor cloud explosions. Table 9-1, derived from his data, classifies the leak by point of origin and shows that pipe failures accounted for half of these incidents— more than half if we exclude transport containers. It is therefore important to know why pipe failures occur. This chapter examines a number of typical failures (or near failures). These and other failures, summarized in references 2 and 3, show that by far the biggest single cause of pipe failures has been the failure of construction teams to follow instructions or to do well what was left to their discretion. Therefore, the most effective way of reducing pipe failures is to do the following:

1. Specify designs in detail.
2. Check construction closely to see that the design has been followed and that details not specified have been constructed according to good engineering practice.

Many publications about pipe failures attribute them to causes such as fatigue or inadequate flexibility. This is not very helpful. It is like saying

TABLE 9-1   Origin of Leaks Causing Vapor Cloud Explosions

Origin of Leak	Number of Incidents	Notes
Transport container	10	Includes 1 zeppelin
Pipeline (incl. valves, flange, sight-glass, and 2 hoses)	34	
Pump	2	
Vessel (incl. 1 internal explosion, 1 foamover, and 1 failure due to overheating)	5	
Relief valve or vent	8	
Drain valve	4	
Error during maintenance	2	
Unknown	2	
Total	67	

a fall was caused by gravity. We need to know why fatigue occurred or why the flexibility was inadequate. To prevent further incidents, should we improve the design, construction, operations, maintenance, inspection, or what? The following incidents, and many others, suggest that improvement should be made at the design/construction interface. That is, we should focus on the detailing of the design and see that it has been followed and that good practice has been followed when details are not specified.

9.1.1   Dead-Ends

Dead-ends have caused many pipe failures. Water, present in traces in many oil streams, collects in dead-ends and freezes, breaking the pipe. Or corrosive materials dissolve in the water and corrode the line.

For example, there was a dead-end branch, 12 in. in diameter and 3 m (10 ft) long, in a natural-gas pipeline operating at a gauge pressure of 550 psi (38 bar). Water and impurities collected in the dead-end, which corroded and failed. The escaping gas ignited at once, killing three men who were looking for a leak [4].

There are other sorts of dead-ends besides pipes that have been blanked. Rarely used valved branches are just as dangerous. The feed line to a furnace (Figure 9-1) was provided with a permanent steam connection for use during de-coking.

The connection was on the bottom of the feed line, and the steam valve was not close to the feed line. Water collected above the steam valve,

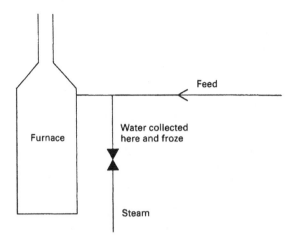

FIGURE 9-1 The steam connection to the furnace formed a dead-end.

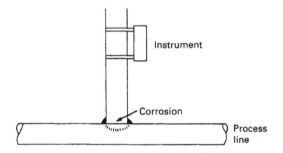

FIGURE 9-2 Water collected in the instrument support and corroded the process line.

froze during cold weather, and ruptured the line, allowing oil at a gauge pressure of 450 psi (30 bar) to escape. If dead-ends cannot be avoided, they should be connected to the *top* of the main pipeline (unless the liquid in the pipeline is denser than water).

An unusual—and unnecessary—dead-end was a length of 2-in. pipe welded onto a process line to provide a support for an instrument (Figure 9-2). Water collected in the support. Four years after it had been installed, the process line corroded through, and a leak of liquefied gas occurred.

Another serious failure occurred because water in a dead-end was suddenly vaporized. A heavy oil was dried by heating it to 120°C (250°F) in a tank filled with steam coils. The oil was circulated while it was being dried. The suction line projected into the conical base of the tank, forming a dead-end, as shown in Figure 9-3.

As long as the circulation pump was kept running, water could not settle out in the dead-end. The foreman knew the pump had to be kept running. When he was transferred to another plant, this information was lost, and the pump was used only for emptying the tank.

This worked satisfactorily for a time until some water collected in the dead-end and gradually warmed up as the oil was heated. When the temperature reached 100°C (212°F), the water vaporized with explosive violence and burst the equipment. The escaping oil caught fire, five men were killed, and the tank ended up in the plant next door.

This incident illustrates the dangers of dead-ends and the pressures developed when water is suddenly vaporized. It also shows how easily knowledge can be lost when people leave. Even if the new foreman was *told* to run the pump all the time or if this was written in the instructions, the reason for doing so might be forgotten, and the circulation might be stopped because it seemed unnecessary or to save electricity.

Other incidents caused by the sudden vaporization of water are described in Sections 12.2 and 12.4.5.

An explosive decomposition in an ethylene oxide (EO) distillation column, similar in its results to that described in Section 7.3.2, may have been set off by polymerization of EO in a dead-end spot in the column base where rust, a polymerization catalyst, had accumulated. Such dead-ends should be avoided. However, it is more likely that a flange leaked; the leaking gas ignited and heated an area of the column above the temperature at which spontaneous decomposition occurs. The source of ignition of the leak may have been reaction with the insulation, as described in Section 7.3.2. When flange leaks are likely or their consequences serious, flanges should be left uninsulated [14].

Dead-ends in domestic water systems can provide sites for the growth of the bacteria that cause Legionnaires' disease [15].

Some vertical drain lines in a building were no longer needed, so they were disconnected and capped but left connected to the horizontal main drain below. The caps were fixed with tape but were not made watertight as there was no way, it seemed, that water could get into them. Fifteen years later, a choke developed in the main drain; water backed up into

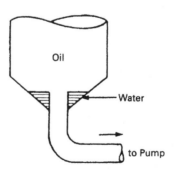

FIGURE 9-3   Water in this dead-end was vaporized by oil.

the disused legs and dripped into an electrical switch box. All power was lost, and some of the switch gear was damaged beyond repair [23].

9.1.2 Poor Support

Pipes have often failed because their support was insufficient and they were free to vibrate. On other occasions they failed because their support was too rigid and they were not free to expand:

(a) Many small-diameter pipes have failed by fatigue because they were free to vibrate. Supports for these pipes are usually run on site, and it is not apparent until startup that the supports are inadequate. It is easy for the startup team, busy with other matters, to ignore the vibrating pipes until the team has more time to attend to them. Then the team gets so used to them that it does not notice them.

Vibration and failure are particularly liable to occur when a small-diameter pipe carries a heavy overhung weight. Within 30 minutes of the start of a new compressor, a pressure gauge fell off for this reason [24].

When equipment receives impulses at its own natural frequency of vibration, excessive vibration (resonance) occurs, and this can lead to rapid failure. A control valve was fitted with a new spindle with slightly different dimensions. This changed its natural frequency of vibration to that of the impulses of the liquid passing through it (the frequency of rotation of the pump times the number of passages in the impeller). The spindle failed after three months. Even a small change in the size of spindle is a modification [24].

(b) A near failure of a pipe is illustrated in Figure 9-4. An expansion bend on a high-temperature line was provided with a temporary support to make construction easier. The support was then left in position. Fortunately, while the plant was coming onstream, someone noticed it and asked what it was for.

(c) After a crack developed in a 22-in.-diameter steam main, operating at a gauge pressure of 250 psi (17 bar) and a temperature of 365°C (690°F), the main was checked against the design drawings. Many of the supports were faulty. Here's an example from four successive supports:
1. On No. 1, the spring was fully compressed.
2. No. 2 was not fitted.

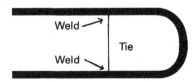

FIGURE 9-4 A construction support on an expansion bend was left in position.

3. No. 3 was in position but not attached to the pipe.
4. No. 4 was attached, but the nuts on the end of the support rod were slack.

Piping with a 12-in. diameter and larger is usually tailored for the particular duty. There is a smaller factor of safety than with smaller sizes. With these large pipes, it is even more important than with smaller ones that the finished pipework is closely inspected, to confirm that the construction team has followed the designer's instructions.

(d) A pipe was welded to a steel support, which was bolted to a concrete pier. A second similar support was located 2 m (6.5 ft) away. The pipe survived normal operating conditions. But when it got exceptionally hot, a segment of the pipe was torn out. The fracture extended almost completely around the weld. The bolts anchoring the support to the concrete pier were bent.

This incident was reported in the safety bulletin of another company. The staff members dismissed the incident. "Our design procedures," they said, "would prevent it happening." A little later it did happen. A reflux line was fixed rigidly to brackets welded to the shell of a distillation column. At startup the differential expansion of the hot column and the cold line tore one of the brackets from the column. Flammable vapor leaked out but fortunately did not catch fire.

(e) A 10-in. pipe carrying oil at 300°C (570°F) was fitted with a 3/4-in. branch on its underside. The branch was located 5 in. from a girder on which the pipe rested. When the pipe was brought into use, the expansion was sufficient to bring the branch into contact with the girder and knock it off. Calculations showed that the branch would move more than 6 in.

(f) On many occasions, pipe hangers have failed in the early stages of a fire, and the collapse of the pipes they were supporting has added to the fire. Critical pipes should therefore be supported from below.

(g) An extension was added to a 30-year-old pipebridge that carried pipes containing flammable liquids and gases. To avoid welding, the extension was joined to the old bridge by bolting. Rust was removed from the joining surfaces, and the extension was painted. Water penetrated the crack between the old and new surfaces and produced rust. As rust is more voluminous than the steel from which it is formed, the rust forced the two parts of the pipebridge apart—a phenomenon known as rust-jacking (see Section 16.3). Some of the bolts failed, and a steam main fractured. Fortunately, the liquid and gas lines only sagged [16].

(h) Eleven pipelines, 50 to 200 mm (2 to 8 in.) in diameter, containing hydrocarbon liquids and gases, were supported on brackets of the type shown in Figure 9-5a, 2.1 m (7 ft) tall and 6 m (10 ft) apart. The pipes

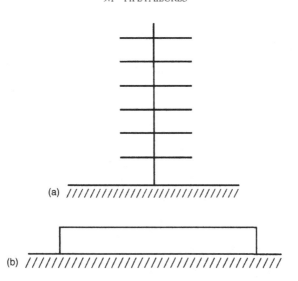

FIGURE 9-5   (a) The original pipe supports. (b) The supports used after the collapse.

were fixed to two of the brackets and rested on the others. The pipe run passed through a tank farm, and the wind flow through the gaps between the tanks caused the upright part of the supports to incline 2 degrees from the vertical. This was noticed when the pipe run was inspected, but no one regarded it as serious.

As the result of a power failure, the flow through many of the pipes suddenly stopped, and the surge caused the angle of inclination to increase to 6 degrees. The tops of the supports were now 5 in. (125 mm) out of line. The supports were now unstable. Eleven hours after the power failure and three hours after the flows had been restored, the pipe run collapsed over a length of 23 m (75 ft); 14 tons of gasoline spilled. Three hours later, a further length collapsed. The pipe supports were replaced by the type shown in Figure 9-5b.

9.1.3   Water Injection

Water was injected into an oil stream using the simple arrangement shown in Figure 9-6. Corrosion occurred near the point shown, and the oil leak caught fire [5]. The rate of corrosion far exceeded the corrosion allowance of 0.05 in. per year.

A better arrangement is shown in Figure 9-7. The dimensions are chosen so that the water injection pipe can be removed for inspection. However, this system is not foolproof. One system of this design was assembled with the injection pipe pointing upstream instead of downstream. This increased corrosion.

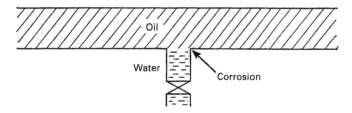

FIGURE 9-6   Water injection—a poor arrangement.

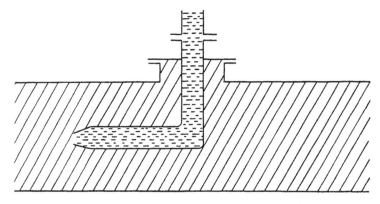

FIGURE 9-7   Water injection—a better arrangement.

As discussed in Section 3.2.1, equipment should be designed so that it is difficult or impossible to assemble it incorrectly or so that the incorrect assembly is immediately apparent.

9.1.4   Bellows

Bellows (expansion joints) are a good example of equipment that is intolerant of poor installation or departure from design conditions. They should therefore be avoided on lines carrying hazardous materials. This can be done by building expansion loops into the pipelines.

The most spectacular bellows failure of all time (Flixborough) was described in Section 2.4. Figure 9-8 illustrates a near failure.

A large distillation column was made in two halves, connected by a 42-in. vapor line containing a bellows. During a shutdown, this line was steamed. Immediately afterward someone noticed that one end of the bellows was 7 in. higher than the other, although it was designed for a maximum difference of 3 in. Someone then found that the design contractor had designed the line for normal operation. But the design contractor had not considered conditions that might be developed during abnormal procedures, such as startup and shutdown.

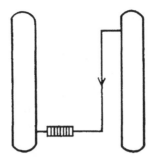

FIGURE 9-8 A large bellows between the two halves of a distillation column.

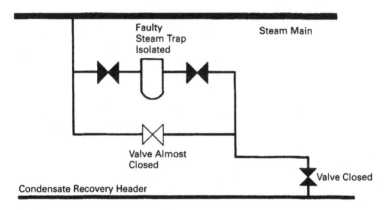

FIGURE 9-9 Arrangement of valves on a steam main that was broken by a water hammer.

9.1.5 Water Hammer

Water hammer (also known as hydraulic shock) occurs in two distinct ways: when the flow of liquid in a pipeline is suddenly stopped, for example, by quickly closing a valve [13], and when slugs of liquid in a gas line are set into motion by movement of gas or condensation of vapor. The latter occurs when condensate is allowed to accumulate in a steam main, because the traps are too few or out of order or in the wrong place. High-pressure mains have been ruptured, as in the following incident:

(a) A 10-in.-diameter steam main operating at a gauge pressure of 600 psi (40 bar) suddenly ruptured, injuring several workers.

The incident occurred soon after the main had been brought back into use after a turnaround. It was up to pressure, but there was no flow along it. The steam trap was leaking and had been isolated. An attempt was made to get rid of condensate through the bypass valve. But steam entered the condensate header, and the line was isolated, as shown in Figure 9-9. Condensate then accumulated in the steam main.

When a flow was started along the steam main by opening a 3/4-in. valve leading to a consuming unit, the movement of the condensate fractured the main [6].

(b) Figure 9-10 shows how water hammer burst another steam main—this time one operating at a gauge pressure of 20 psi (1.4 bar). Two drain points were choked and one isolated. In addition, the change in diameter of the main provided an opportunity for condensate to accumulate. The main should have been constructed so that the bottom was straight and so the change in diameter took place at the top.

(c) An operator went down into a pit to open a steam valve that was rarely operated and had been closed for nine months. Attempts to open the valve with a reach rod, 8 m (26 ft) long, had been unsuccessful. The pit was recognized as a confined space, and so the atmosphere was tested, the operator wore a rescue harness, and a standby co-worker was on duty outside. The steam main was up to pressure on both sides of the valve, and the gauge pressure was 120 psi (8.3 bar) on the upstream side, 115 psi (7.9 bar) on the downstream side. There was a steam trap on the downstream side of the valve but not on the upstream side, and as the valve was on the lowest part of the system, about 5 tons of cold condensate had accumulated on the upstream side.

The operator took about one to two minutes to open the valve halfway; very soon afterward, there was a loud bang as a 6-in. cast-iron valve on a branch (unused and blanked) failed as a result of water hammer. The operator was able to climb out of the pit, but later died from his burns, which covered 65% of his body [17]. Figure 9-11 explains the mechanism.

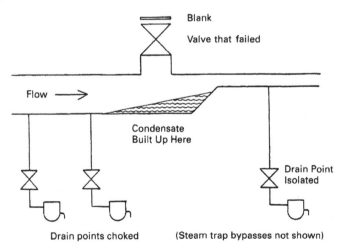

FIGURE 9-10 Arrangement of drains on a steam main that was broken by a water hammer.

The accident would not have occurred (or would have been less serious) if:

- Cast iron had not been used. It is brittle and therefore not a suitable material of construction for steam valves, which are always liable to be affected by water hammer.
- There was a steam trap upstream of the valve.
- The valve had been located in a more accessible place.
- The operator had taken longer to open the valve. On previous occasions operators had taken several hours or even longer, but

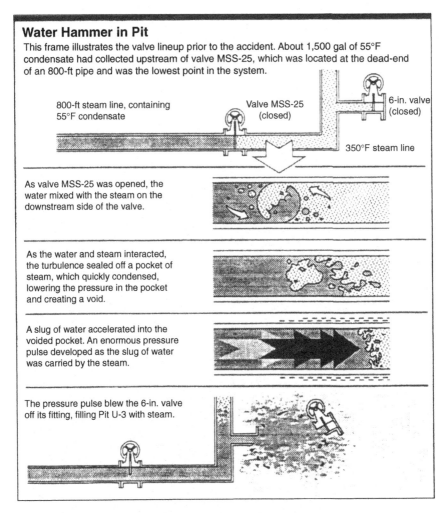

Water Hammer in Pit

This frame illustrates the valve lineup prior to the accident. About 1,500 gal of 55°F condensate had collected upstream of valve MSS-25, which was located at the dead-end of an 800-ft pipe and was the lowest point in the system.

800-ft steam line, containing 55°F condensate

Valve MSS-25 (closed)

6-in. valve (closed)

350°F steam line

As valve MSS-25 was opened, the water mixed with the steam on the downstream side of the valve.

As the water and steam interacted, the turbulence sealed off a pocket of steam, which quickly condensed, lowering the pressure in the pocket and creating a void.

A slug of water accelerated into the voided pocket. An enormous pressure pulse developed as the slug of water was carried by the steam.

The pressure pulse blew the 6-in. valve off its fitting, filling Pit U-3 with steam.

FIGURE 9-11 Condensate collected in a steam main. A valve was opened quickly. Sudden movement of the condensate fractured another valve. The figure explains how this occurred. *(Reproduced by permission of the Office of Environment, Safety, and Health, U.S. Department of Energy.)*

there were no written instructions, and the operator on duty had not been trained or instructed.

- The operating team as a whole had been aware of the well-known hazards of water hammer in steam mains.

For another failure due to water hammer, see Section 10.5.3.

9.1.6   Miscellaneous Pipe Failures

(a) Many failures have occurred because old pipes were reused. For example, a hole 6 in. long and 2 in. wide appeared on a 3-in. pipe carrying flammable gas under pressure. The pipe had previously been used on a corrosive/erosive duty, and its condition was not checked before reuse.

In another case, a 4½-in.-diameter pipe carrying a mixture of hydrogen and hydrocarbons at a gauge pressure of 3,600 psi (250 bar) and a temperature of 350 to 400 degrees C (660 to 750 degrees F) burst, producing a jet of flame longer than 30 m (Figure 9-12). Fortunately, the pipe was located high up, and no one was injured.

The grade of steel used should have been satisfactory for the operating conditions. Investigation showed, however, that the pipe had previously been used on another plant for 12 years at 500°C (930°F). It had used up a lot of its creep life.

FIGURE 9-12   An old pipe was reused and failed by creep.

Old pipes should never be reused unless their history is known in detail and tests show they are suitable (see Section 9.2.1h).

(b) Many failures have occurred because the wrong grade of steel was used for a pipeline. The correct grade is usually specified, but the wrong grade is delivered to the site or selected from the pipe store.

The most spectacular failure of this sort occurred when the exit pipe from a high-pressure ammonia converter was constructed from carbon steel instead of 1.25% Cr, 0.5% Mo. Hydrogen attack occurred, and a hole appeared at a bend. The hydrogen leaked out, and the reaction forces pushed the converter over.

After several such incidents had occurred, many companies insisted that if use of the wrong grade of steel could affect the integrity of the plant, all steel must be checked for composition before use. This applied to flanges, bolts, welding rods, and the like, as well as the raw pipe. However, these identification checks were abandoned when suppliers passed quality checks. Steel can be analyzed easily with a spectrographic analyzer. Other failures caused by the use of the wrong construction material are described in Section 16.1.

(c) Several pipe failures have occurred because reinforcement pads have been welded to pipe walls, to strengthen them near a support or branch, and the spaces between the pads and the walls were not vented. For example, a flare main collapsed, fortunately while it was being stress relieved.

Pipe reinforcement pads can be vented by intermittent rather than continuous welding, or a ⅛-in. or ¼-in. hole can be drilled in the pad.

(d) Corrosion—internal or external—often causes leak-before-break failures, but not always.

A line carrying liquefied butylene at a gauge pressure of about 30 psi (2 bar) passed through a pit where some valves were located. The pit was full of water, contaminated with some acid. The pipe corroded, and a small leak occurred. The line was emptied for repair by flushing with water at a gauge pressure of 110 psi (7.5 bar). The line was designed to withstand this pressure. However, in its corroded state it could not do so, and the bonnet was blown off a valve. The operator isolated the water. This allowed butylene to flow out of the hole in the pipe. Twenty minutes later, the butylene exploded, causing extensive damage [7].

(e) A 1-in. screwed nipple and valve blew out of an oil line operating at 350°C (660°C). The plant was covered by an oil mist, which ignited 15 minutes later. The nipple had been installed about 20 years earlier, during construction, to facilitate pressure testing. It was not shown on any drawing, and the operating team members did not know of its existence. If they had known it was there, they would have replaced it with a welded plug.

Similar incidents are described in Section 7.1.5.

(f) Not all pipe failures are due to inadequacies in design or construction (for example, the one described in Section 1.5.2).

A near failure was also due to poor maintenance practice. A portable, handheld compressed-air grinder was left resting in the space between two live lines. The switch had been left in the On position. So when the air compressor was started, the grinder started to turn. It ground away part of a line carrying liquefied gases. Fortunately, the grinder was noticed and removed before it had ground right through the line, but it reduced the wall thickness from 0.28 in. to 0.21 in.

(g) Figure 9-13 shows the pipework on the top of a reactor. When the pipework was cold, any liquid in the branch leading to the rupture disc drained out; when it was hot, it remained in the branch, where it caused corrosion and cracking [18].

9.1.7   Flange Leaks

Leaks from flanges are more common than those described in Sections 9.1.1 to 9.1.6 but are also usually smaller. On lines carrying LFGs and other flashing liquids, spiral-wound gaskets should be used in place of compressed asbestos fiber (caf) gaskets because they restrict the size of

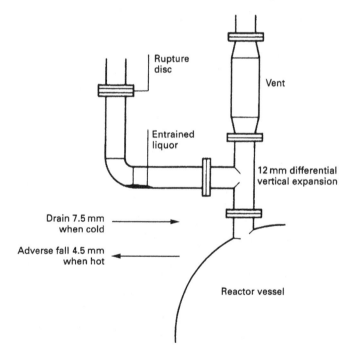

FIGURE 9-13   The vent arrangements at the top of the reactor. Liquid drained out when the pipework was cold but not when it was hot.

any leak to a very small value. A section of a caf gasket between two bolts has often blown out, causing a fair-sized leak. But this will not occur with a spiral-wound gasket.

9.1.8 Catastrophic Failures

The fire and explosions in Mexico City in 1984, which killed more than 500 people (see Section 8.1.4), started with a pipe failure. The cause is not known, but the pipe may have been subjected to excessive pressure. Earlier the same year, in February, at least 508 people, most of them children, were killed in Cubatao, Sao Paulo, Brazil, when a 2-ft-diameter gasoline pipe ruptured and 700 tons of gasoline spread across a strip of swamp. The incident received little publicity, but it seems that, as at Bhopal and Mexico City, a shantytown had been allowed to grow up around the pipeline, on stilts over the swamp. The cause of the failure is not known, but the pipeline was said to have been brought up to pressure in error, and it was also stated that there was no way of monitoring the pressure in the pipeline [10].

9.2 PRESSURE VESSEL FAILURES

Failures of pressure vessels are rare. Many of those that have been reported occurred during pressure testing or were cracks detected during routine examination. Major failures leading to serious leaks are hard to find.

Low-pressure storage tanks are more fragile than pressure vessels. They are therefore more easily damaged. Some failures are described in Chapter 5.

A few vessel failures and near failures are described next—to show that they can occur. Failures of vessels as a result of exposure to fire are described in Section 8.1.

9.2.1 Failures (and Near Failures) Preventable by Better Design or Construction

These are infrequent:

(a) A leak of gas occurred through the weep hole in a multiwall vessel in an ammonia plant. The plant stayed on line, but the leak was watched to see that it did not worsen. Ten days later the vessel disintegrated, causing extensive damage.

The multiwall vessel was made from an inner shell and 11 layers of wrapping, each drilled with a weep hole. The disintegration was attributed to excessive stresses near a nozzle. These had not been recognized when the vessel was designed.

The report on the incident states: "Our reading of the literature led us to believe that as long as the leaking gas could be relieved through the weep holes, it would be safe to operate the equipment. We called a number of knowledgeable people and discussed the safety issue with them. Consensus at the time supported our conclusion. But after the explosion, there was some dispute over exactly what was said and what was meant. Knowing what we know now, there can be no other course in the future than to shut down operations in the event of a leak from a weep hole under similar circumstances" [8].

(b) An ammonia plant vessel disintegrated as the result of low-cycle fatigue, the result of repeated temperature and pressure cycles [9].

(c) An internal ball float in a propane storage sphere came loose. When the tank was overfilled, the ball lodged in the short pipe leading to the relief valve, in which it formed an exact fit. When the sphere warmed up, the rise in pressure caused its diameter (14 m [45 ft]) to increase by 0.15 m (6 in.). The increase in diameter was noticed when someone found that the access stairway had broken loose.

A similar incident occurred in a steam drum in which steam was separated from hot condensate. On this occasion, the operator noticed that the pressure had risen above the set point of the relief valve and tripped the plant [19].

If you use ball-float level indicators, compare the size of the balls with those of the branches on the top of the vessel. If a loose ball could lodge in one of the branches, protect the branch with a metal cage, or use another type of level indicator.

(d) Several vessels have failed, fortunately during pressure testing, adjacent to internal support rings that were welded to the vessel. Expert advice is needed if such features are installed.

(e) N-butane boils at 0°C (32°F) and iso-butane at −12°C (10°F). When the air temperature is below 0°C and a vessel containing butane is being emptied, it is possible to create a partial vacuum and suck in the vessel; this has occurred on several occasions. Vessels used for storing butane and other liquefied gases with boiling points close to 0°C (e.g., butadiene) should be designed to withstand a vacuum. If an existing tank cannot be modified, then warm butane can be recycled, or the butane can be spiked with propane (but the pressure may then be too high in warm weather and the relief valve may lift).

(f) Although I have said that pressure vessel failures are rare, this is not true if vessels are not designed to recognized standards. Davenport [11] has described several liquefied petroleum gas (LPG) vessel failures that were due to poor construction. In the United Kingdom in 1984, no one knew who made 30% of the LPG tanks in use, when, or to what standard [12].

(g) The catastrophic failure of a 34-m^3 vessel storing liquid carbon dioxide killed three people, injured eight, and caused $20 million in damage and a three-month loss in production [25]. There were failures by all concerned:

- The vessel was leased from a supplier of carbon dioxide, and the user company did not check that it conformed to the company's usual standards.
- The supplier modified the vessel, but the workmanship was poor. A weld was weak, as it was not full penetration, and brittle because the weld surface, cut with a torch, was not ground before welding.
- As the result of a heater failure, the temperature of the vessel, designed for $-29°C$ ($-20°F$), fell to $-60°C$ ($-75°F$) by evaporative cooling (see Section 10.5.2); at this temperature, carbon steel becomes brittle, and cracking may have started.
- Five weeks later, the heater failed again, this time in the On position, and the pressure in the tank rose. The two relief valves failed to open because they were fixed to the side of the vessel and connected to the vapor space at the top by an internal line (Figure 9-14)—a most unusual arrangement, presumably adopted so that one nozzle could be used for filling, venting (during filling), and relief. As a result, the relief valve was cooled by the liquid in the vessel and became blocked by ice from condensed atmospheric moisture. There

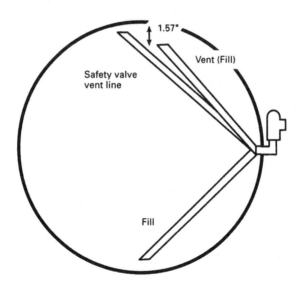

FIGURE 9-14 Unusual arrangement of a relief valve and pipework on a tank truck used to transport liquid carbon dioxide. The relief valve was cooled by the liquid and became blocked by ice from condensed atmospheric moisture. *(Illustration courtesy of the Institution of Chemical Engineers.)*

was no drain hole in the relief valve tailpipe (see Section 10.2.4).

The vessel burst, most of the bits ending up in a nearby river, from which they were salvaged.

After the accident, a search disclosed 11 other failures that had occurred but had received little or no publicity [26]. If they had been publicized, this incident could have been avoided. The company concerned withdrew all its carbon dioxide vessels that could not withstand low temperatures and replaced them with stainless steel ones. The company found that it could manage with 75% fewer vessels than it had used before (see Section 21.2.1).

At least two of the other 11 failures occurred because the plates from which the vessels were made did not get the correct postwelding heat treatment. Once a vessel has been constructed, it is not easy to check that it has had the correct heat treatment. The codes do not ask for microscopic examination of the grain structure, but it has been recommended [20].

The main recommendations from the incident were as follows:

- Leased equipment must meet the same standards as all other equipment.
- Do not say, "It must be safe because we are following the regulations and industry standards." They may be out of date or not go far enough.
- Publicizing accidents can prevent them from happening again.

(h) Designers are sometimes encouraged to use second-hand vessels as they are cheaper or immediately available. As with the pipeline in Section 9.1.6a, designers should do so only when they know the history of the vessel, including its design code, when it was last inspected, and when a materials specialist is satisfied that it is suitable for the new duty. The precautions are particularly important when the vessel is intended for use (1) with hazardous materials, (2) at pressures above atmospheric, or (3) at temperatures above or below atmospheric.

At lunch one day, when I worked in industry, I overheard the chief accountant ask a maintenance engineer if he could let him have a length of old rope to make a swing for his daughter. The engineer refused, as he would not, he said, know the history of the rope. (I am aware that in some companies a length of new rope would be declared scrap.)

A centrifuge was offered for sale. Examination showed that a repair to the bowl, by welding, had not been made by the manufacturer but by a contractor.

Old vessels may not be as cheap as they seem at first sight. Nozzles and manways are often in the wrong place, and the cost of modifying them may make the vessel more expensive than a new one.

A designer who was persuaded to use an old cylindrical pressure vessel ended up using the two dished ends and nothing more!

Penny-pinching can be tragic, for example, when old tab washers, split pins, and pipes are reused (see Section 16.1h).

(i) Alert observation can prevent failure. A welder was asked to weld a flange onto a nozzle on a new vessel. He noticed that the weld attaching the nozzle to the tank appeared to be substandard. Thorough examination showed that the seven other nozzle welds on this tank, and several welds on other tanks supplied as part of the same batch, lacked full penetration along 10% to 40% of their length [27].

9.2.2 Failures Preventable by Better Operation

The incident described in Section 9.2.1a might be classified as one preventable by better operation of equipment:

(a) Low-pressure storage tanks have often been sucked in, as described in Section 5.3. Pressure vessels can also be sucked in if they have not been designed to withstand vacuum, as the following incident shows:

A blowdown drum was taken out of service and isolated. The drain line was removed and a steam lance inserted to sweeten the tank. The condensate ran out of the same opening.

The condensate was isolated, and 45 minutes later the drain valve was closed. Fifteen minutes later the vessel collapsed. Clearly, 45 minutes was not long enough for all the steam to condense.

(b) Contractors installed a redundant pressure vessel, intended for reuse at atmospheric pressure, and decided to pressure-test it. They could not find a water hose to match any of the connections on the vessel. They therefore decided to pressure-test it with compressed air. The vessel reached a gauge pressure of 25 psi (1.7 bar) before it ruptured.

It is possible that the employees concerned did not understand the difference between a pressure test, normally carried out with water, and a leak test, often carried out with compressed air at a pressure well below the test pressure.

This incident shows the need to define the limits within which contractors can work and to explain these limits to contractors' employees.

Another incident in which a pressure vessel was ruptured by compressed air, this time because the vent was choked, is described in Section 2.2a.

(c) A vessel, designed to operate at a gauge pressure of 5 psi (0.3 bar) and protected by a rupture disc, was being emptied by pressurization with compressed air. The operator was told to keep the gauge pressure below 5 psi, but he did not do so, and the vessel burst, spraying him with a corrosive chemical. A valve below the rupture disc was closed and had probably been closed for some time.

It is bad practice (and in some countries illegal) to fit a valve between a vessel and its rupture disc (or relief valve). The valve had been fitted to stop escapes of gas into the plant after the disc had blown and while it was being replaced. A better way, if isolation is required, is to fit two rupture discs, each with its own isolation valve, the valves being interlocked so that one is always open.

If compressed gas has to be blown into a vessel that cannot withstand its full pressure, then it is good practice to fit a reducing valve on the gas supply. This would be possible in the case just described. But it may not be possible if the gas is used to blow liquid *into* a vessel. If the gas pressure is restricted to the design pressure of the vessel, it may not be sufficient to overcome friction and change in height.

A sidelight on the incident is that the operator had worked on the plant for only seven months and during that time had received five warnings for lack of attention to safety or plant operations. However, the incident was not due to the operator's lack of attention but to the poor design of the equipment. Sooner or later, a valve will be shut when it should be open or vice versa, and the design or method of operation should allow for this (see also Section 1.1 on isolation for maintenance).

(d) Failure of a level controller can allow high-pressure gas to enter a storage tank and rupture it (see Section 5.2.2c). Pressure vessels have also been ruptured in this way. In one case, gas at a gauge pressure of about 2,200 psi (150 bar) entered a vessel designed for 150 psi (10 bar). Bits of the vessel, up to 2 tons in weight, were found more than 1 km away. The control system was badly designed, as there were two let-down valves in parallel; failure of either could cause rupture of the downstream vessel. In addition, the signals to the two trip valves had been isolated. If the normal control system failed, something we should expect every few years, the only protection was quick action by the operator [21].

(e) A reactor was overpressured by a runaway reaction. Visual examination showed nothing wrong, so the reactor was allowed to continue in service. Eight weeks later it was again overpressured by another runaway reaction, and this time it burst catastrophically. A thorough examination then showed that the reactor had been damaged by the first runaway. (The control instrumentation may also have been damaged, and this may have led to the second runaway.) [28] Equipment that has been taken outside its design or test range should not be used again until a materials expert has examined it.

9.2.3 Cylinders

Cylinders have been involved in a number of incidents. The following are typical:

(a) A technician was moving a cylinder containing nitrogen, together with some heavier gas, at a gauge pressure of 600 psi (40 bar). The technician accidentally moved the valve operating lever. The cylinder fell over, and the valve was knocked off. The cylinder then became airborne, hit a platform 6 m (20 ft) above, and went through a sheet metal wall into a building. It went through the roof of this building, 15 m (50 ft) above, and then fell back through the roof and landed 40 m (130 ft) from the point where it had started its journey. Remarkably, no one was injured. Four things were wrong:

- The operating lever should have been removed before the cylinder was moved.
- A safety pin, which would prevent accidental operation, was not in place.
- There was no protective cap over the valve, as this was not a part of this particular cylinder design.
- The cylinder should have been moved with a cylinder cart, not by hand.

(b) Several incidents have been due to overfilling. For example, cylinders of uranium hexafluoride (hex) were weighed as they were filled, with the cylinder and the cart that supported it resting on the filling scales. One cylinder was longer than usual. As a result, one wheel of the cart supporting the cylinder overlapped the filling scale and rested on the ground. By the time the operator realized this and moved the cylinder, its weight was above the range of the scales. The operator tried to remove some of the contents of the cylinder by applying a vacuum but without success, probably because some of the hex had solidified. The operator and his supervisor then moved the cylinder into a steam chest and heated it. The contents expanded, and after two hours the cylinder ruptured. One man was killed, and a number were injured by the escaping gas. There were many things wrong:

- An operator was asked to fill a cylinder longer than that for which the filling equipment was designed.
- The normal filled weight was close to the top of the scale, so if a cylinder was overfilled, its weight was unknown.
- There was no equipment for emptying overfilled cylinders.
- Although heating overfilled cylinders was against company rules, the operator and his supervisor may not have known this and probably did not understand the reason for the rule [22].

(c) A chlorine cylinder was left standing, connected to a regulator, for eight months. The valve became rusted and appeared to be tightly

closed though it was not. When someone was disconnecting the regulator, gas spurted into his face. He and three other people who were
in the room at the time were hospitalized [29].

References

1. J. A. Davenport, *Chemical Engineering Progress*, Vol. 73, No. 9, Sept. 1977, p. 54.
2. T. A. Kletz, *Learning from Accidents*, 3rd edition, Butterworth-Heinemann, Boston, MA, 2001, Chapter 16.
3. T. A. Kletz, *Plant/Operations Progress*, Vol. 3, No. 1, Jan. 1984, p. 19.
4. U.S. National Transportation Safety Board, *Safety Recommendations*, P-75-14 and 15, 1975.
5. *The Bulletin, The Journal of the Association for Petroleum Acts Administration*, Apr. 1971.
6. *Explosion from a Steam Line: Report of Preliminary Inquiry No. 3471*, Her Majesty's Stationery Office, London, 1975.
7. C. H. Vervalin, *Fire Protection Manual for Hydrocarbon Processing Plants*, 3rd edition, Gulf Publishing Co, Houston, TX, 1985, p. 122.
8. L. B. Patterson, *Ammonia Plant Safety*, Vol. 21, 1979, p. 95.
9. J. E. Hare, *Plant/Operations Progress*, Vol. 1, No. 3, July 1982, p. 166.
10. *Hazardous Cargo Bulletin*, June 1984, p. 34.
11. J. A. Davenport, "Hazards and Protection of Pressure Storage of Liquefied Petroleum Gases," *Proceedings of the Fifth International Symposium on Loss Prevention and Safety Promotion in the Process Industries*, Société de Chimie Industrielle, Paris, 1986, p. 22-1.
12. A. C. Barrell, *Hazard Assessment Workshop*, Atomic Energy Authority, Harwell, UK, 1984.
13. D. Clarke, *The Chemical Engineer*, No. 449, 1988, p. 44.
14. B. E. Mellin, *Loss Prevention Bulletin*, No. 100, 1991, p. 13.
15. *Loss Prevention Bulletin*, No. 091, Feb. 1990, p. 23.
16. B. M. Hancock, "Preventing Piping Failures," *Safety and Loss Prevention in the Chemical and Oil Processing Industries*, Symposium Series No. 120, Institution of Chemical Engineers, Rugby, UK, 1990, p. 589.
17. *Occupational Safety Observer*, Vol. 2, No. 9, U.S. Dept. of Energy, Washington, DC, Sept. 1993, p. 1.
18. A. B. Smith, *Loss Prevention Bulletin*, No. 102, Dec. 1991, p. 29.
19. M. L. Griffin and F. H. Gurry, "Case Histories of Some Power- and Control-based Process Incidents," Paper presented at AIChE Loss Prevention Symposium, Houston, Tex., Mar./Apr. 1993.
20. T. Coleman, *The Chemical Engineer*, No. 461, June 1989, p. 29.
21. K. C. Wilson, *Process Safety and Environmental Protection*, Vol. 68, No. B1, Feb. 1990, p. 31.
22. *The Safety of the Nuclear Fuel Cycle*, Organization for Economic Cooperation and Development, 1993, p. 206.
23. *Operating Experience Weekly Summary*, No. 97-32, Office of Nuclear and Facility Safety, U.S. Dept. of Energy, Washington, DC, 1997, p. 6.
24. F. K. Crawley, *Loss Prevention Bulletin*, No. 134, 1997, p. 21.
25. W. E. Clayton and M. L. Griffin, *Process Safety Progress*, Vol. 13, No. 4, Oct. 1994, p. 203, and *Loss Prevention Bulletin*, No. 125, Oct. 1995, p. 3.
26. W. E. Clayton and M. L. Griffin, *Loss Prevention Bulletin*, No. 126, Dec. 1995, p. 18.
27. *Operating Experience Weekly Summary*, No. 97-12, Office of Nuclear and Facility Safety, U.S. Dept. of Energy, Washington, DC, 1997, p. 6.
28. S. J. Brown and T. J. Brown, *Process Safety Progress*, Vol. 14, No. 4, Oct. 1995, p. 245.
29. *Safety Management* (South Africa), Nov./Dec. 1990, p. 45.

10

Other Equipment

Occasionally all the valves on a ring main would be closed and the pressure in a pump rise to danger point. No one appeared to realize that there was anything wrong with this state of affairs.

—A U.K. gas works in 1916, described by Norman Swindin,
Engineering without Wheels

Incidents involving storage tanks, stacks, pipelines, and pressure vessels have been described in Chapters 5, 6, and 9. This chapter describes some incidents involving other items of equipment.

10.1 CENTRIFUGES

Many explosions, some serious, have occurred in centrifuges handling flammable solvents because the nitrogen blanketing was not effective.

In one case, a cover plate between the body of the centrifuge and a drive housing was left off. The nitrogen flow was not large enough to prevent air from entering, and an explosion occurred, killing two men. The source of ignition was probably sparking caused by the drive pulley, which had slipped and fouled the casing. However, the actual source of ignition is unimportant. In equipment containing moving parts, such as a centrifuge, sources of ignition can easily arise.

In another incident the nitrogen flow was too small. The range of the rotameter in the nitrogen line was 0 to 60 L/min (0 to 2 ft³/min), although 150 L/min (5 ft³/min) was needed to keep the oxygen content at a safe level.

On all centrifuges that handle flammable solvent, the oxygen content should be monitored continuously. At the very least, it should be checked

doi:10.1016/B978-1-85617-531-9.00010-X

every shift with a portable analyzer. In addition, the flow of nitrogen should be adequate, clearly visible, and read regularly.

These recommendations apply to all equipment blanketed with nitrogen, including tanks (Section 5.4) and stacks (Section 6.1). But the recommendations are particularly important for centrifuges because of the ease with which sources of ignition can arise [1].

Another hazard with centrifuges is that if they turn the wrong way, the snubber can damage the basket. It is therefore much more important than with pumps to make sure this does not occur.

One centrifuge was powered by a hydraulic oil installation 2 to 3 m (6 to 10 ft) away. A leak of oil from a cooler was ignited, and the fire was spread by oil and product spillages and by plastic-covered cables. It destroyed the plastic seal between the centrifuge and its exit chute. There was an explosion in the chute and a flash fire in the drier to which it led. The centrifuge exit valve was closed, but the aluminum valve actuator was destroyed. Fortunately, the exit valve did not leak, or several tons of solvent would have been added to the fire. Aluminum is not a suitable material of construction for equipment that may be exposed to fire.

10.2 PUMPS

10.2.1 Causes of Pump Failures

The main causes of pump failure, often accompanied by a leak, are as follows:

1. Changing the operating temperature or pressure or the composition of the liquid so that corrosion increases. Any such change is a modification and its effects should be reviewed before it is made, as discussed in Chapter 2.
2. Incorrect installation or repair, especially in fitting the bearings or seals; badly fitted pipework can produce large, distorting forces, and sometimes pumps rotate in the wrong direction.
3. Maloperation, such as starting a pump before all the air has been removed, starting with the delivery valve open or the suction valve closed, starting with a choked or missing strainer, or neglecting lubrication.
4. Manufacturing faults. New pumps should be checked thoroughly. Make sure the pump is the one ordered and that the material of construction is the same as that specified (see Section 16.1). ·

10.2.2 Types of Pump Failures

The biggest hazard with pumps is failure of the seal, sometimes the result of bearing failure, leading to a massive leak of flammable, toxic, or corrosive chemicals. Often it is not possible to get close enough to the pump suction and delivery valves to close them. Many companies therefore install remotely operated emergency isolation valves in the suction lines (and sometimes in the delivery lines as well), as discussed in Section 7.2.1. A check valve (nonreturn valve) in the delivery line can be used instead of an emergency isolation valve, provided it is scheduled for regular inspection.

Another common cause of accidents with pumps is dead-heading—that is, allowing the pump to run against a closed delivery valve. This has caused rises in temperature, leading to damage to the seals and consequent leaks. It has caused explosions when the material in the pump decomposed at high temperature. In one incident, air saturated with oil vapor was trapped in the delivery pipework. Compression of this air caused its temperature to rise above the auto-ignition temperature of the liquid, and an explosion occurred—a diesel-engine effect.

Positive pumps are normally fitted with relief valves. These are not usually fitted to centrifugal pumps unless the process material is likely to explode if it gets too hot. As an alternative to a relief valve, such pumps may be fitted with a high-temperature trip. This isolates the power supply. Or a kick-back, a small-diameter line (or a line with a restriction orifice plate) leading from the delivery line back to the suction vessel, may be used. The line or orifice plate is sized so that it will pass just enough liquid to prevent the pump from overheating. Small-diameter lines are better than restriction orifice plates as they are less easily removed.

Pumps fitted with an auto-start will dead-head if they start when they should not. This has caused overheating. Such pumps should be fitted with a relief valve or one of the other devices just described.

A condensate pump was started up by remote operation, with both suction and delivery valves closed. The pump disintegrated, bits being scattered over a radius of 20 m (65 ft). If remote starting must be used, then some form of interlock is needed to prevent similar incidents from occurring.

Pumps can overheat if they run with the delivery valve almost closed. In one incident, a pump designed to deliver 10 tons/hr was required to deliver only ¼ ton/hr. The delivery valve was gagged, the pump got too hot, the casing joint sprang, and the contents leaked out and caught fire.

If a pump is required to deliver a very small fraction of its design rate, a kick-back should be provided.

Many bearing failures and leaks have occurred as the result of lack of lubrication. Sometimes operators have neglected to lubricate the pumps. On one occasion, a bearing failure was traced to water in the lubricating oil. The bearing failure caused sparks, which set fire to some oily residues nearby. Drums of oil in the open-air lubricating-oil storage area were found to be open so that rainwater could get in. This is a good example of high technology—in bearing and seal design—frustrated by a failure to attend to simple things.

More than any other item of equipment, pumps require maintenance while the rest of the plant is on line. Many incidents have occurred because pumps under repair were not properly isolated from the running plant. See Section 1.1. Reference 18 describes other pump failures.

10.3   AIR COOLERS

A pump leak caught fire. There was a bank of fin-fan coolers above the pump, and the updraft caused serious damage to the coolers. There was an emergency isolation valve in the pump suction line. This was soon closed and the fire extinguished but not before the fin-fans were damaged. The damage to them was far greater than the damage to the pumps. It is not good practice to locate fin-fans (or anything else) over pumps or other equipment that is liable to leak.

On several other occasions, the draft from fin-fans has made fires worse. And the fans could not be stopped because the stop buttons were too near the fire. The stop buttons should be located (or duplicated) at least 10 m away.

Another hazard with air coolers is that even though the motor is isolated, air currents have caused the fans to rotate while they were being maintained. Fans should therefore be prevented from moving before any maintenance work is carried out on or near them.

10.4   RELIEF VALVES

Few incidents occur because of faults in relief valves themselves. When equipment is damaged because the pressure could not be relieved, someone usually finds afterward that the relief valve (or other relief device) had been isolated (see Section 9.2.2c), wrongly installed (see Section 3.2.1e), or interfered with in some other way (see Section 9.2.1c). The following incidents are concerned with the peripherals of relief valves rather than the valves themselves.

10.4.1   Location

A furnace was protected by a relief valve on its inlet line (Figure 10-1). A restriction developed after the furnace. The relief valve lifted and took most of the flow. The flow through the furnace tubes fell to such a low level that they overheated and burst. The low-flow trip, which should have isolated the fuel supply to the furnace when the flow fell to a low value, could not do so because the flow through it was normal.

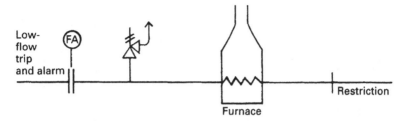

FIGURE 10-1   When the relief valve lifted, the flow through the furnace was reduced.

The relief valve should have been placed after the furnace or, if this was not possible, before the low-flow trip.

Another point on location that is sometimes overlooked is that most relief valves are designed to be mounted vertically and should not be mounted horizontally.

10.4.2   Relief-Valve Registers

All companies keep a register of relief valves. They test them at regular intervals (every one or two years) and do not allow their sizes to be changed without proper calculation and authorization.

However, equipment has been overpressured because the following items were not registered. They had been overlooked because they were not obviously a relief device or part of the relief system:

(a) A hole or an open vent pipe—the simplest relief device possible. Section 2.2a described how two men were killed because the size of a vent hole in a vessel was reduced from 6 in. to 3 in.

(b) A restriction orifice plate limiting the flow into a vessel or the heat input into a vessel should be registered if it was taken into account in sizing the vessel's relief valve. Restriction plates are easily removed. A short length of narrow diameter pipe is better.

(c) A control valve limiting the flow into a vessel or the heat input into a vessel should be registered if its size was taken into account in sizing the vessel's relief valve. The control valve record sheets or database entries should be marked to show that the trim size should not be changed without checking that the relief valve will still be suitable.

(d) Check (nonreturn) valves should be registered *and inspected regularly* if their failure could cause a relief valve to be undersized. Usually two check valves *of different types* in series are used if the check valve forms part of the relief system.

10.4.3   Changing Relief Valves

Some vessels are provided with two full-size relief valves so that one can be changed with the plant on line. On the plant side of the relief valves, isolation valves are usually provided below each relief valve, inter-locked so that one relief valve is always open to the plant (Figure 10-2). If the relief valves discharge into a flare system, it is not usual to provide such valves on the flare side. Instead the relief valve is simply removed and a blank fitted quickly over the end of the flare header before enough air is sucked in to cause an explosion. Later the blank is removed and the relief valve replaced.

On one plant, a fitter removed a relief valve and then went for lunch before fitting the blank. He returned just as an explosion occurred. He was not injured by the explosion but was slightly injured sliding down a pipe to escape quickly.

Removing a valve and fitting a blank is satisfactory if the operators make sure, before the relief valve is removed, that the plant is steady and that this relief valve or any other is unlikely to lift. Unfortunately, such instructions may lapse with the passage of time. This occurred at one plant. The people there were fully aware that air might get into the flare system. They knew about the incident just described. But they were less aware that oil might get out. While an 8-in. relief valve was being changed, another relief valve lifted, and gasoline came out of the open end. Fortunately it did not ignite.

The investigation showed that at the time the operating team members were busy at the main plant, which they operated. A deputy foreman had been left in charge of changing the relief valve. He wanted to get it done while a crane was available.

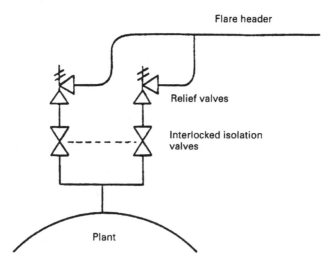

FIGURE 10-2 Two relief valves with interlocked isolation valves. This figure is dia-grammatic. If any liquid might be present, the tailpipe should fall, not rise, after it leaves the relief valve. Otherwise, liquid may collect in the dip and produce a back pressure.

The best way to change a relief valve when a plant is on line is to use the sealing plate shown in Figure 10-3.

All but two of the bolts joining the relief valve to the flare header are removed. The sealing plate is then inserted between the relief valve and the flare header. It is secured by special bolts with small heads that pass through the bolt holes in the relief-valve flange but not through the holes in the sealing plate. The last two bolts can then be removed and the relief valve removed. To replace the relief valve, the procedure is reversed [2]. This system is recommended for changing relief valves on lines greater than 4 in. diameter.

Flare lines should never be slip-plated with ordinary slip-plates because they may be left in position in error. The sealing plate cannot be left in position when the relief valve is replaced.

When replacing relief valves, care must be taken to make sure that the right relief valve is replaced. Section 6.1d describes another incident that occurred when a flare line was broken. Valves of different internal sizes may look alike (see Section 1.2.4).

10.4.4 Tailpipes

Figure 10-4 shows what happened to the tailpipe of a steam relief valve that was not adequately supported. The tailpipe was not provided with a drain hole (or if one was provided, it was too small), and the tailpipe filled with water. When the relief valve lifted, the water hit the curved top of the tailpipe with great force. Absence of a drain hole in a tailpipe also led to the incident described in Section 9.2.1g.

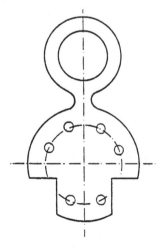

FIGURE 10-3   Sealing plate for changing relief valves.

FIGURE 10-4   This relief-valve tailpipe was not adequately supported.

On other occasions, drain holes have been fitted in relief-valve tail-pipes even though the relief valve discharged into a flare system. Gas has then escaped into the plant area. Sometimes relief-valve exit pipes are not adequately supported and have sagged on exposure to fire, restricting the relief-valve discharge.

10.4.5   Relief-Valve Faults

Here are a few examples of faults in relief valves themselves. These are not the results of errors in design but of poor maintenance practice. The following have all been seen:

1. Identification numbers stamped on springs, thus weakening them (Figure 10-5).
2. The sides of springs ground down so that they fit.
3. Corroded springs.
4. A small spring put inside a corroded spring to maintain its strength. Sometimes the second spring was wound the same way as the first spring so that the two interlocked (Figure 10-6).
5. Use of washers to maintain spring strength.
6. Welding of springs to end caps (Figure 10-7).
7. Deliberate bending of the spindle to gag the valve (Figure 10-8).
8. Too many coils allowing little, if any, lift at set pressure (Figure 10-9).

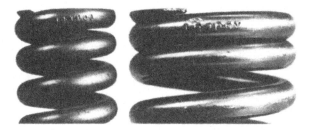

FIGURE 10-5 Identification marks on body coils could lead to spring failure.

FIGURE 10-6 Use of additional inner springs of unknown quality in an attempt to obtain set pressure.

FIGURE 10-7 End caps welded to spring. Failure occurred at weld.

FIGURE 10-8 Deliberate bending of the spindle to gag the valve.

FIGURE 10-9   Example of too many coils.

Do not assume that such things could not happen in your company (unless you have spent some time in the relief-valve workshop). All relief valves should be tested and inspected regularly. Reference 3 describes model equipment and procedures. When a large petroleum company introduced a test program, it was shocked by the results: out of 187 valves sent for testing, 23 could not be tested because they were leaking or because the springs were broken, and 74 failed to open within 10% of the set pressure—that is, more than half of them could not operate as required [4].

Testing, of course, must be thorough. The following incident is described in the form of a conversation between an inspector investigating a boiler explosion and the maintenance foreman [12]:

> *Inspector:* "When was the relief valve last checked?"
> *Foreman:* "After the last overhaul."
> *Inspector:* "How was it checked?"
> *Foreman:* "I set it myself, using the boiler's own pressure gauge."
> *Inspector:* "Why didn't you use a master gauge?"
> *Foreman:* "I didn't need to. The gauge had been checked and found accurate only two weeks before."
> *Inspector:* "Who checked it?"
> *Foreman:* "Mr. X, one of my fitters. He has often done so in the past."
>
> The inspector then spoke to Mr. X.
>
> *Inspector:* "I understand you checked the pressure gauge two weeks before the explosion."
> *Mr. X:* "Yes, the foreman told me to do so."
> *Inspector:* "When was your master gauge last calibrated?"
> *Mr. X:* "I didn't use one."
> *Inspector:* "You didn't use a master gauge? Then how did you check it?"
> *Mr. X:* "I checked it against the relief valve. I knew it was correct because the foreman told me he had adjusted it himself."

This incident occurred in the nineteenth century when boiler explosions were much more frequent than they are today. But are you sure something similar could not occur today? Read Sections 10.7.2b and 10.7.2c before you decide.

Similar incidents have been documented in technical reports. A writes something in a book or paper. B copies it without acknowledgment. A then repeats it in another report, citing B as the source and thus giving it an authenticity it lacked in its first publication.

10.4.6 Disposal of Relief Discharges

Material discharged from relief valves and rupture discs should not be discharged to atmosphere unless the following conditions exist:

- It will have no harmful effects, for example, steam, compressed air, or nitrogen.
- It is a gas at a pressure high enough to disperse by jet mixing. It is necessary to use a pilot-operated relief valve that is either open or shut and is not a type that will hover. Although it is safe to discharge gases such as ethylene and propylene in this way, there may be objections on environmental grounds.
- The amount released is negligible, for example, the relief valves that protect a pipeline that has been isolated.
- A system of trips or interlocks makes the probability that the relief valve will lift very low, say, less than once in 1,000 years for flammable liquids and less than once in 100 years for flammable gases.
- The relief valve will lift only after prolonged exposure of the equipment to fire and will discharge within the fire area so that the discharge will ignite.

Here are some examples of the results of letting relief valves discharge to atmosphere:

- A 6-in. (150-mm) relief valve on a petrochemical plant discharged benzene vapor to atmosphere. It was ignited by a furnace and exploded, rupturing piping, which released more than 100 tons of various flammable liquids. One man was killed, and damage was extensive [5].
- At Seveso in Italy in 1976, a runaway reaction led to the discharge of the reactor contents, including dioxin, a toxic chemical, through a rupture disc direct to the atmosphere. Although no one was killed, many people developed chloracne, an unpleasant skin disease, and an area of about 17 km^2 was made uninhabitable. A catchpot after the relief device would have prevented the reactor contents from reaching the atmosphere. No catchpot was installed as the designers did not foresee that a runaway might occur, although similar runaways had occurred on other plants (see Section 21.2.5) [6].
- A flarestack ignited naphtha vapor from a relief valve on a town gas plant in the United Kingdom. The flame impinged on the naphtha line, which burst, starting a secondary fire [7].

- A relief valve sprayed liquid into the face of a passing operator with such force that it knocked his goggles off.
- A reaction involving concentrated sulfuric acid was carried out at atmospheric pressure in a vessel with an opening to the atmosphere at the top. When a runaway occurred, acid was ejected over the surrounding area [13].
- The rupture discs on some water compressors were allowed to discharge inside a building as the water was clean. However, by the time it had drained down through several floors to the basement of the building, it had dissolved some solid material that had been spilt on one of the intervening floors and became hazardous. Discs had failed on several occasions, for unknown reasons. Possible causes were vibration, hammer pressure, and low-cycle fatigue.

If, despite my advice, you let relief devices discharge to atmosphere, make sure that if the discharge ignites, the flame will not impinge on other equipment and that no one will be in the line of fire.

10.4.7   Vacuum Relief Valves

Some large equipment, though designed to withstand pressure, cannot withstand vacuum and has to be fitted with vacuum relief valves. These usually admit air from the atmosphere. If the equipment contains a flammable gas or vapor, then an explosion could occur with results more serious than collapse of the vessel. Experience shows that a source of ignition may be present even though we have tried to remove all possible sources (see Section 5.4). It is therefore better to protect equipment that cannot withstand vacuum by means of a pressure control valve that admits nitrogen or, if nitrogen is not available, another gas such as fuel gas. Very large amounts may be necessary. For example, if the heat input to a large refinery distillation column stopped but condensation continued, 8,000 m^3/hr of gas, the entire consumption of the refinery, would be required. Instead, a much smaller amount was supplied to the inlet of the condenser, thus blanketing it and stopping heat transfer [14]. The simplest solution, of course, is to design equipment to withstand vacuum.

Protection of storage tanks against vacuum is discussed in Sections 5.3 and 5.4.

10.5   HEAT EXCHANGERS

10.5.1   Leaks into Steam and Water Lines

Hydrocarbons can leak through heat exchangers into steam or condensate systems and appear in unexpected places. Some hydrocarbon gas leaked into a steam line that supplied a heater in the basement of a control building. The gas came out of a steam trap and exploded, killing two men. The operators in the control building had smelled gas but thought it had entered via the ventilation system, so they had switched off the fan. The control gear was ordinary industrial equipment, not suitable for use in a flammable atmosphere, and the sparking ignited the gas. It was fortunate that more people were not killed, as the building housed administrative staff as well as operators.

A leak in another heat exchanger allowed flammable gas to enter a cooling-water return line. Welding, which was being carried out on the cooling tower, ignited the gas. The atmosphere had been tested before work started, five hours earlier (see Section 1.3.2).

10.5.2 Leaks Due to Evaporative Cooling

If the pressure on a liquefied gas is reduced, some of the liquid evaporates, and the rest gets colder. All refrigeration plants, domestic and industrial, make use of this principle. This cooling can affect equipment in two ways: it can make it so cold that the metal becomes brittle and cracks, as discussed in Section 9.2.1f and it can cause water, or even steam, on the other side of a heat exchanger to freeze and rupture a tube or tubes. The leak that caused the explosion in a control building (see the previous section) started this way.

Figure 10-10 illustrates another incident. When a plant was shutting down, the flow of cooling water to the tubes of a heat exchanger was isolated. The propylene on the shell side got colder as its pressure fell. The water in the tubes froze, breaking seven bolts. The operators saw ice forming on the outside of the cooler but did not realize that this was hazardous and took no action. When the plant started up again, propylene entered the cooling-water system, and the pressure blew out a section of the 16-in. (400-mm) line. A furnace located 40 m (130 ft) away ignited the gas, and the fire caused serious damage.

The cooling water should have been kept flowing while the plant was depressured. This would have prevented the water from freezing, provided that depressuring took more than 10 minutes.

10.5.3 Damage by Water Hammer

Water hammer (hydraulic shock) in pipelines is discussed in Section 9.1.5. It can also damage heat exchangers, and Figure 10-11 illustrates such an incident.

The steam supplied to the shell of a distillation column reboiler was very wet, as there was only one steam trap on the supply line although at least three were needed. In addition, condensate in the reboiler drained

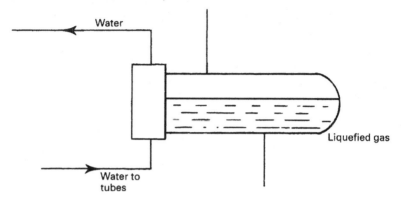

FIGURE 10-10   Evaporative cooling.

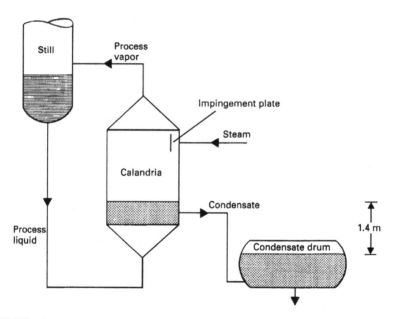

FIGURE 10-11   Condensate in the steam—the result of too few steam traps—knocked
off the impingement plate and damaged the calandria tubes.

away only slowly because the level in the drum into which it drained
was only 1.4 m (4.5 ft) below the level in the reboiler.

An impingement plate was fitted to the reboiler to protect the tubes, but
it fell off, probably as a result of repeated blows by slugs of condensate.

The condensate then impinged on the tubes and squashed or broke 30 of them.

The impingement plate had fallen off several times before and was merely put back with stronger attachments. When something comes apart, we should ask why, not just make it stronger (see Section 1.5.5).

Buildup of condensate in a heat exchanger can cause operating problems as well as water hammer. If the steam supply is controlled by a motor valve and the valve is not fully open, the steam pressure may be too low to expel the condensate, and its level will rise. This will reduce heat transfer, and ultimately the steam supply valve will open fully and expel the condensate. The cycle will then start again. This temperature cycling is bad for the heat exchanger and the plant and may be accompanied by water hammer and corrosion. Proprietary devices are available for overcoming the problem [8].

10.5.4 An Accident during Maintenance

The tube bundle was being withdrawn from a horizontal shell and tube heat exchanger. It was pulled out a few inches and then became stuck. The mechanics decided that the cause was sludge, and to soften it they reconnected the steam supply to the shell. The tube bundle was blown out with some force, causing serious injuries [9].

10.6 COOLING TOWERS

These are involved in a surprisingly large number of incidents; one is described in Section 10.5.1. Wooden packing, after it dries out, is easily ignited, and many cooling towers have caught fire while they were shut down. For example, the support of a force draft fan had to be repaired by welding. An iron sheet was put underneath to catch the sparks, but it was not big enough, and some of the sparks fell into the tower and set the packing on fire.

Corrosion of metal reinforcement bars has caused concrete to fall off the corners of cooling towers.

A large natural-draft cooling tower collapsed in a 70-mph (110-km/hr) wind, probably because of imperfections in the shape of the tower, which led to stresses greater than those it was designed to take and caused bending collapse [10].

An explosion in a pyrolysis gas plant in Romania demolished a cooling tower. It fell on the administration block, killing 162 people. Many people who would not build offices close to an operating plant would consider it safe to build them close to a cooling tower. It is doubtful if this is wise.

10.7   FURNACES

10.7.1   Explosions While Lighting a Furnace

Many explosions have occurred while furnaces were being lit. The two incidents described here occurred some years ago on furnaces with simple manual ignition systems, but they illustrate the principles to be followed when lighting a furnace, whether this is carried out manually or automatically:

(a) A foreman tested the atmosphere inside a furnace (Figure 10-12) with a combustible gas detector. No gas was detected, so the slip-plate (blind) was removed, and two minutes later, a lighted poker

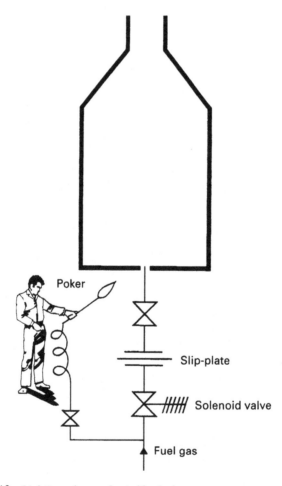

FIGURE 10-12   Lighting a furnace heated by fuel gas.

was inserted. An explosion occurred. The foreman and another man were hit by flying bricks, and the brickwork was badly damaged.

The inlet valve was leaking, and during the two minutes that elapsed after the slip-plate was removed, enough fuel gas for an explosion leaked into the furnace. (Suppose the leak was equivalent to a 1.6-mm [$\frac{1}{16}$-in.] diameter hole, and the gauge pressure of the fuel gas was 0.34 bar [5 psi]. The calculation shows that 80 L [3 ft^3] of gas entered the furnace in two minutes. If this burned in 0.01 second, the power output of the explosion was 100 MW.)

The correct way to light a furnace (hot or cold) that burns gas or burns light oil is to start with a positive isolation, such as a slip-plate, in the fuel line. Other positive isolations are disconnected hoses, lutes filled with water (if the fuel is gas at low pressure), and double block and bleed valves; closed valves without a bleed are not sufficient. Then do the following:

1. Test the atmosphere inside the furnace.
2. If you do not detect gas, light and then insert the poker (or switch on the electric igniter).
3. Remove the slip-plate (or connect the hose, drain the lute, or change over the double block and bleed valves). If the isolation valve is leaking, the poker or igniter will ignite the leaking fuel before it forms an explosive mixture. (The solenoid valve shown in Figure 10-12 should open automatically when the poker is inserted or the igniter is switched on. If it does not, it should be held open until the main burner is lit.)
4. Open the fuel-gas isolation valve.

The furnace had been lit in an incorrect way for many years before the isolation valve started to leak and an explosion occurred. Never say, "It must be safe because we have been doing it this way for 20 years and have never had an accident." Is an accident in the 21st year acceptable?

On furnaces with more than one burner, it may be possible to light a burner from another one if the two are close to each other. If they are not, the full procedure just described should be followed. Explosions have occurred on multiburner furnaces because operators assumed that one burner could always be lit from the next one.

(b) A reduction in fuel oil pressure caused the burner in an oil-fired furnace to go out, and the flame failure device closed the solenoid valve in the fuel oil line (Figure 10-13). The operator closed the two hand-isolation valves and opened the bleed between them. (The group of three valves is equivalent to the slip-plate shown in Figure 10-12.) When the fuel oil pressure was restored, the foreman tested the atmosphere in the furnace with a combustible gas detector. No gas was detected, so he inserted a lighted poker. The fuel oil supply was still positively isolated; nevertheless an explosion occurred, and the foreman was injured, fortunately not seriously.

When the burner went out, the solenoid valve took a few seconds to close, and during this time some oil entered the furnace. In addition, the line between the last valve and the furnace may have drained into the furnace. The flash point of the fuel oil was 65°C (150°F), too high for the oil to be detected by the combustible gas detector. Even though the hot furnace vaporized the oil, it would have condensed in the sample tube of the gas detector or on the sintered metal that surrounds the detector head.

Before relighting a hot furnace that burns fuel oil with a flash point above ambient temperature, sweep it out for a period of time, long enough to make sure that any unburned oil has evaporated. If this causes too much delay, then pilot burners supplied by an alternative supply should be kept alight at all times.

To keep the sweeping-out or purge time as short as possible, the solenoid valve should be close to the burner, and it should close quickly. In addition, the line between the solenoid valve and the burner should not drain into the furnace. As in the previous incident, the furnace had been lit incorrectly for many years before an explosion occurred.

To calculate the purge time, do the following:

1. Calculate the amount of oil between the solenoid valve and the burner.
2. Assume it all drains into the furnace and evaporates. Calculate the volume of the flammable mixture, assuming it is at the lower flammable limit, probably about 0.5% v/v. (If it forms a richer mixture, the volume will be less.)

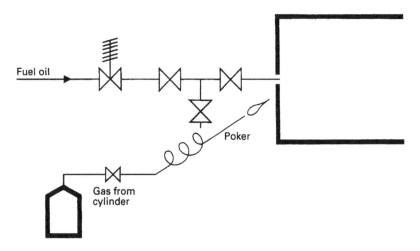

FIGURE 10-13   Lighting a furnace heated by heavy fuel oil.

3. Multiply by four to give a safety margin, and sweep out the furnace with this volume of air [11].

10.7.2 Furnace Tube Ruptures

(a) A heat transfer oil was heated in a furnace. A tube in the convection section ruptured along 8 in. (200 mm) of its length, and the ensuing fire damaged three other furnaces. No one could stop the flow of heat transfer oil into the fire, as the valves in the line and the pump switch were too near the furnace. The fire continued until all the oil was burned.

The tube failure was due to prolonged, though not necessarily continuous, overheating of the furnace tubes at times when maximum output was wanted. This led to a creep failure. There were not enough instruments on the furnace to measure the temperature of the tubes, always a difficult problem, and the operators did not understand the way furnace tubes behave. They are usually designed to last for 10 years, but if they get too hot, they will not last so long. For instance, if the tubes are designed to operate at 500°C (930°F), then the following occurs:

If they are kept at 506°C (940°F), they will last 6 years.

If they are kept at 550°C (1,000°F), they will last 3 months.

If they are kept at 635°C (1,175°F), they will last 20 hours.

If we let the tubes get too hot, however carefully we treat them afterward, they will never be the same again. If we heat them to 550°C, say for six weeks, we will have used up half their creep life, and they will fail after about five years at design temperature. If we find that our pumps, heat exchangers, and distillation columns will handle a greater throughput than design, we can use it. If we try the same with our furnaces, we may be in trouble in the future.

The following recommendations were made after the fire. They apply to all furnaces.

- Provide good viewing ports. (Although this failure occurred in the convection section of the furnace where the tubes are heated by hot flue gas, most failures occur in the radiant section as the result of flames impinging on the tubes.)
- Provide tube temperature measurements (but we can never be sure that we are measuring the temperature at the hottest point).
- Train operators in the principles of furnace operation.
- Look for signs of overheating during overhauls.
- Provide remotely operated emergency isolation valves (see Section 7.2.1).
- Provide remote stop buttons for circulation pumps well away from the furnace.

- Lay out new plants so that circulation pumps are well away from furnaces.
- Do not take a furnace above design output without advice from a materials engineer.
- Examine contractors' proposals critically.

(b) A heat transfer oil was heated in a furnace used only during start-ups to bring reactors up to operating temperature. Startup is always a busy time, and the operator lit the furnace and forgot to open the valves leading from the furnace to the reactors (an example of the sort of lapse of attention we all make from time to time, especially when we are under stress; see Chapter 3). Within 20 to 30 minutes, a furnace tube ruptured, and there was a large fire with flames 15 m tall.

The furnace was fitted with interlocks that should have isolated the fuel supply if the tube wall temperature or the pressure of the heat transfer oil got too high. Neither interlock worked, and neither had been tested or maintained. The set point of the high tube wall interlock had been raised far above its original set point, from 433°C (810°F) to 870°C (1,600°F), a simple way of putting it out of action [15]. Changing the set point of an interlock is a modification and should be allowed only when the equipment is capable of withstanding the new conditions (see Chapter 2).

A similar incident occurred on another furnace when the heat transfer oil froze inside the furnace during unusually cold weather. Outside the furnace, the lines were steam-traced. The operating team decided to thaw the frozen oil by lighting one of the burners in the furnace at a low rate. Later on, someone increased the flow of fuel. About an hour after the furnace was lit, a tube ruptured. There were no instructions on the action to be taken when the oil froze. Lighting a burner had been used before, successfully on that occasion [16].

(c) A feed water pump supplied two boilers. The backup pump also supplied another unit, which was under repair, so the operator on this unit blocked it in. He did not tell the boiler operator what he had done (as in the incident described in Section 17.4).

The online feed water pump tripped, but the operator ignored the alarm signal, presumably because he thought the backup pump would start up automatically.

The smaller of the two boilers became short of water first, and the low water level trip shut it down. The operator was so busy trying to get it back on line that he ignored the low water level and other alarms that were sounding on the other boiler. Unfortunately, the trips on this boiler did not work, as it had been rewired (incorrectly) since it was last checked. Fifteen to 20 minutes later, someone saw flames coming out of the boiler stack. The boiler was then shut down manually. By this time most of the tubes had melted.

After the furnace had been allowed to cool, the operating team, not realizing the extent of the damage, restarted the flow of feed water. They stopped it when they saw water running out of the firebox. It is fortunate they did not start the water flow earlier, or it would have caused explosive vaporization of the water [17]. As stated in Section 9.2.2e, equipment that has been taken outside its design or test range should not be used again until it has been examined.

(d) Another tube failure had an unusual cause. A pipe, sent to an outside workshop for bending, was returned plugged with sand and was welded into the exit line from a furnace. Not surprisingly, the furnace tubes overheated and failed during startup.

The pipe was returned to the plant with a warning that it might contain some sand. The plant staff took this to mean that a few grains might be stuck to the walls, not that the pipe might be full of sand.

Section 14.2.3 describes another failure. For more information on furnace hazards, see reference 19.

References

1. *User's Guide for the Safe Operation of Centrifuges with Particular Reference to Hazardous Atmosphere*, 2nd edition, Institution of Chemical Engineers, Rugby, UK, 1987.
2. T. A. Kletz, *Loss Prevention*, Vol. 6, 1972, p. 134.
3. R. E. Sanders and J. H. Wood, *Chemical Engineering*, Feb. 1993, p. 110.
4. A. B. Smith, *Safety Relief Valves on Pressure Systems*, Reprint No. C454/001/93, Institution of Mechanical Engineers, London, 1993.
5. C. H. Vervalin, *Hydrocarbon Processing*, Vol. 51, No. 12, Dec. 1972, p. 49.
6. T. A. Kletz, *Learning from Accidents*, 3rd edition, Butterworth-Heinemann, Oxford, UK, 2001, Chapter 9.
7. *FPA Journal*, No. 84, Oct. 1969, p. 375.
8. *Loss Prevention Bulletin*, No. 103, Feb. 1992, p. 29.
9. C. Butcher, *The Chemical Engineer*, No. 549, Sept. 16, 1993, p. 24.
10. *Report on the Collapse of the Ardeer Cooling Water Tower, 27 Sept. 1973*, Imperial Chemical Industries Ltd., London, 1974.
11. *Furnace Fires and Explosions*, Hazard Workshop Module No. 005, Institution of Chemical Engineers, Rugby, UK, undated.
12. R. Weaver, *Northern Arrow* (newsletter of the Festiniog Railway Society, Lancashire and Cheshire Branch), No. 147, Oct. 1994.
13. S. J. Brown and T. J. Brown, *Process Safety Progress*, Vol. 14, No. 4, Oct. 1995, p. 244.
14. I. M. Duguid, *Loss Prevention Bulletin*, No. 134, Apr. 1997, p. 10.
15. R. E. Sanders, *Process Safety Progress*, Vol. 15, No. 4, Winter 1996, p. 189.
16. R. E. Sanders, *Chemical Process Safety—Learning from Case Histories*, Butterworth-Heinemann, Boston, MA, 1999, p. 115.
17. S. Mannan, *Process Safety Progress*, Vol. 15, No. 4, Winter 1996, p. 258.
18. S. Grossel, *Chemical Engineering*, Vol. 115, No. 2, Feb. 2008, p. 36.
19. *Safe Furnace and Boiler Firing*, Institution of Chemical Engineers, Rugby, UK, 2005.

Entry to Vessels

A banker described three danger signals:
Ignoring what you know
Forgetting what you know
Believing that this time it will be different.

—[26]

Many people have been killed or overcome because they entered vessels or other confined spaces that had not been thoroughly cleaned or tested. In 1997 about 63 people were killed this way in the United States; about 40 of them were would-be rescuers (see Section 11.6) [25]. A number of incidents are described here. Others are described in Chapter 24 and Section 9.1.5c, and others involving nitrogen are described in Section 12.3. Sometimes it seems that vessels are more dangerous empty than full.

For further details of the procedures that should be followed when preparing vessels for entry, see references 1, 2, and 27.

11.1 VESSELS NOT FREED FROM HAZARDOUS MATERIAL

In these incidents, the vessels were correctly isolated but were not freed from hazardous materials:

(a) A vessel was divided into two halves by a baffle, which had to be removed. The vessel was cleaned out, inspected, and a permit issued for a worker to enter the left-hand side of the vessel to burn out the baffle. It was impossible to see into the right-hand half. But because the left-hand half was clean and because no combustible gas could be detected, it was assumed that the other half was also clean (Figure 11-1). While

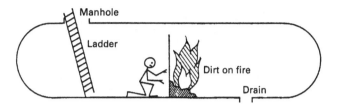

FIGURE 11-1   If part of a vessel cannot be seen, assume it is dirty.

the welder was in the vessel, some deposit in the right-hand half caught fire. The welder got out without serious injury but bruised himself in his haste.

If a part of the vessel cannot be inspected and be seen to be safe, then we should assume the vessel contains hazardous materials.

If the previous contents were flammable, then we should assume there is some flammable material out of sight.

If the previous contents were poisonous, then we should assume there is some poisonous material out of sight, and air masks should be worn for entry.

Gas tests alone are not conclusive. Sludge that gives off gas when heated or disturbed may be present.

(b) The following incidents illustrate that last remark. To clean a paint-mixing tank, it was washed out with xylene. This cleaned the sides, but some residue had to be scraped from the bottom. While an employee was cleaning the tank, wearing neither an air mask nor a life line, he was overcome by xylene, which was trapped in the residue and escaped when it was disturbed [3].

A man was overcome by fumes while removing residues from a tank with a high-pressure water jet. When the entry permit was issued, no one realized that fumes might be given off when the residue was disturbed [14].

(c) After a permit had been issued to weld inside a vessel, a foreman noticed a deposit on the walls. He scraped some off, tested it, and found that it burned. The permit was withdrawn.

(d) A tank had to be entered for inspection. It had contained only water and was not connected to any other equipment, so the usual tests were not carried out. Three men went into the tank and were overcome. Two recovered but one died. The atmosphere inside the tank was tested afterward and found to be deficient in oxygen. It is probable that rust formation used up some of the oxygen. Section 5.3d describes how a similar effect caused a tank to collapse.

Although rusting is normally a slow process, it can be rapid under some conditions. Two men collapsed in an evaporator, which had contained warm, moist magnesium chloride. One of them later died.

Afterward, tests showed that the oxygen content fell to 1% in 24 hours [9, 10]. Other tests showed that corrosion rates increased 10 times when the relative humidity increased from 38% to 52% [11].

Never take shortcuts in entering a vessel. Follow the rules. See also Section 11.6b.

(e) Flammable or toxic liquids have been trapped inside the bearings of stirrers and have then leaked out. In one case, a worker was overcome while working on a bearing although the vessel had been open for entry for 17 days. He disturbed some trapped liquid.

Before issuing an entry permit, look for any places in which liquid might be trapped. Vessels should always be slip-plated as close to the vessel as possible and on the vessel side of isolation valves. Otherwise liquid may be trapped between the valve and the slip-plate. See also Section 11.3h.

(f) On several occasions, young, inexperienced workers have been overcome while cleaning tanks. For example, a 16-year-old boy, on his first day at work, was asked to clean out an oil tank with paraffin. He was not supplied with an air mask or protective clothing or given any supervision. He collapsed on the way home from work [12].

(g) An incident similar to (a) occurred in a distillation column filled with a packing made from corrugated and perforated steel sheets. It was cleaned with hot water and steam and opened for inspection. The distributor above the lowest section of packing was found to need repair. A seized bolt was burned off, and part of it fell into the packing, setting it alight. The packing was destroyed, and a 3-m section of the shell had to be replaced.

The material distilled in the column was known to polymerize, but there was no increased pressure drop, and the top of the packing looked clean, so hot work was allowed. If you cannot see that something is clean, assume it is dirty [20].

11.2 HAZARDOUS MATERIALS INTRODUCED

Sometimes, after a vessel has been freed from hazardous materials, they are then deliberately reintroduced, as in the following incidents:

(a) Two men went into a reactor to carry out a dye-penetrant test on a new weld using trichloroethylene. Because the weld was 8 m long, the solvent was soon used up, and the man who was on duty at the entrance was asked to go for some more. He was away for 10 minutes. When he returned, the two men inside the reactor had collapsed. Fortunately, they were rescued and soon recovered.

The amount of solvent that can be taken into a vessel for dye-penetrant testing or other purposes should be limited so that evaporation of the complete amount will not bring the concentration above the safe concentration, for example, the threshold limit value, making allowance for the air flow if the vessel is force-ventilated.

Standby workers should not leave a vessel when others are inside it.

(b) The Occupational Safety and Health Administration (OSHA) reported a most incredible case [4]. It was decided to shrink-fit a bearing onto a shaft. The shaft was cooled—in a pit—by hosing liquefied petroleum gas onto it while the bearing was heated with an acetylene torch on the floor above the pit. An explosion occurred, killing one man and injuring two others.

(c) The same OSHA report also describes several fatal fires and explosions that occurred while the insides of vessels were being painted, sometimes by spraying. In many cases, the "cause" was said to be unsuitable lighting. But people should never be asked to work in a flammable atmosphere in view of the ease with which sources of ignition can turn up. The concentration of flammable vapor should never exceed 20% of the lower flammable limit while people are in a vessel, and the atmosphere should be monitored continuously. Other fires and explosions were the result of leaks from welding equipment, often ignited when welding started again. Gas tests should always be carried out before welding starts. If oxygen is being used, then the atmosphere should be tested for oxygen as well as for flammable vapors.

(d) A small brick-lined reactor, 5 m long by 1.5 m diameter, had to be repaired. The reactor was cleaned, removed from the plant, and taken outside; the bottom pipework and the whole of the reactor top were removed, leaving holes 0.6 m diameter at one end and the full diameter at the other. It was now no more than a length of wide pipeline with a restriction at one end. A man then went inside the reactor to fill in the cracks between the bricks with a rubber adhesive.

The reactor had been repaired in this way many times without incident. One day the weather was very cold, so the reactor was taken indoors. The man repairing the reactor was overcome by the fumes but fortunately soon recovered when he was taken outside [15].

(e) A contractor's employee was repairing the lining of a tank with acetone when it caught fire and he was badly burned. The source of ignition was an unprotected electric light, supplied by the owner of the tank. Both the owner and the contractor were fined; the owner, a much larger firm, was fined 10 times as much as the contractor. The judge seemed to consider that the provision of an unprotected light was the major offense, but allowing someone to work in a flammable atmosphere was more serious. This should never be permitted, as a source of ignition is always liable to turn up even though we do what we can to remove known sources [16] (see Section 5.4).

(f) An electrician was working in a pit, using a torch fed by a cylinder of liquefied petroleum gas (LPG), which was standing on the edge of the pit. The hose was rather short, and as a result the electrician pulled the cylinder into the pit. The hose connection next to the cylinder valve broke, and the LPG ignited. The electrician was badly burned [17]. Many accidents have simple causes.

11.3   VESSELS NOT ISOLATED FROM SOURCES OF DANGER

Before entry is allowed into a vessel or other confined space, the vessel should be isolated from sources of hazardous material by slip-plating (blinding) or physically disconnecting all pipelines and by isolating all supplies of electricity, preferably by disconnecting the cables. On the whole, these precautions seem to be followed. Accidents as the result of a failure to isolate are less common than those resulting from a failure to remove hazardous materials or from their deliberate reintroduction as described in Sections 11.1 and 11.2. However, the following are typical of the accidents that have occurred.

(a) A reactor had been isolated for overhaul. When maintenance was complete, the slip-plates were removed and the vessel prepared for startup. It was then realized that an additional job had to be done, so employees were allowed to enter the vessel without the slip-plates being put back and without any gas tests being carried out. An explosion occurred, killing two and injuring two others. It was later found that hydrogen had entered the vessel through a leaking tube [5].

(b) The same report describes a number of incidents in which steam lines failed, as the result of corrosion, while employees were working in a pit or other confined space from which they could not escape quickly. In general, steam lines, heating coils, and the like should be depressured and isolated before entry is permitted to a confined space [6].

(c) On a number of occasions, people have been injured because machinery was started up while they were inside a vessel. For example, two men were fixing new blades to the No. 2 unit in a pipe-coating plant. A third man wanted to start up another unit. By mistake, he pressed the wrong button. No. 2 unit moved, and one of the workers was killed [7]. The power supply should have been disconnected (see item [g]).

(d) Contractors were installing a heating coil in a small tank, 2.4 m tall by 1.8 m diameter, which was entered through an opening in the top. A nitrogen line entered the tank, and the nitrogen valve was near the opening. When the job was nearly complete, one of the workmen entered the tank alone. It is believed that while doing so he accidentally knocked the lever-operated valve open, as he was found dead inside the tank, with the nitrogen valve open. The nitrogen supply should have been slip-plated or disconnected. The report said that there were no facilities for isolating the supply, but the valve was not even locked shut [13].

(e) This incident occurred in 1910, but its lessons are still relevant. Two men entered a revolving filter to examine the inside. The inlet line was disconnected, the blow-back gas line was slip-plated, and the 5-in. outlet valve was wide open. Nevertheless the two men were affected by gas but were fortunately able to get out through the manhole.

The liquid, after passing through the outlet valve, joined the line from another filter that was in use. It is believed that carbon dioxide gas from the filter liquid passed up the outlet line into the filter that was being inspected (Figure 11-2). A test showed that "contamination was not sufficient to prevent a candle burning."

The manhole was smaller than those used today, and if the men had been overcome, rescue would have been difficult. Before allowing people to enter a vessel or other confined space, always ask how they will be rescued if they are overcome.

(f) Forty-five years later, in the same company, the accident was repeated, and this time a man was killed. While a man was working inside a boiler, the process foreman noticed that the water level in another boiler was too high. He asked an operator to lower the level through the blowdown (drain) valve, which discharged into a common drain line. Steam entered the boiler under repair from this common line. None of the lines had been slip-plated or disconnected, and the blowdown valve had been left open (Figure 11-2).

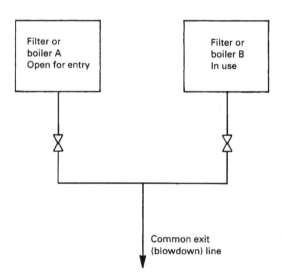

FIGURE 11-2 Liquid or gas traveled backward from the common line into the vessel that was open for entry.

To quote from the accident report,

> On previous occasions men have entered the boiler without complete isolation. It seems that this first occurred in an emergency when it was thought essential to get a boiler back on line with the minimum of delay, although it is admitted that it does not take long (about 1½ hours) to isolate this type of boiler. Since everything went satisfactorily, the same procedure has apparently been adopted on other occasions, even when, as in this case, there was no particular hurry. Gradually, therefore, the importance of the correct procedure seems to have been forgotten, and on more than one occasion complete reliance seems to have been placed on the presence and reliability of a boiler fitter and his pair of keys. On this occasion, unfortunately, his memory let him down.

The blowdown valves on the boilers were operated by a special key, which had a lug on it so that it could not be removed when the valve was open. It was therefore impossible, in theory, for two blowdown valves to be open at the same time. However, the boiler fitter "kept and jealously guarded" a private key without a lug and had used this one to open the blowdown valve on the boiler that was under repair. He forgot to tell the process foreman what he had done or to close the valve. "The presence of this key would appear to have been of little moment as long as the correct procedure of complete isolation was maintained, but as soon as it was departed from, the additional key became a menace, which eventually enabled the present tragedy to occur," the accident report said.

Any system based on the use of a single key is error prone, as it is so easy to acquire a spare.

(g) Two men were asked to clean out a mixer, which was fitted with large internal mixing blades. Before entering the vessel, they pressed the start button, as usual, to confirm that the power was isolated. The blades did not move, but the power supply was not isolated; it had failed. The switch was closed, and the fuses had not been removed. When power was restored, the mixer started to turn and both men were killed [18].

Checking the starter before working on electric equipment is a useful final check that should always be made, but it does not prove that the equipment has been isolated; there may be an interruption in the power supply. Make sure the fuses have been removed.

(h) A vessel stirrer was fitted with a double mechanical seal supplied with a barrier liquid from a small tank, which was blanketed with nitrogen. The barrier liquid leaked away, and the nitrogen entered the vessel and reduced the oxygen level to 15%. A man who was working in the vessel felt unwell and fortunately left at once.

The atmosphere in the vessel was tested before the man started work, and then a portable blower was installed to keep the atmosphere clean. However, it disturbed the dust produced by grinding, so the man switched it off [19].

11.4   UNAUTHORIZED ENTRY

(a) Contractors unfamiliar with a company's rules have often entered vessels without authority. For example, a contractor's foreman was found inside a tank, which was disconnected and open, ready for entry, but not yet tested. He had been asked to estimate the cost of cleaning the tank. The foreman said that he did not realize that a permit was needed just for inspection. He had been given a copy of the plant rules but had not read them.

If vessels are open but entry is not yet authorized, the manhole should be covered by a barrier. Do not rely on contractors reading rules. Explain the rules to them.

(b) It is not only contractors who enter vessels without authorization, as Section 11.1d and the following incident show: A process foreman had a last look in a vessel before it was boxed up. He saw an old gasket lying on the floor. He decided to go in and remove it. Everyone else was at lunch, so he decided to go in alone. On the way out, while climbing a ladder, he slipped and fell and was knocked out. His tongue blocked his throat, and he suffocated.

(c) The incident described in Section 12.3.2d shows that you do not have to get inside a confined space to be overcome. Your head is enough. People should never put their heads inside a vessel unless entry has been authorized.

(d) Two teenagers were employed during school holidays as casual laborers at a flour mill, sweeping out wheat left in rail trucks after off-loading. They jumped or fell into a truck before it was empty, were sucked to the base of the truck, became covered in wheat, and were asphyxiated [21].

(e) The most horrendous entry accident I have seen reported occurred in a pulp and paper mill. Two welders were asked to repair the tines in a repulper. This is an open-topped vessel in which large bundles of wood pulp or recycled paper are macerated by a high-energy (40 to 400 kW) impeller, fitted with tines or blades, to form a 1% to 5% suspension in water. At 10 a.m., the operators left for a tea break after having drained and cleaned the repulpers. A few minutes later, the welders arrived. As No. 3 repulper, the one they had to repair, was clean and empty, they thought it had been cleaned for them and started work.

The operators returned at 10:15 a.m. They saw the welding machine, but the lighting was dim, and they did not see the leads entering the vessel. They filled the vessel with water, turned on the impeller, and fed dry pulp into the vessel.

At 2:30 p.m., the welders were reported missing. Their job card led to No. 3 repulper. It was stopped and drained. Inside were the remains of a rope ladder, welding equipment, and human bones [22].

The welders should have known that they should never enter a vessel until they or their foreman has checked that the vessel is isolated, by slip-plates or disconnection, and has accepted (and signed) an entry permit. Power supplies should have been disconnected, or if merely locked off, each person entering should have added his or her personal lock and kept the key. It is hard to believe that this was the first time normal good practice had not been followed. The management system was at fault.

For another incident, see Section 12.3.2f.

11.5   ENTRY INTO VESSELS WITH IRRESPIRABLE ATMOSPHERES

A man was standing on a ladder, ready to go down into a drain manhole to plug one of the inlet lines. The drain contained some hydrogen sulfide, so he had an air mask ready. But he had not yet put it on because he was well outside the manhole. His feet were at ground level (Figure 11-3). He was about to put on a safety harness when his two companions heard a shout and saw him sliding into the manhole. They were unable to catch him, and his body was recovered from the outfall. He had been

FIGURE 11-3   A man was overcome by fumes from the manhole.

overcome by hydrogen sulfide arising from the drains, although his face was 1.5 m above ground level.

This incident shows that if a vessel or confined space contains a toxic gas, people can be overcome a meter or more from the opening. Similar incidents involving nitrogen are described in Section 12.3.2b.

The incident and the one described in Section 12.3.2c show that special precautions are necessary when entering vessels containing atmospheres that contain so much toxic gas or so little oxygen that there is immediate danger to life. In most cases of entry in which an air mask is worn, it is used because the atmosphere is unpleasant to breathe or will cause harm if breathed for several hours. Only on very rare occasions should it be necessary to enter vessels containing an irrespirable atmosphere. In such cases, two persons trained in rescue and resuscitation should be on duty outside the vessel. And they should have available any equipment necessary for rescuing the person inside the vessel, who should always be kept in view.

11.6   RESCUE

If we see another person overcome inside a vessel, there is a strong natural impulse to rush in and rescue the individual, even though no air mask is available. Misguided bravery of this sort can mean that other people have to rescue two people instead of one, as Section 12.3.3 and the following incidents show:

(a) A contractor entered the combustion chamber of an inert-gas plant while watched by two standby men but without waiting for the air masks to arrive. While he was climbing out of the chamber, he lost consciousness halfway up. His body was caught between the ladder and the chamber wall. The standby men could not pull him out with the lifeline to which he was attached, so one of them climbed in to try to free him, without an air mask or a lifeline. The standby man also lost consciousness. The contractor was finally pulled free and recovered. The standby man was rescued by the fire service, but by this time he was dead.

(b) Three men were required to inspect the ballast tanks on a barge tied up at an isolated wharf 20 km from the plant. No tests were carried out. One tank was inspected without incident. But on entering the second tank, the first man collapsed at the foot of the ladder. The second man entered to rescue him and also collapsed. The third man called for assistance. Helpers who were asked to assist in recovering the two men were partly overcome themselves. Representatives of the safety department 20 km away set out with air masks. One man died before he could be rescued.

Tests on other tanks showed oxygen contents as low as 5%. It is believed that rust formation had used up the oxygen, as in the incident described in Section 11.1d [8].

(c) Waste zinc cyanide solution dripped into an open-topped tank from equipment that had been electroplated. The tank was about 5 ft by 4 ft by 5 ft deep (1.5 m by 1.2 m by 1.5 m deep) and was located in a building. The liquid had been pumped out, and the next job was to remove sludge from the bottom. An operator sprayed hydrochloric acid into the tank, thus producing hydrogen cyanide, and then went down a ladder into the tank. He was killed, and so were five others who tried to help him. Thirty people were injured, including police and firefighters called to the scene. No one had any awareness of the hazards. There were no plans for handling emergencies, no one had clearly defined responsibilities, onlookers and relatives milled around, and everyone who entered the building needed medical treatment [23]. It is not clear from the report who authorized the addition of hydrochloric acid and if it had been used before to clean the tank.

(d) A man had to be lowered into a confined space. Great care was taken when choosing the harness, rope, and hoist. The test records were checked to make sure that the equipment chosen had been inspected as required by the company's procedures. But no one noticed that the man put the harness on from back to front. As a result, when he was lifted he could not control his position by holding the rope in front of himself. Instead he hung helpless on the rope, though he came to no harm.

(e) On one occasion, a man was locked inside a confined space after his co-workers failed to check that the space was clear before locking the door [25].

11.7   ANALYSIS OF VESSEL ATMOSPHERE

Portable oxygen analyzers that sound an alarm if the oxygen concentration falls are available. All who enter confined spaces should carry them, as they could have prevented many of the accidents described in this chapter.

The following incidents show how errors in analysis nearly resulted in people entering an unsafe atmosphere. In both cases, the laboratory staff members were asked to test the atmosphere inside a vessel. In the first case, they checked the oxygen content with a portable analyzer, in which a sample of the atmosphere being tested is drawn through the apparatus with an aspirator. There was a blockage in the apparatus, so it merely registered the oxygen content of the air already inside it. A bubbler or other means of indicating flow should have been fitted.

In the second case, the sample was taken near the manhole instead of being taken near the middle of the vessel. Samples should always be taken well inside the vessel. Long sample tubes should be available so that this can be done. In large vessels and in long, tortuous places like flue gas ducts, samples should be taken at several points.

On both these occasions, vigilance by the operating staff members prevented what might have been a serious incident. They suspected something was wrong with the analysis results and investigated the way the samples were taken.

Inevitably, a book like this one is a record of failures. It is pleasant to be able to describe accidents prevented by the alertness of operating staff.

11.8   WHAT IS A CONFINED SPACE?

If we are building a tank or digging a hole in the ground, when does it become a confined space? Guidance is needed. I suggest that when the depth becomes greater than the diameter (and the space is large enough for someone to enter), the space is considered confined. At this time we should consider what precautions, if any, are necessary. In oil and chemical factories, chemicals may drain out of the ground and give off fumes or catch fire; acid has been known to react with limestone in the ground and produce carbon dioxide. On construction sites, the atmosphere in tanks under construction has been contaminated with oxygen or welding gas, or well-meaning fellow workers have connected up nitrogen supplies (see Section 12.3.3).

Section 11.2d describes an incident in which no tests were carried out and the vessel was not recognized as a confined space because ventilation was good. When the vessel was moved indoors, ventilation was no longer good, and a man was overcome.

11.9   EVERY POSSIBLE ERROR

Earlier sections have described how people were killed because vessels were not freed from hazardous materials, atmospheres were not tested and were not respirable, no thought was given to methods of rescue, the correct equipment was not used, or rescue was bungled. This section describes an incident in which all these things were wrong.

A man was asked to clean a 45-m^3 tank that had contained toluene. It was 6m (20 ft) tall and 3m (10 ft) in diameter. It had not been gas-freed, and the atmosphere had apparently not been tested. He entered the tank through the 16-in.-diameter top opening, using a rope for descent. A self-contained breathing apparatus was available on the job, but he did not wear it. He was overcome by the vapor and lack of air and collapsed on the floor of the tank.

The fire department was called. In an attempt to rescue him, the firefighters started to cut an opening in the side of the tank. The tank exploded, killing one firefighter and injuring 15 others. The man who had entered the tank also died, probably from asphyxiation [24].

More reports on accidents involving entry to confined spaces are presented in Chapter 24.

References

1. T. A. Kletz, in D. A. Crowl and S. S. Grossel (editors), *Handbook of Toxic Materials Handling and Management*, Marcel-Dekker, New York, 1994, Chapter 12.
2. M. S. Mannan (editor), *Lees' Prevention in the Process Industries*, 3rd edition, Elsevier, Boston, MA, 2005, Chapter 21.
3. *Health and Safety at Work*, July 1982, p. 38.
4. W. W. Cloe, *Selected Occupational Fatalities Related to Fire and/or Explosion in Confined Work Spaces as Found in Reports of OSHA Fatality/Catastrophe Investigations*, Report No. OSHA/RP-82/002, U.S. Dept. of Labor, Washington, DC, Apr. 1982, p. 32.
5. Reference 4, p. 22.
6. Reference 4, pp. 65–71.
7. *Occupational Safety and Health, Supplement*, Oct. 1975.
8. *Petroleum Review*, May 1977, p. 49.
9. *Chemical Safety Summary*, Vol. 55, No. 219, Chemical Industries Association, London, 1984, p. 66.
10. R. G. Melchior, *American Industrial Hygiene Association Journal*, Vol. 48, No. 7, 1987, p. 608.
11. *The Sentinel*, Vol. XLIV, No. 2, Industrial Risk Insurers, Hartford, Conn., 2nd quarter 1987, p. 8.
12. *Health and Safety at Work*, Vol. 6, No. 8, Apr. 1984, p. 13.
13. *Chemical Safety Summary*, Vol. 55, No. 219, Chemical Industries Association, London, 1984, p. 63.
14. R. A. Carson and C. J. Mumford, *Loss Prevention Bulletin*, No. 091, Feb. 1990, p. 7.
15. *Loss Prevention Bulletin*, No. 110, Apr. 1993, p. 8.
16. *Health and Safety at Work*, Vol. 15, No. 5, May 1993, p. 9.
17. *Safety Management* (South Africa), Vol. 16, No. 9, Sept. 1990, p. 79.
18. *Safety Management* (South Africa), Vol. 16, No. 4, Apr. 1990, p. 24.
19. *Loss Prevention Bulletin*, No. 112, Aug. 1993, p. 20.
20. *Loss Prevention Bulletin*, No. 122, Apr. 1995, p. 7.
21. *Occupational Safety and Environmental News* (South Africa), Vol. 2, No. 1, 1996, p. 1.
22. *Safety Management* (South Africa), Apr. 1997, p. 36.
23. *Occupational Safety and Health Observer*, Vol. 3, No. 10, U.S. Dept. of Energy, Washington, DC, Oct. 1994, p. 2.
24. An OSHA report quoted in *Operating Experience Weekly Summary*, No. 97-40, Office of Nuclear and Safety Facility, U.S. Dept. of Energy, Washington, DC, 1997, p. 7.
25. *Operating Experience Weekly Summary*, No. 97-46, Office of Nuclear and Safety Facility, U.S. Dept. of Energy, Washington, DC, 1997, p. 5.
26. Quoted by R. Ford in *Modern Railways*, Feb. 2001, p. 25.
27. *Confined Space Entry*, Institution of Chemical Engineers, Rugby, UK, 2005.

Hazards of Common Materials

I deplore the phrase 'Near Miss', because it has such happy-go-lucky con-
notations. A near miss is an accident that, solely by chance, did not happen,
Near Hit is better.... For every 400 near-hits, there is a fatal or serious
injury. Railways must find a way of capturing this information, and turning
it into part of the learning process. If you cover up a near hit (which is so
easy) the elephant trap stays in place waiting for next time.

—Neil Howard [30]

This chapter is not concerned with the hazards of obviously dangerous materials, such as highly flammable liquids and gases, or toxic materials. Rather, the focus is on accidents involving those common but dangerous substances: air, water, nitrogen, and heavy oils.

12.1 COMPRESSED AIR

Many operators find it hard to grasp the power of compressed air. Section 2.2a describes how the end was blown off a pressure vessel, killing two men, because the vent was choked. Compressed air was being blown into the vessel to prove that the inlet line was clear. It was estimated that the gauge pressure reached 20 psi (1.3 bar) when the burst occurred. The operators found it hard to believe that a pressure of "only twenty pounds" could do so much damage. Explosion experts had to be brought in to convince them that a chemical explosion had not occurred.

Unfortunately, operators often confuse a force (such as 20 lbs) with a pressure (such as 20 psi) and forget to multiply the 20 lbs by the number of square inches in the end of the vessel.

Section 13.5 describes a similar accident, whereas Section 5.2.2 describes other incidents in which equipment was damaged by compressed air. Because employees do not always appreciate the power of compressed air, it has sometimes been used to remove dust from workbenches or clothing. Consequently, dust and metal splinters have been blown into people's eyes or into cuts in the skin. Worse still, compressed air has been used for horseplay. A man was killed when a compressed air hose was pushed up his rectum [1].

Fires have often occurred when air is compressed. Above 140°C (284°F), lubricating oil oxidizes and forms a carbonaceous deposit on the walls of air compressor delivery lines. If the deposit is thin, it is kept cool by conduction through the pipework. But when deposits get too thick, they can catch fire. Sometimes the delivery pipe has got so hot that it has burst or the aftercooler has been damaged. In one case the fire vaporized some of the water in the aftercooler and set up a shock wave, which caused serious damage to the cooling-water lines.

To prevent fires or explosions in air compressors, we should do the following:

1. Keep the delivery temperature below 140°C (284°F). It is easier to do this if the inlet filters are kept clean and the suction line is not throttled. On some rotary air compressors, a large oil surface is exposed to the air, deposits readily form and ignite, and the temperatures should be kept lower.
2. Install a high-temperature alarm or trip on the delivery line.
3. Avoid long periods of operation at low rate, as this can increase oil deposition.
4. Avoid traps in the delivery pipework in which oil can collect.
5. Clean the pipework regularly so that deposits do not get more than ⅛ in. (3 mm) thick. One fire occurred in a compressor on which it was impossible to clean the pulsation dampers.
6. Use special lubricants that reduce the formation of deposits.
7. Use nonlubricated compressors. However, oil is still needed for bearings and gear boxes and may leak into the compressors unless special attention is paid to their design and maintenance [19].

After passing through the aftercooler, the compressed air is usually too cool for deposits to form or catch fire but not always. On one plant an instrument air drier became contaminated with oil and caught fire during the drying cycle.

One company experienced 25 fires or explosions in air-compressor discharge pipework within 35 years. In one of the worst, the fire heated the air going forward into an air receiver, which was lined with bitumen to prevent corrosion. On heating, the bitumen gave off flammable vapors, which exploded, toppling the receiver and demolishing part of a building.

Thin films of oil in pipework can explode without a previous fire if subjected to sudden shock, for example, by rapid opening of a valve [20].

Unexpected concentration of oxygen can occur when compressed air is dried or purified by passing it over certain types of molecular sieves. Nitrogen is absorbed preferentially after regeneration, and the air first produced may be rich in oxygen. This can widen flammability limits and lower auto-ignition temperatures. At least one explosion has occurred as a result. If possible, use Type 3A molecular sieves [21].

Another hazard of compressed air is that it contains dust (organic and inorganic), water, and traces of hydrocarbons, which if they are not removed can cause excessive wear of tools or contamination of products. Morris writes,

> Those who use air for pneumatic tools or even paint spray seem to have an inbuilt resistance to any idea that the quality of their compressed air is of any serious consequence. The fact that it transmits concentrated quantities of abrasive particles and water into the finely machined orifices and cylinders of their tools seems to pass them by. [12]

At one time it was believed that hydrocarbon vapor and air in the form of a foam could not explode, and it was even suggested that tanks containing flammable vapor could be made safe for welding or other hot work by filling them with firefighting foam. It is now known that this is incorrect and that such foams can explode. In fact, a method proposed for exploding antipersonnel mines laid during the Falkland Islands War is to cover the ground with foam, with a hydrocarbon-air mixture in the bubbles, and then ignite it [13]. (Tanks can, of course, be made safe for welding by filling them with foam made from nitrogen instead of air. This method is often used if the tank contains openings through which nitrogen gas would rapidly disperse.)

Other hazards of compressed air are described in reference 2.

12.2 WATER

The hazards of water hammer are described in Section 9.1.5 and the hazards of ice formation in Section 9.1.1. This section describes some accidents that have occurred as the result of the sudden vaporization of water, incidents known as foamovers, boilovers, slopovers, frothovers, or puking. Boilover is used if the tank is on fire and hot residues from the burning travel down to the water layer. Slopover is often used if water from fire hoses vaporizes as it enters a burning tank. Sections 9.1.1 and 12.4.5 describe incidents in which vessels burst because water that had collected in a trap was suddenly vaporized. But most foamovers have occurred when a water layer in a tank was suddenly vaporized, as in the following incidents:

(a) Hot oil, the residue from a batch distillation, was being moved into a heavy residue storage tank. There was a layer of water in the tank—the result of steaming the oil transfer line after previous movements—and this vaporized with explosive violence. The roof of the tank was lifted, and structures taller than 20 m (65 ft) were covered with black oil. A man who saw the incident said the tank exploded, though the sudden release of energy had a physical rather than a chemical cause.

To prevent similar incidents from happening, if heavy oil is being transferred into a tank, incoming oil should be kept *below* 100°C (212°F), and a high-temperature alarm should be installed on the oil line. Alternatively, water should be drained from the tank, the tank kept *above* 100°C, and the tank contents circulated before the movement of oil into the tank starts. In addition, the movement of oil into the tank should start at a low rate.

(b) In other cases, a water layer has vaporized suddenly when it was heated by conduction from a hotter oil layer above. For example, to clean a tank that had contained heavy oil, some lighter oil was put into it and heated by the steam coil. There was a layer of water below the oil. The operators were told to keep the temperature of the oil below 100°C, but they did not realize that the height of the thermocouple (1.5 m [5 ft]) was above that of the top of the oil (1.2 m [4 ft]). Although the thermocouple was reading 77°C (170°F), the oil was above 100°C, the water vaporized, and the roof was blown off the tank. As the water started to boil and lift up the oil, the hydrostatic pressure on the water was reduced, and this caused the water to boil with greater vigor.

(c) Some paraffin that had been used for cleaning was left in a bucket. There was some water under the paraffin. Some hot equipment set fire to some cleaning rags, and the fire spread to the paraffin in the bucket.

To put out the fire, a man threw a shovelful of wet scale into the bucket. The water became mixed with the oil, turned to steam, and blew the oil over the man, who was standing 1 to 2 m (3 to 5 ft) away. He died from his burns.

1. Never mix water and hot oil.
2. Do not use flammable solvents for cleaning.
3. Do not carry flammable liquids in buckets. Use a closed can (see Section 7.1.3).

Water can be trapped behind heat exchanger baffles and then suddenly be vaporized by circulation of hot oil. It can also be trapped in dead-ends and U-bends in pipework (see Section 9.1.1). Such U-bends can form when one end of a horizontal pipe is raised by thermal expansion. The trays in a distillation column were damaged during startup when hot gas met water, from previous steaming, dripping down the column [3]. Section 17.12 describes an incident somewhat similar to a foamover.

Accidents have occurred because hot water was not treated with respect. Five men were killed when a plastic hot-water tank split along a seam [14]. On another plant, a man, about to make some tea, caught his sleeve on the tap of an electric water heater. The heater fell over, 2 gal of hot water fell on him, and he died in the hospital five days later [15]. The heater should have been fixed to the wall. If it had contained a hazardous chemical, it would have been secured, but no one thought hot water was hazardous. Chemicals are not the only hazards on a plant.

Other hazards of water are described in reference 3.

12.3 NITROGEN [4, 29]

Nitrogen is widely used to prevent the formation of flammable mixtures of gas or vapor and air. Flammable gases or vapors are removed with nitrogen before air is admitted to a plant, and air is removed with nitrogen before flammable gases or vapors are admitted.

There is no doubt that without nitrogen (or other inert gas), many more people would be killed by fire or explosion. Nevertheless, we have paid a heavy price for the benefits of nitrogen. Many people have been asphyxiated by it. In one group of companies in the period 1960–1978, 13 employees were killed by fire or explosion, 13 by toxic or corrosive chemicals, and 7 by nitrogen. It is our most dangerous gas.

This section describes some accidents in which people were killed or overcome by nitrogen. Some of the accidents occurred because nitrogen was used instead of air. In others, people were unaware of the dangers of nitrogen or were not aware that it was present.

The name *inert gas*, often used to describe nitrogen, is misleading. It suggests a harmless gas. Nitrogen is not harmless. If people enter an atmosphere of nitrogen, they can lose consciousness, without any warning symptoms or distress, in as little as 20 seconds. Death can follow in three or four minutes. A person falls, as if struck down by a blow on the head. In German, nitrogen is known as stickstoff ("suffocating gas"). Perhaps we would have fewer incidents if we called it choking gas instead of inert gas.

12.3.1 Nitrogen Confused with Air

Many accidents have occurred because nitrogen was used instead of compressed air. For example, on one occasion a control room operator noticed a peculiar smell. On investigation it was found that a hose, connected to a nitrogen line, had been attached to the ventilation intake. This had been done to improve the ventilation of the control room, which was rather hot. On other occasions, nitrogen has been used by mistake to freshen the atmosphere in vessels in which employees were working. And in another incident, nitrogen was used by mistake to power an air-driven light, used during entry to a vessel. In this case, the error was discovered in time. More serious are incidents in which nitrogen has been connected to air masks.

To prevent these errors, we should always use different fittings for compressed air and nitrogen. Nevertheless, confusion can still occur, as the following story shows:

An operator donned a fresh-air hood to avoid breathing harmful fumes. Almost at once he felt ill and fell down. Instinctively he pulled off the hood and quickly recovered. It was then found that the hood had been connected by mistake to a supply of nitrogen instead of compressed air.

Different connectors were used for nitrogen and compressed air, so it was difficult at first to see how a mistake had been made. However, the place where the man was working was a long way from the nearest compressed air connection, so several lengths of hose had to be joined together. This was done by cutting off the special couplings and using simple nipples and clamps. Finally, the hoses were joined to one projecting through an opening in the wall of a warehouse. The operator then went into the warehouse, selected what he thought was the other end of the projecting hose, and connected it to the air line. Unfortunately, there were several hoses on the floor of the warehouse, and the one to which he had joined the air line outside was already connected to a nitrogen line.

To prevent incidents similar to those described, we should do the following:

1. Use cylinder air for breathing apparatus.
2. Label all service points.
3. Use different connectors for air and nitrogen, and publicize the difference so that everyone knows.

Another incident occurred on a plant where the pressure in the instrument air system was maintained with nitrogen when the instrument air compressor failed. Two operators who were required to wear air masks attached them to the instrument air system. Unknown to them, the compressor had broken down, and the system was full of nitrogen. They both died [16].

A third incident occurred at a U.S. government facility. An employee connected his air mask onto a nitrogen line and immediately blacked out, fell, and hit his head. Fortunately, a standby man came to his assistance, and he recovered without serious injury. The compressed air and nitrogen lines used the same couplings, and the nitrogen lines, which should have been a distinctive color, had not been painted [22].

When possible, air from cylinders or a dedicated system should be used instead of general-purpose compressed air. If the latter has to be used, it should be tested at the point of use immediately before use, every time.

12.3.2 Ignorance of the Dangers

(a) A member of a cleaning crew decided to recover a rope, which was half inside a vessel and was caught up on something inside. While kneeling down, trying to disentangle the rope, he was overcome by nitrogen. Afterward he admitted that if necessary, he would have entered the vessel.

(b) On several occasions, people who were working on or near leaky joints on nitrogen lines have been affected. Although they knew nitrogen is harmful, they did not consider that the amount coming out of a leaky joint would harm them (Figure 12-1).

Two maintenance workers had just removed the cover from a manhole near the top of a distillation column, which had been swept out with nitrogen, when one of them collapsed. The other pulled him free, and he soon recovered. The bottom manhole had already been removed, and it seems that a chimney effect caused nitrogen to come out of the upper manhole [23].

TAKE CARE –
Don't go too near nitrogen leaks!

FIGURE 12-1

(c) Two men without masks were killed because they entered a vessel containing nitrogen. Possibly they had removed their masks on other occasions, when the atmosphere was not harmful to breathe, for a moment or two and did not appreciate that in a 100% nitrogen atmosphere they would be overcome in seconds. It is believed that one man entered the vessel, removed his mask, and was overcome and that the second man then entered, without a mask, to rescue him.

Entry should not normally be allowed to vessels containing irrespirable atmospheres. Special precautions are necessary if entry is permitted (see Section 11.5).

(d) You do not have to get right inside a confined space to be overcome. Your head is enough. When a plant was being leak-tested with nitrogen after a shutdown, a leak was found on a manhole joint on the side of a vessel. The pressure was blown off, and a fitter was asked to remake the joint. While he was doing so, the joint ring fell into the vessel. Without thinking, the fitter squeezed the upper part of his body through the manhole so that he could reach down and pick up the joint. His companion saw his movements cease, realized he was unconscious, and pulled him out into the open air, where he soon recovered.

(e) In another incident, the cover of a large converter was removed, but nitrogen was kept flowing through it to protect the catalyst. An inspector did not ask for an entry permit, as he intended only to "peep in." Fortunately, someone noticed that he had not moved for a while, and he was rescued in time.

(f) A contract welder was asked to repair some cracks near the manhole on top of a vessel that had been swept out with nitrogen. To gain access, he removed the plastic sheet that covered the open manhole and placed a ladder inside the vessel, protruding through the manhole. He then stood on the ladder, in a position similar to that shown in Figure 11-3. He dropped the tip of his torch into the vessel, went part way down the ladder to see if he could see it, and collapsed. By the time he was found, he was dead [24].

As stated in Section 11.4, if a manhole has been removed but entry has not been authorized, the manhole should be covered by a fixed barrier, not just a plastic sheet. A ladder inside a manhole that is protected by only a loose cover is almost an invitation to enter.

12.3.3 Nitrogen Not Known to Be Present

Some of the incidents described in Section 12.3.2 may fall into this category. Most of the incidents of this type, however, have occurred during construction when one group of workers has, unknown to others, connected up the nitrogen supply to a vessel. The following is an account of a particularly tragic accident of this type.

Instrument personnel were working inside a series of new tanks, installing and adjusting the instruments. About eight weeks earlier, a nitrogen manifold to the tanks had been installed and pressure-tested; the pressure was then blown off and the nitrogen isolated by a valve at the plant boundary. The day before the accident, the nitrogen line was put back up to pressure because the nitrogen was required on some of the other tanks.

On the day of the accident, an instrument mechanic entered a 2-m^3 tank to adjust the instruments. There was no written entry permit because the people concerned believed, mistakenly, that entry permits were not required in a new plant until water or process fluids had been introduced. Although the tank was only 2 m (6 ft) tall and had an open manhole at the top, the mechanic collapsed. An engineer arrived at the vessel about five minutes later to see how the job was getting on. He saw the mechanic lying on the bottom, climbed in to rescue him, and was overcome as soon as he bent down.

Another engineer arrived after another five or ten minutes. He fetched the process supervisor and then entered the vessel. He also collapsed. The supervisor called the plant fire service. Before they arrived, the third man recovered sufficiently to be able to climb out of the vessel. The second man was rescued and recovered, but the first man died. It is believed that an hour or two before the incident, somebody opened the nitrogen valve leading to the vessel and then closed it.

What can we learn from this incident?

1. If someone is overcome inside a vessel or pit, we should never attempt to rescue the individual without an air mask. We must curb our natural human tendency to rush to the person's aid, or there will be two people to rescue instead of one (see Section 11.6).
2. Once a vessel has been connected up to any process or service line, the full permit-to-work and entry procedure should be followed. In the present case, this should have started eight weeks before the incident. And the nitrogen line should have been disconnected or slip-plated where it entered the vessel.

    There should be a formal handover from construction so that everyone is aware when it has taken place. The final connection to process or service lines is best made by plant fitters rather than by the construction team. In each plant, the procedure for handover should be described in a plant instruction.
3. When the plant is still in the hands of construction, the normal permit-to-work procedure is not necessary, but an entry permit system should be in force. Before anyone enters a vessel, it should be inspected by a competent, experienced person who will certify that it is isolated

and free from danger. While a tank is being built, when the walls reach a certain height (say, greater than the diameter) the tank should be deemed to be a confined space, and the entry procedure should apply.
4. All managers and supervisors should be aware of the procedure for handover and entry to vessels.

12.3.4 Liquid Nitrogen

Supplies of liquid nitrogen should always be tested before they are offloaded into the plant. Suppliers of liquid nitrogen often say there is no need to test, as they use different fittings on liquid nitrogen and liquid air (or oxygen) trucks and confusion is impossible. However, in several cases the impossible has happened, and liquid air (or oxygen) has been supplied instead of liquid nitrogen. One incident is described in Section 4.1f. Sometimes the mistake has been discovered by testing, but I know of two cases in which liquid air or oxygen was fed to a plant. Fortunately, in one case a high-oxygen-concentration alarm operated and in the other case a high-temperature alarm. The first incident occurred on a plant where the crew always tested the regular consignments of nitrogen but did not test a special extra delivery.

If a high-temperature or high-oxygen-concentration alarm will detect a wrong delivery, is there a need to test before acceptance? The alarms are our last layer of protection; if they fail, a fire or explosion is likely, and so we should never deliberately rely on them. Our preventive measures should lie as far as possible from the top event [17].

Nitrogen boils at a lower temperature than oxygen, so oxygen will condense on materials that are cooled with liquid nitrogen. If these materials are flammable, a fire or explosion can occur. Some pork rind that had to be ground was cooled with liquid nitrogen. When the grinder was started up, it exploded, and two men were killed.

Other hazards of liquid nitrogen and liquid air are due to their low temperature:

- Many materials become very brittle. Vehicle tires can explode, and carbon steel equipment can fail if exposed to the liquid or its vapor. A steel pressure vessel designed for use at a gauge pressure of 450 psi (30 bar) broke into 20 pieces when it was filled with cold nitrogen gas. The liquid nitrogen vaporizer should have been fitted with a low-temperature trip.
- Liquid trapped between valves will produce a large rise in pressure as it warms up.
- Spillages produce a fog, which restricts visibility [25, 26].

12.4 HEAVY OILS (INCLUDING HEAT TRANSFER OILS)

This term is used to describe oils that have a flash point above ambient temperature. They will therefore not burn or explode at ambient temperature but will do so when hot. Unfortunately, many people do not realize this and treat heavy oils with a disrespect that they would never apply to gasoline, as shown by the incidents described next. Another incident was described in Section 12.2c. Heavy oils are widely used as fuel oils, solvents, lubricants, and heat transfer oils, as well as process materials.

12.4.1 Traces of Heavy Oil in Empty Tanks

Repairs had to be carried out to the roof of a storage tank, which had contained heavy oil. The tank was cleaned out as far as possible, and two welders started work. They saw smoke coming out of the vent and flames coming out of the hole they had cut. They started to leave, but before they could do so the tank's roof lifted, and a flame 25 m (80 ft) long came out. One of the men was killed, and the other was badly burned. The residue in the tank continued to burn for 10 to 15 minutes [5].

Though the tank had been cleaned, traces of heavy oil were stuck to the sides or behind rust or trapped between plates. These traces of oil were vaporized by the welding and ignited.

Some old tanks are welded along the outside edge of the lap only, thus making a trap from which it is hard to remove liquids. Even light oils can be trapped in this way (see Section 5.4.2c and Figure 5-10).

A similar incident is described in an official report [6]. A tank with a gummy deposit on the walls and roof had to be demolished. The deposit was unaffected by steaming but gave off vapor when a burner's torch was applied to the outside. The vapor exploded, killing six firefighters who were on the roof at the time.

It is almost impossible to completely clean a tank (or other equipment) that has contained heavy oils, residues or polymers, or material that is solid at ambient temperature, particularly if the tank is corroded. Tanks that have contained heavy oils are more dangerous than tanks that have contained lighter oils, such as gasoline. Gasoline can be completely removed by steaming or sweeping with nitrogen. Note also that while light oils, such as gasoline, can be detected with a combustible gas detector, heavy oils cannot be detected. Even if a heavy oil is heated above its flash point, the vapor will cool down in the detector before it reaches the sensitive element.

Before welding is allowed on tanks that have contained heavy oils, the tanks should be filled with inert gas or with firefighting foam generated with inert gas, *not* with firefighting foam generated with air (see Section 12.3.2). Filling the tank with water can reduce the volume to be inerted.

Another incident occurred when an old 45-m^3 diesel oil tank was being cut up by acetylene welding. The top half was removed, and four holes were being cut in the lower half so that it could be picked up and moved. A piece of hot slag fell onto sludge on the bottom of the tank and set it alight. The fire could not be extinguished with handheld extinguishers, and the fire department had to be called. Cold-cutting methods should be considered when equipment that cannot be cleaned has to be cut up. Other fires have been started by falling welding slag; it can fall farther than expected [28].

An unusual case of an explosion in a "vessel" containing traces of heavy oil occurred when welding was carried out on the brakes of a tractor. The heat vaporized and ignited the lubricant used in fitting the tires, and the resulting explosion killed three men.

12.4.2   Traces of Heavy Oil in Pipelines

Some old pipelines had to be demolished. They were cleaned as far as possible and then tested with a combustible gas detector. No gas or vapor was detected, so a burner was given permission to cut them up. While doing so, sitting on the pipes 4 m (13 ft) above the ground, a tarry substance seeped out of one of the pipes and caught fire. The fire spread to the burner's clothing, and he ended up in the hospital with burns to his face and legs. The deposit did not give off enough vapor when cold for the combustible gas detector to have detected it.

It is almost impossible to completely clean pipes that have contained heavy oils or polymers. When demolishing old pipelines, there should be as many open ends as possible so that pressure cannot build up. And good access should be provided so that the burner or welder can escape readily if he or she needs to do so.

12.4.3   Pools of Heavy Oil

An ore-extracting process was carried out in a building with wooden floors. But this was considered safe because the solvent used had a flash point of 42°C (108°F), and it was used cold. Leaks of solvent drained into a pit inside the building. While welding was taking place, a burning piece of rag fell into the pit, and in a few seconds the solvent film, which covered the water in the pit, was on fire. The rag acted as a wick and set fire to the solvent, although a spark or a match would not have done so. The fire spread to the wooden floor; some glass pipes burst and these added more fuel to the fire. In a few minutes, the building was ablaze and two-thirds of the contents were destroyed [7].

12.4.4 Spillages of Heavy Oil, Including Spillages on Insulation

The heat transfer section of a plant was filled with oil after mainte-nance by opening a vent at the highest point and pumping oil into the system until it overflowed out of the vent. The overflow should have been collected in a bucket, but sometimes a bucket was not used or the bucket was overfilled. Nobody worried about small spillages because the flash point of the oil was above ambient temperature and its boiling point and auto-ignition temperature were both above 300°C (570°F).

A month after such a spillage, the oil caught fire. Some of it might have soaked into insulation, and if so, this would have caused the oil to degrade, lowering its auto-ignition temperature so that it ignited at the temperature of the hot pipework. The oil fire caused a leak of process gas, which exploded causing further localized damage and an oil fire.

All spillages, particularly those of high-boiling liquids, should be cleaned up promptly. Light oils will evaporate, but heavy oils will not. Besides the fire hazard, spillages produce a risk of slipping.

Insulation that has been impregnated with heavy oil—or any other organic liquid—should be removed as soon as possible before the oil ignites. If oil is left in contact with insulation materials, the auto-ignition temperature is low-ered by 100 to 200 degrees C (180 to 360 degrees F) [8] (see Section 7.3.2).

12.4.5 Heavy Oil Fireballs

Sections 9.1.1 and 12.2 describe incidents that occurred when heavy oils, at temperatures above 100°C (212°F), came into contact with water. The water vaporized with explosive violence, and a mixture of steam and oil was blown out of the vessel, after rupturing it.

In another incident of the same nature, the oil caught fire. A furnace supplied heat transfer oil to four reboilers. One was isolated for repair and then pressure-tested. The water was drained out of the shell, but the drain valve was 8 in. above the bottom tube plate, so a layer of water was left in the reboiler (Figure 12-2).

When the reboiler was brought back on line, the water was swept into the heat transfer oil lines and immediately vaporized. This set up a liquid hammer, which burst the surge tank. It was estimated that this required a gauge pressure of 450 psi (30 bar). The top of the vessel was blown off in one piece, and the rest of the vessel was split into 20 pieces. The hot oil formed a cloud of fine mist, which ignited immediately, forming a fire-ball 35 m in diameter. (Mists can explode at temperatures below the flash point of the bulk liquid; see Section 19.5.)

This incident led to the following recommendations:

1. Adequate facilities must be provided for draining water from heat transfer and other hot oil systems.
2. Oil rather than water should be used for pressure testing.

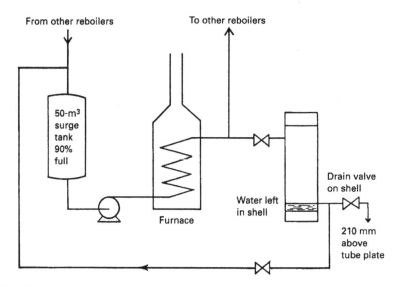

FIGURE 12-2   Water left in the heat exchanger was vaporized by hot oil.

3. Surge vessels should operate about half full, not 90% full as in this case.
4. In new plants, water should be considered as a heat transfer medium instead of oil. A decision to use water has to be made early in the design because the operating pressure will be higher. Although this will add to the cost, there will be savings in lower fire protection costs. In some plants, the heat transfer oil is a bigger fire hazard than the process materials [9–11].

12.4.6   A Lubricating Oil Fire

An ethylene plant compressor was lubricated by a pump, which took suction from a sump. The sump was originally topped up by hand, but to save labor a pump was installed to supply oil from a storage tank some distance away. An operator forgot to shut down this pump when the sump was filled to the required level, and it was overfilled. The pump had a greater capacity than the vent on the sump, so the sump was overpressured. The pressure backed up the oil line from the gearbox, which failed. Oil spewed out and ignited. The material damage was $500,000, but the consequential loss was many times greater [18].

On chemical plants and oil refineries, steam, nitrogen, compressed air, lubricating oil, and other utility systems are responsible for a disproportionately large number of accidents. Flammable oils are recognized as a hazard, but services are given less attention. If the modification to the lubricating system had been systematically studied before it was made, as recommended in Chapter 2, a larger vent could have been installed, or a pipe-break and funnel could have been installed at the inlet to the sump.

12.4.7   Degradation of Heavy Oils

Degradation of heavy oils spilled on insulation has already been described in Sections 7.3.2 and 12.4.4. Heat transfer oils can degrade in normal use, producing both light and heavy ends. The light ends accumulate at high points and can further degrade into a mixture of carbon and rust, known as "coffee grounds," which forms hard deposits in dead-end nozzles, such as those leading to relief valves. To prevent blockages, we should vent light ends frequently and inspect relief valve nozzles whenever the relief valves are removed for routine examination.

Heavy ends can further degrade into carbon deposits on the insides of furnace tubes and lead to tube failure. Sometimes the tube blocks completely and prevents a serious spillage, but at other times spillages have produced costly and spectacular fires, as in the incident described in Section 10.7.2 (though that one was not due to accumulation of heavy ends). To prevent tube failures, keep the concentration of heavy ends below 5%, and follow the recommendations on furnace operation in Section 10.7.2 [27].

References

1. *Safety Management* (South Africa), Apr. 1982, p. 30; and Feb. 1993, p. 36.
2. *Hazards of Air and Oxygen*, Institution of Chemical Engineers, Rugby, UK, 2004.
3. *Hazards of Water* and *Hazards of Steam*, Institution of Chemical Engineers, Rugby, UK, 2004.
4. T. A. Kletz, "Nitrogen—Our Most Dangerous Gas," *Proceedings of the Third International Symposium on Loss Prevention and Safety Promotion in the Process Industries*, Swiss Society of Chemical Industries, 1980, p. 1518.
5. T. A. Kletz, *Journal of Hazardous Materials*, Vol. 1, No. 2, 1976, p. 165.
6. A. W. M. Davies, *Public Enquiry into a Fire at Dudgeon's Wharf on 17 July 1969*, Her Majesty's Stationery Office, London, 1970.
7. R. Hoy-Petersen, *Proceedings of the First International Symposium on Loss Prevention and Safety Promotion in the Process Industries*, Elsevier, Amsterdam, 1984, p. 325.
8. P. E. Macdermott, *Petroleum Review*, July 1976.
9. J. W. Boley, *A Guide to Effective Industrial Safety*, Gulf Publishing Co, Houston, TX, 1977.
10. L. Pilborough, *Inspection of Chemical Plants*, Gulf Publishing Co, Houston, TX, 1977.
11. K. Gugan, *Unconfined Vapor Cloud Explosions*, Gulf Publishing Co, Houston, TX, 1979.
12. N. Morris, *The Chemical Engineer*, No. 437, June 1987, p. 51.
13. *Chemistry in Britain*, Vol. 22, No. 7, July 1986, p. 610.
14. *Chemical Engineering*, Dec. 2, 1977, p. 67.
15. *Bulletin*, Royal Society for the Prevention of Accidents, Birmingham, UK, May 1980.
16. *Chemical Safety Summary*, Vol. 56, No. 221, Chemical Industries Association, London, 1985, p. 4.
17. T. A. Kletz, *Learning from Accidents*, 3rd edition, Butterworth-Heinemann, Boston, MA, 2001.
18. V. G. Geishler, *Loss Prevention*, Vol. 12, 1979, p. 10.
19. A. Jacob, *The Chemical Engineer*, No. 503, Sept. 12, 1991, p. 19.
20. Y. Guo and C. W. Kauffman, "An Experimental Investigation of Air Line Explosions Caused by Film Detonation," *Paper presented at Combustion Institute Eastern Section Fall Meeting*, Albany, NY, Oct./Nov. 1989.
21. G. R. Schoofs, *AIChE Journal*, Vol. 38, No. 9, Sept. 1992, p. 1385.
22. *Employee Suffers Oxygen Deprivation*, Safety Note No. DOE/EH-0110, U.S. Dept. of Energy, Washington, DC, Oct. 1989.

23. *Loss Prevention Bulletin*, No. 110, Apr. 1993, p. 8.
24. *Loss Prevention Bulletin*, No. 098, Apr. 1991, p. 19.
25. *Loss Prevention Bulletin*, No. 097, Feb. 1993, p. 1.
26. British Cryogenics Council, *Cryogenics Safety Manual*, 3rd edition, Butterworth-Heinemann, Oxford, UK, 1991.
27. J. W. Bowman and R. P. Perkins, *Plant/Operations Progress*, Vol. 9, No. 1, Jan. 1990, p. 39.
28. *Occupational Health and Safety Observer*, Vol. 3, No. 12, U.S. Dept. of Energy, Washington, DC, Dec. 1994, p. 6.
29. *Hazards of Nitrogen and Catalyst Handling*, Institution of Chemical Engineers, Rugby, UK, 2006.
30. N. Howard, *Railway World*, March 2002, p. 69.

13

Tank Trucks and Cars

A cloud of chlorine drifted into a swimming pool after a tanker pumping sodium hypochlorite into a storage tank sprang a leak. It is thought that the hypochlorite came into contact with hydrochloric acid, causing a release of chlorine.... In total 30 children, six firemen and seven others were taken to hospital after the incident in Leicestershire.

—*Health and Safety at Work*, October 1984

The quotation shows us that though we use a safer substance (hypochlorite) instead of a hazardous one (chlorine), accidents are still possible.

This chapter is not concerned with accidents on the road. Rather, it describes some of the many incidents that have occurred while tank trucks and cars (known in Europe as road and rail tankers) were being filled or emptied. Section 18.8 shows how hazard and operability studies have been used to spot potential hazards in filling systems, and Section 22.3 describes some runaway reactions in tank trucks and cars.

13.1 OVERFILLING

Tank trucks and cars have been overfilled on many occasions, both when filled automatically and when filled by hand.

In automatic systems, the filler sets the quantity to be filled on a meter, which closes a valve when this required quantity has been delivered. Overfilling has occurred because the wrong quantity was set on the meter, because there was already some liquid in the tank (left over from the previous load), and because the filling equipment failed. For these reasons, many companies now fit their tank trucks with high-level trips, which automatically close a valve in the filling line [8].

Tank trucks and cars that are filled by hand have been overfilled because the filler left the job for a few minutes and returned too late. On one occasion, an operator thought a tank truck had a single-compartment tank when in fact there were two compartments. He tried to put the full load into one compartment.

On another occasion, after a tank truck had been filled during the night, the operator completed a filling certificate—a very small piece of paper—and slipped it inside the dispatch papers. This was the usual practice. When the next shift came on duty, the driver had not returned to get the truck. The overnight record sheets had all been sent to the plant office. So the operator shook the dispatch papers to see if there was a filling certificate among them. Nothing fell out because the certificate was caught up in the other papers. The operator therefore started to fill the tanker again.

In contrast, a case of overfilling, which was the subject of an official report [1], was due to the poor design of complex automatic equipment at a large terminal for loading gasoline and other hydrocarbons.

The grade and quantity of product required were set on a meter. The driver inserted an authorization card and pressed the start button. The required quantity was then delivered automatically. The filling arm had to be lowered before filling could start.

One day the automatic equipment broke down, and the foreman decided to change over to manual filling. He asked the drivers to check that the hand valves on the filling lines were shut, but he did not check himself. He then operated the switches that opened the automatic valves. *Some of the hand valves were open.* Gasoline and other products came out, overfilled the tanker (or splashed directly on the ground), and caught fire. Three men were killed, 11 were injured, and the whole row of 18 filling points was destroyed.

To quote from the official report,

> The decision to override the individual controls on the loading arms by means of a central switchboard, without the most rigid safeguards, was a tragic one. After its installation an accident from that moment on became inevitable sooner or later.
>
> That this switchboard was installed, with the approval of the terminal management ... in a switchroom from which the loading stands were not visible, suggests some failure to take into account the basic fundamentals of safety in operation of plant....
>
> [H]ad the same imagination and the same zeal been displayed in matters of safety as was applied to sophistication of equipment and efficient utilization of plant and men, the accident need not have occurred.

13.2   BURST HOSES

Hoses have failed while tank trucks or cars were being filled or emptied for all the reasons listed in Section 7.1.6, in particular because damaged hoses or hoses made from the wrong material were used. However, the most common cause of hose failure is the tanker driving away before the hose is disconnected.

The following incidents are typical:

(a) A tank truck was left at a plant for filling with liquefied flammable gas. Some hours later, the transport foreman assumed that it would be ready and sent a driver to get it. There was no one in the plant office, so the driver went to the loading bay. He found that the truck was grounded and that the grounding lead had been looped through the steering wheel—the usual practice—to prevent the driver from driving away before disconnecting it. He removed the lead and drove off, snapping off the filling branch and tearing the hose that was connected to the vent line. Fortunately, there was no flow through the filling line at the time though the valves were open, and the spillage was relatively small. It did not catch fire.

Plant instructions stated that a portable barrier should be put in front of tank trucks being filled, but the barrier was not being used. However, if it had been in use, the driver might have removed it.

A device that can be fitted to a tank truck to prevent anyone from driving it away while a hose is connected is described in reference 2. A plate is fixed in front of the hose connection. To connect the hose, this plate has to be moved aside, and this applies the brakes. Reference 3 describes a special type of hose that seals automatically if it breaks; there are also other types.

Remotely operated emergency isolation valves (see Section 7.2.1) should be fitted on filling lines. If the hose breaks for any reason, pressing a button located at a safe distance can stop the flow. A check valve can prevent reverse flow from the tank truck or car.

Note that it is not necessary to ground tank trucks containing liquefied flammable gases because no air is present in the tank.

(b) Gasoline was being discharged at a service station from a tank truck, which was carrying diesel fuel in one compartment. To save time, the driver decided to discharge the diesel fuel while discharging the gasoline. To do this he had to move the tank truck about 1 to 2 m.

He drove the truck slowly forward, while the discharge of fuel continued. The hose caught on an obstruction and was pulled part way out of its fastening. Gasoline escaped and caught fire. The service station and tank truck were destroyed [4].

(c) Similar incidents have occurred at gasoline filling stations when motorists have driven off before removing the filling nozzles from their cars. In one case, the pump and nozzle were damaged and sparking ignited the spilled gasoline.

13.3   FIRES AND EXPLOSIONS

A number of explosions or fires have occurred in tank trucks or cars while they were being filled. The most common cause is "switch filling." A tank contains some flammable vapor, such as gasoline vapor, from a previous load and is then filled with a safer, higher-boiling liquid, such as gas oil. The gas oil is not flammable at ambient temperature. So no special precautions are normally necessary to prevent the formation of static electricity. The tank may be filled quickly—may even be splash-filled—a static charge is formed, and a spark jumps from the liquid to the wall of the tank, igniting the gasoline vapor.

A similar incident occurred in a tank truck used to carry waste liquids. While it was being filled with a nonflammable liquid and the driver was standing on the top, smoking, an explosion occurred, and the manhole cover was thrown 60 m. On its previous journey the tank truck had carried a waste liquid containing dissolved flammable gas. Some of the gas was left in the tank and was pushed out when it was filled with the next load. For other examples see reference 10.

Flammable liquids should never be splash-filled, even though they are below their flash points. The splash filling may form a mist, which can be ignited by a static discharge. Mists, like dusts, can be ignited at any temperature (see Section 19.5).

On one occasion a tank truck was being splash-filled with gas oil, flash point 60°C (140°F). The splashing produced a lot of mist, and it also produced a charge of static electricity on the gas oil. This discharged, igniting the mist. There was a fire with flames 10 m (30 ft) high, but there was no explosion. The flames went out as soon as the mist had been burned.

Many thousands of tank trucks had been splash-filled with gas oil at this installation before conditions were exactly right for a fire to occur. When handling flammable gases or liquids, we should never say, "It's okay. We've been doing it this way for 20 years and have never had a fire." Such a statement should be made only if an explosion in the 21st year is acceptable.

Note that grounding a tank truck will not prevent ignition of vapor by a discharge of static electricity. Grounding prevents a discharge from the tank to earth, but it does not prevent a discharge from the liquid in the tank to the tank or to the filling arm.

There is more information on static electricity in Chapter 15.

13.4   LIQUEFIED FLAMMABLE GASES

Tank trucks or cars that carry liquefied gases under pressure at ambient temperature present additional hazards.

When the tanks are filled, the vapor is vented to a stack or back to the plant through a vapor return line, which is fitted to the top of the tank. An official report [5] described a fire that occurred because the fillers had not bothered to connect up this vapor return line. Vapors were discharged into the working area. Seven people were injured.

Following this incident, a survey at another large installation showed that the fillers there were also forgetting to connect up the vapor lines. Reference 5 also reports that at another plant the vapor return line was connected in error to another filling line. The vapor could not escape, the pressure in the tank rose, and the filling hose burst. There was no emergency isolation valve in the filling line, no check valve on the tank (see Section 13.2a), and no excess flow valve on either, so the spillage was substantial.

Vapor return lines and filling lines should be fitted with different sizes or types of connections.

13.5   COMPRESSED AIR

Compressed air is often used to empty tank trucks and cars. Plastic pellets are often blown out of tank trucks. When the tank is empty, the driver vents the tank and then looks through the manhole to check that the tank is empty. One day a driver who was not regularly employed on this job started to open the manhole before releasing the pressure. When he had opened two out of five quick-release fastenings, the manhole blew open. The driver was blown off the tank top and killed.

Either the driver forgot to vent the tank or thought it would be safe to let the pressure (a gauge pressure of 10 psi or 0.7 bar) blow off through the manhole. After the accident, the manhole covers were replaced by a different type in which two movements are needed to release the fastenings. The first movement allows the cover to be raised about ¼ in. while still capable of carrying the full pressure. If the pressure has not been blown off, this is immediately apparent, and the cover can be resealed or the pressure allowed to blow off. In addition, the vent valve was repositioned at the foot of the ladder [6].

Many of those concerned were surprised that a pressure of "only 10 pounds" could cause a man to be blown off the top of the tank. They forgot that 10 psi is not a small pressure. It is 10 lbs of force on every square inch (see Section 12.1).

A similar incident is described in Section 17.1.

13.6   TIPPING UP

On several occasions, tank trailers have tipped up because the rear compartments were emptied first, as shown in Figure 13-1.

FIGURE 13-1   A tank trailer may tip up if the rear compartments are emptied first.

It is not always possible to keep the trailer connected to the truck's unit during loading/unloading. If it is not connected, the front compartments should be filled last and emptied first or a support put under the front of the trailer.

Some tank trailers are fitted with folding legs. Some designs are difficult to lubricate adequately and difficult to maintain, and as a result, a number have collapsed.

13.7   EMPTYING INTO OR FILLING FROM THE WRONG PLACE

On many occasions, tank trucks have been discharged into the wrong tank. The following incident is typical of many.

A tank truck containing isopropanol arrived at a plant during the night. It was directed to a unit that received regular supplies by tank trucks. The unit was expecting a load of ethylene glycol. So without looking at the label or the delivery note, unit staff members emptied the tank truck into the ethylene glycol tank and contaminated 100 tons of ethylene glycol.

Fortunately, in this case the two materials did not react. People who have emptied acid into alkali tanks have been less fortunate. A plant received caustic soda in tank cars and acid in tank trucks. One day a load of caustic soda arrived in a tank truck. It was labeled "Caustic Soda," the delivery papers said it was caustic soda, and the hose connections were unusual. But the operators had a mind-set (see Section 3.3.5) that anything in a tank truck was acid, and they spent two hours making an adaptor to enable them to pump the contents of the tank truck into the acid tank.

A plant received tetra-ethyl lead (TEL) and hydrogen fluoride (HF) in tank cars of different shapes, colors, and markings. One day a load of HF arrived in a tank car of the type normally used for TEL. It was therefore put into the siding alongside the TEL off-loading point, and an operator started to transfer the contents into the TEL tank. He stopped when he noticed white fumes coming out of the vent on the tank car. The contents of the TEL tank were ruined, but fortunately the reaction did not run away [9]. It is difficult not to sympathize with the operators. The delivery of HF in a tank car of the type used for TEL set a trap for them (see Chapter 3). The supplier might reasonably have been expected to draw attention to the change.

On other occasions, tank trucks have been filled with the wrong material. In particular, liquid oxygen or liquid air has been supplied instead of liquid nitrogen. One incident, the result of confusion over labeling, was described in Section 4.1f.

I do not know of any case in which delivery of liquid oxygen instead of liquid nitrogen caused an explosion. But, as stated in Section 12.3.1, in one case the "nitrogen" was used to inert a catalyst bed, and the catalyst got hot; in another case a high-oxygen-concentration alarm in the plant sounded, and in several cases check analyses showed that oxygen had been supplied.

Many suppliers of liquefied gases state that they use different hose connections for liquid oxygen and liquid nitrogen so mistakes cannot arise. However, mistakes *have* occurred, possibly because of the well-known tendency of operators to acquire adaptors.

Liquid nitrogen should always be analyzed before it is off-loaded. The same applies in other cases where delivery of the wrong material could have serious unwanted results, such as a fire or runaway reaction, as in the two incidents that follow. If analysis causes too much delay, the new load should be put in a holding tank.

- After a load of diesel fuel intended for standby generators had been off-loaded into the stock tank, it was found to contain too much particulate matter. This could have affected the performance of the generators [11].
- As the result of a mixup at a distribution center, two tank truck drivers received each other's papers. One of the trucks carried a load of sodium chlorite solution, and the other carried epichlorohydrin. The chlorite truck went to the customer who was expecting epichlorohydrin and was off-loaded into a tank that already contained some epichlorohydrin. The result was an explosion and a serious fire; fumes and smoke led to the closure of the bridges over the Severn Estuary in the United Kingdom [12, 13].

Suppliers' papers tell us what they intend to deliver, not what is in the tank truck or car. We can find that out either by analyzing the contents or by seeing what happens.

The following incident involved cylinders rather than bulk loads, but it shows how alertness to an unusual observation can prevent an accident.

A plant used nitrogen in large cylinders. One day a cylinder of oxygen, intended for another plant, was delivered in error. The foreman noticed that the cylinder had an unusual color and unusual fittings, and he thought it strange that only one was delivered. Usually several cylinders were delivered at a time. Nevertheless he accepted the cylinder. He did not notice that the invoice said "Oxygen."

The invoice was sent as usual to the purchasing department for payment. The young clerk who dealt with it realized that oxygen had been delivered to a unit that had never received it before. She told her supervisor, who telephoned the plant, and the error came to light.

For another success story, see Section 11.7.

13.8   CONTACT WITH LIVE POWER LINES

The manhole covers on tank cars are sometimes sealed with wires. Loose ends of wires protruding above the manhole cover have come into contact with the overhead electric wire, which supplies power to the train, and caused a short circuit.

In the United Kingdom, there is normally a gap of 4 in. between the highest point of the tank car and the lowest point of the cables, but if the gap falls below 2 in., arcing may occur [7].

Somewhat similar incidents have occurred on railway lines powered by a third electrified rail. The cap covering the discharge pipe has vibrated loose, the retaining chain has been too long, and the cap has contacted the third rail [7].

For more information on the safety of loading and unloading, see reference 14.

References

1. H. K. Black, *Report on a Fatal Accident and Fire at the West London Terminal on 1 April 1967*, Her Majesty's Stationery Office, London, 1967.
2. T. A. Kletz, *Loss Prevention*, Vol. 10, 1976, p. 151.
3. *Petroleum Review*, July 1976, p. 433.
4. *Petroleum Review*, July 1976, p. 428.
5. *Annual Report of H. M. Inspectors of Explosives for 1967*, Her Majesty's Stationery Office, London, 1968.
6. T. A. Kletz, *Lessons from Disaster: How Organizations Have No Memory and Accidents Recur*, Institution of Chemical Engineers, Rugby, UK, 1993, Chapter 2.

7. *Petroleum Review,* Apr. 1976, p. 241.

8. J. N. Baker, *Measurement and Control,* Vol. 18, Apr. 1985, p. 104.

9. *Petroleum Review,* June 1985, p. 36.

10. T. H. Pratt, *Process Safety Progress,* Vol. 15, No. 3, Fall 1996, p. 177.

11. *Operating Experience Weekly Summary,* No. 97-06, Office of Nuclear and Safety Facility, U.S. Dept. of Energy, Washington, DC, 1997, p. 3.

12. *The Chemical Engineer,* No. 620, Oct. 10, 1996, p. 10; and No. 621, Oct. 24, 1996, p. 7.

13. *The Industrial Emergency Journal,* Vol. 1, No. 2, Oct./Nov. 1996, p. 14.

14. *Safe Tank Farms and (Un)loading Operations,* Institutions of Chemical Engineers, Rugby UK, 2006.

14

Testing of Trips and Other Protective Systems

The driver of a ramshackle Maputo taxi that had holes in the floor and kept breaking down was asked if all the gears worked. "Yes," he said, "but not all at the same time."

—*Daily Telegraph* (London), January 7, 1995

Many accidents have occurred because instrument readings or alarms were ignored (see Sections 3.2.8, 3.3.1, and 3.3.2). Many other accidents, including Bhopal (see Section 21.1), have occurred because alarms and trips were not tested or not tested thoroughly or because alarms and trips were made inoperative or their settings altered, both without authority. These and some related accidents are described here.

Microprocessor-based control systems are being increasingly used in place of traditional instrumentation. Some accidents that have occurred on these systems are described in Chapter 20.

In many companies, especially in the United States, the trips described in this chapter would be called interlocks.

14.1   TESTING SHOULD BE THOROUGH

All protective equipment should be tested regularly, or it may not work when required. Whereas it is sufficient to test relief valves every year or every two years, instrumented alarms and trips are less reliable and should be inspected every month or so.

Testing must be thorough and as near as possible to real-life situations, as shown by the following incidents:

(a) A high-temperature trip on a furnace failed to operate. The furnace was seriously damaged. The trip did not operate because the pointer touched the plastic front of the instrument case, and this prevented it from moving to the trip level. The instrument had been tested regularly—by injecting a current from a potentiometer—but to do this *the instrument was removed from its case and taken to the workshop.*

(b) A reactor was fitted with a high-temperature trip, which closed a valve in the feed line. When a high temperature occurred, the trip valve failed to close although it had been tested regularly.

Investigation showed that the pressure drop through the trip valve—a globe valve—was so high that the valve could not close against it. There was a flow control valve in series with the trip valve (Figure 14-1), and the trip normally closed this valve as well. However, this valve failed in the open position—this was the reason for the high temperature in the reactor—and the full upstream pressure was applied to the trip valve.

Emergency valves should be tested against the maximum pressure or flow they may experience and, whenever possible, should be installed so that the flow assists closing.

(c) If the response time of protective equipment is important, it should always be measured during testing. For example, machinery is often interlocked with guards so that if the guard is opened, the machinery stops. Brakes are often fitted so that the machinery stops quickly. The actual stopping time should be measured at regular intervals and compared with the design target.

Another example: a mixture of a solid and water had to be heated to 300°C (570°F) at a gauge pressure of 1,000 psig (70 bar) before the solid would dissolve. The mixture was passed through the tubes of a heat exchanger while hot oil, at low pressure, was passed over the outside of the tubes. It was realized that if a tube burst, the water

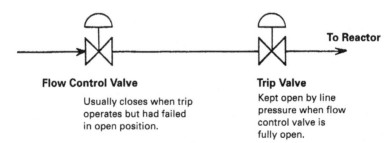

Flow Control Valve

Usually closes when trip
operates but had failed
in open position.

Trip Valve

Kept open by line
pressure when flow
control valve is
fully open.

FIGURE 14-1   When the control valve was open, the pressure prevented the trip valve from closing.

would come into direct contact with the hot oil and would turn to steam with explosive violence. An automatic system was therefore designed to measure any rise in the oil pressure and to close four valves, in the water and oil inlet and exit lines. The heat exchanger was also fitted with a rupture disc, which discharged into a catch-pot. The system was tested regularly; nevertheless, when a tube actually burst, most of the oil was blown out of the system and caught fire, as the valves had taken too long to close. They had been designed to close quickly but had got sluggish; the time of response was not measured during the test, so no one knew that they were not responding quickly enough.

Procedures, like equipment, also take time to operate. For example, how long does it take to empty your building when the fire alarm sounds? Is this quick enough?

(d) A large factory could be supplied with emergency power from a diesel-driven generator. It was tested regularly to ensure that the diesel engine started up when required. When the power supply actually failed, the diesel generator started up, but the relay that connected it to the distribution system failed to operate.

The emergency supply was tested when the distribution system was live. No one understood how the emergency circuits worked and did not realize that they were not being thoroughly tested [2].

(e) Here is an example from another industry. For many years railway carriage doors in the United Kingdom opened unexpectedly from time to time, and passengers fell out. Afterward, the locks were removed from the doors and sent for examination. No faults were found, and it was concluded that passengers had opened the doors. However, it was not the locks that were faulty but the alignment between the locks and the recesses in the doors. This was faulty and allowed them to open [3].

(f) A plant was pressure-tested before startup, but the check valves (nonreturn valves, NRV) in the feed lines to each unit (Figure 14-2) made it impossible to test the equipment to the left of them. A leak of liquefied petroleum gas (LPG) occurred during startup at the point indicated. The three check valves were then replaced by a single one in the common feed line at the extreme left of the diagram.

(g) Before testing an interlock or isolation to make sure it is effective, ask what will happen if it is not. For example, if a pump or other item of equipment has been electrically isolated by removing the fuses, it should be switched on to check that the correct fuses have been withdrawn. Suppose they have not; will the pump be damaged by starting it dry?

A radioactive source was transferred from one container to another by remote operation in a shielded cell. A radiation detector, inter-locked with the cell door, prevented anyone from opening the cell

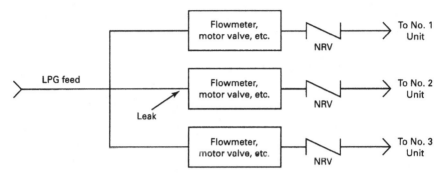

FIGURE 14-2   The check valves (nonreturn valves, NRV) prevented a leak test of the equipment to the left of them. During startup, a leak occurred at the point indicated.

door when radiation could be detected inside it. To make sure the interlock was working, an operator tried to open the cell door, by remote control, during a transfer. He found he could open it. He then found that the closing mechanism would not work. Fortunately, he had not opened the door very far.

(h) Do not test a trip or interlock by altering the set point. The trip or interlock may operate at the altered set point, but that does not prove it will operate at the original set point.

14.2   ALL PROTECTIVE EQUIPMENT SHOULD BE TESTED

This section lists some protective equipment that has often been overlooked and not included in testing schedules.

14.2.1   Leased Equipment

After a low-temperature trip on a nitrogen vaporizer failed to operate, it was found that the trip was never tested. The equipment was rented, and the user assumed—wrongly—that the owner would test it.

14.2.2   Emergency Valves

A pump leaked and caught fire. It was impossible to reach the suction and delivery valves. But there was a second valve in the suction line between the pump and the tank from which it was taking suction, situated in the tank dike. Unfortunately, this valve was rarely used and was too stiff to operate.

All valves—whether manual or automatic—that may have to be operated in an emergency should be tested regularly (weekly or monthly). If completely closing a valve will upset production, it should be closed halfway during testing and closed fully during shutdowns.

Emergency blowdown valves are among those that should be tested regularly. Reference 5 describes in detail the measures necessary to test emergency isolation valves when very high reliability is needed.

14.2.3 Steam Tracing

A furnace feed pump tripped out. The flowmeter was frozen, so the low-flow trip did not operate. Two tubes burst, causing a long and fierce fire. The structure and the other tubes were damaged, and the stack collapsed.

In cold weather, the trace heating on instruments that form part of trip and alarm systems should be inspected regularly. This can be part of the test routine, but more frequent testing may be necessary.

14.2.4 Relief Valves, Vents, Flame Arrestors, and Similar Items

Section 10.4.2 lists some items that should be registered for inspection as part of the relief valve register. Section 2.2a described an accident that killed two men. A vent was choked, and the end of the vessel was blown off by compressed air.

Open vents, especially those on storage tanks, are often fitted with flame arrestors. If the vents, and in particular the flame arrestors, are not kept clean, they are liable to choke, and the tanks maybe sucked in (see Section 5.3a). If the flame arrestors are ineffective, a lightning strike or other external source of ignition may ignite the flammable mixture often present inside the tank, above the liquid level, and produce an explosion. According to a 1989 report, in the Province of Alberta, Canada, alone, failures of flame arrestors were responsible for 10 to 20 tank explosions every year. Some of the failures were due to damage not detected during inspection, others were due to unsuitable design [4].

14.2.5 Other Equipment

Other equipment, in addition to that already mentioned, that should be tested regularly includes the following:

- Check valves and other reverse-flow prevention devices, if their failure can affect the safety of the plant.
- Drain holes in relief valve tailpipes. If they choke, rainwater will accumulate in the tailpipe (see Section 10.4).
- Drain valves in tank dikes. If they are left open, the dike is useless.

- Emergency equipment, such as diesel-driven fire water pumps and generators.
- Filters for both gases and liquids, including air filters. Their performance should be checked.
- Fire and smoke detectors and firefighting equipment.
- Grounding connections, especially the movable ones used for grounding trucks.
- Labels (see Chapter 4) are a sort of protective equipment. They vanish with remarkable speed, and regular checks should be made to make sure they are still there.
- Mechanical protective equipment, such as overspeed trips.
- Nitrogen blanketing (on tanks, stacks, and centrifuges).
- Passive protective equipment, such as insulation. If 10% of the fire insulation on a vessel is missing, the rest is useless.
- Spare pumps, especially those fitted with auto-starts.
- Steam traps.
- Trace heating (steam or electrical).
- Trips, interlocks, and alarms.
- Valves, remotely operated and hand-operated, that have to be used in an emergency.
- Ventilation equipment (see Section 17.6).
- Water sprays and steam curtains.
- Finally, equipment used for carrying out tests should itself be tested.

If equipment is not worth testing, then you don't need it.

Trips and interlocks should be tested after a major shutdown, especially if any work has been done on them. The following incidents demonstrate the need to test all protective equipment:

(a) A compressor was started up with the barring gear engaged. The barring gear was damaged. The compressor was fitted with a protective system that should have made it impossible to start the machine with the barring gear engaged. But the protective system was out of order. It was not tested regularly.

(b) In an automatic firefighting system, a small explosive charge cut a rupture disc and released the firefighting agent, halon. The manufacturers said it was not necessary to test the system. To do so, a charge of halon, which is expensive, would have to be discharged.

The client insisted on a test. The smoke detectors worked, and the explosive charge operated, but the cutter did not cut the rupture disc. The explosive charge could not develop enough pressure because the volume between it and the rupture disc was too great. The volume had been increased as the result of a change in design: installation of a device for discharging the halon manually.

(c) A glove box on a unit that handled radioactive materials was supposed to be blanketed with nitrogen, as some of the materials handled were combustible. While preparing to carry out a new operation, an operator discovered that the nitrogen supply was disconnected and that there was no oxygen monitor. The supply was disconnected several years before when nitrogen was no longer needed for process use, and the fact that it was still needed for blanketing was overlooked. Disconnecting a service was not seen as a modification and was not treated as such. The oxygen analyzer had apparently never been fitted [6].

One sometimes comes across a piece of protective equipment that is impossible to test. All protective equipment should be designed so that it can be tested easily.

14.3 TESTING CAN BE OVERDONE

An explosion occurred in a vapor-phase hydrocarbon oxidation plant, injuring 10 people and seriously damaging the plant, despite the fact that it was fitted with a protective system that measured the oxygen content and isolated the oxygen supply if the concentration approached the flammable limit.

It is usual to install several oxygen analyzers, but this plant was fitted with only one. The management therefore decided to make up for the deficiency in numbers by testing it daily instead of weekly or monthly.

The test took more than an hour. The protective system was therefore out of action for about 5% of the time. There was a chance of 1 in 20 that it would not prevent an explosion because it was being tested. It was, in fact, during testing when the oxygen content rose.

14.4 PROTECTIVE SYSTEMS SHOULD NOT RESET THEMSELVES

(a) A gas leak occurred at a plant and caught fire. The operator saw the fire through the window of the control room and operated a switch, which should have isolated the feed and opened a blowdown valve. Nothing happened. He operated the switch several times, but still nothing happened. He then went outside and closed the feed valve and opened the blowdown valve by hand.

The switch operated a solenoid valve, which vented the compressed air line leading to valves in the feed and blowdown lines (Figure 14-3). The feed valve then closed, and the blowdown valve opened. This did not happen instantly because it took a minute or so for the air pressure to fall in the relatively long lines between the solenoid valve and the other valves.

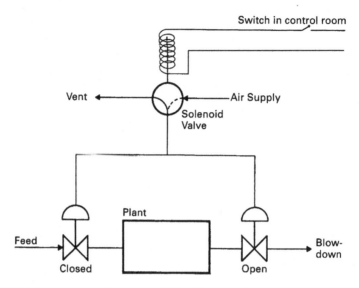

FIGURE 14-3   An automatic system, which will take a minute or so to operate.

The operator expected the system to function as soon as he operated the switch. When it did not, he assumed it was faulty. Unfortunately, after operating the switch several times, he left it in its normal position.

The operator had tested the system on several occasions, as it was used at every shutdown. However, it was tested in conditions of no stress, and he did not notice that it took a minute or so to operate. The solenoid valve should have been fitted with a latch so that once the switch had been operated, the solenoid valve could not return to its normal position until it was reset by hand.

(b) A liquid-phase hydrocarbon oxidation plant was fitted with a high-temperature trip, which shut off the air and opened a drain valve that dumped the contents of the reactor in a safe place (Figure 14-4). If the air valve reopened after a dump, a flammable mixture could form in the reactor.

One day the temperature-measuring device gave a false indication of high temperature. The air valve closed, and the drain valve opened. The temperature indication fell, perhaps because the reactor was now empty. The drain valve stayed open, but the air valve reopened, and a flammable mixture was formed in the reactor. Fortunately, it did not ignite.

The air valve reopened because the solenoid valve in the instrument air line leading to the air valve would not stay in the tripped position. It should have been fitted with a latch.

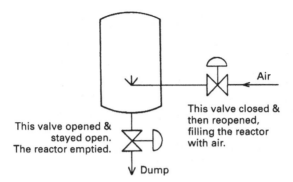

This valve closed &
then reopened,
filling the reactor
with air.

This valve opened &
stayed open.
The reactor emptied.

Dump

Air

FIGURE 14-4   When the air valve reopened after a dump, a flammable mixture formed in the reactor.

14.5   TRIPS SHOULD NOT BE DISARMED WITHOUT AUTHORIZATION

Many accidents have occurred because operators made trips inoperative (that is, disarmed, blocked, or deactivated). The following incidents are typical:

(a) Experience shows that when autoclaves or other batch reactors are fitted with drain valves, the valves may be opened at the wrong time and the contents tipped onto the floor, often inside a building. To prevent this outcome, the drain valves on a set of reactors were fitted with interlocks so that they could not be opened when the pressure was above a preset value. Nevertheless, a drain valve was opened when a reactor was up to pressure, and a batch emptied onto the floor. The inquiry disclosed that the pressure-measuring instruments were not reliable. So the operators had developed the practice of defeating the interlocks either by altering the indicated pressure with the zero adjustment screw or by isolating the instrument air supply.

One day the inevitable happened. Having defeated the interlock, an operator opened a drain valve in error instead of a transfer valve.

Protective equipment may have to be defeated from time to time, but this should only be done after authorization in writing by a responsible person. And the fact that the equipment is out of action should be clearly signaled—for example, by a light on the panel.

(b) Soon after a startup, part of a unit was found to be too hot. Flanged joints were fuming. It was then found that the combined temperature controller and high-temperature trip had been unplugged from the power supply.

Trips should normally be designed so that they operate if the power supply is lost. If this will cause a dangerous upset in plant operation, then an alarm should sound when power is lost. Trips should be tested at startup if they have been worked on during a shutdown. Particularly important trips, such as those on furnaces and compressors and high-oxygen concentration trips, should always be tested after a major shutdown.

The most common cause of a high temperature (or pressure, flow, level, etc.) is a fault in the temperature measuring or control system.

(c) Trips and interlocks may have to be disarmed (that is, made inoperative) so that equipment can be maintained. The operators or maintenance workers may then forget to re-arm the trip or interlock. For example, to maintain an emergency diesel generator, the auto-start mechanism was blocked. According to the procedure, when work is complete, one electrician should remove the block, and another should verify that it has been removed. Both signed the procedure to indicate that the block was removed. Nevertheless, a week later a routine test found that the block was still in position [7].

As stated in Sections 1.2.7e and 3.2.7b, checking procedures often break down, as the first person assumes the checker will spot anything missed; after a while the checker, having never found anything wrong, stops checking. When safety equipment has to be blocked or disarmed, this should be clearly signaled by a light or prominent notice on the panel.

(d) On computer-controlled plants, it may be possible to override an interlock by means of a software block. On one plant passwords and codes were needed for access to the program. They were kept under lock and key and issued only to electricians and engineering staff, but nevertheless 40 people had access to them. When an interlock was found, by routine tests, to be blocked, all 40 denied any knowledge. A secret shared by 40 people is no secret.

(e) At Gatwick airport in the United Kingdom, an employee put his head through the hatch in the driver's cab of a cargo transfer vehicle. He thought the vehicle had stopped, but it was still moving slowly, and he became trapped between the vehicle and a nearby pillar. Fortunately, he was only bruised. An interlock, which should have stopped the vehicle when the hatch was opened, had been taped over to improve the ventilation of the cab. According to the report, the company should have checked the safety equipment regularly, and a systematic assessment of the operation could have identified the risk. The company was fined [8].

(f) Alarms were deactivated, by reprogramming a data logger, to prevent them from sounding during the routine monthly test of an emergency generator. Afterward those involved forgot to reactivate the alarms. This was not discovered until nine days later, when someone looked at the data logger printout and noticed the alarms were still listed as deactivated. There were no written logs, policies, or procedures for deactivating the alarms.

In another similar case, the deactivation was noted in the plant log book, but few people look at old logs. The deactivation was discovered during an upset, when someone realized that an alarm had not sounded. As stated in (c), if an alarm is temporarily out of action, this should be prominently signaled [9].

(g) If disarming an interlock is occasionally necessary, the procedure for doing so should not be too easy, as the railways discovered long ago. Interlocks prevent a signal from being set at Go if another train is already in the section of track that it protects. An interlock occasionally has to be bypassed, for example, when a train has broken down or when the equipment for detecting the presence of a train has failed. Originally a single movement of a key was all that was necessary, and this caused several accidents. A change was then made. To get the key, the signalman (dispatcher) had to break a glass and then send for a technician to repair it. Everyone knew he had used the key, and he was less ready to use it. In an alternative system, a handle had to be turned 100 times. This gave ample time for him to consider the wisdom of his action [10].

Many of these incidents show the value of routine testing.

14.6   INSTRUMENTS SHOULD MEASURE DIRECTLY WHAT WE NEED TO KNOW

An ethylene oxide plant tripped, and a light on the panel told the operator that the oxygen valve had closed. Because the plant was going to be restarted immediately, he did not close the hand-operated isolation valve as well. Before the plant could be restarted, an explosion occurred. The oxygen valve had not closed, and oxygen continued to enter the plant (Figure 14-5).

The oxygen valve was closed by venting the air supply to the valve diaphragm, by means of a solenoid valve. The light on the panel merely said that the solenoid had been de-energized. Even though the solenoid is de-energized, the oxygen flow could have continued for the following reasons:

1. The solenoid valve did not open.
2. The air was not vented.
3. The trip valve did not close.

Actually the air was not vented. A wasp's nest choked the 1-in. vent line on the air supply. Whenever possible, we should measure directly

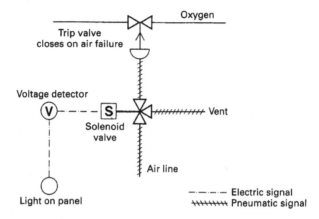

FIGURE 14-5   The light shows that the solenoid is de-energized, not that the oxygen flow has stopped.

what we need to know and not some other parameter from which it can be inferred [1].

Other incidents in which operators relied on automatic valves and did not back them up with hand valves are described in Sections 17.3b and 17.5c.

14.7   TRIPS ARE FOR EMERGENCIES, NOT FOR ROUTINE USE

(a) Section 5.1.1 described how a small tank was filled every day with sufficient raw material to last until the following day. The operator watched the level in the tank and switched off the filling pump when the tank was 90% full. This system worked satisfactorily for several years before the inevitable happened and the operator allowed the tank to overfill. A high-level trip was then installed to switch off the pump automatically if the level exceeded 90%. To everyone's surprise, the tank overflowed again after about a year.

When the trip was installed, the following was assumed:

1. The operator will occasionally forget to switch off the pump in time, and the trip will then operate.
2. The trip will fail occasionally (about once in two years).
3. The chance that both will occur at the same time is negligible.

However, it did not work out like this. The operator decided to rely on the trip and stopped watching the level. The manager and foreman knew this but were pleased that the operator's time was being utilized better. A simple trip fails about once every two years, so the tank was bound to overflow after a year or two. The trip was being used as a process controller and not as an emergency instrument.

After the second spillage, the following options were considered:

1. Persuade the operator to continue to watch the level. This was considered impracticable if the trip was installed.
2. Remove the trip, rely on the operator, and accept an occasional spillage.
3. Install two trips, one to act as a process controller and the other to take over if the first one fails.

(b) When a furnace fitted with a low-flow trip has to be shut down, it is common practice to stop the flow and let the low-flow trip isolate the fuel supply to the burners. In this way, the trip is tested without upsetting production.

On one occasion the trip failed to operate, and the furnace coils were overheated. The operator was busy elsewhere on the unit and was not watching the furnace.

All trips fail occasionally. So if we are deliberately going to wait for a trip to operate, we should watch the readings and leave ourselves time to intervene if the trip fails to work.

14.8   TESTS MAY FIND FAULTS

Whenever we carry out a test, we may find a fault, and we must be prepared for one.

After changing a chlorine cylinder, two workers opened the valves to make sure there were no leaks on the connecting pipework. They did not expect to find any, so they did not wear air masks. Unfortunately, there were some small leaks, and they were affected by the chlorine.

The workers' actions were not very logical. If they were sure there were no leaks, there was no need to test. If there was a need to test, then leaks were possible, and air masks should have been worn.

Similarly, pressure tests (at pressures above design, as distinct from leak tests at design pressure) are intended to detect defects. Defects may be present—if we were sure there were no defects, we would not need to pressure-test—and therefore we must take suitable precautions. No one should be in a position where he or she may be injured if the vessel or pipework fails (see Section 19.2).

14.9   SOME MISCELLANEOUS INCIDENTS

(a) A radioactive-level indicator on the base of a distillation column was indicating a low level although there was no doubt that the level was normal. Radiography of pipewelds was in operation 60 m (200 ft) away, and the radiation source was pointing in the direction of the radiation detector on the column. When the level in the column is high, the liquid absorbs radiation; when the level is low, more radiation falls on the detector. The detector could not distinguish between radiation from the normal source and radiation from the radiographic source and registered a low level.

(b) As pointed out in Section 1.5.4d, on several occasions fitters have removed thermowells—pockets into which a temperature-measuring device is inserted—without realizing that this would result in a leak.

(c) Section 9.2.1c describes an incident in which a float came loose from a level controller in a sphere containing propane and formed a perfect fit in the short pipe below the relief valve. When the sphere was filled completely and isolated, thermal expansion caused the 14-m (46-ft)-diameter sphere to increase in diameter by 0.15 m (6 in.).

14.10   SOME ACCIDENTS AT SEA

Rudyard Kipling wrote, "What do they know of England who only England know?" In the same way, what do we know about process safety if we know nothing about accidents in other industries? Here are some shipping accidents with lessons for the process industries.

More than 30 years have passed since the U.S. nuclear submarine *Thresher* sank, with the loss of 129 lives, and the reasons may have been forgotten. The immediate cause was a leak of seawater from a silver-brazed joint in the engine room. This, it is believed, short-circuited electrical equipment, causing the reactor to shut down. As a result, the submarine was unable to empty its ballast tanks and rise to the surface.

According to a 1994 report [11], the "nuclear power plant was the focus of the designers' attention; the standards used for the nuclear power plant were more stringent than those for the rest of the submarine." In the process industries' utilities, storage areas and offplots often get less attention than the main units and are involved in disproportionately more incidents. The report continues:

> The Navy had experienced a series of failures with silver-brazing, which resulted in several near-misses, indicating that the traditional quality assurance method, hydrostatic testing, was inadequate. Therefore, the Navy instructed the shipyard to use ultrasonic testing … on the *Thresher's* silver-brazed joints. However, the Navy failed to specify the extent of the testing required and did not confirm that the testing program was fully implemented. When ultrasonic testing proved burdensome and time-consuming, and when the pressures of the schedule became significant, the shipyard discontinued its use in favor of the traditional method. This action was taken despite the fact that 20 out of 145 joints passing hydrostatic testing failed to meet minimum bonding specifications when subject to ultrasonic testing.

In the process industries, many incidents have shown the need to tell contractors precisely what they should do and then check that they have done it. It is easy to forget this at a time of recession and economizing.

Another incident occurred on a British submarine. At the time, small drain valves were used to check that the torpedo outer doors were closed; if water came out of the drain valve, then the outer door was open. The reverse, however, was not true. On one occasion the drain valve was plugged; the inner door was opened when the outer door was also open; the submarine sank, and many sailors drowned. Many similar incidents have occurred in the process industries, for example, when testing for trapped pressure, though with less serious results. Before testing, ask what will happen if the result is not what we expect it to be (Section 14.1g).

The loss of the *Titanic* in 1912 has been the subject of many books. The loss of another luxury ship, the *Ville du Havre*, off the Newfoundland coast in 1873, as the result of a collision, is less well known. The lifeboats were difficult to detach, as the ship was newly painted and everything was stuck fast; many could not be detached in time. The life preservers, along the sides of the deck, were also stuck fast. Fifty-seven people were rescued, but 226 drowned. On chemical plants, painters have been known to paint everything in sight [12].

This disaster, like the loss of the *Thresher*, shows the importance of checking the work of contractors. It also shows the need to try out all emergency equipment from time to time, especially after maintenance, whether it is a diesel generator, an interlock, an alarm, or a lifeboat. On the *Titanic*, the most serious deficiency was lack of sufficient boats for all the passengers, but failure to try out emergency equipment added to the loss of lives. The crew had difficulty removing covers from the boats and cutting them loose. There had been no lifeboat drills, and some of the crew members did not know where to go [13].

Overheard from a woman leaving a movie theater after seeing James Cameron's *Titanic:* "You know, that could really happen."

—*Daily Telegraph* (London), Mar. 2, 1998

There is more on protective systems in Chapter 33.

References

1. W. H. Doyle, "Some Major Instrument Connected CPI Losses," *Paper presented at Chemical Process Industry Symposium*, Philadelphia, PA, 1972.
2. J. A. McLean, *Loss Prevention Bulletin*, No. 110, Apr. 1993, p. 1.
3. Health and Safety Executive, *Passenger Falls from Train Doors*, Her Majesty's Stationery Office, London, 1993.
4. R. Roussakis and K. Lapp, "A Comprehensive Test Method for Inline Flame Arresters," *Paper presented at AIChE Loss Prevention Symposium*, San Diego, CA, Aug. 1990.
5. R. A. McConnell, *Process Safety Progress*, Vol. 16, No. 2, Summer 1997, p. 61.

6. *Operating Experience Weekly Summary*, No. 97-10, Office of Nuclear and Safety Facility, U.S. Dept. of Energy, Washington, DC, 1997, p. 1.
7. *Operating Experience Weekly Summary*, No. 97-12, Office of Nuclear and Safety Facility, U.S. Dept. of Energy, Washington, DC, 1997, p. 5.
8. *Health and Safety at Work*, Vol. 19, No. 5, May 1997, p. 5.
9. *Operating Experience Weekly Summary*, No. 97-27, Office of Nuclear and Safety Facility, U.S. Dept. of Energy, Washington, DC, 1997, p. 4.
10. S. Hall, *Railway Detectives*, Ian Allen, Shepperton, UK, 1990, p. 117.
11. *Occupational Safety Observer*, Vol. 3, No. 6, U.S. Dept. of Energy, Washington, DC, June 1994, p. 4.
12. B. S. Vester, *Our Jerusalem*, Ariel, Jerusalem, 1950, 1988, p. 32.
13. Lord Mersey, *Report on the Loss of the S.S. Titanic*, St. Martin's Press, New York, 1990 (reprint).

Static Electricity

We cannot carry on inspiration and make it consecutive. One day there is no electricity in the air, and the next the world bristles with sparks like a cat's back.

—Ralph Waldo Emerson (1803–1882)

Static electricity (static for short) has been blamed for many fires and explosions, sometimes correctly. Sometimes, however, investigators have failed to find any other source of ignition. So they assume that it must have been static even though they are unable to show precisely how a static charge could have been formed and discharged.

A static charge is formed whenever two surfaces are in relative motion, for example, when a liquid flows past the walls of a pipeline, when liquid droplets or solid particles move through the air, or when someone walks, gets up from a seat, or removes an article of clothing. One charge is formed on one surface—for example, the pipe wall—and an equal and opposite charge is formed on the other surface—for example, the liquid flowing past it.

Many static charges flow rapidly to earth as soon as they are formed. But if a charge is formed on a nonconductor or on a conductor that is not grounded, it can remain for some time. If the level of the charge, the voltage, is high enough, the static will discharge by means of a spark, which can ignite any flammable vapors that may be present. Examples of nonconductors are plastics and nonconducting liquids, such as most pure hydrocarbons. Most liquids containing oxygen atoms in the molecule are good conductors.

Even if a static spark ignites a mixture of flammable vapor and air, it is not really correct to say that static electricity caused the fire or explosion.

The real cause was the leak or whatever event led to the formation of a flammable mixture. Once flammable mixtures are formed, experience shows that sources of ignition are likely to turn up. The deliberate formation of flammable mixtures should never be allowed except when the risk of ignition is accepted—for example, in the vapor spaces of small fixed-roof tanks containing flammable nonhydrocarbons (see Section 5.4).

15.1   STATIC ELECTRICITY FROM FLOWING LIQUIDS

Section 5.4.1 described explosions in storage tanks, and Section 13.3 described explosions in tank trucks, ignited by static sparks. The static was formed by the flow of a nonconducting liquid, and the spark discharges occurred *between the body of the liquid and the grounded metal containers* (or filling arms).

If a conducting liquid such as acetone or methanol flows into an ungrounded metal container, the container acquires a charge from the liquid, and a spark may occur *between the container and any grounded metal that is nearby*, as in the following incidents:

(a) Acetone was regularly drained into a metal bucket. One day the operator hung the bucket on the drain valve instead of placing it on the metal surface below the valve (Figure 15-1).

The handle of the bucket was covered with plastic. When acetone was drained into the bucket, a static charge accumulated on the acetone and on the bucket. The plastic prevented the charge from flowing to earth via the drainpipe, which was grounded. Finally a spark passed between the bucket and the drain valve, and the acetone caught fire.

Even if the bucket had been grounded, it would still have been bad practice to handle a flammable (or toxic or corrosive) liquid in an open container. It should have been handled in a closed can to prevent

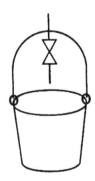

FIGURE 15-1   The bucket was not grounded and acquired a charge.

spillages (see Sections 7.1.3 and 12.2c). Closed cans, however, will not prevent ignition by static electricity, as the following incidents show.

(b) A man held a 10-L metal container while it was being filled with acetone. When he tried to close the valve in the acetone line, the acetone ignited, and the fire spread to other parts of the building. The man was wearing insulating (crepe rubber) shoes, and it is believed that a static charge accumulated on the acetone, the can, and the man. When he put his hand near the valve, a spark jumped from him to the valve, which was grounded, and ignited the acetone vapor.

(c) Metal drums were occasionally filled with vinyl acetate via a 2-in. diameter rubber hose. There was no means for grounding the drum, and the rubber hose did not reach to the bottom of the drum; the liquid splashed down from a height of 0.6 m (2 ft). A few minutes after filling started, a violent explosion occurred, and the ends of the drum were blown out. One end hit a man in the legs, breaking both of them, and the other end broke another man's ankle. He was burned in the ensuing fire and died a few days later.

Note that, as in the incident described in Section 13.3, the operation had been carried out a number of times before conditions were right for an explosion to occur.

(d) Explosions have occurred because external paint prevented grounding of a drum or internal linings prevented grounding of the contents [4].

As with tanks (Section 5.4.1), explosions can also occur in grounded drums containing liquids of low conductivity if a static charge accumulates on the liquid and passes to a grounded conductor, such as a filling pipe. Reference 4 describes some incidents that have occurred. They are most likely when the following occurs:

- The liquid has a low conductivity (less than $50\,pS/m$) (S = siemen) and a low minimum ignition energy (less than $1\,mJ$).
- The vapor-air mixture in the drum is close to the optimum for an explosion. This usually occurs about midway between the lower and upper explosive limits.
- The liquid acquires a high charge by flowing through a filter, rough-bore hose, or other obstruction.

If these conditions are unavoidable, it may be necessary to inert the drum with nitrogen before filling.

15.2 STATIC ELECTRICITY FROM GAS AND WATER JETS

On a number of occasions, people have received a mild electric shock while using a carbon dioxide fire extinguisher. The gas jets from the extinguishers contain small particles of solid carbon dioxide, so a charge will collect on the horn of the extinguisher and may pass to earth via the hand of the person who is holding the horn.

A more serious incident of the same sort occurred when carbon dioxide was used to inert the tanks of a ship, which had contained naphtha. An explosion occurred, killing four men and injuring seven. The carbon dioxide was added through a plastic hose 8 m (25 ft) long, which ended in a short brass hose (0.6 m [2 ft] long) that was dangled through the ullage hole of one of the tanks. It is believed that a charge accumulated on the brass hose and a spark passed between it and the tank (see Section 19.4) [1].

A few years later, carbon dioxide was injected into an underground tank containing jet fuel as a tryout of a firefighting system. The tank blew up, killing 18 people who were standing on top of the tank. In this case, the discharge may have occurred from the cloud of carbon dioxide particles.

The water droplets from steam jets are normally charged, and discharges sometimes occur from the jets to neighboring grounded pipes. These discharges are of the corona type rather than true sparks and may be visible at night; they look like small flames [2].

Discharges from water droplets in ships' tanks (being cleaned by high-pressure water-washing equipment) have ignited flammable mixtures and caused serious damage to several supertankers [3]. The discharges occurred from the cloud of water droplets and were thus "internal lightning."

A glass distillation column cracked, and water was sprayed onto the crack. A spark was seen to jump from the metal cladding on the insulation, which was not grounded, to the end of the water line. Although no ignition occurred in this case, the incident shows the need to ground all metal objects and equipment. They may act as collectors for charges from steam leaks or steam or water jets.

Most equipment is grounded by connection to the structure or electric motors. But this may not be true of insulation cladding, scaffolding, pieces of scrap or tools left lying around, or pieces of metal pipe attached by nonconducting pipe or hose (see next item). In one case, sparks were seen passing from the end of a disused instrument cable; the other end of the cable was exposed to a steam leak.

15.3   STATIC ELECTRICITY FROM POWDERS AND PLASTICS

A powder was emptied down a metal duct into a plant vessel. The duct was replaced by a rubber hose, as shown in Figure 15-2. The flow of powder down the hose caused a charge to collect on it. Although the hose was reinforced with metal wire and was therefore conducting,

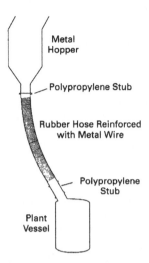

FIGURE 15-2 The flow of powder caused a static charge to collect on the insulated hose.

it was connected to the plant at each end by short polypropylene pipes that were nonconducting. A charge therefore accumulated on the hose, a spark occurred, the dust exploded, and a man was killed.

A nonconducting hose would have held a charge. But a spark from it would not have been as big as from a conducting hose and might not have ignited the dust, though we cannot be certain. It would have been safer than an ungrounded conducting hose but less safe than a grounded conducting hose.

Hoses and ducts used for conveying explosive powders should be made from conducting material and be grounded throughout. Alternatively (or additionally) the atmosphere can be inerted with nitrogen, the ducts can be made strong enough to withstand the explosion, or an explosion vent can be provided.

Electrostatic discharges can ignite a chemical reaction even when no air is present. For example, when a powder was dried under vacuum, electrostatic discharges produced, in the powder, a network of channels of increased conductivity. When the vacuum was broken, with nitrogen, the rise in pressure produced sudden increased sparking and a runaway decomposition of the powder. Operation under a lower vacuum prevented the ignitions, as the discharges were then more frequent and therefore less energetic and less damaging [12].

Another incident occurred in a storage bin for a granular material. The level in the bin was measured by the change in the capacity of a vertical steel cable. The measuring device was disconnected, and the cable thus became an ungrounded conductor. A charge accumulated on it, and a spark passed between the cable and the wall, about 0.3 m (1 ft) away. At the time, the level in the bin was low, and the whole of the cable was uncovered. An explosion occurred in the bin, but it was vented through a relief panel, and there was no damage. The granules were considered difficult to ignite, but the fines in them accumulated on the cable [9].

The first and third incidents are examples of hazards introduced by simple modifications (see Chapter 2). Many dust explosions caused by other sources of ignition are reviewed in reference 10.

Note that introducing a plastic section in a pipeline so that the metal pipe beyond the section is no longer grounded can be a hazard with liquids as well as powders. On several occasions, to prevent splashing when tank trucks are filled, plastic extension pieces have been fitted to the filling arms. The extension pieces included ungrounded metal parts; charge accumulated on them and then discharged, igniting the vapor in the tank trucks [13].

Several fires have occurred when powders were added manually to vessels containing flammable atmospheres, and the use of mechanical methods of addition is recommended [5, 11]. It is better to prevent the formation of explosive mixtures by blanketing with inert gas or by lowering the temperature of the liquid. Reference 5 also describes several discharges that have occurred from plastic surfaces. For example, an operator wiped the plastic cover of an inspection lamp, approved for use in flammable atmospheres, with his glove. The cover became charged, and when it was inserted into a vessel containing a flammable atmosphere—it was an aluminum vessel that had been cleaned with sodium hydroxide solution so that hydrogen was produced—an ignition occurred. Electrical equipment for use in flammable atmospheres should have a surface resistance of less than 1 G ohm at 50% relative humidity. The vessel should not, of course, have been inspected until it had been gas-freed.

A gasoline spillage ignited when someone attempted to sweep it up with a broom that had plastic bristles. The spillage should have been covered with foam.

Although ignitions have occurred as a result of static discharges from plastic surfaces, "the number of incidents is extremely small in relation to the widespread use of plastic material" [6]. If plastic surfaces are liable to become charged and flammable mixtures are likely to be present, then the exposed area of plastic should not exceed 20 cm² if the ignition energy of the mixture is 0.2 mJ; it should be less if the ignition energy is lower.

15.4   STATIC ELECTRICITY FROM CLOTHING

(a) An operator slipped on a staircase, twisted his ankle, and was absent for 17 shifts. The staircase was in good condition, and so were the operator's boots.

The reaction of many people would have been that this is another of those accidents that we can do nothing about, another occasion when "man told to take more care" appears on the accident report.

However, in the plant where the accident occurred, they were not satisfied with this easy way out. They looked into the accident more thoroughly. The injured man was asked why he had not used the handrails.

It then came to light that the handrails were covered with plastic and that anyone using them *and wearing insulating footwear* acquired an electric charge. When he touched the metal of the plant, he got a mild electric shock. The spark, of course, was not serious enough to cause any injury. But it was unpleasant. People therefore tended not to use the handrails.

For a spark to be felt, it must have an energy of at least 1 mJ. The minimum energy required to ignite a flammable mixture is 0.2 mJ, so a spark that can be felt is certainly capable of causing ignition if flammable vapor is present.

(b) We have all acquired a static charge by walking across a man-made fiber carpet (or just by getting up from our chairs) and then felt a mild shock when we touched a metal object, such as a filing cabinet. Similar charges can be acquired by walking across a plant floor wearing nonconducting footwear. And sparks formed in this way have been known to ignite leaks of flammable gas or vapor, especially in dry climates. However, the phenomenon is rare. It does not justify insistence on the use of conducting footwear unless leaks are common [7]. If leaks are common, action to prevent them from occurring is more effective than action to prevent them from igniting.

(c) A driver arrived at a filling station, removed the cap from the end of the filler pipe, and held it in his hand while an attendant filled the car with gasoline. The driver took off his pullover sweater, thus acquiring a charge and leaving an equal and opposite charge on the pullover, which he threw into the car. He was wearing nonconducting shoes, so the charge could not leak away to earth.

When he was about to replace the cap on the end of the filler pipe, a spark jumped from the cap to the pipe, and a flame appeared on the end of the pipe. It was soon extinguished. The flame could not travel back into the gasoline tank, as the mixture of vapor and air in the tank was too rich to explode.

At one time there was concern that man-made fiber clothing might be more likely than wool or cotton clothing to produce a charge on the wearer. The incident just described shows that the static charge was produced only when the clothing was removed. When dealing with a leak, we do not normally start by removing our clothing. There is therefore no need to restrict the types of cloth used, so far as static electricity is concerned. Electrostatic sparks from people are reviewed in reference 8.

References

1. *Fire Journal,* Nov. 1967, p. 89.
2. A. F. Anderson, *Electronics and Power,* Jan. 1978.
3. *S. S. Mactra (ON 337004)—Report of Court No. 8057 Formal Investigation,* Her Majesty's Stationery Office, London, 1973.
4. L. G. Britton and J. A. Smith, "Electrostatic Hazards of Drum Filling," Paper presented at AIChE Loss Prevention Symposium, Minneapolis, Minn., Aug. 1987.
5. Health and Safety Executive, *Electrostatic Ignition,* Her Majesty's Stationery Office, London, 1982.
6. N. Gibson and D. J. Harper, *Journal of Electrostatics,* Vol. 11, 1981, p. 27.
7. R. W. Johnson, *Loss Prevention,* Vol. 14, 1981, p. 29.
8. B. D. Berkey, T. H. Pratt, and G. M. Williams, "Review of Literature Related to Human Spark Scenarios," Paper presented at AIChE Loss Prevention Symposium, Minneapolis, Minn., Aug. 1987.
9. L. G. Briton and D. C. Kirby, *Plant/Operations Progress,* Vol. 8, No. 3, July 1989, p. 177.
10. R. K. Eckhoff, *Dust Explosions in the Process Industries,* 2nd edition, Butterworth-Heinemann, Oxford, UK, 1997.
11. J. Bond, *Loss Prevention Bulletin,* No. 088, Aug. 1989, p. 21.
12. D. K. Davies, *Electrostatics News,* No. 5, Southampton University, UK, Spring 1966, p. 3.
13. T. H. Pratt and J. G. Atharton, *Process Safety Progress,* Vol. 15, No. 3, Fall 1996, p. 173.

Materials of Construction

For him iron is as flimsy as straw, and bronze as soft as rotten wood.

—Job 41:27, Good News Bible

16.1 WRONG MATERIAL USED

Many incidents have occurred because the wrong material of construction was used. This has usually been the result of errors by maintenance or construction personnel or suppliers, who did not use or did not supply the materials specified. Few failures have been the result of errors by materials specialists who incorrectly specified the materials to be used.

The following incidents are typical:

(a) A titanium flange was fitted by mistake on a line carrying dry chlorine. The flange caught fire. Titanium is ideal for wet chlorine but catches fire on contact with dry chlorine. (Burning in this case means rapid combination with chlorine, not oxygen.)

In another incident, on a new plant, two flanged joints leaked an hour after chlorine was introduced. The gaskets were removed and analyzed and were found to be made from titanium although they were stamped Hastelloy, the material specified [5].

PTFE gaskets were specified for a section of plant that handled acid. As they are fragile and expensive and as an extensive series of tests using water had to be carried out during startup, temporary nitrile rubber gaskets were used during this period. You can guess what happened. One of them was left in position and corroded, causing an acid leak. Subsequent checks showed that many more gaskets were made of the wrong material.

(b) A carbon steel valve painted with aluminum paint was used instead of a stainless steel valve. It corroded rapidly.

(c) A plug valve was supplied with a pure nickel plug instead of one made from 304L stainless steel. The valve body was made from the correct material. The valve was installed in a nitric acid line. Five hours later the plug had disappeared, and acid was escaping through the stem seal.

The manufacturers had provided a test certificate stating that the valve was made from 304L steel.

(d) During the night, a valve had to be changed on a unit that handled a mixture of acids. The fitter could not find a suitable valve in the workshop, but on looking around he found one on another unit. He tested it with a magnet and, finding it nonmagnetic, assumed it was similar to the stainless steel valves normally used. He therefore installed it. Four days later, the valve was badly corroded and there was a spillage of acids.

The valve was made of Hastelloy, an alloy suitable for use on the unit where it was found but not suitable for use with the mixture of acids on the unit on which it had been installed.

(e) A tank truck, used for internal transport, looked as if it was made of stainless steel. It was therefore filled with 50% caustic soda solution. Twelve hours later, the tank was empty. It was made of aluminum, and the caustic soda created a hole and leaked out.

The material of construction has now been stenciled on all tank trucks used for internal transport in the plant where the incident occurred.

(f) A small, new tank was installed with an unused branch blanked off. A month later the branch was leaking. It was then discovered that the tank had arrived with the branch protected by a blank flange made of wood. The wood was painted the same color as the tank, and nobody realized that it was not a steel blank.

(g) A leak on a refinery pump, which was followed by a fire, was due to incorrect hardness of the bolts used. Other pumps supplied by the same manufacturer were then checked, and another was found with off-specification bolts. The pump had operated for 6,500 hours before the leak occurred.

If the pump had been fitted with a remotely operated emergency isolation valve as recommended in Section 7.2.1, the leak could have been stopped quickly. Damage would have been slight. As it was, the unit shut down for five weeks.

(h) Section 9.1.6b describes what happened when the exit pipe of a high-pressure ammonia converter was made from carbon steel instead of 1.25% Cr, 0.5% Mo. Hydrogen attack occurred, a hole appeared at a bend, and the reaction forces from the escaping gas pushed the converter over.

A return bend on a furnace failed after 20 years of service. It was then found that it had been made from carbon steel instead of the alloy specified.

(i) After some new pipes were found to be made of the wrong alloy, further investigation showed that many of the pipes, clips, and valves in store were also made of the wrong alloys. The investigation was extended to the rest of the plant, and the following are some examples of the findings:

1. The wrong electrodes had been used for 72 welds on the tubes of a fired heater.
2. Carbon steel vent and drain valves had been fitted on an alloy steel system.
3. An alloy steel heat exchanger shell had been fitted with two large carbon steel flanges. The flanges were stamped as alloy.

(j) Checks carried out on the materials delivered for a new ammonia plant showed that 5,480 items (1.8% of the total) were delivered in the wrong material. These included 2,750 furnace roof hangers; if the errors had not been spotted, the roof would probably have failed in service:

> [V]endors often sent without notice what they regarded as "superior" material. Thus, if asked to supply 20 flanges in carbon steel of a given size, the vendor, if he had only 19 such flanges available, was quite likely to add a 20th of the specified size in "superior" 2.25% Cr. When challenged the vendor was often very indignant because he had supplied "superior," i.e., more expensive, material at the original price. We had to explain that the "superior" material was itself quite suitable, *if we knew about it*. If we didn't, we were quite likely to apply the welding procedures of carbon steel to 2.25% Cr steel with unfortunate results [1].

As the result of incidents such as those described in (c) and (g) through (j), many companies now insist that if the use of the wrong grade of steel can affect the integrity of the plant, all steel (flanges, bolts, welding rods, etc., as well as pipes) must be checked for composition before use. The analysis can be carried out easily with a spectrographic analyzer. The design department should identify which pipelines and so on need to be checked and should mark drawings accordingly.

I was present at a meeting where the unexpected corrosion of a pipe was discussed. A materials expert gave a long, complex description, above the heads of most of those present, of possible reasons. I then asked if the material of construction had been checked before installation. We were told it had not, as the company did not have amaterials identification program at the time of construction. Several of those present recalled such incidents that had occurred elsewhere, including a case where a scaffold pole, which looked similar to the process piping, was installed in a boiler.

According to Quinion [17], writing in another context, "The better [explanations] sound, the more circumstantial and detailed the background, the neater the conclusion, the less likely they are to be true. Conversely, if a story is mundane and boring, it is likely to be correct."

Anecdotal evidence exists that some companies have relaxed their material identification programs when their suppliers' systems comply with quality standards. In view of the serious results of occasional minor errors—minor from the suppliers' point of view—it is doubtful if this is wise.

(k) The recycle of scrap to produce stainless steel has led to increases in the concentration of trace elements not covered in the steel specifications. This may lead to poorer corrosion resistance and weld quality, although so far only the nuclear power industry has reported problems. One report says, "Work on understanding the basic processes of impurity segregation in steels and the resulting embrittlement has been very important in understanding component failure problems on plant" [6].

(l) Sometimes the wrong steel has been supplied as the result of misunderstanding rather than wrong labeling. Thus, suppliers have delivered CS (carbon steel) instead of C5 steel (5% Cr, 1.5% Mo, 1% W). 5% Cr, 0.5% Mo is sometimes called P5, but this name is also used to describe 2.5% Cr steel.

(m) The U.S. Department of Energy has reported that some imported nuts and bolts were substandard and failed in service, causing 61 crashes of private planes between 1984 and 1987 and a fire in a U.S. Navy destroyer [7].

(n) Creep failures have been described in furnaces (Section 10.7.2) and in a pipe (Section 9.1.6a). After 28 years of service at 540 to 600 degrees C (1,000 to 1,100 degrees F) and a gauge pressure of 900 psi (60 bar), the studs holding a bonnet of a 28-in. valve expanded as the result of creep. The effect was similar to that produced when a nut is forced onto a stud with a thread of a different pitch. The load was held by only a few threads, the studs failed, and the bonnet separated from the body. Once one stud failed, the load on the others increased, and there was a rapid cascade of failures [14].

During design the life expectancy, due to creep or other forms of corrosion, should be estimated and examination or replacement planned. Cheap fittings, such as studs, bolts, and nuts, should be replaced in good time. Not to do so is penny-pinching and expensive in the end.

Here are two more examples of penny-pinching. The piston of a reciprocating engine was secured to the piston rod by a nut, which was locked in position by a tab washer. When the compressor was overhauled, the tightness of this nut was checked. To do this, the tab on the washer had to be knocked down and then knocked up again. This weakened the washer so that the tab snapped off in service, the nut worked loose, and the piston hit the end of the cylinder, fracturing the piston rod.

The load on a 30-ton hoist slipped, fortunately without injuring anyone. It was then found that a fulcrum pin in the brake mechanism had worked loose, as the split pin holding it in position had fractured and fallen. The bits of the pin were found on the floor.

Split pins and tab washers should not be reused but replaced every time they are disturbed. Perhaps we cannot be bothered to go to the store for a fresh supply. Perhaps there is none in the store.

16.2   HYDROGEN PRODUCED BY CORROSION

Hydrogen produced by corrosion can turn up in unexpected places, as shown by the following incidents:

(a) An explosion occurred in a tank containing sulfuric acid. As the possibility of an explosion had not been foreseen, the roof/wall weld was stronger than usual, and the tank split at the base/wall weld. The tank rose 15 m (50 ft) into the air, went through the roof of the building, and fell onto an empty piece of ground nearby, just missing other tanks. Fortunately, no one was hurt. If the tank had fallen on the other side of the building, it would have fallen into a busy street.

Slight corrosion in the tank had produced some hydrogen. The tank was fitted with an overflow pipe leading down to the ground, but it had no vent. So the hydrogen could not escape, and it accumulated under the conical roof. Welders working nearby ignited the hydrogen. (presumably, some found its way out of the overflow) [2].

The tank should have been fitted with a vent at the highest point, as shown in Figure 16-1.

Many suppliers of sulfuric acid recommend that it is stored in pressure vessels designed to withstand a gauge pressure of 30 psi (2 bar). The acid is usually discharged from tank trucks by compressed air, and if the vent is choked the vessel could be subjected to the full pressure of the compressed air.

(b) Hydrogen produced by corrosion is formed as atomic hydrogen. It can diffuse through iron. This has caused hydrogen to turn up in unexpected places, such as the insides of hollow pistons. When holes have been drilled in the pistons, the hydrogen has come out and caught fire [3].

In another case, acidic water was used to clean the inside of the water jacket that surrounded a glass-lined vessel. Some hydrogen

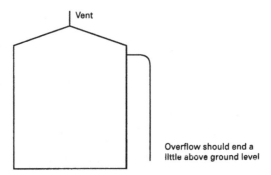

FIGURE 16-1 Acid tanks should be fitted with a high-point vent, as well as an overflow, so that hydrogen can escape.

diffused through the wall of the vessel and developed sufficient pressure to crack the glass lining.

Corrosion uses up oxygen, and this has caused tanks to collapse (see Section 5.3d) and persons to be overcome when entering a vessel (see Section 11.1d).

(c) The sudden failure of six bonnet studs on an 8-in. valve caused a release of hydrogen fluoride, which killed two men and hospitalized 10 others. The failure was the result of hydrogen-assisted stress corrosion cracking. In this phenomenon, hydrogen, produced by corrosion, migrates to flaws in areas of high tensile stress, where it lowers the energy needed for cracks to grow. When the cracks reach a critical size, the equipment fails suddenly. The grade of steel used in this case was unsuitable; reference 8 lists the types that should be used.

16.3   OTHER EFFECTS OF CORROSION

Corrosion usually results in a leak or failure of a support because a vessel or support gets too thin. It is then not strong enough to withstand the pressure or load. However, rust can cause failure in another way. It occupies about seven times the volume of the steel from which it was formed. When rust occurs between two plates that have been bolted or riveted together, a high pressure develops. This can force the plates apart or even break the bolts or rivets (see Section 9.1.2g). Corrosion of the reinforcement bars in concrete can cause the concrete to crack and break away.

16.4   LOSS OF PROTECTIVE COATINGS

Aluminum pump impellers are often used to pump fluorinated hydrocarbon refrigerants. If the impeller rubs against the casing, the protective film of aluminum oxide is removed, and combined with the local heating produced by the rubbing, which allows the aluminum to react with the refrigerant, the impeller may disappear. Contact between the impeller and the casing may be a result of worn bearings, which in turn are the result of compressor surges, so the reasons for any surging should be investigated [4].

A special type of high-pressure joint incorporated copper gaskets. A change was made to aluminum after laboratory tests showed no sign of reaction with the process material. The gaskets normally lasted for many years, but one failed after a few days. It was then found that the man who installed it, anxious to do a good job, had cleaned the gasket immediately before installing it. In doing, so he removed the film of oxide, and the aluminum now dissolved in the process liquid. It was usual to clean the gaskets a few days before they were installed. Though it was not realized at the time, this allowed a fresh oxide film to form.

A change was made back to copper. It is more user friendly than aluminum and will tolerate cleaning or scratching of the surface.

16.5   SOME OTHER INCIDENTS CAUSED BY CORROSION

(a) An oil company took a section of plant out of use and, because of an oversight, did not remove process materials from all the pipework. For 18 years a pipe was left with a mixture of hydrogen fluoride and benzene boxed up inside it. Finally, the walls became so thin that they burst, and 10 men were taken to the hospital suffering from the effects of acid gas [9].

(b) A plant made an evaporator for liquid nitrogen by running hundreds of meters of copper piping through a steel tank filled with water. Although the steel was painted, it corroded right through in six months as the result of galvanic corrosion—that is, the steel and the copper formed an electrolytic cell. Paint never gives 100% cover, and if 1% of the steel was uncovered, all the current would have passed though this area, and its corrosion rate would have been increased 100 times. Painting the copper, which did not corrode, would be more effective than painting the steel [10]!

This incident illustrates the hazards of do-it-yourself engineering by people who do not fully understand the properties of the materials they are using.

(c) Minute amounts (up to $300\,\mu/m^3$) of mercury in natural gas have caused brittle failure of certain alloys. Valves have failed as a result. In addition, reaction of the mercury with ammonia can produce explosive compounds [11].

(d) Some catalyst tubes in a reactor failed as a result of chloride-induced stress corrosion cracking soon after startup. A materials expert, called in to investigate, found that all the failures had occurred in one corner of the reactor; that men had been working on the roof, day and night, for several weeks after the tubes had been fitted; that this area of the roof could not be seen from the rest of the plant; and that to reach the nearest restroom the men had to negotiate three ladders [12].

Reference 13 describes some other corrosion problems.

16.6   FIRES

We know that metals, especially aluminum (see Section 10.1), can be affected by fire, but we do not usually consider the possibility that they will burn. However, some metals, including titanium, will burn when powdered or finely divided, and bulk titanium will also burn. Three titanium heat exchangers were set alight and destroyed by burning operations. In one case, ignition was started by direct contact with the torch and in the other two cases by contact with hot slag [15]. Great care is needed if welding or burning is carried out anywhere near titanium equipment.

16.7   CHOOSING MATERIALS

In choosing materials of construction, we have to compromise between various factors. Kirby [16] uses the acronym SHAMROCK to summarize and remember them:

S = Safety: What are the consequences of failure? If they are serious, a more resistant material than usual may be justified. For example, on a plant where leaking water would react violently with process materials, the water lines were made from a grade of steel resistant to stress corrosion cracking (from the chloride in the cooling water) as well as rust.

H = History: If a plant has used material successfully for many years, and the staff members know its strengths and limitations, how to weld it, and so on, they may hesitate before making a change. For example, a fiberglass-reinforced plastic had given excellent service for many years; when another composite from the same company, with the same name but a different number, was used instead, it failed overnight.

A = Availability: Before a salesperson sells you the latest wonder-working material, ask how easy it will be to get replacement supplies in a hurry.

M = Maintenance: A plant engineer saved $10,000 per year by no longer neutralizing the slightly acidic cooling water. In time, rust formation in 30 jacketed reactors increased reaction times by 25%. I have known several engineers who gained a reputation for efficiency by similar measures, including neglecting maintenance, and then left their successors to pick up the tab.

R = Reparability: A plant bought some vessels with a new type of plastic lining instead of the one they had used for many years. The new material had better temperature resistance than the old, but when it did need repair, the patches would not stick. In time the problems were overcome, but reparability should have been considered before the change was made.

O = Oxidizing/reducing nature of process fluids: In acidic solutions, this affects the choice of alloys.

C = Cost: An important consideration, but look at lifetime costs, including maintenance, not just at initial costs. Penny-pinching (Section 16.1n) is rarely worthwhile.

K = Kinetics of corrosion mechanisms: Unless we understand these, we will not know which materials will be suitable and which will not.

References

1. G. C. Vincent and C. W. Gent, *Ammonia Plant Safety*, Vol. 20, 1978, p. 22.
2. *Chemical Safety Summary*, No. 192, Chemical Industries Association, London, Oct./Dec. 1977.
3. *Case Histories of Accidents in the Chemical Industry*, No. 1807, Manufacturing Chemists Association, Washington, DC, Apr. 1975.
4. R. Stevens, *Plant/Operations Progress*, Vol. 4, No. 2, 1985, p. 68.
5. *Loss Prevention Bulletin*, No. 097, Feb. 1991, p. 9.
6. B. Eyre, *Atom*, No. 407, Oct. 1990, p. 11.
7. *DOE Quality Alert*, Bulletin No. DOE/EH-0266, U.S. Dept. of Energy, Washington, DC, Aug. 1992.
8. *Loss Prevention Bulletin*, No. 089, Oct. 1989, p. 27.
9. *Health and Safety at Work*, Vol. 14, No. 6, June 1990, p. 4.
10. M. Turner, *The Chemical Engineer*, No. 468, Jan. 1990, p. 28.
11. S. M. Williams, *Plant/Operations Progress*, Vol. 10, No. 4, Oct. 1991, p. 189.
12. M. Turner, *The Chemical Engineer*, No. 492, 1991, p. 40.
13. *Corrosion Awareness—A Three-Part Videotape Series*, Gulf Publishing Co., Houston, Tex., 1991.
14. S. J. Brown, *Plant/Operations Progress*, Vol. 6, No. 1, Jan. 1987, p. 20.
15. G. E. Mahnken and M. T. Rook, *Process Safety Progress*, Vol. 16, No. 1, Spring 1997, p. 54.
16. G. N. Kirby, *Chemical Engineering Progress*, Vol. 92, No. 6, June 1996, p. 38.
17. M. Quinion, *Port Out, Starboard Home and Other Language Myths*, Penguin Books, London, 2004, p. 228.

CHAPTER

17

Operating Methods

[P]eople place their faith in systems either because they're new (so they simply must be good) or because they're old and have worked a long time.

—Wendy Grossman, *Daily Telegraph* (London), July 29, 1997

This chapter describes some accidents that occurred because operating procedures were poor. It does not include accidents that occurred because of defects in procedures for preparing equipment for maintenance or vessels for entry. These are discussed in Chapters 1, 11, 23, and 24.

17.1 TRAPPED PRESSURE

Trapped pressure is a familiar hazard in maintenance operations and is discussed in Section 1.3.6. Here we discuss accidents that have occurred as a result of process operation.

Every day, in every plant, equipment that has been under pressure is opened up. This is normally done under a work permit. One employee prepares the job, and another opens up the vessel. And it is normally done by slackening bolts so that any pressure present will be detected before it can cause any damage—provided the joint is broken in the correct way, described in Section 1.5.1.

Several fatal or serious accidents have occurred when one worker has carried out the whole job—preparation and opening up—and has used a quick-release fastening instead of nuts and bolts. One incident, involving a tank truck, is described in Section 13.5. Here is another:

A suspended catalyst was removed from a process stream in a pressure filter. After filtration was complete, the remaining liquid was blown

doi:10.1016/B978-1-85617-531-9.00017-2

out of the filter with steam at a gauge pressure of 30 psi (2 bar). The pressure in the filter was blown off through a vent valve, and the fall in pressure was observed on a pressure gauge. The operator then opened the filter for cleaning. The filter door was held closed by eight radial bars, which fitted into U-bolts on the filter body. The bars were withdrawn from the U-bolts by turning a large wheel fixed to the door. The door could then be withdrawn.

One day an operator started to open the door before blowing off the pressure. As soon as he opened it a little, it blew open and he was crushed between the door and part of the structure and was killed instantly.

In situations such as this, it is inevitable that sooner or later an operator will forget that he has not blown off the pressure and will attempt to open up the equipment while it is still under pressure. On this particular occasion, the operator was at the end of his last shift before starting his vacation.

As with the accidents described in Section 3.2, it is too simple to say that the accident was due to the operator's error. The accident was the result of a situation that made it almost inevitable.

Whenever an operator has to open up equipment that has been under pressure, the following should be in place:

(a) The design of the door or cover should allow it to be opened about ¼ in. (6 mm) while still capable of carrying the full pressure, and a separate operation should be required to release the cover fully. If the cover is released while the vessel is under pressure, then this is immediately apparent, and the pressure can blow off through the gap, or the cover can be resealed.

(b) Interlocks should be provided so that the vessel cannot be opened up until the source of pressure is isolated and the vent valve is open.

(c) The pressure gauge and vent valve should be visible to the operator when he or she is about to open the door or cover [1].

Pressure can develop inside drums, and then when the lid is released, it may be forcibly expelled and injure the person releasing it. Most of the incidents reported have occurred in waste drums where chemicals have reacted together. For example, nitric acid has reacted with organic compounds. Acids may corrode drums and produce hydrogen. Rotting organic material can produce methane. Materials used for absorbing oil spillages can expand to twice their original volume. Some absorbent was placed in drums with waste oil; the drums were allowed to stand for two days before the lids were fitted, and 10% free space was left, but nevertheless pressure developed inside them. If drums are bulging, lid-restraining devices should be fitted before they are opened or even moved [9].

17.2 CLEARING CHOKED LINES

(a) A man was rodding out a choked ¼-in. line leading to an instrument (Figure 17-1a). When he had cleared the choke, he found that the valve would not close and he could not stop the flow of flammable liquid. Part of the unit had to be shut down.

Rodding out *narrow bore* lines is sometimes necessary. But before doing so, a ball valve or cock should be fitted on the end (Figure 17-1b). It is then possible to isolate the flow when the choke has been cleared, even if the original valve will not close.

(b) Compressed air at a gauge pressure of 50 psi (3.4 bar) was used to clear a choke in a 2-in. line. The solid plug got pushed along with such force that when it reached a slip-plate (spade), the slip-plate was knocked out of shape, rather like the one shown in Figure 1-6.

On another occasion, a 4-in.-diameter vertical U-tube, part of a large heat exchanger, was being cleaned mechanically when the cleaning tool, which weighed about 25 kg, stuck in the tube. A supply of nitrogen at a gauge pressure of 3,000 psi (200 bar) was available, so it was decided to use it to try to clear the choke. The tool shot out of the end of the U-tube and came down through the roof of a building 100 m away.

Gas pressure should never be used for clearing choked lines.

(c) A 1-in. line, which had contained sulfuric acid, was choked. It was removed from the plant, and an attempt was made to clear it with water from a hose. A stream of acid spurted 5 m into the air, injuring one of the men working on the job. Those concerned either never knew or had forgotten that much heat is evolved when sulfuric acid and water are mixed.

(d) When clearing chokes in drain lines, remember that there may be a head of liquid above the choke. The following incident illustrates the hazards:

The drain (blowdown) line on a boiler appeared to be choked. It could not be cleared by rodding (the choke was probably due to

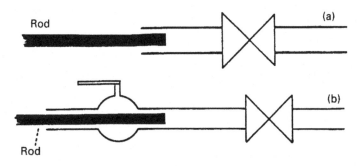

FIGURE 17-1 The wrong (a) and right (b) ways of clearing a choked line.

scale settling in the base of the boiler), so the maintenance foreman pushed a water hose through the drain valve and turned on the water. The choke cleared immediately, and the head of water left in the boiler pushed the hose out of the drain line and showered the foreman with hot water. Although the boiler had been shut down for 15 hours, the water was still at 80 to 90 degrees C (175 to 195 degrees F) and scalded the foreman.

Clearing the choke should not have been attempted until the temperature of the water was below 60°C (140°F), the foreman should have worn protective clothing, and if possible a second valve should have been fitted to the end of the drain line as described in part (a). The accumulation of scale suggests that the water treatment was not adequate [3].

(e) An acid storage tank was emptied so that the exit valve could be changed. The tank was then filled with acid, but the new valve seemed to be choked. After the tank had been emptied again (quite a problem, as the normal exit line was not available), the staff found that the gasket in one of the flanged joints on the new valve had no hole in it!

(f) An operator who tried to clear a choke in a pump with high-pressure steam was killed when the seal gave way and sprayed him with a mixture of steam and a corrosive chemical (2,4-dichlorophenol). He was not wearing protective clothing. The seal was the wrong type, was badly fitted, and had cracked. When the company was prosecuted, its defense was that the operator should have notified the maintenance department and not attempted to clear the choke himself; had the managers known that operators tried to clear blockages by themselves, they would not have condoned the practice. However, this is no excuse; it is the responsibility of managers to keep their eyes open and know what goes on.

(g) The company had set up a computer system designed to pinpoint any equipment that needed replacing, but eight months before the accident it was found to be faulty and was shut down. The judge said, "You don't need an expert armed with a computer to know what will happen when the wrong type of seal is mixed with high-pressure steam" [4].

17.3   FAULTY VALVE POSITIONING

Many accidents have occurred because operators failed to open (or close) valves when they should have. Most of these incidents occurred because operators forgot to do so, and such incidents are described in Sections 3.2.7, 3.2.8, 13.5, and 17.1. In this section we discuss incidents

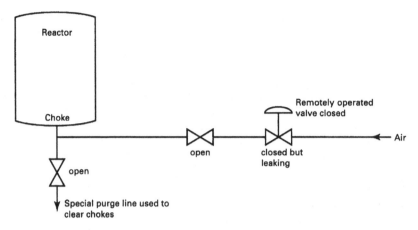

FIGURE 17-2 Liquid purge burned in the drain line.

that occurred because operators did not understand why valves should be open (or closed):

(a) As described in Section 3.3.4c, the emergency blowdown valves on a plant were kept closed by a hydraulic oil supply. One day the valves opened, and the plant started to blow down. It was then discovered that, unknown to the manager and contrary to instructions, the foreman had developed the practice of isolating the oil supply valve "in case the supply pressure in the oil system failed." This was a most unlikely occurrence and much less likely than the oil pressure leaking away from an isolated system.

(b) The air inlet to a liquid-phase oxidation plant became choked from time to time. To clear the choke, the flow of air was isolated, and some of the liquid in the reactor was allowed to flow backward through the air inlet and out through a purge line, which was provided for this purpose (Figure 17-2).

One day the operator closed the remotely operated valve in the air line but did not consider it necessary to close the hand valve as well, although the instructions said he should. The remotely operated valve was leaking, the air met the reactor contents in the feed line, and reaction took place there. The heat developed caused the line to fail, and a major fire followed.

The air line should have been provided with remotely operated double block and bleed valves, operated by a single button.

Other incidents in which operators relied on automatic valves and did not back them up with hand valves are described in Sections 14.6 and 17.5c.

(c) An engineer flew from Japan to Korea to investigate a customer's complaint: there must be something wrong with the crude oil supplied, as no distillate was produced. Within 30 minutes he found that a valve in the vacuum system had been closed incorrectly [10].

17.4 RESPONSIBILITIES NOT DEFINED

The following incident shows what can happen when responsibility for plant equipment is not clearly defined and operators in different teams, responsible to different supervisors, are allowed to operate the same valves.

The flarestack shown in Figure 17-3 was used to dispose of surplus fuel gas, which was delivered from the gasholder by a booster through valves C and B. Valve C was normally left open because valve B was more accessible.

One day the operator responsible for the gasholder saw that it had started to fall. He therefore imported some gas from another unit. Nevertheless, a half-hour later the gasholder was sucked in.

Another flarestack at a different plant had to be taken out of service for repair. An operator at this plant therefore locked open valves A and B so that he could use the "gasholder flarestack." He had done this before, though not recently, and some changes had been made since he last used the flarestack. He did not realize that his action would result in the gasholder emptying itself through valves C and B. He told three other men what he was going to do, but he did not tell the gasholder operator. He did not know that this man was concerned.

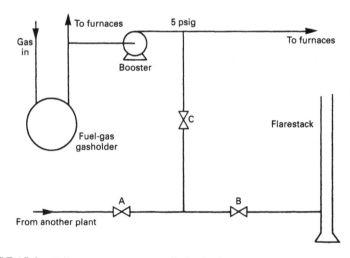

FIGURE 17-3   Different operators controlled valve B.

Responsibility for each item of equipment should be clearly defined at the supervisor, foreman, and operator levels, and only the people responsible for each item should operate it. If different teams are allowed to operate the same equipment, then sooner or later an incident will occur. Section 10.7.2c describes a similar incident.

17.5 COMMUNICATION FAILURES

This section describes some incidents that occurred because of failures to tell people what they needed to know, because of failures to understand what had been told, and because of misunderstandings about the meanings of words:

(a) A maintenance foreman was asked to look at a faulty cooling water pump. He decided that, to prevent damage to the machine, it was essential to reduce its speed immediately. He did so, but did not tell any of the operating team members straight away. The cooling water rate fell, the process was upset, and a leak developed on a cooler.

(b) A tank truck, which had contained liquefied petroleum gas, was being swept out before being sent for repair. The laboratory staff was asked to analyze the atmosphere in the tanker to see if any hydrocarbon was still present. The laboratory staff regularly analyzed the atmosphere inside liquefied petroleum gas (LPG) tank trucks to see if any oxygen was present. Owing to a misunderstanding, they assumed that an oxygen analysis was required on this occasion and reported over the telephone, "None detected." The operator assumed that no hydrocarbon had been detected and sent the tank truck for repair.

Fortunately, the garage had its own check analysis carried out. This showed that LPG was still present—actually more than 1 ton of it.

For many plant control purposes, telephone results are adequate. But when analyses are made for safety reasons, results should be accepted only in writing.

(c) A batch vacuum still was put on standby because there were some problems in the unit that took the product. The still boiler was heated by a heat transfer oil, and the supply was isolated by closing the control valve. The operators expected that the plant would be back on line soon, so they did not close the hand isolation valves, and they kept water flowing through the condenser. However, the vacuum was broken, and a vent on the boiler was opened.

The problems at the downstream plant took much longer than expected to correct, and the batch still stayed on standby for five days. No readings were taken, and when recorder charts ran out, they were not replaced.

The heat transfer oil control valve was leaking. Unknown to the operators, the boiler temperature rose from 75 to 143 degrees C (167 to 290 degrees F), the boiling point of the contents. Finally, bumping in the boiler caused about 0.2 ton of liquid to be discharged through the vent.

Other incidents that occurred because operators relied on automatic valves and did not back them up with hand valves are described in Sections 14.6 and 17.3b. In this incident, the point to be emphasized in addition is that the operators were not clear on the difference between a standby and a shutdown. No maximum period for standby was defined. And no readings were taken during periods on standby. Plant instructions should give guidance on both these matters.

(d) Designers often recommend that equipment is "checked" or "inspected" regularly. But what do these words mean? Designers should state precisely what tests should be carried out and what they hope to determine by the test.

In 1961, a brake component in a colliery elevator failed, fortunately without serious consequences. An instruction was issued that all similar components should be examined. It did not say how or how often. At one colliery the component was examined in position but was not removed for complete examination and was not scheduled for regular examination in the future.

In 1973 it failed, and 18 men were killed [2].

(e) Under the U.K. Ionizing Radiation (Sealed Sources) Regulations, all sealed radioactive sources must be checked by an authorized person "each working day" to make sure that they are still in position. Following an incident at one plant, it was found that the plant took this to mean that the authorized person must check the presence of the sources on Mondays to Fridays but not on weekends. However, "each working day" means each day the radioactive source is working, not each day the authorized person is working!

(f) Teams develop their own shorthand. It is useful, but it can also lead to misunderstandings. On a new unit, the project team had to order the initial stocks of materials. One member of the team, asked to order some TEA, ordered some drums of triethylamine. He had previously worked on a plant where triethylamine was used, and it was called TEA. The manager of the new unit ordered a continuing supply of drums of triethanolamine, the material actually needed and called TEA on the plant where he had previously worked. An alert storeman discovered the confusion when he noticed that two different materials with similar names had been delivered for the same unit, and he asked if both were really required.

On other occasions, the wrong material has been delivered because prefixes such as *n*- or *iso*- were left off when ordering.

(g) A low pumping rate was needed during startup, so the designer installed a kick-back line. For unknown reasons, it fell out of use—perhaps it was not possible to operate at a low enough rate even with the kick-back in use—and instead the operators controlled the level in the suction vessel by switching the pump on and off. The control room operator watched the level and asked the outside operators over a loudspeaker to start up and shut down the pump as required. The two outside operators worked as a team; both could do every job, and they shared the work. One day the control room operator asked for the pump to be shut down. Both outside operators were some distance away; each assumed that the other would be nearer and would shut it down. Neither shut it down, the suction vessel was pumped dry, and the pump overheated and caught fire.

Teamworking, in which everybody can do a job, can easily deteriorate into a system where nobody does it.

17.6   WORK AT OPEN MANHOLES

It was the practice on one plant to remove the manhole cover from a vessel containing warm toluene, inside a building, in order to add a solid. A change in the composition of the feedstock, not detected by analysis, resulted in the emission of more vapor than usual, and the operator was killed. Afterward it was found that the ventilation system was "poorly designed, badly installed, and modified somewhat ineffectively. In addition there appeared to have been no scheduled maintenance of the ventilation system, which was subsequently in an ineffective condition."

It is bad practice to carry out operations at open manholes when flammable or toxic vapors may be present. (Another incident was described in Section 3.3.4a.) Whenever possible, operations should be carried out in the open air or in open-sided buildings. Gas detectors should be installed if vapors are liable to leak into closed buildings.

Many ventilation systems are part of the protective equipment of the plant (see Chapter 14), and like all protective equipment, they should be tested regularly against agreed performance criteria.

17.7   ONE LINE, TWO DUTIES

The following incident shows the hazards of using the same line for different materials. The cost of an extra line is well repaid if it prevents just one such incident.

An operator made up a solution of hydrogen peroxide (1% to 3%) in a makeup tank. His next job was to pump the solution into another vessel.

A branch of the transfer line led to a filter, and the valve in this line had been left open (following an earlier transfer of another material). Some of the solution went into the filter. When the operator realized what was happening, he closed the filter inlet valve but did not remove the solution that was in the filter; he did not know that it would decompose on standing. There was no relief valve on the filter, and about 12 hours later the pressure broke the head bolts and blew the head off the filter. After the explosion, separate lines and pumps were installed for the two duties, a relief valve was fitted to the filter, and the hazards of hydrogen peroxide were explained to the operators [5]—all actions that could have been taken beforehand (see also Section 20.2.1).

17.8   INADVERTENT ISOLATION

(a) The compressed air supply to a redundant tank was isolated so that the tank could be removed. No one realized that the compressed air supply to a sampling device on a vent stack came from the same supply.

   When a service line supplies plant items that have no obvious connection with each other, it is good practice to fix a label on or near the valves, listing the equipment that is supplied. Alternatively, each item can be supplied by independent lines.

(b) A manganese grinding mill was continually purged with nitrogen to keep the oxygen content below 5%; an oxygen analyzer sounded an alarm if the oxygen content was too high. A screen became clogged with fine dust, and before clearing it the maintenance team members isolated the power supply. They did not know that the switch also isolated the power supply to the nitrogen blanketing equipment and to the oxygen analyzer. Air leaked into other parts of the plant undetected, and an explosion occurred.

   As this incident shows, operators and maintenance workers may know how individual items of equipment work but may not understand the way they are linked together. In addition, air entered the plant because a blind flange had not been inserted (a common failing; see Section 1.1), and the screen became clogged because it was finer than usual. Changing the screen size was a modification, but its consequences had not been considered beforehand—another common failing (see Chapter 2) [6].

17.9   INCOMPATIBLE STORAGE

Two incompatible chemicals were kept in the same store; if mixed they became, in effect, a firework, easily ignited. One of the chemicals was stored in cardboard kegs on a shelf close to a hot condensate pipe. As it was known to decompose at 50°C (120°F) the electrical department staff members were asked to disconnect the power supply to the steam boiler, but instead of doing so they merely turned the thermostat to zero. The kegs ruptured, and the chemical fell onto the second chemical, which was stored in bags immediately below. A fire occurred, followed by an explosion. The source of ignition was uncertain, but a falling lid may have been sufficient. Firefighting was hampered by a shortage of water, which had been known to the company for four years.

The company had received advice on the storage of incompatible chemicals, but no chemist or chemical engineer was involved, and one of the chemicals was classified incorrectly [7].

17.10   MAINTENANCE: IS IT REALLY NECESSARY?

Suppose I found that my car alternator was not charging and took the car to a garage with an instruction to change the alternator. When I got it back, with a new alternator fitted, the fault would probably be cured. But the fault might have been a slack fan belt, a sticking or worn brush, or something else that could be put right for a fraction of the cost of a replacement alternator. These minor faults would probably have been put right when the alternator was changed and would have hidden the real cause of the fault. I would have paid a high and unnecessary price, and the unnecessary maintenance may have introduced a few new faults.

In the same way, if we do not carry out some simple diagnostic work first, some of the maintenance work we carry out on our plants may be unnecessary. Process operators, with the best of intentions, often say what they think is wrong; for example, if a pump is not working correctly, they ask the maintenance team to check or clean the suction strainer. Sometimes the strainer is found to be clean, or the pump is no better after the strainer has been cleaned. We then find that there is a low level in the suction tank, the suction temperature is too high, the impeller is corroded, or a valve is partially shut.

In another example, a high-level alarm sounds. The tank could not possibly be full, so the operators ask the instrument maintenance department staff to check the level measurement. After they have done so and shown that it is correct, further investigation shows that an unforeseen flow has taken place into the tank (and perhaps the tank overflows; see Section 3.3.2a).

In a third example, a heat exchanger is not giving the heat transfer expected. The maintenance team is asked to clean the tubes. When it withdraws the bundle, there is only a sprinkling of dust. We then find that the inlet temperatures or flows have changed, but no one calculated the effect on heat transfer, and no one expected that it would be so great.

Maintenance is expensive (and hazardous). A little questioning before work is carried out might save money, reduce accidents, and get the plant back on line sooner. It might also show a need for more diagnostic information: a pressure gauge here, a temperature point there [8].

17.11   AN INTERLOCK FAILURE

Interlocks can fail because they have been disarmed (that is, made inoperative), their set points have been changed, or they are never tested, as described in Section 14.5. They can also fail as the result of errors in operation and design, as in the following incident.

A vessel was fitted with a simple mechanical interlock: a horizontal pin fitted into a slot in the vessel lid; the lid could not be moved sideways until the pin was withdrawn (Figure 17-4). A solenoid controlled movement of the pin. The solenoid could not be activated and the pin withdrawn until various measurements, including the temperature and level of the liquid in the vessel, were within specified ranges.

Nevertheless the lid was moved, although the measurements were not correct. Several possible explanations were considered:

1. The pin might have been seized inside the solenoid. Unfortunately, the operator, believing this to be the case, had squirted a lubricant into the solenoid chamber before any investigation could be carried out. A vertical pin would have been less likely to stick.
2. The operator, believing that all measurements were correct, might have assumed that the system was faulty and inserted a thin strip of metal into the end of the slot and moved the pin back into the solenoid. He denies doing this but admits that he did not check the temperature and level readings to make sure they were correct before trying to move the lid.

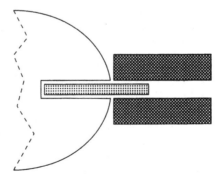

FIGURE 17-4   A simple mechanical interlock: the lid could not be moved until the pin was withdrawn from the slot.

In the original design, the pin fitted into a hole, but the hole was changed to a slot so the operator could see the position of the pin. At the time no one realized that this made it possible for someone to move the pin by hand, another example of the unseen results of a plant modification. To prevent this, a sheet of transparent plastic could have been used to cover the slot.

3. The connection between the temperature and level measurements and the solenoid was not hardwired but went through the plant control computer. A software error might have caused the solenoid to be activated when it should not have been. The system had been in use without problems for many years, but a slight change in, for example, the order in which signals are received and processed can result in a fault that has been lying in waiting like a time bomb for many years. Many people believe that safety interlocks should be hard wired rather than software based (see Chapter 20). If they are software based, they should be independent of the control system.

17.12   EMULSION BREAKING

In 1968, there was a discharge of oil vapor and mist followed by a devastating explosion at Pernis in the Netherlands. The release of vapor that caused the explosion was due to a sort of foamover (Section 12.2), but the mechanism was not the usual one. In a normal foamover, a layer of heavy oil, above a water layer, is heated above 100°C (212°F). The heat gradually travels through the oil to the water. When the water boils, the steam lifts up the oil, thus reducing the pressure on the water so that it boils more vigorously. The mixture of steam and oil may blow the roof off the storage tank. Foamovers can also occur if oil, above 100°C, is added to a tank containing a water layer.

In the Netherlands incident, there were two layers in the slops tank, which was almost full. The lower layer was a stable emulsion of water in heavy oil; the upper layer was a mixture of oils with an initial boiling point of 60°C (140°F). The steam supply to the heating coils was cracked open, and the temperature of the emulsion gradually rose. When it reached 100°C (212°F), the emulsion split into water and oil layers. The oil mixed with the upper oil layer and heated it rapidly. The lighter components vaporized, and a mixture of oil vapor and mist was expelled from the tank. The escaping cloud was ignited, probably by one of the plant furnaces, and the resulting explosion caused extensive damage. Two people were killed, 10 were hospitalized, and about 70 were slightly injured. There was some damage outside the plant site.

According to the official report [11], no one had ever realized before that an emulsion layer could suddenly split and give rise to a sudden eruption of hydrocarbon mist. No recommendations were made in the report. The authors presumably assumed that the recommendations were obvious and now that the cause of the explosion is known, everyone will check any tanks in which emulsion layers might form, and if they find any they will either segregate the emulsion layers, keep them at the same temperature as the overlying oil layers (by circulating the tanks), or keep them well below the temperature at which the emulsions will split. It is also clear that slops tanks should not be heated unless it is essential to do so.

17.13   CHIMNEY EFFECTS

Chimneys are common, and we all know how they work, but chimney effects in plants often take us by surprise. We fail to apply familiar knowledge because it seems to belong to a different sphere of thought, as in the following incidents:

(a) A distillation column was emptied, washed out, and purged with nitrogen. A manhole cover at the base was removed. While two men were removing the manhole cover at the top of the column, one of them was overcome. The other pulled him clear, and he soon recovered. It seems that due to a chimney effect, air entered the base of the column and displaced the lighter nitrogen [12].

(b) A hydrogen line, about 12-in. diameter, had to be repaired by welding. The hydrogen supply was isolated by closing three valves in parallel (one of which was duplicated) (Figure 17-5). The line was purged with nitrogen and was tested at a drain point before welding started, to confirm that no hydrogen was present. When the welder struck his arc, an explosion occurred, and he was injured. The investigation showed that two of the isolation valves were leaking. It also showed why the hydrogen was not detected at the drain point: the drain point was at a low level, and air was drawn through it into the plant to replace gas leaving through a vent. The source of ignition was sparking, which occurred because the welding return lead was not securely connected to the plant (another familiar problem) [13].

(c) A flarestack and its associated seal vessel were being prepared for maintenance. The seal vessel was emptied, and all inlet lines were slip-plated (blinded). A control valve located in one of the inlet lines, between the vessel and one of the spades, was removed (Figure 17-6). Five minutes later, an explosion occurred inside the equipment. Thirty seconds later, there was a second explosion, and flames came out of the opening where the control valve had been. As the result of the chimney effect, air had entered the system, and a mixture of air and vapor had moved up the stack. The source of ignition was probably another flarestack nearby [14].

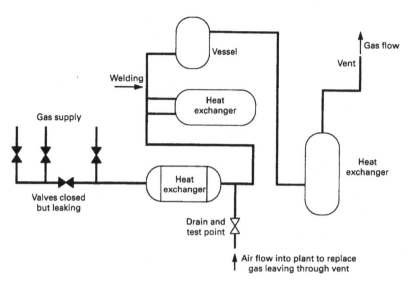

FIGURE 17-5 A simplified diagram of a plant, showing why hydrogen gas was not detected at the drain point. *(Reproduced with permission of the American Institute of Chemical Engineers. Copyright © 1995 AIChE. All rights reserved.)*

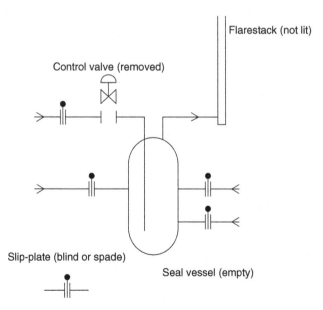

FIGURE 17-6 When the control valve was removed, a chimney effect caused air to enter the system, and an explosion occurred. *(Reproduced with permission of the American Institute of Chemical Engineers. Copyright © 1995 AIChE. All rights reserved.)*

The flare system should have been purged with nitrogen before the lines were spaded or the control valve removed. At the very least, the open end should have been blanked as soon as the valve was removed.

The incident also shows the importance of placing spades as near as possible to the equipment that is to be isolated, particularly when a vessel is to be entered. Valves should not be left between a slip-plate and the vessel, as liquid can then be trapped between the valve and the slip-plate and enter the vessel if the valve leaks or is opened.

References

1. T. A. Kletz, *An Engineer's View of Human Error*, 3rd edition, Institution of Chemical Engineers, Rugby, UK, 2000, Chapter 2.
2. *Accident at Markham Colliery, Derbyshire*, Her Majesty's Stationery Office, London, 1974.
3. *Loss Prevention Bulletin*, No. 092, Apr. 1990, p. 9.
4. *Health and Safety at Work*, Vol. 14, No. 12, Dec. 1992, p. 10.
5. J. S. Arendt and D. K. Lorenzo, "Investigation of a Filter Explosion," Paper presented at AIChE Loss Prevention Symposium, San Diego, Aug. 1990.
6. J. A. Senecal, *Journal of Loss Prevention in the Process Industries*, Vol. 4, No. 5, Oct. 1991, p. 332.
7. Health and Safety Executive, *The Fire at Allied Colloids Ltd.*, Her Majesty's Stationery Office, London, 1994.
8. E. H. Frank, private communication.
9. *Operating Experience Weekly Summary*, No. 97-03, Office of Nuclear and Safety Facility, U.S. Dept. of Energy, Washington, D.C., 1997, p. 1; No. 97-19, p. 2; No. 97-22, p. 1; and No. 97-30, p. 1. See also *Mixing and Storing Incompatible Chemicals*, Safety Note No. 97-1, Office of Nuclear and Safety Facility, U.S. Dept. of Energy, Washington, D.C., 1997.
10. A. Wilson, *Well Oiled*, Northgate, London, 1979, p. 76.
11. Ministry of Social Affairs and Public Health, *Report Concerning an Inquiry into the Cause of the Explosion on 20th January 1968 at the Premises of Shell Nederland Rafftnaderiji NV in Pernis*, State Publishing House, The Hague, The Netherlands, 1968.
12. *Loss Prevention Bulletin*, No. 107, Oct. 1992, p. 24.
13. P. J. Nightingale, *Plant/Operations Progress*, Vol. 8, No. 1, Jan. 1989, p. 28.
14. *Loss Prevention Bulletin*, No. 107, Oct. 1992, p. 24.

18

Reverse Flow, Other Unforeseen Deviations, and Hazop

[To] divide each of the difficulties under examination into as many parts as possible, as might be necessary for its adequate solution.

—René Descartes (1596–1650),
Discourse on Method

This chapter describes some incidents that occurred because of deviations from the design intention, as expressed in the process flow diagrams (also known as line diagrams or process and instrumentation diagrams). The fact that these deviations could occur was not spotted during the design stage, and they had unfortunate unforeseen results. Ways of spotting these deviations by conducting hazard and operability studies are discussed in Section 18.7.

Errors introduced during modifications are discussed in Chapter 2, whereas designs that provided opportunities for operators to make errors are discussed in Chapter 3.

One of the most common errors made at the process-flow-diagram stage is failure to foresee that flow may take place in the reverse direction to that intended, as discussed next [1].

18.1 REVERSE FLOW FROM A PRODUCT RECEIVER OR BLOWDOWN LINE BACK INTO THE PLANT

(a) Accidents have occurred because gas flowed from a product receiver into a plant that was shut down and depressured. In one incident,

doi:10.1016/B978-1-85617-531-9.00018-4

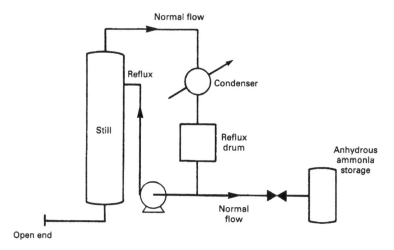

FIGURE 18-1   Reverse flow occurred from the storage vessel to the open end.

ammonia flowed backward from a storage vessel, through a leaking valve, into a reflux drum, into a still, and out of an open end in the bottoms line, which was open for maintenance (Figure 18-1).

If the possibility of reverse flow had been foreseen, then a slip-plate could have been inserted in the line leading to the ammonia storage vessel, as described in Section 1.1.

(b) In another incident, a toxic gas in a blowdown header flowed through a leaking blowdown valve into a tower and out of the drain valve. The operator who was draining the tower was killed (Figure 18-2).

(c) The contents of a reactor were pumped out into another vessel for further treatment. The pump delivery valve should have closed automatically when the reactor was empty, but on this occasion the automatic system was not working correctly and was on manual control. Fifteen minutes passed before it was closed, and during this time, steam traveled backward up the transfer line between the treatment vessel and the reactor and heated a heel of reaction product (about 150 kg), left behind in the reactor, from about 100°C (212°F) to about 175°C (365°F). At this temperature, the reaction product, a nitro compound, decomposed explosively, causing extensive damage [12].

A hazard and operability study (see Section 18.7) was carried out during design, but flow of steam from the treatment vessel to the reactor was never considered as a possible deviation, perhaps because the team thought that prompt closing of the valve between the two vessels would prevent it. If it had been considered, a check valve might have been inserted in the line. Hazard and operability studies are only as good as the knowledge and experience of

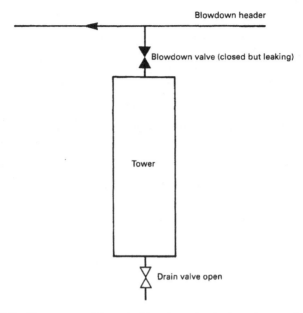

FIGURE 18-2   Flow occurred from the blowdown header into the vessel and out of the drain valve.

the team. In hazard and operability studies, the team members do not always ask what will happen if the automatic equipment is on manual control. This question should always be asked, as safety may depend on the correct operation of such equipment.

(d) Reverse flow into vessels open for entry is discussed in Sections 11.3e and 11.3f.

18.2   REVERSE FLOW INTO SERVICE MAINS

This occurs when the pressure in the service line is lower than usual or when the pressure in the process line is higher than usual. Many plants have experienced incidents such as the following:

1. Liquefied gas leaked into a steam line that had been blown down. Ice then formed on the outside of the steam line.
2. A leak on a nitrogen line caught fire.
3. The paint was dissolved in a cabinet that was pressurized with nitrogen; acetone had leaked into the nitrogen [2].
4. A compressed air line was choked with phenol.
5. Toxic fumes in a steam system affected a man who was working on the system (see Section 1.1.4).

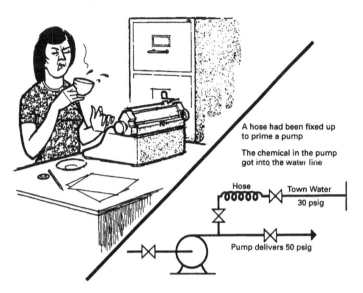

FIGURE 18-3   Never connect town water to process equipment.

Another incident is illustrated in Figure 18-3. Town water should never be directly connected to process lines by a hose [19] or permanent connections. A break tank should be provided.

A service that is used intermittently should be connected to process equipment by a hose, which is disconnected when not in use, or by double block and bleed valves. If a hose is used, it should be provided with a vent so that it can be depressured before it is disconnected.

If a service is used continuously, it may be connected permanently to process lines. If the service pressure is liable to fall below the normal process pressure, then a low-pressure alarm should be provided on the service line. If the process pressure is liable to rise above the normal service pressure, then a high-pressure alarm should be provided on the process side.

In addition, check valves should be fitted on the service lines.

18.3   REVERSE FLOW THROUGH PUMPS

If a pump trips (or is shut down and not isolated), it can be driven backward by the pressure in the delivery line and damaged. Check valves are usually fitted to prevent reverse flow, but they sometimes fail.

When the consequences of reverse flow are serious, then the check valve should be scheduled for regular inspection. The use of two, preferably

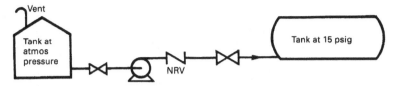

FIGURE 18-4   The check valve was relied on to prevent backflow, and the isolation valves were not used. A spillage was inevitable.

of different types, in series, should be considered. The use of reverse rotation locks should also be considered.

When lines are being emptied by steaming or by blowing with compressed air or nitrogen, care should be taken that pumps are not turned so fast in reverse (or even forward) that they are damaged.

In one plant, light oil was pumped at intervals from a tank at atmospheric pressure to one at a gauge pressure of 15 psi (1 bar). For many years the practice was not to close any isolation valves but to rely on the check valve in the pump delivery. One day a piece of wire got stuck in the check valve, oil flowed backward, and the atmospheric tank overflowed (Figure 18-4).

This is a good example of an accident waiting to happen. Sooner or later the check valve was bound to fail, and a spillage was then inevitable.

In this case, the design was not at fault. The operators did not understand the design philosophy. Would this have been foreseen in a hazard and operability study (Section 18.7), and would special attention have been paid to the point in operator training?

18.4   REVERSE FLOW FROM REACTORS

The most serious incidents resulting from reverse flow have occurred when reactant A (Figure 18-5) has passed from the reactor up the reactant B feed line and reacted violently with B.

In one incident, paraffin wax and chlorine were reacted at atmospheric pressure. Some paraffin traveled from the reactor back up the chlorine line and reacted with liquid chlorine in a catchpot, which exploded with great violence. Bits were found 30 m (100 ft) away [3].

A more serious incident occurred at a plant in which ethylene oxide and aqueous ammonia were reacted to produce ethanolamine. Some ammonia got back into the ethylene oxide storage tank, past several check valves in series and a positive pump. It got past the pump through the relief valve, which discharged into the pump suction line. The ammonia reacted with 30 m^3 of ethylene oxide in the storage tank. There was a

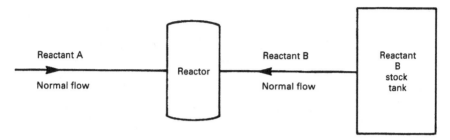

FIGURE 18-5   Reverse flow of A occurred from the reactor into the B stock tank.

violent rupture of the tank, followed by an explosion of the vapor cloud, which caused damage and destruction over a wide area [4].

Another somewhat similar incident occurred when butadiene from a reactor flowed in the wrong direction up a line used for adding emulsifier. The check valve, which should have prevented the flow, was obstructed. The emulsifier tank was in a building and had an open vent. Butadiene came out and exploded. The explosion was heard 15 km away, but damage was minimized by the light construction of the building, which ruptured at the junction of the roof and walls [13].

When such violent reactions can occur, it is not sufficient to rely on check valves. In addition, either of the following should occur:

1. The reactant(s) should be added via a small break tank so that if reverse flow occurs only a small quantity will react and not the main stock, or
2. The pressure drop in the pipeline should be measured, and if it gets too low, a trip valve should be closed automatically. A reliable, duplicated system may be necessary [16].

A seemingly minor backflow from a reactor occurred on an ammonium nitrate plant, but it led to an explosion. The two reactants, nitric acid and ammonia gas, entered the titanium reactor through separate spargers, and a corresponding amount of ammonium nitrate solution overflowed into a rundown tank. When the plant was shut down for a minor repair, some ammonium nitrate solution flowed backward into the nitric acid sparger and mixed with the acid. Most of it was blown out when the nitric acid line was emptied with compressed air, but a small amount of the ammonium nitrate solution was trapped and left behind. Steam was blown through the acid line to keep the vessel warm. After about nine hours, the sparger exploded, and the explosion spread to the rest of the reactor and the rundown tank. The blast damaged an ammonia tank, and people living within seven miles had to leave their homes. The main recommendations in the report were to redesign the sparger so that liquid could not be trapped in it and avoid the use of titanium, as it increases the sensitivity of ammonium nitrate [17, 18].

Reference 14 reviews other ways of preventing backflow.

18.5   REVERSE FLOW FROM DRAINS

This has often caused flammable liquids to turn up in some unexpected places. For example, construction had to be carried out next to a compound of small tanks. Sparks would fall onto the compound. Therefore, all flammable liquids were removed from the tanks while the construction took place. Nevertheless, a small fire occurred in the compound.

Water was being drained from a tank on another part of the plant. The water flow was too great for the capacity of the drains, so the water backed up into the compound of small tanks, taking some light oil with it. Welding sparks ignited this oil.

Another incident occurred on a plant that handled liquefied vinyl chloride (VC) (boiling point −14°C [7°F]). Some of the liquid entered a vessel through a leaking valve, and the operator decided to flush it out to drain with water. As the VC entered the drain it vaporized, and the vapor flowed backward up the drainage system; white clouds came out of various openings. Some of the VC came out inside a laboratory 30 m (100 ft) away, as the pressure was sufficient to overcome the level of liquid in the U-bends. The VC exploded, injuring five people and causing extensive damage. The amount that exploded was estimated as about 35 kg [15].

18.6   OTHER DEVIATIONS

(a) Figure 18-6 shows part of an old unit. Valve A could pass a higher rate than valve B. Inevitably, in the end the lower tank overflowed.

(b) Raw material was fed to a unit from two stock tanks, A and B. Tank A was usually used; tank B was used infrequently. The raw material was pumped to a head tank from which excess flowed back, as shown in Figure 18-7. The system was in use for several years before the inevitable happened. Tank B was in use; tank A was full, and the flow from the head tank caused it to overflow.

(c) A funnel was installed below a sample point so that excess liquid was not wasted but returned to the process (Figure 18-8). What will happen if a sample is taken while the vessel is being drained?

The design errors in these cases may seem obvious, but the diagrams have been drawn so that the errors are clear. Originally they were hidden among the detail of a "spaghetti bowl" drawing. To bring the errors to

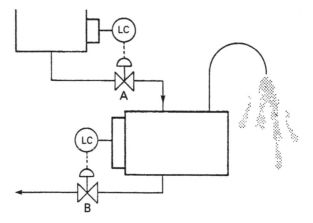

FIGURE 18-6   Valve A could pass a higher rate than valve B, thus making a spillage inevitable.

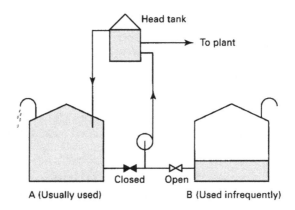

FIGURE 18-7   If A is full and suction is taken from B, A will overflow.

light, it is necessary to go·through line diagrams systematically, line by line and deviation by deviation, as described in the next section.

18.7   A METHOD FOR FORESEEING DEVIATIONS

The incidents listed earlier in this chapter and many others could have been foreseen if the design had been subjected to a hazard and operability study (Hazop). This technique allows people to let their imaginations go free and think of all possible ways in which hazards or operating problems might arise. But to reduce the chance that something is missed,

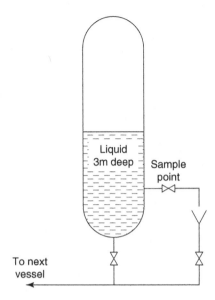

FIGURE 18-8   What will happen if the vessel is drained while a sample is being taken?

Hazop is done in a systematic way, each pipeline and each sort of hazard being considered in turn.

A pipeline for this purpose is one joining two main plant items—for example, we might start with the line leading from the feed tank through the feed pump to the first feed heater. A series of guide words are applied to this line in turn, the words being as follows:

NONE
MORE OF
LESS OF
PART OF
MORE THAN
OTHER

NONE, for example, means no forward flow or reverse flow when there should be forward flow. We ask the following questions:

Could there be no flow?
If so, how could it arise?
What are the consequences of no flow?
How will the operators know that there is no flow?
Are the consequences hazardous, or do they prevent efficient operation?
If so, can we prevent no flow (or protect against the consequences) by changing the design or method of operation?
If so, does the size of the hazard or problem justify the extra expense?

The same questions are then applied to "reverse flow," and we then move on to the next guide word, MORE OF. Could there be more flow than design? If so, how could it arise? And so on. The same questions are asked about "more pressure" and "more temperature," and, if they are important, about other parameters, such as "more radioactivity" or "more viscosity."

PART OF prompts the team to ask if the composition of the material in the pipeline could differ from design, MORE THAN prompts them to ask if additional substances or phases could be present, and OTHER THAN reminds them to consider startup, shutdown, maintenance, catalyst regeneration, services failure, and other abnormal situations. For more detailed accounts of Hazop, see references 5 through 10.

18.8   SOME PITFALLS IN HAZOP

The success of a Hazop in identifying hazards depends on the knowledge and experience of the team members. If they lack knowledge and experience, the exercise is a waste of time. The following incidents show how an inexperienced team can miss hazards:

(a) Figure 18-9 shows a floating-roof tank located in a dike. Rainwater can be drained from the roof into the dike and from the dike into a waterway. The team members are considering whether any substance other than water can get into the waterway. For this to occur, there would have to be a hole in the hose, and both valves would have to be left open. An inexperienced team may decide that a triple failure is so improbable that there is no need to consider it further.

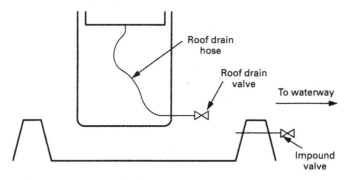

FIGURE 18-9   Liquid other than rainwater can reach the waterway only if there is a hole in the hose and both valves are left open. This is not as unlikely as it seems at first sight. *(Used with permission from Hydrocarbon Processing, Apr. 1992. Copyright © 1992 Gulf Publishing Co. All rights reserved.)*

Someone with knowledge of the practicalities of plant operation would realize that during prolonged rain the operators may leave both drain valves open, whatever the instructions say, to avoid frequent visits to the tank. Any hole in the hose will then contaminate the waterway with oil [20].

(b) According to a design, an explosive powder had to be transferred in a scoop. The Hazop team realized that this could lead to the formation of an electrostatic charge on the powder and scoop and decided that a metal scoop would be safer than a plastic one. No one realized that if the operator was not grounded, a conducting scoop would increase the risk of ignition, as the charge could pass as a spark from the scoop to ground. A spark from a nonconducting plastic scoop would be less likely to occur and less energetic if it did occur [21]. The best solution is not to use an open scoop.

(c) During the final purification of a product, a small amount of an oxidizing agent had to be added to a much larger amount of hydrocarbon. The reaction between the two substances was known to be highly exothermic and is listed as such in the standard work on the subject, *Bretherick's Handbook of Reactive Chemical Hazards* [22]. However, not one member of the team knew this, and none of them was sufficiently aware to consult this standard work. (Like those who designed the temporary pipe at Flixborough (Section 2.4a), they did not know what they did not know.) An explosion occurred after a few months of operation [21].

In all three examples, the senior managers of the companies involved were committed to safety, but the staff lacked the necessary knowledge and experience. It was not necessary for the whole team to have been aware of the hazard. One member's awareness would have been enough, so long as the other team members were willing to listen. It was not necessary for him or her to be fully conversant with the details of the hazard, so long as concerns were followed up.

18.9   HAZOP OF BATCH PLANTS

When studying a batch plant, the guide words should be applied to the instructions as well as the pipelines. For example, if an instruction says that 1 ton of A should be charged to a reactor, the Hazop team should consider the effects of the following deviations:

DON'T CHARGE A
CHARGE MORE (OR LESS) A
CHARGE AS WELL AS A

CHARGE PART OF A (if A is a mixture)
CHARGE OTHER THAN A
REVERSE CHARGE A (That is, can flow occur from the reactor to the
    A storage vessel [see Section 18.4]?)
A IS ADDED EARLY (OR LATE)
A IS ADDED TOO QUICKLY (OR SLOWLY)

Here are three examples of hazards uncovered during Hazops of batch processes:

- During the Hazop of a batch reaction, when discussing the guide words "AS WELL AS A," someone asked what contaminants could lead to a runaway reaction. Another member said organic acids could do so. Other members remarked that organic acids were used in another process and were stored in similar drums in the same warehouse. This example shows how Hazop is able to combine the knowledge and thoughts of different team members [23].
- During the Hazop of another batch process, when discussing services failure, the team members realized that a power failure would result in the loss of both agitation and cooling and that at certain stages of the process this could lead to a runaway reaction. They decided to use town water for emergency cooling and nitrogen injection for emergency agitation [23].
- During the Hazop of a proposed experimental rig, it came to light that one of the reactants was hydrogen cyanide, supplied in cylinders, and the designers expected the operators to convey the cylinders to the top floor of the building in the elevator. Toxic or flammable gases and people should never be together in a confined space.
- A large distillation column in a refinery operated at high vapor loads, just above atmospheric pressure. It was not designed for vacuum and so had to be protected if the heat input from the reboiler failed but condensation continued. An inexperienced Hazop team might have accepted without comment the original design intention, which was to break the vacuum with fuel gas (or nitrogen if available). A more experienced team might have realized that the volume of gas required was enormous but that it could be reduced to a manageable figure by locating the vacuum breaker valve at the inlet to the condensers, thus blanketing them and reducing heat transfer [24].

18.10   HAZOP OF TANK TRUCKS

Hazop has been applied mainly to fixed plants, but application of the technique to tank trucks used for carrying anhydrous ammonia and liquid carbon dioxide disclosed a number of hazards [11].

18.10.1 "More of Pressure"

Use of this guide word brought out the fact that if there was a leak on the filling line, there was no way of preventing the contents of the tank truck from flowing backward into the filling line and out to the atmosphere unless the leak was so big that the excess flow valve on the tank truck would operate. This will not occur unless the flow is at least 1½ to 2 times the normal flow. A remotely operated emergency isolation valve prevents flow from the plant. It was therefore decided to install compressed air cylinders on the tank trucks to operate their internal valves; the cylinders were connected to the plant emergency valve system so that when this was operated, the emergency valves on the tank truck also closed. As a bonus, the internal valves also close if the tanker is driven away while still filling.

The tank trucks were not fitted with relief valves—normal European practice for toxic liquids. The study showed that the plant was designed for a higher pressure than the tank trucks and that in certain circumstances they could be overpressured. Modifications were made.

18.10.2 "Less of Temperature"

Some of the older tank trucks were made from grades of steel that are brittle at low temperatures, and they are never moved at temperatures below 0°C (32°F). It was discovered that some customers wanted liquid carbon dioxide delivered at less than the usual pressure, and arrangements had to be made for them to be supplied only by selected tank trucks. (All new tank trucks are capable of withstanding the lowest temperatures that can be reached.)

18.10.3 "More Than"

Some customers complained that there was oxygen in the ammonia. It was found that the road transport maintenance department was preparing tanks for repair by washing them out with water and then returning them to the plant full of air. The oxygen could cause stress corrosion cracking. Arrangements were made for the plant staff to take over responsibility for preparing tank trucks for repair.

18.11 HAZOP: CONCLUSIONS

The benefits of Hazop go far beyond a simple risk of recommendations for a safer plant. The interactions between team members can bring about a change in individual and departmental attitudes. Employees become more likely to seek one another out and discuss possible consequences of proposed changes; departmental rivalries and barriers recede. The dangers of working in isolation and the consequences of ill-judged and hasty actions become better appreciated, particularly by those who have spent their whole careers in the same department [25].

Learning from experience, the main subject of this book, is a lantern on the stern, illuminating the hazards the ship has passed through. It is essential to do so, as we will be repeating the journey in the future. Hazop, however, is a lantern on the bow, illuminating the hazards that lie ahead.

The disadvantage of Hazop is that it takes place late in design, at a time when it is too late to make major changes. We need similar studies at earlier stages, when we are deciding which process to use and before the process design is passed to the engineers who will prepare the line diagrams [26].

References

1. T. A. Kletz, *Hydrocarbon Processing*, Vol. 55, No. 3, Mar. 1976, p. 187.
2. T. A. Kletz, *Learning from Accidents*, 3rd edition, Butterworth-Heinemann, Boston, Mass, 2001, Chapter 2.
3. D. B. de Oliveria, *Hydrocarbon Processing*, Vol. 52, No. 3, Mar. 1973, p. 113.
4. J. E. Troyan and R. Y. Le Vine, *Loss Prevention*, Vol. 2, 1968, p. 125.
5. H. G. Lawley, *Chemical Engineering Progress*, Vol. 70, No. 4, Apr. 1974, p. 45.
6. B. Tyler, F. Crawley, and M. Preston, *HAZOP: Guide to Best Practice*, 2nd edition, Institution of Chemical Engineers, Rugby, UK, 2008.
7. R. E. Knowlton, *An Introduction to Hazard and Operability Studies*, Chemetics International, Vancouver, Canada, 1981, and *A Manual of Hazard and Operability Studies*, Chemetics International, Vancouver, Canada, 1992.
8. T. A. Kletz, *Hazop and Hazan—Identifying and Assessing Process Industry Hazards*, 4th edition, Institution of Chemical Engineers, Rugby, UK, 1999.
9. M. S. Mannan (editor), *Lees' Loss Prevention in the Process Industries*, 3rd edition, Elsevier, Boston, Mass, 2005, Chapter 8.
10. T. A. Kletz, *Chemical Engineering*, Vol. 92, No. 7, Apr. 1985, p. 48.
11. E. A. George, *Loss Prevention*, Vol. 14, 1981, p. 185.
12. K. Bergroth, *Loss Prevention Bulletin*, No. 109, Feb. 1993, p. 1.
13. M. S. Mannan (editor), *Lees' Loss Prevention in the Process Industries*, 3rd edition, Elsevier, Boston, Mass, 2005, Appendix B10.
14. S. M. Englund, J. L. Mallory and D. J. Grinwis, *Chemical Engineering Progress*, Vol. 88, No. 2, Feb. 1992, p. 47.
15. J. Easterbrook and D. V. Gagliardi, *Plant/Operations Progress*, Vol. 3, No. 1, Jan. 1984, p. 29.
16. H. G. Lawley, *Hydrocarbon Processing*, Vol. 55, No. 4, Apr. 1976, p. 247.
17. *Chemical Process Safety Report*, Vol. 5, No. 11, Sept. 1995.
18. *Energetic Events*, Vol. 3, No. 3, Wilfred Baker Engineering, Aug. 1995, p. 4.
19. *Loss Prevention*, No. 124, Aug. 1995, p. 13.
20. D. W. Jones, *Hydrocarbon Processing*, Vol. 71, No. 4, Apr. 1992, p. 78.
21. G. S. Melville, *Journal of Loss Prevention in the Process Industries*, Vol. 7, No. 5, 1994, p. 387.

22. P. G. Urben (editor), *Bretherick's Handbook of Reactive Chemical Hazards*, 5th edition, Butterworth-Heinemann, Oxford, UK, 1995.

23. R. L. Collins, *Chemical Engineering Progress*, Vol. 91, No. 4, 1995, p. 48.

24. I. M. Duguid, *Loss Prevention Bulletin*, No. 134, Apr. 1997, p. 10.

25. N. Carling, Hazop Study of BAPGO's FCCU Complex, *American Petroleum Institute Committee on Safety and Fire Protection Spring Meeting*, Denver, Colorado, April 8–11, 1986.

26. T. A. Kletz, Don't Just Pass the Parcel; Accidents That Would Not Have Occurred if Those Involved Had Talked Together, *Mary Kay O'Connor Process Safety Center Symposium*, October 23–24, 2007.

I Didn't Know That . . .

I know that the multitude walk in darkness. I would put into each man's hand a lantern, to guide him; and not have him set out upon his journey depending for illumination on abortive flashes of lightning, or the coruscations of transitory meteors.

—William Wordsworth (1770–1850)

This chapter describes some accidents that occurred because people were unaware of accidents that had happened many times before.

19.1 AMMONIA CAN EXPLODE

In reports on ammonia explosions, the authors often say that they were surprised to find that ammonia can explode. For example, a leak of ammonia from the 50-year-old refrigeration system of an ice cream plant in Houston, Texas, ignited and severely damaged the building. The ignition source was not identified, but there were several possible sources. The chief of the Houston Fire Department Hazardous Materials Response Team wrote, "The hazards, it was believed, were limited to health; never had much thought been given to the flammability of ammonia" and "It is hard to find any of the old, experienced ammonia refrigeration men who believe it possible for ammonia to explode" [1].

Another explosion occurred in Brazil. Welding had to be carried out on the roof of an aqueous ammonia tank. The tank was emptied but not gas-freed. When a welder applied his torch to the roof, the tank blew up. The welder survived but was crippled for life.

Ammonia explosions are not common, as the lower explosive limit (LEL) of ammonia is unusually high: 16%; the upper limit is 25%. Typical limits for hydrocarbons are propane, 2% to 9.5%, and cyclohexane, 1.3% to 8.3%. In addition, the auto-ignition temperature of ammonia is high, about 650°C (1,200°F), compared with about 480°C (900°F) for propane and about 270°C (520°F) for cyclohexane, so ammonia is harder to ignite. Nevertheless, there was little excuse for the ignorance of the responsible people in Texas and Brazil, as ammonia explosions have occurred from time to time and the explosibility of ammonia has been known since at least 1914 [2]. In a paper presented in Houston in 1979, Baldock said that a number of ammonia leaks had exploded, although some reported incidents may not have been caused by ammonia at all. He gave no details. He added that there had been 11 explosions in aqueous ammonia tanks and several explosions in nitric acid plants when the ammonia/air ratio became too high [3].

A series of incidents in one nitric acid plant has been described in detail [21]. Rust passed through the ammonia gas filter and catalyzed oxidation of the ammonia in the ammonia/air mixer and in the pipe leading from it to the platinum catalyst. This occurred even though the ammonia concentration was below the normal flammable limit. The temperature of the pipe rose from 220°C (430°F) to more than 1,000°C (1,830°F) in an hour, and then the pipe ruptured. This damaged the platinum catalyst, and some dust from it ended up in the ammonia/air mixer. As a result, several further ignitions occurred after the plant was repaired and restarted. On these occasions it was shut down at once, before the pipework failed. The report admits that the mixer had not been cleaned for years as "it was so time consuming to remove it." (Compare the tank that was sucked in because the flame arrestors had not been cleaned for two years [Section 5.3a].) The report recommended installation of high-temperature alarms as well as regular cleaning of the mixer.

In 1968, an explosion in a sausage plant in Chicago killed 9 people, including 4 firefighters, injured 72, and destroyed the plant. The incident started when a gasoline tank truck hit an obstruction. The gasoline leaked into a basement and caught fire. The fire heated a 136-kg (300-lb) ammonia cylinder, which discharged its contents through the relief device. The ammonia rose into the ground floor area and the floor above where it exploded. It was assumed that, because of its high LEL, the ammonia was able to pass through the fire zone without igniting and then accumulate in a confined space until the concentration reached 16%.

Brief reports have appeared of several other ammonia fires or explosions:

• An explosion occurred in 1976 while a refrigeration plant in Hexham, England, was being demolished [4].
• A fire broke out in 1977 at Llandarcy, South Wales, fed by leaking ammonia valves [5].

- A fire and explosion occurred in 1978 in a disused cold store in Southwark, London [6].
- In Enid, Oklahoma, in 1978, the refrigeration system on an ammonia storage vessel failed. The ammonia warmed up, its pressure rose, and some ammonia was discharged through a relief valve and ignited by a nearby flare [14].
- A welder was killed by an explosion in New Zealand in 1991 while working on an empty 28-m^3 tank, which contained a flammable mixture of ammonia vapor and air.

A feature of ammonia explosions is that any ammonia that continues to leak out after the explosion may not burn, as its concentration may be too low.

So far as I am aware, ammonia has never exploded in the open air, and it is doubtful if a concentration as high as 16% could be attained out of doors. In 1989, at Jovona, Lithuania, a storage tank split from top to bottom, and 7,000 tons of liquid ammonia were spilled. The pool caught fire [15, 16], but according to later reports, the fire was due to rupture of a natural gas line that passed through the area [17].

In 2008, the U.S. media reported that a fire that destroyed a meat-packing plant was caused by the ignition of a leak of ammonia gas from a freezer. The fire actually started when a welder set fire to some plastic panels. Some ammonia leaked as a result, but there was no evidence that it caught fire.

What can we do to prevent ammonia explosions? The action required is much the same as for other flammable gases:

1. Use equipment of sound design and construction. (The Houston ice cream plant was not up to today's standards but had been "grandfathered.")
2. Use nonflammable refrigerants instead of ammonia.
3. If ammonia is used, see that the ventilation is adequate. (It does not have to be all that good to prevent the ammonia concentration from reaching 16%, but it has to be reasonably good if we wish to prevent the ammonia concentration from reaching 10,000 ppm, the concentration that is fatal to about 50% of people in 30 minutes.)
4. Gas-free and test before introducing a source of ignition.

19.2   HYDRAULIC PRESSURE TESTS CAN BE HAZARDOUS

As water is incompressible, hydraulic pressure tests are often considered safe. If the vessel fails, the bits will not fly far.

Hydraulic pressure testing is safer than pneumatic testing, as much less energy is released if the equipment fails. Nevertheless, some spectacular failures have occurred during hydraulic tests. In 1965, a large pressure vessel (16 m [52 ft] long by 1.7 m [5.6 ft] in diameter), designed for operation at a gauge pressure of 350 bar, failed during pressure test at the manufacturer. The failure, which was of the brittle type, occurred at a gauge pressure of 345 bar, and four large pieces were flung from the vessel. One piece weighing 2 tons went through the workshop wall and traveled nearly 50 m (160 ft). Fortunately, there was only one minor injury. The failure occurred during the winter, and the report recommends that pressure tests should be carried out above the ductile-brittle transition temperature for the grade of steel used. It also states that the vessel was stress-relieved at too low a temperature [7]. Another similar failure is described in reference 8. Substandard repairs and modifications were contributory factors.

When carrying out pressure tests, remember that the equipment may fail, and take precautions accordingly. If we were sure that the equipment would not fail, we would not need to test it (see Section 14.8). Reference 22 gives advice on the measures necessary. Remember also that equipment may fail during on line pressurization with process materials if the temperature is too low [8]. I do not know of any vessels that have burst for this reason, but rupture discs have failed because they were too cold.

19.3   DIESEL ENGINES CAN IGNITE LEAKS

Most companies do not allow spark ignition (gasoline) engines to enter areas where flammable gases or liquids are handled, except under strict control, as a leak of gas or vapor might be ignited by the spark mechanism. Many companies, however, allow uncontrolled access by diesel engines, believing that they cannot ignite gas or vapor. This is incorrect, as the following incident shows.

Four tons of hot, flammable hydrocarbon leaked out of a plant while maintenance work was in progress. A diesel engine was operating in the area. The hydrocarbon vapor was sucked into the air inlet, and the engine started to race. The driver tried to stop it by isolating the fuel supply, the usual way of stopping a diesel engine, but without success, as the fuel was reaching the engine through the air inlet. Finally flashback occurred, and the hydrocarbon was ignited. Two men were killed [9].

Another incident occurred when a tank truck drove underneath a loading arm that was dripping gasoline. The engine started to race and emitted black smoke, but fortunately no ignition occurred [10].

In yet another incident, a hydraulic hose leaked, and an oil mist was sucked into the air inlet of a diesel engine. It continued to run for three to five minutes after the normal fuel supply was isolated. The air filter on the engine was missing. Had it been present, it would probably have trapped the oil mist [23].

Proprietary devices that shut off the air supply as well as the fuel supply are available for protecting diesel engines that have to operate in areas in which leaks of flammable gas or vapor may occur [11]. However, diesel engines can ignite leaks of flammable gas or vapor in other ways. Sparks or flames can be emitted by the exhaust, the exhaust pipe can be hot enough to ignite the vapor directly [23], and ancillary equipment, such as electrical equipment, can produce sparks. One explosion occurred because an engine was stopped by use of the decompression control. Spark arrestors and flame arrestors should therefore be fitted to the exhaust, its temperature should be below the auto-ignition temperature of the materials handled, electrical equipment should be protected, and if a decompression control is fitted, it should be disconnected.

The degree of protection adopted in any particular case will depend on the length of time the diesel engine is present and the degree of supervision [12]. A truck delivering goods does not need any special protection but should not be allowed to enter the plant area unless conditions are steady and leaks unlikely. Plants should be laid out so that such vehicles do not normally have to enter areas where flammable gases or liquids are handled. A diesel pump that is permanently installed or a tow motor (forklift truck) in everyday use requires the full treatment. An intermediate level of protection is suitable for a crane or pump used occasionally. It should be fitted with a device for shutting off the air supply, and it should never be left unattended with the engine running. Pumps driven by compressed air are safer than diesel pumps. Flooded drains and sumps can be emptied with ejectors powered by a water supply.

An entirely different diesel hazard is compression of a pocket of air and flammable vapor trapped in a vessel or pipeline by a column of liquid. If the pressure of the liquid rises, the air is compressed, and the heat developed may heat the vapor above its auto-ignition temperature [13].

19.4 CARBON DIOXIDE CAN IGNITE A FLAMMABLE MIXTURE

In 1966, a naphtha tanker, the *Alva Cape,* was involved in a collision near New York and was severely damaged. Some naphtha was spilled, and the rest was pumped out into another vessel. The owners wanted to move the ship to a shipyard where it could be gas-freed and the damage could be surveyed, but the New York Fire Department said that the ship's tanks should be inerted before it was moved. The salvage company, therefore, ordered some carbon dioxide cylinders and hoses. Two tanks were inerted without incident, but when carbon dioxide was discharged into a third tank, an explosion occurred, followed by a fire. Four men were killed, and further explosions occurred in other tanks when the fire heated them.

When the carbon dioxide was discharged, the adiabatic cooling caused particles of solid carbon dioxide to form, and these collected a charge of static electricity (see Section 15.2). The charge discharged as a spark and ignited the mixture of naphtha vapor and air in the tank. The company that supplied the carbon dioxide did not know how it would be used but warned the salvage company that it was hazardous to inert tanks with carbon dioxide. The vessel was towed out to sea and sunk by gunfire [18].

A similar incident occurred in France a year or two later. Carbon dioxide was injected into a tank containing jet fuel during a tryout of a firefighting system. The tank exploded, killing 18 people who were standing on the top.

19.5   MISTS CAN EXPLODE

Most everyone knows that dusts—fine particles of solid—can explode, but not everyone is aware that mists—fine droplets of liquid—can explode just as easily and that they can explode at temperatures far below the flash point of the bulk liquid or vapor [19].

For example, a material had been oxidized many times without incident in 1- and 4-L vessels, in an oxygen atmosphere, at a temperature of 80°C (175°F), and at a gauge pressure of 225 psi (15.5 bar). The flash point of the solvent at this temperature and pressure was 130°C (265°F). The next step was to scale up to a 48-L vessel. The rate of reaction was limited by the rate at which the oxygen and the material to be oxidized could be brought into contact, so a highly efficient gas-dispersing and agitation system was installed. This filled the vapor space of the reaction vessel with a fine mist, and several hours after startup, the vessel exploded. The pressure or temperature did not rise beforehand, so the explosion could not have been caused by a runaway reaction; it was a mist explosion. The source of ignition was a small amount of a catalyst left over from an earlier set of experiments [20]. The introduction of the efficient agitation system was a process modification (see Section 2.6), but its consequences were not foreseen.

Another incident occurred when contractors were employed to clean several black oil tanks, with a 4,500-m³ capacity, so they could be used for the storage of kerosene. The details of the contract were agreed verbally. After removing solid residues and the heater coils, the tanks would be sprayed with hot water and a detergent. Floodlights were suspended through roof manholes and were to be removed before the tanks were sprayed.

The first tank was cleaned without incident. By the time the contractors started on the second tank, a new foreman was in charge. To soften the deposits, he disconnected the steam coil, blew live steam into the tank, and then sprayed kerosene onto the walls, section by section, using a spinner mounted on a tripod. The report does not make it clear whether or not the kerosene was heated. While men were moving the tripod, a fire started in the tank, followed by an explosion. Three men were killed, one by the fire and two by falling bits of the external concrete cladding.

Survivors said that conditions in the tank resembled a thick fog, which tests confirmed. The source of ignition was either the floodlights, which had been left in position and had a surface temperature of 300°C (570°F), or a discharge of static electricity generated by the steam cloud. Tests showed that the oil mist could ignite at 11°C (52°F), 60 degrees C (110 degrees F) below the flash point of the oil. Though not suggested in the report, the hot steam pipe could also have been a source of ignition. The contractors were fined. It seems that they had no idea that mists could explode, or if they did know, they failed to tell their foreman [24].

Oil mist explosions have often occurred in the crankcases of reciprocating engines. They can be prevented by installing relief valves.

The incidents described in Sections 12.4.5 and 17.12 were also mist explosions.

19.6   THE SOURCE OF THE PROBLEM LAY ELSEWHERE

The cause of a problem may be difficult to find when it lies in another part of the plant. One example was described in Section 2.6a. Here are two more.

The product from a new plant was purified in a vacuum distillation column. Soon after startup, the column developed a high-pressure drop. It was opened up for inspection, and the lower trays were found so full of solid that the pressure drop had caused them to buckle. Analysis showed that the solid was a polymer of the product; traces of the reaction catalyst were present and had presumably caused the polymerization. The obvious solution was to remove the traces of catalyst before distillation; this would have been expensive, so changing the operating conditions was tried first. Lower throughput was tried, then lower boilup. Next on the list was a lower pressure, which would give lower temperatures. It proved impossible to lower the pressure, though the vacuum system should have been able to pull a harder vacuum. Was there a leak of air into the plant? A pressure test showed that there was, in the seal of the bottoms pump. When this was repaired, the polymerization problem disappeared. Oxygen as well as traces of catalyst were needed for polymerization.

The pump seal was padded with nitrogen so that if it leaked again, only nitrogen would enter the column.

A molten salt cooled a reactor. At startup, an electric heater was used to heat the salt to reaction temperature. During one startup, the temperature of the salt rose at only half the usual rate. Obviously one of the heaters was faulty, but no fault could be found. The problem was finally traced to a nitrogen valve, which had been left open. The flow of nitrogen through the reactor was taking away half the heat.

Commenting on these two incidents, Gans and colleagues noted that big failures usually have simple causes, whereas marginal failures usually have complex causes. If the product bears no resemblance to the design, look for something simple, like a leak of water into the plant. If the product is slightly below specification, the cause may be hard to find. Look for something that has changed, even if there is no obvious connection between the change and the fault [25].

More hazards that you might not be aware of are discussed in Chapter 32.

References

1. M. H. McRae, *Plant/Operations Progress*, Vol. 6, No. 1, Jan. 1987, p. 17.
2. Reis, *Zeitschrift Physikal. Chem.*, Vol. 88, 1914, p. 513.
3. P. J. Baldock, *Loss Prevention*, Vol. 13, 1980, p. 35.
4. *Northern Echo*, Darlington, UK, May 15, 1976.
5. *Lloyds List*, London, June 20, 1977.
6. *Lloyds List* and *The Times*, London, Jan. 11, 1978.
7. *British Welding Research Association Bulletin*, Vol. 7, Part 6, June 1966, p. 149.
8. P. G. Snyder, "Brittle Fracture of a High Pressure Heat Exchanger," Paper presented at AIChE Loss Prevention Symposium, Minneapolis, Aug. 1987.
9. *Chemical Age*, Dec. 12, 1969, p. 40; and Jan. 9, 1970, p. 11.
10. *Petroleum Review*, Apr. 1984, p. 36.
11. *Hazardous Cargo Bulletin*, Jan. 1983, pp. 22, 24.
12. Oil Companies Materials Association, *Recommendations for the Protection of Diesel Engines Operating in Hazardous Areas*, Wiley, Chichester, UK, 1977.
13. O. A. Pipkin, in C. H. Vervalin (editor), *Fire Protection Manual for Hydrocarbon Processing Plants*, 3rd edition, Vol. 1, Gulf Publishing Co, Houston, Tex., 1985, p. 95.
14. M. Kneale, The Safe Disposal of Relief Discharges, Paper 7, *Selection & Use of Pressure Relief Systems for Process Plants*, Institution of Chemical Engineers North Western Branch, Manchester, UK, 1989.
15. B. O. Andersson, "Lithuania Ammonia Accident," *Hazards XI—New Directions in Process Safety*, Symposium Series No. 124, Institution of Chemical Engineers, Rugby, UK, 1991.
16. B. O. Andersson and J. Lindley, *Loss Prevention Bulletin*, No. 107, Oct. 1992, p. 11.
17. T. A. Kletz, *Journal of Loss Prevention in the Process Industries*, Vol. 5, No. 4, 1992, p. 255.
18. *Fire Journal*, Nov. 1967, p. 89.
19. J. Eichhorn, *Petroleum Refiner*, Vol. 34, No. 11, Nov. 1955, p. 194.
20. H. T. Kohlbrand, "Case History of a Deflagration Involving an Organic Solvent/Oxygen System below Its Flash Point," Paper presented at AIChE Loss Prevention Symposium, San Diego, Calif., Aug. 1990.
21. W. D. Verduijn, *Plant/Operations Progress*, Vol. 15, No. 2, Summer 1996, p. 89.

22. G. Saville, B. Skillerne de Bristowe, and S. M. Richardson, "Safety in Pressure Testing," *Hazards XIII—Process Safety: The Future*, Symposium Series No. 141, Institution of Chemical Engineers, Rugby, UK, 1997.

23. *Loss Prevention Bulletin*, No. 133, Feb. 1997, p. 17.

24. Health and Safety Executive, *Report on Manufacturing and Service Industries, 1982*, Her Majesty's Stationery Office, London, 1983, p. 13.

25. M. Gans, D. Kohan, and B. Palmer, *Chemical Engineering Progress*, Vol. 87, No. 4, Apr. 1992, p. 25.

Problems with Computer Control

In … human-machine interaction, the human and the automated system may both be assigned control tasks. However, unless the partnership is carefully planned, the operator may simply end up with the tasks that the designer cannot figure out how to automate. The number of tasks that the operator must perform is reduced but, surprisingly, the error potential may be increased.

—Nancy Leveson, *Safeware*, Addison-Wesley Publishing,
Reading, Mass, 1995

The use of computers and microprocessors (also known as programmable electronic systems [PES]) in process control continues to grow. They have brought about many improvements but have also been responsible for some failures. If we can learn from these failures, we may be able to prevent them from happening again. A number of them are therefore described here. Although *PES* is the most precise description of the equipment used, I refer to it as a *computer,* as this is the term the nonexpert usually uses.

20.1   HARDWARE AND SOFTWARE FAULTS

Most hardware faults occur in the measurement and control systems attached to the computer rather than in the computer hardware itself, but computer faults do occur and are more common than was once thought. Their effects can be reduced by installing "watchdogs," devices that detect computer failures. However, one accident in which valves were opened at the wrong time and several tons of hot liquid were spilled was due to an error in a watchdog card [1]. Some incidents have been due to voltage variations, including those caused by lightning [2], which in one case caused a computer to leave out a step in a sequence. Systems should be designed so that the effects of foreseeable failures of power or equipment are minimized, and hazard and operability studies (see Section 18.7) should include a check that this has been done.

Software faults can occur in the systems software that comes with the computer or in the applications software written for the particular application. Systems software faults can be reduced by using only well-tested systems—not always easy in a rapidly changing field. Applications software faults can be reduced by thorough testing to make sure that the program behaves as we want it to during abnormal as well as normal conditions. Such testing can take longer than design, and even then some faults may be missed and may lie like time bombs, waiting for a particular combination of unusual conditions to set them off. It is impossible to test every possible pathway in a computer system.

The view is therefore growing that we should try to design plants so that they are safe even if there is a fault in the software. We do this by adding on independent safety systems, such as relief valves and hard-wired trips and interlocks, or by designing inherently safer plants that remove the hazards instead of controlling them (see Chapter 21).

There is an important difference in the failure modes of hardware and software. Computer hardware is similar to other hardware. Once initial faults have been removed and before wear becomes significant, failure can be considered random and treated probabilistically. In contrast, failure of software is systemic. Once a fault is present, it will always produce the same result when the right conditions arise, wherever and whenever that piece of software is used. There is no agreed method for estimating the type and number of faults that might occur, and it is impossible to be 100% confident that we have found them all by testing. For this reason many people, including some authorities, are reluctant to use a computer for the last line of protection on a high-hazard plant. If a computer is used, it should be independent of the control computer.

Errors in written instructions are also systemic, but it is easy for the author to check them, and readers can understand what is meant even though they contain errors in spelling or grammar or are ambiguous. We know what is meant if we are told to save soap and wastepaper.

20.2   TREATING THE COMPUTER AS A BLACK BOX

The most common types of errors are probably those that occur because operators treat the computer as a "black box," that is, something that will do what we want it to do without the need to understand what goes on inside it. There is no fault in the hardware or software, but nevertheless the system does not perform in the way that the designer or operators expected it to perform. The fault is in the specification and arises because the software engineer did not understand the designer's or operator's requirements or was not given sufficiently detailed instructions covering all eventualities. Operators can do what we want them to do even though we have not covered the point precisely; they can decode vague instructions. A computer, however, can do only what it is told to do.

Errors of this type can be reduced by carrying out hazard and operability studies, or Hazops (see Section 18.7), on the instructions given to the computer as well as on the process lines. We should ask what the computer will do for all possible deviations (no flow, reverse flow, more flow, more pressure, more temperature, etc.), for all operating modes, and for all stages of a batch process. The software engineer should be a member of the Hazop team. This will give him or her a better understanding of the process requirements and will give the process engineers and the people who will operate the plant a better understanding of the capabilities of the control system.

20.2.1 The Hazards of Complexity

A pump and various pipelines were used for three different duties: for transferring methanol from a tank truck to storage, for charging it to the plant, and for moving recovered methanol back from the plant (Figure 20-1). A computer set the various valves and monitored their positions.

A tank truck was emptied. The pump had been started from the control room but had been stopped by means of a local button. The next job was to transfer some methanol from storage to the plant. The computer set the valves, but as the pump had been stopped manually, it had to be started manually. When the transfer was complete, the computer told the pump to stop; but as it had been started manually, it did not stop, and a spillage occurred [3].

A thorough Hazop probably would have revealed that this error could have occurred. The control system could have been modified, or better still, separate lines could have been installed for the various different movements, thus greatly reducing the opportunities for error. The incident shows how easily errors in complex systems can be overlooked if the system is not thoroughly analyzed. In addition, it illustrates the paradox that we are willing to spend money on complexity but are less willing to spend it on simplicity [4]. Yet the simpler solution, independent lines (actually installed after the spillage), makes errors less likely and may not be more expensive if lifetime costs are considered. Control systems need regular testing and maintenance, which roughly doubles their lifetime

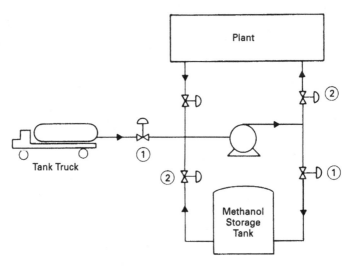

FIGURE 20-1   A pump and lines, controlled by a computer, were used for several different jobs. The pump could also be started and stopped locally. (1) Valves open for first job; others closed. (2) Valves open for second job; others closed.

cost (even after discounting), whereas extra pipelines involve little extra operating cost.

20.2.2   Unforeseen Effects of a Small Leak

A computer controlled a pressure filter. It circulated the liquor through the filter for two hours. As more solid was deposited on the filter, the pressure drop increased. To measure the pressure drop, the computer counted the number of times that the pressure of the air in the filter needed to be topped up in 15 minutes. It had been told that if fewer than five top-ups were needed, filtration was complete, and it could move on to the next phase: smoothing the filter cake. If more than five top-ups were needed, the liquor was circulated for an additional two hours.

There was a leak of compressed air into the filter, which misled the computer into calculating that filtration was complete. The computer signaled completion to the operator, who opened the filter door; the entire batch—liquid and solid—was spilled.

To be fair to the computer, or rather to the programmer, the computer had detected that something was wrong—there was no increase in power consumption during smoothing—and had signaled this finding by stopping the operation, but the operator ignored this warning sign or did not appreciate its significance [3].

Again, a Hazop would probably have disclosed the weakness in the system for detecting the pressure drop through the cake, and changes could have been made. In particular, the filter should have been fitted with a device to prevent the operator from opening it more than a crack while it was full of liquid. Many accidents have occurred because operators opened up autoclaves or other pressure vessels while they were up to pressure (see Sections 13.5 and 17.1). Opening up a vessel while it is full of liquid is not as dangerous but, nevertheless, dangerous enough.

20.2.3 Unforeseen Effects of a Measurement Failure

The furnace on a steel plant was started up from cold shutdown after repair. The temperature indicator was out of order and continually registered a low temperature. The computer, therefore, supplied the maximum fuel-gas rate to the furnace and continued to supply it after an hour, when the furnace was hot. After four hours, the furnace was seriously damaged [5].

Instrument failure is a foreseeable event and should be considered during design by means of a Hazop, a failure mode and effect analysis (FMEA), or some other way. One wonders what the operators were doing. Did they have such confidence in the computer that they never bothered to look at the information displays? If the furnace had not been controlled by a computer, even the most inept operator would have suspected that something was wrong if the temperature had not changed after an hour. The computer, of course, could have been programmed to sound an alarm if the temperature did not change after a period of time, but no one recognized the need to tell it to do so. A Hazop or FMEA could have shown the need.

This illustrates a point that applies to all the incidents described in this chapter: computers do not introduce new errors, but they can provide new opportunities for making old errors; they allow us to make more errors faster than ever before. Incidents will occur on any plant if we do not check readings from time to time or if instructions do not allow for foreseeable failures of equipment.

20.2.4 Changing Trends May Not Be Noticed

This incident occurred on a plant where the possibility of a leak of liquid had been foreseen; a sump had been provided into which any leaks would drain, and a level alarm would then sound. Unfortunately, when a leak occurred, it fell onto a hot surface; most of it evaporated, leaving a solid residue. Because no liquid entered the sump, the leak was not detected for several hours.

The operators could have detected that something was wrong by a careful comparison of trends in a number of measurements, but they saw no need to make such a comparison, as they were not aware of any problem. They did not normally display the relevant measurements on the computer screens and had not called them up. Afterward, the operators said the spillage would have been detected earlier if the chart recorders had not been removed from the control room when the computer was installed in place of the original control equipment.

The computer could have been programmed to carry out mass balances, compare readings for consistency, or sound an alarm (or, better, display a message advising the operators that something was amiss) when unexpected measurements were received, but no one had foreseen the need to program it to do so. It is also possible to install additional screens that can continuously display the trends of selected measurements, in much the same way as old-fashioned chart recorders did. It is, however, easier to change the scale or the measurements displayed (though not relevant to this case, a computer can also detect the absence of noise in measuring instruments that are "stuck") [6].

20.2.5 An Error That Would Not Be Made without a Computer

A computer was controlling a batch reaction on a chemical plant during the night, when daylight saving time ended, and the clocks had to be put back one hour. The operator reset the computer's clock so that it indicated 2 a.m. instead of 3 a.m. The computer then shut the plant down for an hour until the clock indicated 3 a.m. again [7]. Perhaps Hazops should consider reverse flow of time as well as reverse flow of liquids!

20.3   MISJUDGING THE WAY OPERATORS WILL RESPOND

This comes close to the last category as a source of error, and there is much scope for improving the operator/computer interface. The following are some of the incidents that have occurred:

(a) When a power failure occurred, the computer printed a long list of alarms. The operator did not know what had caused the upset, and he did nothing. After a few minutes, an explosion occurred. Afterward, the designer admitted that he had overloaded the operator with too much information but asked the operator why he did not assume the worst and trip the plant. Unfortunately, when people are overloaded by too much information, they tend to switch off (themselves, not the computer) and do nothing. Computers make it easy to overload people with too much information.

(b) The information that operators need for dealing with an alarm is often distributed between several pages of the display. It should be possible to bring together on a special page for each major alarm the information needed for dealing with it [8].

Different display pages often look alike, which saves development time and costs. But during an emergency, an operator may turn to the wrong page and not realize immediately that he has done so.

(c) To reduce the chance that operators will enter the wrong data or instructions, computers are often programmed so that when an operator presses the Enter button, the data or instructions are displayed for the operator to check, and then the operator has to press the Enter button again. Unfortunately, operators soon get into the habit of pressing the Enter button twice in rapid succession after entering data or instructions. It is better if operators have to carry out two distinct operations after entering data, for example, moving a cursor before pressing Enter the second time.

(d) A computer was taken off line so that the program could be changed. At the time, it was counting the revolutions on a metering pump, which was feeding a batch reactor. When the computer was put back on line, it continued counting where it had left off, and the reactor was overcharged.

(e) A computer collected spot values of each instrument reading every minute and then wrote them onto a hard disk every five minutes. The hard disk survived a runaway reaction followed by an explosion, but the explosion occurred toward the end of one of the five-minute periods, and all the data for the five minutes immediately before the explosion were lost. The highest pressure recorded was 60 psi (4 bar), although the bursting pressure of the ruptured reactor was about 900 psi (60 bar) [9].

(f) Some operators seem to expect computers to behave like humans and cannot understand why they make mistakes no human would make. These people instinctively trust seemingly intelligent machines. According to one report, even when alarms were sounding, the operator did not believe it was a real emergency; "the computer can cope," he believed [13].

(g) Two girls hired a taxi to take to take them from their home near Northampton, about 80 miles north of London, to a football ground in London called Stamford Bridge. The taxi driver entered "Stamford Bridge" in his satellite navigation system and then followed its instructions. He arrived at the village of Stamford Bridge, 150 miles to the North of Northampton. Neither the driver nor his passengers suspected that anything was wrong until they arrived there [18].

20.4   OTHER PROBLEMS

20.4.1   Errors in the Data Entered in the Computer

An operator wanted to reduce the temperature on a catalytic cracker from 528°C (982°F) to 527°C (980°F). Unfortunately, he pressed the keys in the wrong order (908) and immediately pressed the Enter key. The computer responded with impressive speed, slamming slide valves shut and causing a flow reversal along the riser. Fortunately, there were no injuries and only a small fire at a leaking joint.

Standards should be written and vendors should be chosen so that the computer will reject or query data or instructions that are outside specified ranges, that deviate more than a specified amount from the existing value, or that fail consistency tests.

On another occasion, an operator was changing a feed rate from 75 to 100 gal per minute. She entered 1,000 in error. The computer opened the feed valve to the full extent, raising the pressure in the plant. There was no damage, however, because the relief valve lifted [10]. A second line of defense, as recommended at the end of Section 20.1, countered an error in the design of the software—failure to foresee and allow for a slip that operators can easily make.

It is possible for errors to occur because data are entered in the wrong units. (See Section 6.2.) A plane became short of fuel and had to make a forced landing because x lbs of fuel were loaded instead of x kg.

20.4.2   Failures to Tell Operators of Changes

I do not know of any incidents in the process industries that have occurred because operators were not told of changes in data or programs, but this has caused an aircraft accident. In 1979, the destination way-point of an Air New Zealand sightseeing flight to Antarctica was moved 2 degrees to the east, but the crew members were not told. The inertial navigation system guided the plane, which was flying low so the passengers could see the scenery, along a valley that ended in a cliff. It looked similar to the open-ended valley that the crew members expected to follow. They did not realize they were on the wrong course, and they flew into the cliff. All 257 people on board were killed [11, 12].

20.4.3   Modifications

Chapter 2 stressed the need to consider the results of plant modifications before they are made and to prevent unauthorized ones. This applies to computers as well as traditional plant. No change should be made to hardware or software unless authorized by a professionally competent person who has carried out a systematic survey of possible consequences. It is easier to change a software control system than a traditional one and therefore harder to control the changes, but it is just as important to do so. Section 20.5 describes an unauthorized change to hardware that could have had serious results.

20.4.4 Old Software

Sections 9.1.6c and 9.2.1h drew attention to the hazards of using old equipment. Similar remarks apply to old software except that, unfortunately, it never wears out. I do not know of any incidents in the process industries due to this cause, but it was responsible for the loss of the European space rocket *Ariane 5*. A function that no longer served any purpose was left in "for commonality reasons," and the decision to do so "was not analyzed or fully understood" [14]. In another incident, cancer patients received excessive doses of radiation because operators were able to enter data faster than the computer could process them. This had always been the case, but originally a hardwired interlock had prevented the excessive dose. The interlock was removed when the old software was "improved" [15, 16].

20.5 UNAUTHORIZED INTERFERENCE

Unauthorized interference with computer hardware is usually difficult, but interference with peripheral equipment may be more serious than on a traditional plant, as the computer will not know that interference has occurred. For example, the leads on a limit switch on a valve were interchanged to carry out some tests. The plant was on manual control at the time but was switched back to computer control before the leads were restored to their correct positions. The computer "thought" the valve was open (when it was shut) and elected to close it. It actually opened it, releasing flammable material [2].

If the plant had been controlled conventionally, the operators involved may have known of the temporary interchange of the leads, or a notice could have been placed on the panel informing them. However, it would be difficult to tell the computer that the valve is open when the signal says it is shut! A computer provides new opportunities for familiar communication errors.

Files had to be transferred from a control computer to a training simulator. At first, there was no direct connection; the files were first transferred to a freestanding workstation and then from there to the simulator. To simplify the transfer, a direct cable connection was made between the control computer and the simulator.

Unfortunately, the address of the gateway in the control computer used for the data transfer was the same as that used to connect to the distributed control system (dcs). As a result data flowed from the simulator through the control computer to the dcs and replaced the current input data by historic data. Some conditions on the plant started to change; fortunately, alert operators soon noticed these changes, and the plant was brought back under control.

Connecting a control computer to another system is a modification and should only be carried out after systematic study of possible consequences (see Section 20.4.3). If made, data flow should be possible only in the outward direction (see Section 24.3). All systems should be secure. Houses need doors. The doors on control systems are less tangible than those on houses but just as important.

If an instrument reading is faulty, operators are sometimes able to override the instrument and type in an estimated reading. Sometimes they are right, and production continues; sometimes they are wrong, and an incident occurs. Operators are usually reluctant to believe unusual readings and rush to the conclusion that the instrument is faulty, whatever the type of control (see Section 3.3.2).

Today it is usually harder than in the early days of computer control for operators to interfere with the software, override interlocks, or type in "correct" readings. However, many operators acquire keys or passwords that they should not have, in much the same way as operators have always unofficially acquired and secreted an assortment of tools and adaptors. On one plant, an interlock was found to be illegally blocked; the password had been disclosed to 40 people, all of whom denied responsibility (see Section 14.5d).

I have seen only one report of a virus in process control software and none of access by hackers. The virus was found on a Lithuanian nuclear reactor and is said to have been introduced by someone who wanted the credit for detecting and removing it. However, this does not mean virus infection or hacking will never occur, and their consequences could be much more serious than loss of accountancy data. As long as a control PES stands alone and is not connected to other systems, infection is impossible (unless a virus is present in the original software), but networking has become increasingly common.

Computer viruses are rather like AIDS. To avoid infection, do not promiscuously share data or disks, and keep the covers on your disks and other storage equipment in the presence of computers whose background is unknown.

20.6 NEW APPLICATIONS

Permits-to-work could be prepared and stored on a computer. The saving in effort would not be great, but additional functions are now possible. Consider these examples:

- The computer could remind the user of any special hazards associated with this piece of equipment and its contents and the actions that should be taken.
- The computer could also remind the user of any problems encountered when the equipment was being prepared or maintained on earlier occasions.
- If a vessel is being prepared for entry, the computer could check that the number of slip-plates (blinds) to be fitted (or pipes disconnected) is the same as the number of connections shown on the drawing.
- If someone tries to take out a second permit on the same item of equipment, this would be instantly apparent, and the computer could refuse to issue it.
- Suppose a fitter has to replace a gasket during a night shift. On some plants it is easy; only one sort is used, and all the fitter has to do is select the right size. On other plants, many types are used. The fitter has to get out a line diagram, find the line number, and then look up the details in a bulky equipment list. It should be possible for him to view the line diagram on a computer screen, select the line with a cursor, and have details of the line displayed, including the location of spare parts and any distinguishing marks, such as the color of the gaskets. The line diagram and equipment list will have been prepared on a computer; all that is needed is a link between the design system and the maintenance system. (Of course, we should, if possible, reduce the number of types of gaskets, nuts, bolts, etc., required even though we may use more expensive types than strictly necessary on some duties.)

Another new application under development is to give operators more information about approaching hazards. For example, if hot oil, over 100°C (212°F), is added to a storage tank containing a water layer or the oil in the tank is heated above 100°C, the water may be vaporized with explosive violence; a mixture of steam and oil will be expelled through the tank vent and may even blow the roof off the tank (see Section 12.2). If the temperature of the incoming oil or the oil in the tank approaches 100°C, then the screen could display a warning message, not merely announcing a high temperature but reminding the operator of the consequences. The reminder message could also be displayed if the operator starts up or increases the heat supply to a tank that contains a water layer. On request the system could explain why the consequences may occur and refer the operator to a plant instruction, accident report, or other document, accessible on the screen, from which the operator could find more information.

The number of possible incidents that might occur and warnings that might be given is enormous, and each plant would have to make a selection based on its own experience and that of the industry. The information would also be accessible to designers and Hazop teams, though they will probably require access to the whole accident database [17].

20.7   CONCLUSIONS

If we can learn from the incidents that have occurred in process plants controlled by computers, we may be able to prevent them from happening again. Familiar errors caused the incidents that have occurred. Accidents or process upsets will occur in any plant, whatever the method of control, if we do not allow for foreseeable slips or equipment failures if modifications are not controlled, if operators are overloaded by too much information, if information display is poor, if controllers are set incorrectly, if warnings are ignored, or if operators are not told of changes that have been made. However, some of these errors are more likely to occur on plants controlled by computers than on conventional plants. This is because different departments may be responsible for operation of the plant and design and operation of the control system, and operating staff members may have exaggerated views of the power of the computer and a limited understanding of what it can and cannot do.

One way of improving communication between chemical and software engineers would be to combine the jobs. There is a need for engineers who are equally at home in the two fields.

References

1. I. Nimmo, S. R. Nunns, and B. W. Eddershaw, "Lessons Learned from the Failure of a Computer System Controlling a Nylon Polymer Plant," Paper presented at Safety and Reliability Society Symposium, Altrincham, UK, Nov. 1987.
2. B. W. Eddershaw, *Loss Prevention Bulletin*, No. 088, p. 3.
3. *Chemical Safety Summary*, Vol. 56, No. 221, Chemical Industries Association, London, UK, 1985, p. 6.
4. T. A. Kletz, *Process Plants: A Handbook for Inherently Safer Design*, 2nd edition, Taylor and Francis, Washington, DC, 1998, Chapter 7.
5. N. G. Leveson, *IEEE Software*, Vol. 7, No. 6, Nov. 1990, p. 55.
6. S. M. Englund and D. J. Grinwis, *Chemical Engineering Progress*, Vol. 88, No. 10, Oct. 1992, p. 36.
7. A. M. Wray, *New Scientist*, Vol. 119, Sept. 8, 1988, p. 61.
8. L. Bodsberg and O. Ingstad, "Technical and Human Implications of Automatic Safety Systems," Paper presented at Sixth International Symposium on Loss Prevention and Safety Promotion in the Process Industries, Oslo, Norway, 1989.

9. D. G. Mooney, "An Overview of the Shell Fluoroaromatics Explosion," *Hazards XI—New Directions in Process Safety*, Symposium Series No. 124, Institution of Chemical Engineers, Rugby, UK, 1991.

10. D. K. Lorenzo, *A Manager's Guide to Reducing Human Errors*, Chemical Manfacturers Association, Washington, DC, 1990, p. 18.

11. P. Mahon, *Verdict on Erebus*, Collins, Auckland, New Zealand, 1984.

12. M. Shadbolt, *Reader's Digest*, Nov. 1984, p. 164.

13. R. E. Eberts, *Chemical Engineering Progress*, Vol. 82, No. 12, Dec. 1985, p. 30.

14. J. L. Lions, *Ariane 5—Flight 501 Failure*, European Space Agency, Paris, 1996.

15. N. G. Leveson, *Safeware—System Safety and Computers*, Addison-Wesley, Reading, MA, 1996, Appendix A.

16. T. A. Kletz, et al, *Computer Control and Human Error*, Institution of Chemical Engineers, Rugby, UK, 1995, pp. 13 and 107.

17. M. Jefferson, P. W. H. Chung, and T. A. Kletz, "Learning the Lessons of Past Accidents," *Hazards XIII—Process Safety: The Future*, Symposium Series No. 141, Institution of Chemical Engineers, Rugby, UK, 1997.

18. N. Allen and N. Britain, *Daily Telegraph* (London), April 4, 2008.

Additional Reading

References 15 and 16

Inherently Safer Design

[A]ll great controversies depend on both sides sharing one false premise.

—A 4th-century theologian

Those who want to spend more money to make a plant safer and those who think enough has been spent share a false premise: they both assume more safety will cost more money.

Many of the incidents in this book were the result of leaks of hazardous materials, and the recommendations describe ways of preventing leaks by providing better equipment or procedures. As we have seen, equipment can fail or can be neglected, and procedures can lapse. The most effective methods, therefore, of preventing leaks of hazardous materials are to use so little that it hardly matters if it all leaks out (intensification or minimization) or to use a safer material instead (substitution). If we cannot do this and have to store or handle large amounts of hazardous material, we should store or handle it in the least hazardous form (attenuation or moderation). Plants in which this is done are said to be inherently safer because they are not dependent on added-on equipment or procedures that might fail; the hazard is avoided rather than controlled, and the safety is inherent in the design.

Because hazards are avoided, there is less need to add on protective equipment, such as interlocks, alarms, emergency isolation valves, fire insulation, water spray, and the like, and the plants are therefore usually cheaper as well as safer.

The principles of inherently safer design may seem obvious, but until the explosion at Flixborough in 1974 (see Section 2.4), little thought was given to ways of reducing inventories of hazardous materials. We simply designed a plant and accepted whatever inventory was needed for that design, confident of our ability to keep it under control. Flixborough weakened our own and the public's confidence in this ability, and 10 years later Bhopal almost destroyed it. The first incident described in this chapter on inherently safer design is therefore the toxic gas release at Bhopal.

My book *Plant Design for Safety: A User-Friendly Approach* [1] and references 12 through 15 describe many examples of ways in which plants can be made inherently safer. Note that we use the term *inherently safer*, not *inherently safe*, as we cannot avoid every hazard.

21.1 BHOPAL

The worst disaster in the history of the chemical industry occurred in Bhopal, in the state of Madhya Pradesh in central India, on December 3, 1984. A leak of methyl isocyanate (MIC) from a chemical plant, where it was used as an intermediate in the manufacture of the insecticide carbaryl, spread beyond the plant boundary and caused the death by poisoning of more than 2,000 people. The official figure was 2,153, but some unofficial estimates were much higher. In addition, about 200,000 people were injured. Most of the dead and injured were living in a shantytown that had grown up next to the plant.

The immediate cause of the disaster was the contamination of an MIC storage tank by several tons of water and chloroform. A runaway reaction occurred, and the temperature and pressure rose. The relief valve lifted, and MIC vapor was discharged into the atmosphere. The protective equipment, which should have prevented or minimized the release, was out of order or not in full working order: the refrigeration system that should have cooled the storage tank was shut down, the scrubbing system that should have absorbed the vapor was not immediately available, and the flare system that should have burned any vapor that got past the scrubbing system was out of use.

The contamination of the MIC may have been the result of sabotage [2], but, as we shall see, the results would have been much less serious if less MIC had been stored, if a shantytown had not grown up close to the plant, and if the protective equipment had been kept in full working order.

21.1.1 "What You Don't Have Can't Leak"

Most commentators missed the most important lesson to be learned from Bhopal: the material that leaked was not a product or raw material but an intermediate, and although it was convenient to store it, it was not essential to do so. Following Bhopal, the company concerned, Union Carbide, and other companies decided to greatly reduce their stocks of MIC and other hazardous intermediates. A year after the disaster, Union Carbide reported that stocks of hazardous intermediates had been reduced by 75% [3].

The product, carbaryl, was manufactured by reacting phosgene and methylamine to produce MIC, which was then reacted with alpha-naphthol. The same product can be made from the same raw materials by reacting them in a different order and avoiding the production of MIC. Phosgene is reacted with alpha-naphthol, and then the intermediate is reacted with methylamine.

If the MIC had been manufactured on the site, there would have been be no need for any storage as it could have been passed down a pipe and used as it was made. The worst possible leak would have been a few kilograms from a broken pipe.

21.1.2 Plant Location

If materials that are not there cannot leak, people who are not there cannot be killed. The death toll at Bhopal—and at Mexico City (see Section 8.1.4) and Sao Paulo (see Section 9.1.8)—would have been lower if a shantytown had not been allowed to grow up near the plant. It is, of course, much more difficult to prevent the spread of shantytowns than of permanent dwellings, but nevertheless we should try to do so by buying and fencing land if necessary (or removing the need to do so, as described earlier).

21.1.3 Keep Incompatible Materials Apart

The MIC storage tank was contaminated by substantial quantities of water and chloroform—up to a ton of water and 1½ tons of chloroform—and this led to a complex series of runaway reactions [4]. The precise route by which water entered the tank is unknown; several theories have been put forward, and sabotage seems likely [2], though whoever deliberately added the water may not have realized how serious the consequences would be. Hazard and operability studies (Section 18.7) are a powerful tool for identifying ways in which contamination and other unwanted deviations can occur, and because water was known to react violently with MIC, it should not have been allowed anywhere near it.

21.1.4 Keep Protective Equipment in Working Order—and Size It Correctly

As already stated, the refrigeration, flare, and scrubbing systems were not in full working order when the leak occurred. In addition, the high temperature and pressure on the MIC tank were at first ignored because the instruments were known to be unreliable. The high-temperature alarm did not operate, as the set point had been raised and was too high. One of the main lessons of Bhopal is, therefore, the need to keep protective equipment in working order. Chapter 14 describes some other accidents that illustrate this theme.

It is easy to buy safety equipment. All we need is money, and if we make enough fuss we get the equipment in the end. It is much more difficult to make sure the equipment is kept in full working order when the initial enthusiasm has faded. All procedures, including testing and maintenance procedures, are subject to a form of corrosion more rapid than that which affects the steelwork and can vanish without a trace once managers lose interest. A continuous auditing effort is needed to make sure that procedures are maintained.

Sometimes managers and supervisors lose interest, and unknown to them, operators stop carrying out procedures. However, shutting the flare system down for repair and taking the refrigeration system out of use were not decisions operators would make on their own. Managers must have made these decisions and thus showed a lack of understanding or commitment.

The refrigeration, scrubbing, and flare systems were probably not big enough to have prevented a discharge of MIC of the size that occurred, but they would have reduced the amount discharged to the atmosphere. The relief valve was not big enough to handle the two-phase flow of liquid and vapor it was called upon to handle, and the tank was distorted by the rise in pressure, although it did not burst. Protective systems cannot be designed to handle every conceivable eventuality, but nevertheless Bhopal does show the need to consider a wide range of circumstances, including contamination, when highly toxic materials such as MIC are handled. It also shows the need, when sizing relief valves, to ask if two-phase flow will occur.

21.1.5   Joint Ventures

The Bhopal plant was half-owned by a U.S. company and half-owned locally. The local company was responsible for the operation of the plant as required by Indian law. In such joint ventures, it is important to be clear who is responsible for safety—in both design and operation. The technically more sophisticated partner has a special responsibility and should not go ahead unless it is sure that the operating partner has the knowledge, experience, commitment, and resources necessary for handling hazardous materials. It cannot shrug off responsibility by saying that it is not in full control.

21.1.6 Training in Loss Prevention

Bhopal—and many of the other incidents described in this book—leads us to ask if those who designed and operated the plant received sufficient training in loss prevention, as students and from their employers. In the United Kingdom, all chemical engineering undergraduates get some training in loss prevention, but this is not the case in most other countries, including the United States. Loss prevention should be included in the training of all engineers; it should not be something added onto a plant after design, like a coat of paint, but an integral part of design. Whenever possible, hazards should be removed by a change in design, such as a reduction in inventory, rather than by adding on protective equipment. Although we may never use some of the skill and knowledge we acquire as students, every engineer will have to make decisions about loss prevention, such as deciding how far to go in removing a hazard [5].

At Bhopal, there had been changes in staff and reductions in manning, and the new recruits may not have been as experienced as the original team. However, I do not think that this contributed significantly to the cause of the accident. The errors that were made, such as taking protective equipment out of commission, were basic ones that cannot be blamed on the inexperience of a particular plant.

21.1.7 Public Response

Bhopal showed the need for companies to collaborate with local authorities and emergency services in drawing up plans for handling emergencies. Inevitably, Bhopal produced a great deal of public reaction throughout the world, but especially in India and the United States. There have been calls for greater control (a paper titled "A Field Day for the Legislators" [6] listed 32 U.S. government proposals or activities and 35 international activities that had been started by the end of 1985) and attempts to show that the industry can put its own house in order (for example, the setting up of the Center for Chemical Process Safety by the American Institute of Chemical Engineers and of the Community Awareness and Response program by the Chemical Manufacturers Association).

Terrible though Bhopal was, we should beware of overreaction or suggestions that insecticides, or the whole chemical industry, are unnecessary. Insecticides, by increasing food production, have saved more lives than were lost at Bhopal. But Bhopal was not an inevitable result of insecticide manufacture. By better design and operations and by learning from experience, further Bhopals can be prevented. Accidents are not due to lack of knowledge but failure to use the knowledge we have. Perhaps this book will help spread some of that knowledge.

21.2   OTHER EXAMPLES OF INHERENTLY SAFER DESIGN

21.2.1   Intensification

The most effective way of designing inherently safer plants is by intensification—that is, using or storing smaller amounts of hazardous material so that the effects of a leak are less serious. When choosing designs for heat exchangers, distillation columns, reactors, and all other equipment, we should, whenever possible, choose designs with a small inventory or holdup of hazardous material. References 1 and 12 through 15 describe some of the changes that are possible. Intensification is easy to apply on a new plant, but its application to existing plants is limited unless we are prepared to replace existing equipment. However, as we have seen, stocks of hazardous intermediates can be reduced on existing plants. When the product of one plant is the raw material of another, stocks can be reduced by locating both plants on the same site, and this also reduces the amount of material in transit.

One company found that it could manage without 75% of its product storage tanks, though in this case the tanks, not the product, were hazardous (see Section 9.2.1g).

Nitroglycerin (NG) production provides a good example of the reductions in inventory that can be achieved by redesign. It is made from glycerin and a mixture of concentrated nitric and sulfuric acids. The reaction is very exothermic; if cooling and stirring do not remove the heat, an uncontrollable reaction is followed by explosive decomposition of the NG.

The reaction was originally carried out batchwise in large stirred pots each containing about a ton of material. The operators had to watch the temperature closely, and to make sure they did not fall asleep, they sat on one-legged stools next to the reactors.

If we were asked to make this process safer, most of us would add onto the reactor instruments for measuring temperature, pressure, flows, rate of temperature rise, and the like and then use these measurements to operate valves that stopped flows, increased cooling, opened vents and drains, and so on. By the time we had finished, the reactor would hardly be visible beneath the added-on protective equipment. However, when the NG engineers were asked to improve the process, they asked why the reactor had to contain so much material. The obvious answer was because the reaction is slow. But the chemical reaction is not slow. Once the molecules come together, they react quickly. It is the chemical engineering—the mixing—that is slow. They therefore designed a small well-mixed reactor, holding only about a kilogram of material, which achieves about the same output as the batch reactor. The new reactor resembles a laboratory water pump. The rapid flow of acid through it creates a partial vacuum, which sucks in the glycerin through a side arm. Very rapid mixing occurs, and by the time the mixture leaves the reactor, the reaction is complete. The residence time in the reactor was reduced from 120 minutes to 2 minutes, and a blast wall of reasonable size could then protect the operator. Similar changes were made to the later stages of the plant where the NG is washed and separated. The reactor, cooler, and centrifugal separator together contain 5 kg of NG [16].

Whatever the method of production, the product, NG, is still hazardous, and safer explosives are now replacing it. Increasingly, quarries now use explosives prepared at the point of use by mixing two nonexplosive ingredients.

Intensification, when practicable, is the preferred route to inherently safer design, as the reduction in inventory results in a smaller and thus cheaper plant. This is, in addition to the reduction in cost, achieved by reducing the need for added-on protective equipment.

21.2.2   Substitution

If intensification is not possible, we should consider substitution—that is, replacing a hazardous material by a less hazardous one. For example, benzene, once widely used as a solvent, is immediately toxic in high concentrations and produces long-term toxic effects in low concentrations. Other solvents, such as cyclohexane, can often be used instead. Better still, nonflammable or high-flashpoint solvents may be suitable. In the food industry, supercritical carbon dioxide is now widely used instead of hexane for decaffeinating coffee and extracting hops. It can also be used for degreasing equipment [7]. Similarly, nonflammable or high-flashpoint heat transfer fluids should be used whenever possible instead of those that have low flash points.

Suppliers of helium operate a recovery service. Helium contaminated with air or nitrogen is returned to them and is purified by passing the gas through a carbon bed cooled to $-196°C$ ($-320°F$) by a jacket of boiling liquid nitrogen. The carbon absorbs the oxygen and nitrogen. The returned helium usually contains far more nitrogen than oxygen. However, one batch of returned helium contained 1.3% nitrogen and 2.2% oxygen. As the oxygen has a higher boiling point ($-183°C$ [$-297°F$]), it condensed out preferentially, and the gas absorbed on the top of the carbon bed was 85% oxygen. The carbon-oxygen mixture exploded, causing extensive blast and missile damage but, fortunately, no injuries. Minor mechanical shock could have detonated the mixture.

None of the people on the plant realized that a change in feed composition could be hazardous. However, when they told other helium suppliers what had happened, two earlier unpublished incidents came to light. This third incident might not have occurred if the people who failed to publicize the earlier incidents had been more open.

Analysis procedures are now more rigorous. In addition, an example of substitution, silica gel, which does not react with oxygen, is used as the absorbent instead of carbon. Silica gel is less efficient but inherently safer [8].

As in the case of Bhopal (Section 21.1.1), a different chemical route can sometimes be substituted for one that has hazardous consequences. Here is an example from another industry. Pastry dough is made, at home and industrially, by mixing flour and fat and then adding water. If used to cover a pie, it is liable to crack or slump. According to International Patent Application No. 96/39852, these disasters can be prevented by emulsifying the water and fat, with the help of an emulsion stabilizer, and then adding the flour [17].

21.2.3   Attenuation

A third method of making plants inherently safer is attenuation—using hazardous materials in the least hazardous form. For example, while small quantities of liquefied toxic or flammable gases such as chlorine, ammonia, and propane are usually stored under pressure at atmospheric temperature, large quantities are usually stored at low temperature at or near atmospheric pressure. Because the pressure is low, the leak rate through a hole of a given size is smaller, and because the temperature is low, evaporation is much less (see Section 8.1.5). The possibility of a leak from the refrigeration equipment has to be considered as well as the possibility of a leak from the storage vessel, and for this reason only large quantities are refrigerated.

Another example of attenuation is storing or transporting a hazardous material in a solvent. Thus, acetylene has been stored and transported for many years as a solution in acetone, and many organic peroxides, which are liable to decompose spontaneously, are stored and transported in solution.

21.2.4   Limitation of Effects

A fourth road to inherently safer design is limitation of effects, by equipment design or by limiting the energy available, rather than by adding on protective equipment. For example, many spillages of liquefied petroleum gas are due to the overfilling of storage vessels and discharge of liquid from a relief valve that is not connected to a flare system. If the filling pumps can be rated so that their closed-head delivery pressure is below the set point of the relief valves (or the vessels designed so that they can withstand the delivery pressure), then the relief valves will not lift when the vessels are filled.

Similarly, overheating can be prevented by using a heating medium at a temperature too low to be hazardous. For example, corrosive liquids are often handled in plastic (or plastic-coated) tanks heated by electric immersion heaters. If the liquid level falls, exposing part of the heater, the tank wall may get so hot that it catches fire. One insurance company reported 36 such fires in two years, many of which spread to other parts of the plants. Five were due to failure of a low-level interlock.

The inherently safer solution is to use a source of heat that is not hot enough to ignite the plastic, for example, hot water, low-pressure steam, or low-energy electric heaters [9].

If unstable chemicals have to be kept hot, the heating medium should be incapable of overheating them. Some acidic dinitrotoluene should have been kept at 150°C (300°F), as it decomposes at higher temperatures. It was heated by steam at 210°C (410°F) for 10 days in a closed pipeline and decomposed explosively [10].

21.2.5   Seveso

The use of an unnecessarily hot heating medium led to the runaway reaction at Seveso, Italy, in 1976, which caused a fallout of dioxin over the surrounding countryside, making it unfit for habitation. Although no one was killed, it became one of the best-known chemical accidents, exceeded only by Bhopal, and had far-reaching effects on the laws of many countries.

A reactor containing an uncompleted batch of 2,4,5-trichlorophenol (TCP) was left for the weekend. Its temperature was 158°C (316°F), well below the temperature at which a runaway reaction can start (believed at the time to be 230°C [645°F] but possibly as low as 185°C [365°F]). The reaction was carried out under vacuum, and the reactor was heated by an external steam coil supplied with exhaust steam from a turbine at 190°C (374°F) and a gauge pressure of 12 bar (Figure 21-1). The turbine was on reduced load, as various other plants were also shutting down for the weekend (as required by Italian law), and the temperature of the steam rose to about 300°C (570°F). The temperature of the bulk liquid could not

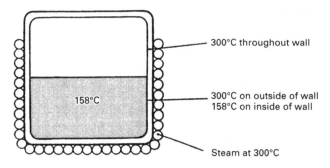

FIGURE 21-1   The Seveso reactor.

get much above 158°C (316°F) because of its heat capacity, and so below the liquid level there was a temperature gradient through the walls of the reactor, 300°C (570°F) on the outside, 158°C (316°F) on the inside. Above the liquid level, the walls were at 300°C (570°F) right through.

When the steam was isolated and, 15 minutes later, the stirrer was switched off, heat passed by conduction and radiation from the hot wall above the liquid to the top 10 cm or so of the liquid, which became hot enough for a runaway to start. If the steam had been cooler, 185°C (365°F) or less, the runaway could not have occurred [11] (see also Section 10.4.6).

21.2.6   Existing Plants

As already stated, it is difficult to introduce inherently safer designs on existing plants, though storage can often be reduced. However, one company improved the inherent safety of an acquired pant, which contained a unit for the manufacture of phosgene. Improvements to the reliability of the control equipment made it possible to reduce in-plant storage; buying purer carbon monoxide, one of the raw materials, made it possible to eliminate parts of the purification section (thus simplifying as well as intensifying); and changing the chlorine supply from liquid to gas reduced the chlorine inventory by 90% [18].

21.3   USER-FRIENDLY DESIGN

A related concept to inherently safer design is user-friendly design: designing equipment so that human error or equipment failure does not have serious effects on safety (and also on output or efficiency). While we try to prevent human errors and equipment failures, only very low failure rates are acceptable when we are handling hazardous materials, and, as this book has shown, it is hard to achieve them. We should, therefore, try to design so that the effects of errors are not serious. The following are some of the ways in which we can accomplish this:

- By simplifying designs: complex plants contain more equipment that can fail, and there are more ways in which errors can occur.
- By avoiding knock-on effects: for example, if storage tanks have weak seam roofs, an explosion or overpressuring may blow the roof off, but the contents will not be spilled (see Section 5.2).
- By making incorrect assembly impossible (for an example, see Section 9.1.3).
- By making the status of equipment clear. Thus, figure-8 plates are better than slip-plates, as the position of the former is obvious at a glance, and valves with rising spindles are better than valves in which

the spindle does not rise. Ball valves are friendly if the handles cannot be replaced in the wrong position.

- Using equipment that can tolerate a degree of misuse. Thus, fixed pipework is safer than hoses (see Section 7.1.6), and fixed pipework with expansion loops is safer than expansion joints (bellows).

Reference 1 gives more examples. The phrase "user-friendly design" has not been widey adopted and many writers describe the proposals suggested in this section as examples of inherently safer design.

References

1. T. A. Kletz, *Process Plants: A Handbook for Inherently Safer Design*, 2nd edition, Taylor and Francis, Washington, D.C., 1998.
2. A. Kalelkar, "Investigation of Large Magnitude Incidents: Bhopal as a Case Study," *Preventing Major Chemical and Related Process Accidents*, Symposium Series No. 110, Institution of Chemical Engineers, Rugby, UK, 1988.
3. *Chemical Insight*, late Nov. 1985, p. 1.
4. *Bhopal Methyl Isocyanate Incident: Investigation Team Report*, Union Carbide Corporation, Danbury, Conn., Mar. 1985.
5. T. A. Kletz, *Plant/Operations Progress*, Vol. 7, No. 2, Apr. 1988, p. 95.
6. W. Stover, "A Field Day for the Legislators," *The Chemical Industry after Bhopal*, Proceedings of a symposium, IBC Technical Services, London, Nov. 7–8, 1985.
7. *Swedish Work Environment Fund Newsletter*, No. 2, 1992.
8. J. W. Hempseed and R. W. Ormsby, *Plant/Operations Progress*, Vol. 10, No. 3, July 1991, p. 184.
9. *Sentinel*, International Risk Insurers, 3rd quarter, 1981.
10. *Loss Prevention Bulletin*, No. 088, p. 13.
11. T. A. Kletz, *Learning from Accidents*, 3rd edition, Elsevier, Boston, MA, 2001, Chapter 9.
12. Center for Chemical Process Safety, *Inherently Safer Chemical Processes—A Life Cycle Approach*, 2nd edition, Wiley, San Francisco, CA, 2009.
13. *International Conference and Workshop on Process Safety Management and Inherently Safer Processes*, AIChE, New York, 1996.
14. *Guidelines for Engineering Design for Process Safety*, AIChE, New York, 1993, Chapter 2.
15. S. M. Englund, Process and Design Opportunities for Inherently Safer Plants. In: V. M. Fthenakis (editor), *Prevention and Control of Accidental Releases of Hazardous Gases*, Van Nostrand Reinhold, New York, 1993.
16. N. A. R. Bell, "Loss Prevention in the Manufacture of Nitroglycerin," *Loss Prevention in the Process Industries*, Symposium Series No. 34, Institution of Chemical Engineers, Rugby, UK, 1971.
17. *Chemistry and Industry*, No. 14, July 21, 1997, p. 548.
18. R. Gowland, *1996 Process Safety Progress*, Vol. 15, No. 1, Spring 1996, p. 52.

Additional Reading on Bhopal

S. Bajpai, N. Jain, H. P. K. Warrier, and J. P. Gupta, *Bhopal Gas Tragedy and Its Effects on Process Safety*, Indian Institute of Technology, Kanpur, India, 2004.

P. Shrivastava, *Bhopal—Anatomy of a Crisis*, Ballinger, Cambridge, MA, 1987.

S. Varadarajah, *Report on Scientific Studies on the Factors Related to Bhopal Toxic Gas Leakage*, Indian Planning Commission, New Delhi, India, Dec. 1985.

S. Hazarika, *Bhopal—The Lessons of a Tragedy*, Penguin Books, London, 1987.

T. A. Kletz, *Learning from Accidents*, 3rd edition, Elsevier, Boston, MA, 2001, Chapter 10.

22

Reactions—Planned and Unplanned

Famous last words: "I didn't see an exotherm in the lab"; "I only saw a little bit of foaming"; "It goes a bit brown if you leave it in the oven for too long."

—Chilworth Technology

Many accidents, particularly on batch plants, have been due to runaway reactions—that is, reactions that get out of control. The reaction becomes so rapid that the cooling system cannot prevent a rapid rise in temperature, or the relief valve or rupture disc cannot prevent a rapid rise in pressure, and the reactor ruptures. Examples are described in the chapter on human error (Sections 3.2.1e and 3.2.8), although the incidents were really due to poor design, which left traps into which someone ultimately fell.

The number of reactions that can run away is enormous, *Bretherick's Handbook of Reactive Chemical Hazards* [1] lists about 4,700 chemicals that have been involved in hazardous reactions of one sort or another, and there are more than 20,000 cross-references to entries involving more than one chemical. It is an essential work of reference for the chemist, the process engineer, and everyone involved in process safety. All I can do here is give a few examples to illustrate the reasons why runaways occur.

22.1 LACK OF KNOWLEDGE

This is, or was at one time, the major cause of runaway reactions. After many years of safe operation, a chemical or a reaction mixture gets a little hotter than usual or is kept warm for a little longer than usual, and a runaway occurs. Today there is little excuse for such runaways, as many methods are available for testing both pure substances and reaction mixtures. They include accelerating rate calorimetry (ARC), differential scanning calorimetry, and reaction calorimetry. There are also methods for determining the size of the relief valve, rupture disc, or vent required. Expert advice is needed on the most suitable technique for each case. If process conditions are changed, then further testing may be necessary. Section 2.6d describes a slight change in operating conditions, which led to a violent explosion.

Safe operation for many years does not prove that a reaction will not run away. Unknown to the operators, the plant may be close to the conditions under which it becomes unstable, and a slight change in pressure, temperature, or concentration, too small to cause concern, may take it over the brink. The operators are blind men walking along the edge of a precipice, as the following incidents illustrate:

(a) Some zoalene (3,5-dinitro-ortho-toluamide), a poultry food additive, was left standing for longer than usual after drying. After 27 hours, a devastating explosion occurred. Tests by accelerating rate calorimetry (ARC) showed that zoalene could self-heat to destruction if held at 120 to 125 degrees C (248 to 257 degrees F) for 24 hours.

The company had started to test by ARC, then a new technique, all the chemicals it handled. At the time of the explosion, the company had tested 5% of them. It had no reason to give zoalene priority, as other tests and 17 years of manufacturing experience had given no inkling of its instability. The official report on the explosion concluded, "There appears to be no substantial grounds for criticizing the management or operating personnel for undertaking and conducting the operation that led to the explosion in the way they did" [2]. The conclusion would, of course, have been different if the zoalene had blown up several years later, after the company had time to test all its chemicals.

(b) As the result of a steam leak into a reactor jacket, some nitrobenzene sulfonic acid was held for 11 hours at 150°C (302°F). Decomposition occurs above 145°C (293°F), and a violent explosion expelled the reactor from the building. At a time, decomposition was believed to occur only above 200°C (390°F) [3].

(c) A solution of ferric chloride in a solvent was manufactured by suspending iron powder and adding chlorine. The process was tested by ARC and on a pilot scale and then transferred to a full-scale reactor (1.2 m diameter by 2 m tall). During the third batch, a devastating explosion occurred, killing 2 people and injuring 50.

The control of the reaction was based on the assumption that stopping the flow of chlorine would stop all reaction; this was true on the pilot unit but not on the full-scale plant. On the pilot unit, there was no stirrer, as the incoming chlorine gave sufficient mixing. When chlorine addition stopped, mixing also stopped and so did the reaction. On the full-scale plant, a stirrer was necessary, and this continued in operation after chlorine addition stopped. In addition, on the pilot unit the cooling was sufficient to hide any continuing reaction that did occur.

Stirring was attempted during the ARC tests, but the iron powder interfered with the mechanically coupled stirrer [4].

(d) Many incidents have occurred because designers failed to realize that as the volume of a reactor is increased, the surface area and thus the heat loss do not increase in proportion. If the height of a reactor is doubled and the shape stays the same, then the volume increases eight times, but the surface area increases only four times.

A reaction was believed to be thermally neutral, as no rise in temperature was observed in the laboratory. No cooling was provided on the pilot plant, and the first batch developed a runaway. Fortunately, the relief valve was able to handle it. Subsequent research showed that the reaction developed 2 watts/kg/°C. Laboratory glassware has a heat loss of 3 to 6 watts/kg/°C, so no rise in temperature occurred. On the 2.5-m^3 pilot plant reactor, the heat loss was only 0.5 watt/kg/°C [21]. Reference 22 lists heat losses and cooling rates for vessels of various sizes.

(e) In an unusual cause of a runaway reaction, thieves tried to burn their way through the steel door of a fireworks store with an oxyacetylene torch. The building, made of concrete 12 in. thick, was reduced to rubble [23].

22.2   POOR MIXING

(a) Section 3.2.8 described a runaway that occurred because a valve in a reactor circulation line was closed. As a result there was no mixing, and the incoming reactant formed a separate layer. When someone opened the valve, the two layers were suddenly mixed, with a catalyst, and reacted violently. Even though two liquids are miscible, they may still layer.

Many similar incidents have occurred when a stirrer or circulating pump stopped. For example, an acidic waste stream in a tank was neutralized with chalk slurry. The operator realized that the liquid going to drain was too acidic. Looking around, he then found that the stirrer had stopped. He switched it on again. The acid and chalk, which had formed two separate layers, reacted violently, and the gas produced blew the bolted lid off the tank.

(b) An aromatic hydrocarbon was being nitrated by slowly adding a mixture of nitric and sulfuric acids. After several hours, the operator discovered the temperature had not risen. He noticed the stirrer had stopped and switched it on. Almost at once he realized that this could start a runaway reaction; he switched off the stirrer, but it was too late. Within 20 seconds fumes were coming out of the vent on the reactor. He went down to the floor below to open the drain valve, but the fumes were so thick he could not see what he was doing. Wisely, he decided to leave the area. Five minutes later the vessel burst. The lid ended up 15 m away, and the rest of the reactor was propelled several meters onto the ground [5].

Failure of a reactor stirrer or circulation pump should automatically isolate the supply of incoming liquid. If all the reactants have already been added and mixing failure is still hazardous, then the action to be taken should be agreed in advance, for example, increasing cooling, adding a quenching agent, or dumping the contents.

(c) If there is no mixing in a reactor, the temperature measurements will indicate the temperature at the point of measurement, but it may be different in other parts of the bulk liquid. A reactor was provided with a quench water system; if the contents got too hot, water could be added from a hose. A power failure caused the stirrer to stop. The operator watched the temperature. As it was falling he did nothing. After a while it started to rise; before he could connect up the water supply, the rupture disc failed, and soon afterward the reactor blew up.

No accident has a simple cause. In this case, a contributory factor was the inexperience of the operator, who had to ask someone to show him how the quench system worked. It had not been used for several years. Emergency equipment is usually used infrequently, and without regular training, people forget (or never learn) how to use it [6].

(d) A devastating explosion, which killed 46 people including 19 members of the public, destroyed a trinitrotoluene (TNT) factory in Ashton, near Manchester in the United Kingdom, during the first World War. During the final stage of production, the addition of nitric acid to convert dinitrotoluene to TNT, the nitrator pan 1.5 m [5 ft] tall and 1.5 m diameter) started to give off nitric acid fumes. Acid flow was stopped and so was the stirrer, but nevertheless the contents boiled over. Hot acid fell onto the wooden staging around the pan, starting a fire. Soon afterward the stocks of TNT in surrounding equipment and in drums exploded.

At the time of the explosion, the wooden stagings were being replaced by iron ones, but the work was going slowly, as the materials needed were rationed and the Munitions Works Board had classified the change as desirable but not absolutely necessary.

Why did the pan boil over? To improve efficiency the operator in charge had reduced the amount of sulfuric acid added, along with nitric acid, during the final stage of the process. As less acid was added, the pan was not as full as usual, and for a time the top stirring blades were above the surface of the liquid. In addition, the use of less sulfuric acid made the reaction less stable. When the level rose and covered the blades, the unreacted acid started to react, the temperature rose, and a runaway reaction occurred [24, 25]. We thus see once again the unforeseen result of a process change that was not properly thought through (see Chapter 2).

(e) A similar incident occurred more recently during the sulfonation of a nitroaromatic compound. Some product from a previous batch was put into the reactor and heated to 85°C (185°F); the melted nitroaromatic and oleum were then added simultaneously and the temperature allowed to rise to 110°C (230°F).

The reactor was also used for other processes. The cooler was not adequate for one of them, so an additional coil was fitted. To make room for it, the stirrer was removed and replaced by a turbine agitator, located higher up the vessel. At a later date, when the process that needed the extra cooling was no longer used, the extra coil, which had corroded, was removed. This lowered the level in the reactor.

Six months later, during a routine maintenance check, the position of the agitator was inadvertently set at the highest position possible. At the start of the reaction, the agitator was now uncovered. The two reactants formed separate layers; the temperature fell, and the control system increased the steam supply to the reactor jacket. After a while, the rising liquid level covered the agitator, the accumulated quantities of the two hot reactants were mixed, and reaction began. The cooling system could not control the unusually high rate of reaction, the temperature rose, decomposition set in, the pressure rose, the reactor cover ruptured, and the contents overflowed like a stream of lava [26].

This incident and the previous one show how easy it is to reduce mixing inadvertently and how serious the results can be.

22.3   CONTAMINATION

The most famous case of a runaway reaction caused by contamination is Bhopal (see Section 21.1). In this case, the reaction occurred in a storage vessel. It did not burst but was distorted, and the discharge of vapor was larger than the scrubbing and flare systems could have handled, even if they had been in operation.

Here are some other examples of unwanted reactions caused by contaminants [7]:

- Storage tanks containing ethylene oxide are usually inerted with nitrogen. One plant used nitrogen made by cracking ammonia. The nitrogen contained traces of ammonia, which catalyzed an explosive decomposition of the ethylene oxide. Similar decompositions have been set off by traces of other bases, chlorides, and rust.
- A storage tank containing acrolein was kept cool by circulating the liquid through a water-cooled heat exchanger. Demineralized water was normally used, but the supply failed, and water from an underground borehole was used instead; it contained numerous minerals. There was a slight leak in the heat exchanger, some water contaminated the acrolein, and the minerals catalyzed rapid polymerization. The tank exploded.
- A high-pressure air compressor drew its air from an area where oxyacetylene welding was taking place. Small amounts of copper acetylide formed on a bronze valve and exploded.
- Unstable impurities may concentrate on certain trays in a distillation column.
- Traces of oxygen in nitrogen used for inerting can react with some products, such as butadiene and acrolein, and cause explosive polymerization. In one case, unknown to the acrolein plant, a trace of oxygen was deliberately added to the nitrogen supply at the request of another plant.
- Inhibitors are usually added to butadiene and acrolein to prevent polymerization, but the system is not foolproof. Several runaways have occurred in tank trucks or tank cars containing acrolein. As it cooled, some of the liquid crystallized, leaving the inhibitor in solution. In other cases, impurities have been left behind in the bulk liquid, and their concentration has risen sufficiently to start a runaway.

Vessels are sometimes contaminated by material left over from a previous use. For example, tank trucks were filled with a waste sludge containing particles of aluminum. One day the tank truck contained some caustic soda left over from the previous load. The caustic soda reacted with the aluminum, producing hydrogen. The increase in pressure blew open an inspection port and knocked an operator onto the ground.

In another example, a solvent was put into a small reactor to remove some polymer, which was stuck to the walls. Some monomer, which was trapped behind the polymer, reacted with the solvent, and the pressure rose. Bits of polymer plugged the relief valve, and the pressure broke a glass connecting line [8].

Reactions have often taken place in waste containers because assorted substances were added indiscriminately [8]. Leaking heat exchangers can also cause contamination as described previously.

The management at a top-quality London hotel was horrified to find that its house champagne, for which it charged $60 per bottle, lost its bubbles a few moments after it was poured. At first the hotel blamed the suppliers, but the trouble was found to be due to a trace of the detergent used to wash the glasses [27].

22.4   REACTIONS WITH AUXILIARY MATERIALS

As well as testing the reactants, as described in Section 22.1, we should also test auxiliary materials. For example, nitrogen trifluoride reacted with silica, which was used as a drying agent. Whenever a new batch of silica was installed, there was a rise in temperature, which the operators never reported and in time accepted as normal. They were walking along the precipice described in Section 22.1, and one day the temperature rise got out of control.

22.5   POOR TRAINING OR PROCEDURES

An operator was told to add a reactant over a certain period of time. He started to add it too slowly. Finding that he was getting behind, he added the rest too quickly, and a runaway occurred. Fortunately, in this case, the relief device controlled the situation, and the reactor did not rupture, though product was wasted. It may be necessary to specify the rate of addition as well as the time of addition.

Operators were told to add a reactant at 45°C (113°F) over a period of 1 to 1½ hours. They believed this to be impossible, as the heater was not powerful enough, so they decided to add it at a lower temperature and heat the material in the reactor. They did not tell anyone. This went on for a long period of time and, unknown to the supervisor, became the accepted practice. Again they were on the edge of a precipice, and ultimately a runaway reaction occurred with emission of toxic fumes.

Unfortunately, if people are given instructions that are impossible, or that they think are impossible, to carry out, they do not like to tell their supervisors, and so they often just do the best they can. However, in this case if proper records had been kept and the supervisor examined them, he would have noticed that the addition temperature was wrong.

Runaways have also occurred when operators added the wrong material to a reactor, often because different materials had similar names, were stored in similar drums, or were poorly labeled (see Chapter 4).

A batch distillation column, used for distilling nitrotoluene, had not been cleaned for 30 years. A buildup of sludge caused some problems, or so it was believed, and those in charge decided to clean the column with live steam. The operators were told not to let the sludge get above 90°C (194°F), but there was no way they could measure its temperature; all they could measure was the temperature of the vapor in the still. The sludge got much hotter, and a runaway reaction occurred. A ball of flame came out of an open manhole on the base of the still, engulfed the control room—a wooden building (!)—25 m away, crossed a parking lot, and reached the office block. Five men were killed; four of them were in the control room. The company had to pay fines and costs of $600,000 [9].

When solids have to be added to a reactor containing a volatile liquid, this is sometimes done by opening a cover and adding the solid quickly before much vapor has time to escape. Such an operation may be carried out many times without incident, but a fire or explosion is always possible. If the reactor is inerted with nitrogen the inerting may be lost. Reference 30 describes many lock systems that allow a powder to be added to a liquid without any escape of vapor or inerting gas.

Nitration has been described as the "most widespread and powerfully destructive industrial unit process operation" [10].

22.6   USE-BY DATES

We are used to seeing sell-by or use-by dates on food. Some chemicals also have, or ought to have, use-by dates on them. The best known are ethers; on standing, they form peroxides, which can explode if subjected to shock. Ethers should not be kept for more than limited periods—six months in the case of dimethyl and other low-molecular-weight ethers. This has been known for many years, and Bretherick [11] gives references to a number of explosions that occurred because ethers were kept for too long. A particularly tragic accident befell a research chemist. He tried to open a bottle of isopropyl ether by holding it against his stomach and twisting the cap. The bottle exploded, injuring him so severely that he died two hours later [12]. Nevertheless, according to a recent report from the U.S. Department of Energy [13], 21 containers of dimethyl ether more than 21 months old were found in one of its laboratories.

The U.S. Department of Energy also points out that polyethylene bottles containing corrosive chemicals may deteriorate with prolonged use [14].

Other limited-life chemicals listed by Bretherick are bleaching powder ("Material which has been stored for a long time is liable to explode on exposure to sunlight or on overheating of tightly packed material in closed containers" [15]) and aqua regia, a 1:4 mixture of nitric and sulfuric acids used for cleaning ("Aqua regia decomposes with evolution of gas and should not be stored in tightly closed bottles [and preferably not at all]" [16]).

A dilute solution of hydroxylamine nitrate and nitric acid was left in a vented tank for four years. Evaporation caused the concentration of the chemicals to rise until they started to react together. They produced so much steam and gas that they blew the lid off the tank. In another incident, the same mixture, plus hydrazine, was trapped between two valves. Decomposition ruptured a gasket. Chemicals should be removed from vessels that are no longer in use [28, 29].

Oil spillages onto warm, absorbent materials, such as insulation, also have a limited life (see Sections 7.3.2 and 12.4.4). The oil soon decomposes to materials with a low auto-ignition temperature and self-ignites. As many insulation fires have started in this way, oil-soaked insulation should be removed without delay. Linseed oil ignites particularly easily. This has been known since at least 1925; nevertheless, in 1965 some cloths used to apply linseed oil to laboratory benches were not burned as directed but dropped into a waste bin. A fire started after a few hours and destroyed the laboratory [17]. Reference 18 lists substances that are liable to self-heat, and reference 19 includes references to a number of incidents that have occurred involving substances as diverse as wood shavings, tobacco, milk powder, and soap powder.

A manufacturer of ethylene oxide received some old returned cylinders in which the ethylene oxide had partly polymerized, sealing the valves. The cylinders were taken to an explosives testing site and blown up [20].

References

1. P. G. Urben (editor), *Bretherick's Handbook of Reactive Chemical Hazards*, 6th edition, Butterworth-Heinemann, Oxford, UK, 1999.
2. Health and Safety Executive, *The Explosion at the Dow Chemical Factory, King's Lynn, 27 June 1976*, Her Majesty's Stationery Office, London, Mar. 1977.
3. *Case Histories of Accidents in the Chemical Industry*, Vol. 3, Manufacturing Chemists Association, Washington, DC, 1970, p. 111.
4. E. S. De Haven and T. J. Dietsche, *Plant/Operations Progress*, Vol. 9, No. 2, Apr. 1990, p. 131.
5. T. Kotoyori, *Journal of Loss Prevention in the Process Industry*, Vol. 4, No. 2, Apr. 1989, p. 120.
6. *Loss Prevention Bulletin*, No. 098, Apr. 1991, p. 7.
7. R. Grollier Baron, "Hazards Caused by Trace Substances," *Seventh International Symposium on Loss Prevention and Safety Promotion in the Process Industries*, Taormina, Italy, May 4–8, 1992.
8. M. A. Capraro and J. H. Strickland, *Plant/Operations Progress*, Vol. 8, No. 4, Oct. 1989, p. 189.
9. Health and Safety Executive, *The Fire at Hickson and Welch Ltd.*, HSE Books, Sudbury, UK, 1994.

10. Reference 1, Vol. 2, p. 246.
11. Reference 1, Vol. 2, p. 125.
12. *Case Histories of Accidents in the Chemical Industry*, Vol. 2, Manufacturing Chemists Association, Washington, DC, 1966, p. 6.
13. *Safe Chemical Storage*, Bulletin 91-2, DOE/EH-0168, U.S. Dept. of Energy, Washington, DC, Feb. 1991.
14. *Polyethylene Bottles Containing Corrosive Chemicals May Deteriorate with Prolonged Use*, Bulletin 89-1, DOE/EH-0094, U.S. Dept. of Energy, Washington, DC, Aug. 1989.
15. Reference 1, Vol. 2, p. 57.
16. Reference 1, Vol. 2, p. 43.
17. Reference 1, Vol. 2, p. 186.
18. J. Bond, *Sources of Ignition*, Butterworth-Heinemann, Oxford, UK, 1990, Appendix 2.6.
19. Reference 1, Vol. 2, p. 361.
20. E. S. Hunt, *Loss Prevention*, Vol. 6, 1972, p. 140.
21. *Process Safety News*, No. 3, Chilworth Technology, Southampton, UK, Autumn 1996/Winter 1997, p. 1.
22. J. Singh, *Chemical Engineering*, May 1997, p. 92.
23. *The Times* (London), 1996 (precise date unknown).
24. J. Billings and D. Copland, *The Ashton Munitions Explosion*, Tameside Leisure Services, Stalybridge, UK.
25. *Explosion of TNT at Ashton-under-Lyne*, Ref. No. MUN7/37 XC/B/8909, Public Records Office, London.
26. *Thermal Process Safety*, Expert Commission for Safety in the Swiss Chemical Industry, Basle, Switzerland, 1993.
27. *Daily Telegraph Business News* (London), Nov. 5, 1994.
28. *Operating Experience Weekly Summary*, No. 97-21, Office of Nuclear and Safety Facility, U.S. Dept. of Energy, Washington, DC, 1997, p. 1.
29. *Chemical Explosion at Hanford*, Safety Alert No. 97-1, U.S. Dept. of Energy, Washington, DC, 1997.
30. M. Glor, *Chemical Engineering*, Vol. 114, No. 10, Oct. 2007, p. 88.

Additional Reading on Runaway Reactions

J. Barton and R. Rogers, *Chemical Reaction Hazards*, 2nd edition, Institution of Chemical Engineers, Rugby UK, 1997.

STILL GOING WRONG

Introduction to Part B

As explained in the preface, there is an important difference between Part A (*What Went Wrong?*) and Part B (*Still Going Wrong*). Part A is mainly concerned with engineering matters. It describes the changes in design, maintenance, and operations necessary to prevent the accidents described from happening again. Part B also does this, but in addition, whenever possible, it looks for the underlying causes of the accidents, such as weaknesses in organization, "custom and practice," and culture. For this reason, a few items in Part A are discussed further in Part B.

This book is written primarily for all those involved in design, maintenance, and operations. But fundamental changes in organization, "custom and practice," and culture require action by senior managers, many of whom do not realize that their involvement is necessary. They frequently and rightly say that safety is everybody's business, but in practice they often exclude themselves and leave it to the technical staff. They may comment on the lost-time accident rate, but that does not measure process safety. This part of the book describes a number of serious accidents that occurred because of serious errors of judgment by the leaders of large companies. For examples, see Section 24.8 and Chapter 26.

Maintenance

People should have to take a class on this information before they receive their undergraduate degrees in engineering. Nobody really tells us this stuff.

—A message from a chemical engineering student
who found *What Went Wrong?* in a library

The longest chapter in Part A is Chapter 1, Preparation for Maintenance. This is still the source of many accidents, and more are described in the following pages. They have been chosen to emphasize that the need for good practice and its enforcement is as great as ever and to draw attention to features not discussed in Part A.

23.1 INADEQUATE PREPARATION ON A DISTANT PLANT

This accident occurred in a large, responsible international company but in a distant plant many thousands of miles away from the United States and Europe. Pipework connected to a tank that had contained a flammable liquid was being modified. The tank was "washed clean with water," to quote the report. The foreman checked that the tank looked clean and that there was no smell. The valves on the tank and the manway cover were all closed, or so it was thought, and a permit was issued for welding on the pipework. One of the pipes was cut with a hacksaw and a section removed. When a welder started to weld the replacement section, an explosion occurred in the tank. The welder was hit by the manway cover and hurled 5 m (16 ft) to the ground. He died from his injuries.

doi:10.1016/B978-1-85617-531-9.00023-8

23.1.1   What Went Wrong?

- Water washing may remove all the liquid from a tank, but it cannot remove all the vapor. Tests for flammable vapor should have been carried out *inside and outside* the tank before work started, and it is good practice to place a portable gas detector alarm near the welding site in case conditions change.
- Two of the valves on the tank were found to be open to atmosphere. One of them, on the top of the tank, probably provided the flame to ignition path as the welder was working several meters away. The foreman should have checked these valves before issuing the permit-to-work.
- It is possible that the valve between the tank and the line being welded was leaking. The lines between the tank and the welding operations should have been blinded.
- The job was completed by removing all the pipework and modifying it in the workshop. That could have been done before the explosion.
- The procedures for preparing equipment for maintenance were grossly inadequate or ignored (or both). It is most unlikely that this was the first time that a job had been prepared in such a slipshod way, so more senior and professional staff should have noticed what was going on.

Before you say, "This couldn't happen in my company," remember Bhopal (see Section 21.1) (or Longford; see Section 26.2). Do you know what goes on in your overseas plants or in that little faraway plant that you recently acquired, not because you really wanted it but as part of a larger deal? Do you circulate all your recommended practices and accident reports to these outstations? Do you audit their activities?

23.2   PRECAUTIONS RELAXED TOO SOON

When a whole unit is shut down for an extended overhaul, the usual practice is to isolate the unit at the battery limits by inserting blinds in all pipe lines, to remove all hazardous materials, and to check that any remaining concentrations are low enough for safety. Many publications [1] describe how this can be done. It is then not necessary to isolate individually every piece of equipment that is going to be inspected or maintained. (However, equipment that is going to be entered should still be individually isolated by blinding or disconnection.)

After a long shutdown, there is obviously a desire to get back on line as soon as possible. A few jobs are not quite finished. Can we remove the battery limit isolations, or some of them, and start warming up a section of the plant where all the work is complete?

The correct answer is "Yes, but first the equipment that is still being worked on must be individually blinded. Do not depend on valve isolations. Valves can leak" (see Section 23.3). The following incident occurred because this advice was not followed.

A fluid coker was starting up after a four-week shutdown. Work on some items of equipment including the main fractionation column was not quite finished and its vent line was still open to the atmosphere. Some, but not all, of the lines leading to this column were blinded to support this work, so it was decided to start removing the battery limit blinds. When the blind on the low-pressure natural gas supply line was removed, passing gas was detected in the plant, as the natural gas isolation valve was leaking. The blind was replaced but removed again the next day. The leak then seemed small. Six hours later there was an explosion in the fractionation column. The trays were displaced and damaged, but the shell was unharmed.

The precise route by which the gas got into the column is uncertain and is not described in the report [2]. It probably came from the leaking valve just described. However, the next level of cause is clear: before the battery limit blinds were removed, every line leading to equipment that was still being worked on or was open to the atmosphere should have been individually blinded. The underlying causes were taking chances to get the plant back on line quickly and insufficient appreciation of the hazards.

23.2.1 Lessons Learned

Under this heading the report describes with commendable frankness some well-known information that was apparently not known to those in charge.

A small quantity of flammable gas or vapor can cause a large explosion with severe consequences, especially when the fuel is confined. As little as 5 to 15 kg (10 to 30 lb) of methane could have caused the damage as it is not necessary for the whole of the vessel to be filled with the flammable mixture. Vessels should be inerted if there is any possibility of flammable gas entering, through leaking valves or in other ways (but it is better to prevent gas entering by adequate blinding). To bring home to people the power of hydrocarbons, remind them that a gallon (4 liters, ≈3 kg, or 7 lb), burned in a rather inefficient engine, can accelerate a ton of car to 70 mph (110 km/h) and push it 30 mi (50 km). Looked at this way, the damage to the column seems less surprising. Most of us get practical experience of the energy in hydrocarbons every day, but we do not relate it to the hydrocarbons we handle at work.

The quantity of gas that might leak through a closed valve is significantly more than most people realize.

Consideration should be given to double blocks with a drain (bleed) between them on isolating installations that are troublesome or highly sensitive to leakage. Note that the double block and drain (or bleed) does not remove the need for more positive isolation by blinding. It makes the fitting of blinds safer and is adequate for quick jobs but not for extended ones such as turnarounds.

Valves are the mainstay of any plant. Trying to stop leaks with excessive torque will damage them. Any that are troublesome should be noted for change at the next shutdown.

The rigor with which commissioning activities are carried out is often less than that which is applied to normal operating procedures ... all procedures should be written in a clear, concise, and consistent manner.

23.3   FAILURE TO ISOLATE RESULTS IN A FIRE

In the last incident, the equipment under maintenance was not isolated from a source of danger, natural gas, because blinds were removed prematurely and the consequences not thought through. In this incident, there was not only a leaking valve but no blinds were (or could be) inserted.

A pin-hole leak occurred on a 6-in. diameter naphtha draw-off line from a fractionation column at a height of 34 m (112 ft) above ground level. Many attempts were made to isolate and drain the line but without success as the valve between the line and the column was passing intermittently when it was supposed to be closed and the bottom of the line was plugged with debris. Nevertheless, it was decided to replace a corroded 30 m (100 ft) length of it with the plant on line, despite the fact that the workers doing so would be working at a height, with limited means of escape, and with hot pipework nearby. This decision was made at the operator level and professional staff were not involved.

Two cuts were made in the pipe with a pneumatic saw. When naphtha leaked from the second cut, it was decided to open a flange and drain the line. As the line was being drained, there was a sudden release of naphtha from the first cut. It was ignited, probably by the hot surface of the column, and quickly engulfed the column. Four men were killed and another seriously injured.

The immediate cause of the fire was the grossly unsafe method of working. The plant should have been shut down. (If the line had been narrower and not corroded, it might have been possible to run a new line alongside the existing one and carry out an underpressure connection.)

The underlying causes were the following:

- The technical and managerial staff members were rarely seen on the site, did not take sufficient interest in the details of plant operation, and, in particular, allowed an operator to authorize and control an obviously hazardous job.

- Employees at all levels had a poor understanding of the hazards.
- They did not recognize the need for a systematic evaluation of the hazards of specific jobs and the need to prepare a detailed plan of work [3].

In cases like this, managers have been known to say afterward, "I didn't know this sort of thing was going on. If I had known, I would have stopped it." This is a poor excuse. It is a manager's job to know what is going on, and this knowledge cannot be learned by sitting in an office but by visiting the site, carrying out audits, and generally keeping one's eyes open. When an accident discloses a poor state of affairs, it is stretching credulity too far to claim that it was the first time that risks or shortcuts had been taken. They are usually taken many times before the result is an accident.

23.4 UNINTENTIONAL ISOLATION

Many incidents have occurred because someone isolating a flow or an electricity supply has not realized that he or she was also isolating the supply to other equipment besides the equipment intended for isolation. If this is not obvious from the position of the isolation valve, then a label should indicate which equipment or unit is supplied via the valve. Similarly, labels on fuse boxes and main switches should indicate which equipment or unit is supplied.

The flow of compressed air to a sampling system was isolated unintentionally. This was not discovered for some time as the bulb in the alarm light had failed. The operator canceled the audible alarm but with no indicator light to remind him he forgot that the alarm had sounded, or perhaps he assumed that flow had been restored. The alarm was checked weekly to make sure that the set point was correct but the alarm light was not checked.

Sometimes an unintentional isolation is the result of a slip. An operator was asked to switch a spare transformer on line in place of the working one. This was done remotely from the computer in the control room. He inadvertently isolated the working transformer before switching on the spare one. He realized his error almost immediately and the supply was restored within a minute. The report on the incident blamed distraction:

It is apparent that the Control room is used as a gathering area for personnel, as well as a general thoroughfare for persons moving about the building, to the detriment of the Control room operator's concentration.

The report also suggested greater formality in preparing and follow-ing instructions when equipment is changed over. Though not suggested in the report, it should be simple for the computer program, when the computer is asked to isolate a transformer, to display a warning message such as "Are you sure you want to shut down the electricity supply?" We get such messages on our computers when we wish to delete a file. There is no need for control programs to be less user-friendly than word processors.

Notice that the default action of the people who wrote the report was to describe ways of changing the operator's behavior rather than to look for ways of changing designs or methods of working (see Chapter 27).

23.5   BAD PRACTICE AND POOR DETAILED DESIGN

A reciprocating air compressor was shut down for repair. The process foreman closed the suction and delivery valves and isolated the electric-ity supply. He then tried to vent any pressure left in the machine by the method normally used, opening the drain valve on the bottom of the pulsation damper. It was seized and he could not move it. So instead he vented the pressure by operating the unloading devices on both cylin-ders. Unfortunately, this did not vent all parts of the machine, though the foreman and most of the workforce did not realize this. He then issued a permit-to-work for the repair of the machine.

Two men started to dismantle the machine. They noticed that the han-dle of the drain valve on the pulsation damper was vertical and assumed that the valve was open. They therefore assumed that the pressure had been blown off. After they had unbolted one component, it flew off, injur-ing one of them.

We can learn a number of lessons from this incident:

- Members of process teams often do not always understand in detail the construction of mechanical equipment or the way it works. They should therefore be given detailed instructions on the action to be taken when preparing such equipment for maintenance and, of course, encouraged to learn more about the equipment they operate (see also Section 23.9).
- When handing over the permit to the maintenance worker or foreman, the process foreman should have explained exactly what he had done. The report [4] does not state whether or not he handed it over in person but if he had done so he would presumably have mentioned that he was unable to blow off the pressure in the usual way. Unfortunately, it is all too common for people to leave permits on a table for others to pick up. This is bad practice. When permits are

being issued or handed back on completion of a job, this should be done person to person.

- It is possible on many cocks and ball valves to remove the handle and turn it through to 90 degrees before replacing it. Such valves are accidents waiting to happen. We should use valves that tell us at a glance whether they are open or shut. Rising spindle valves are better than those with nonrising spindles. Ball valves and cocks should have handles that cannot be replaced wrongly.
- Drain valves often become plugged with scale or dirt. Valves used to blow off pressure should therefore be on the top rather than the bottom of a vessel.
- People dismantling equipment should always assume that it may contain trapped pressure and should proceed cautiously. They should loosen all bolts and prize the joint open so that any trapped pressure can blow off through the crack or, if the leak is serious, the joint can be remade.

23.6   DISMANTLING

23.6.1   Wrong Joint Broken

A supervisor decided to remove a number of redundant pipes and branches from a service trench. They had not been used for many years and were rusty and unsightly. An experienced worker went around with a spray can and marked with green paint the sections of pipe that were to be removed and then a permit-to-work was issued to remove the marked sections.

One of the sections was a short vertical length of pipe, 75 mm (3 in.) in diameter and ≈1.5 m (5 ft) long, sticking up above a compressed air main that was still in use (Figure 23-1). The valve between the pipe and the main was tagged to show that it was closed to protect equipment under maintenance. The short length of pipe was marked with several green patches. Unfortunately, there was also some green paint on the flange below the isolation valve. This green paint might have been the remains of an earlier job or it might have accidentally got onto the flange while the pipe above it was being sprayed. The mechanic who had been asked to remove the pipe broke this flange. There was a sudden release of compressed air at a gauge pressure of ≈7 bar (100 psig). Fortunately, the mechanic escaped injury.

The mechanic did not, of course, realize that the compressed air line was still in use. Like the old pipes he was removing, it was rusty and he assumed it was out of use.

This incident displays several examples of poor practice. Each job should have its own permit-to-work, which should make it quite clear

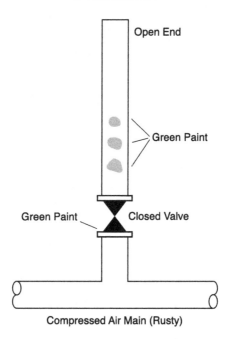

FIGURE 23-1   The lower flange was unbolted in error.

which joint or joints should be broken. The report on the incident stated that in the future, maintenance workers should be shown precisely which piece of equipment is to be maintained, which joint to break, and so on. However, experience shows that this is not enough. Before starting work, the maintenance worker may go for tools or for spares and then come back and break the wrong joint or remove the wrong valve. Equipment to be worked on should be numbered or labeled and the number or name put on the permit-to-work. If there is no permanent number, a numbered tag should be tied to the flange that is to be broken, the valve that is to be changed, or the pipe that is to be cut (and at the point at which it has to be cut).

These identifying tags should be distinct in appearance from tags used to show that valves are isolating equipment under repair. For this purpose, padlocks and chains, or other locking devices, are better than tags as they prevent the valve being opened in error.

During the investigation, someone suggested that the job did not need a permit-to-work as it was noninvasive. However, the purposes of a permit-to-work are to define precisely the work to be done, list the hazards that are present, and describe the necessary precautions. If a job is not defined precisely, it may become invasive.

Though it did not contribute to the accident, rainwater will have collected above the valve and caused corrosion. The open end should have been blanked.

23.6.2 Trapped Pressure in Disused Equipment

Equipment that is no longer used or is not going to be used for some time should be emptied and made safe. If you are sure it will not be needed again, dismantle it as soon as you can. Leave knowledge, not problems, for your successors. They may not know what was in the line or how to handle it. Nevertheless, if you have to dismantle old equipment, do not assume that it has been made safe. Assume that pressure may be trapped behind solid plugs. Here are some examples:

(a) An old disused pipeline was being dismantled by cutting it into lengths with a hacksaw and lowering them to the ground. Both ends of the pipeline were open. When a mechanic cut into the pipe, a spray of sulfuric acid hit him in the face. Fortunately, he was wearing goggles. There were two closed valves in the line, but no one had noticed them. The acid had attacked the metal, forming hydrogen, which pressurized the line [5].

(b) A stainless steel pipe was isolated at both ends and left for six years. After this time, no one remembered what it was last used for. One end was still connected to the plant; the other, lower, end had a valve fitted on it. A mechanic was asked to dismantle the pipe. He opened this valve. Nothing came out. He then unbolted the joint between the valve and the pipe and prized the flanges apart. A little liquid dribbled out. He prized the flanges farther apart. A large and forceful escape of gas, liquid, and dirt sprayed the fitter and his assistant. The pipe had contained acetic acid, and over the years it had corroded the pipe sufficiently to produce a pressure of hydrogen.

(c) A unit was "abandoned in place" to save the cost of demolition. A pump that handled a 50% solution of caustic soda was isolated by closing both valves but was not drained and the fuses were not removed. A contractor was asked to switch on a ventilation fan that served an adjoining area. The switch was next to that for the caustic pump, though 15 m away from the pump and the labels on the switches were very small. The contractor switched on the caustic pump in error. It ran between closed valves and overheated. There was a loud boom, which rattled windows 60 m (200 ft) away. The pump was damaged and dislodged from its baseplate [6].

There were five elementary errors. The incident would not have occurred if one of the following five tasks had been carried out:

- The pump had been drained.
- The pump had been defused.
- The switches were near the equipment they served. (Additional switches for emergency use could have been provided some distance away.)

- The labels were easy to read.
- Someone familiar with the unit had been asked to switch on the ventilation fan.

Incidentally, the same source describes how two other pumps were damaged because they were operated while isolated. One was switched on remotely; the casing was split in two. A power failure caused the other to stop. An operator closed the isolation valves, not realizing that the pump would restart automatically when power was restored. When it did, a bit of the pump was found 120 m (400 ft) away and local damage was extensive.

23.7   COMMISSIONING

A new unit was being commissioned. It was bigger and more complex than any of the other units on the site so the project and engineering teams had checked and double-checked everything, or so they thought. To make sure there were no leaks and that the instruments worked correctly, they operated the plant with water, except for a vessel that was intended for the storage of a water-sensitive reagent. To avoid contaminating this vessel, it was left isolated by two closed valves, a manual valve on the vessel and a control valve below it.

This vessel was later filled with the reagent, and commissioning started. When an operator, standing on a ladder, opened the manual valve, a cloud of dense white fumes surrounded him. Fortunately, he was able to close the valve and escape without injury. There was no gasket in one of the joints between the two valves [7].

As it was impracticable to leak-test this part of the unit with water, it should have been tested with compressed air, by either checking whether or not it retained the pressure or checking for leaks with soapy water. Note that during such a leak test with compressed air, the design pressure of equipment should not be exceeded. Pressure tests to check the integrity of equipment should normally be carried out with water—or other liquid—as then much less energy is released if the equipment fails.

Valves that are operated only occasionally, say, once per year at a planned turnaround, may be operable only from a ladder but valves that may be required during process upsets, such as leaks, should be easy to reach.

23.8   OTHER HIDDEN HAZARDS

In new plants, and extensions to old ones, we often find welding rods, tools, and odd bits of metal left in pipework. Even small bits of

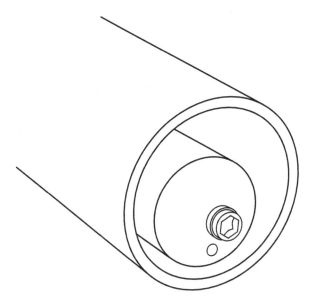

FIGURE 23-2   When a length of new pipe was cut open, a propane cylinder was found inside it. Fortunately, the cutting tool just missed it.

rubbish can harm machinery, and most companies make sure that pipes are clean before new equipment is started up. The most extreme example of unwanted contamination occurred on a U.S. refinery when a length of new pipe, complete with plastic end caps, was being prepared for installation. Welders fitted a bend on one end of the pipe and then, with the end cap still in place, cut a length off the other end. They then found a propane cylinder just inches away from the cut (Figure 23-2). Had it been a few inches nearer the end, there would have been a very nasty accident [8].

Therefore, one never knows what suppliers and construction teams have left inside new equipment. Have a good look before boxing it up or working on it for the first time (see also Section 23.10).

23.9   CHANGES IN PROCEDURE

An instrument probe in a tank truck used to carry gasoline had to be replaced. An experienced mechanic regularly did the job. After the new probe had been inserted, some electrical connections were made and secured with a heat-activated shrink-wrap sleeve. A propane torch was used to seal the shrink-wrap. This was hardly the most suitable tool to use on a vessel containing flammable vapor, but as the probe was always replaced before the shrink-wrap was fitted, the vapor was not open to the atmosphere.

One day the tank truck was wanted back in use as soon as possible so the mechanic used a different type of electrical connection that could be made more quickly but required the application of heat. The mechanic used the torch; he had been using it for several years and the only difference, as he probably saw it, was that he was now using it at a slightly earlier stage of the job, before the probe had been replaced and while the tank was still open to the atmosphere. It exploded and killed the mechanic [9].

The mechanic and his supervisor did not understand the hazards and thought they were making only a minor change in the way the job was done. Many maintenance people do not understand process hazards (just as many process people do not understand equipment hazards; see Section 23.5). The company that owned or used the tank truck should have removed all flammable vapor from the tank before sending it for repair or at the very least should have made clear the nature of the hazards and the precautions to be taken.

When equipment is owned by one company, rented by another, and repaired by a third, responsibilities for maintenance, pressure testing, and inspection should be agreed upon and made clear.

23.10   DEAD-ENDS

Dead-ends in pipes have caused many pipe failures. Traces of water can accumulate in them and freeze, or corrosive materials can dissolve in them (see Section 9.1.1). Other materials can also accumulate in them and remain there when the rest of the equipment has been emptied for maintenance (or for any other reason).

23.10.1   A Disused Pipe Becomes a Dead-End

A furnace was taken out of use. The 10-in. diameter pipe that supplied coke oven gas to the burners was disconnected at the lower (furnace) end and closed by a valve, but the other end was left connected to the main. Ten years later, a crack appeared in the top flange of the valve and gas leaked out. The freezing of water that had collected in the pipe probably caused the crack. Water and other liquids were normally removed from the main coke oven gas line via a number of drain lines, but missing insulation had allowed the water in these lines to freeze. As an immediate measure, a blind was inserted immediately above the cracked valve. While this was being done, some tar oozed from the pipe.

A few days later, a maintenance team started to replace the cracked valve with a blank flange fitted with a drain. When the flange was loosened, the valve and blind dropped down several inches while hanging on the remaining bolts. A large amount of liquid sprayed out and soaked three of the workers. It was ignited either by an infrared lamp used to warm the line or by a gas-fired space heater. Poor access and junk lying around prevented the workers from making a quick escape; two were killed—one of them fell over the junk—and another was seriously injured [10].

There were at least eight things wrong, and putting just one of them right could have prevented the accident:

- The redundant pipe should have been removed when it became clear that it was no longer needed. It is not good practice to leave disused pipes in position in the belief that one day someone might find a use for them.
- If this was not possible for any reason, the pipe should have been regularly drained to remove any liquid that accumulated. This should have been through a small valve fitted below or in place of the 10-in. valve but not so small that it would be liable to choke, perhaps of 2 in. internal diameter, but smaller for less viscous drainings. (The liquid should, of course, be drained into a closed container, not into a bucket and not onto the floor.)
- The missing insulation on the lines that drained the main coke oven gas lines should have been replaced.
- No sources of heat should have been allowed anywhere near equipment that might contain flammable material.
- All employees should have been told about other incidents in which liquids in coke oven gas lines had caught fire, including several that had occurred only days before as a result of the freezing of the drain lines. Many employees thought these liquids were not flammable.
- There should have been regular surveys of the unit to look for dead-end pipes, missing insulation, and other defects.
- The material that leaked out when the blind was being fitted should have been checked for flammability.
- The job should have been properly planned. The company's procedures were frequently ignored.

The dead and injured might have escaped had there been less junk lying around.

23.10.2   A Dead-End inside a Vessel

Paint had to be removed from the manway of a reactor that had contained ethylene oxide. The reactor was swept out with nitrogen, and tests showed that no oxygen or combustible gas or vapor could be detected. Unfortunately, the people who prepared the reactor overlooked a disused line on the base of the reactor that was permanently blinded. Some ethylene oxide that accumulated in this line evaporated and was ignited by sparks from the grinder used to remove the paint. A flash fire killed the man using the grinder, but there was no explosion.

Why was it that the tests did not detect the ethylene oxide? According to the report, the sample tube used was 3.4 m (11 ft) long, not quite long enough to reach to the bottom of the reactor, which was 3.7 m (12 ft) deep. In addition, the ethylene oxide may have been at the bottom of the disused line and as its vapor is heavier than nitrogen, it would diffuse out only slowly. There are other possibilities not mentioned in the report [7]. The sample tube might have absorbed the ethylene oxide so that it never reached the detector head (see Sections 24.4 and 32.4); and combustible gas detectors will not detect flammable gas unless air is present. The report does not say how much time elapsed between the tests and the grinding. A test at 8 a.m., say, does not prove that the plant is still safe several hours later. Tests should be carried out just before work starts, and it is good practice to use a portable gas detector alarm, which gives an audible warning if conditions change.

Finally, there was no need to use a grinder. The paint could have been removed by chipping or with a paint-removal solvent. When flammable materials are handled, it is good practice to add an extra layer of safety by not using sources of ignition when safer methods are practicable.

Ethylene oxide can be ignited and decompose, producing both heat and a rise in pressure, in the absence of oxygen. However, some oxygen will have been present in the incident described because the manway was open.

There are more reports on maintenance accidents in Chapters 1 and 7.

References

1. T. A. Kletz, Equipment maintenance, Chapter 11. In: S. S. Grossel and D. A. Crowl (editors), *Handbook of Highly Toxic Materials Handling and Management*, Dekker, New York, 1995.
2. J. L. Woodward and J. K. Thomas. Lessons learned from an explosion in a large fractionator, *Proceedings of the AIChE Annual Loss Prevention Symposium*, March 2002.
3. Anon. (2001). *Refinery Fire Incident*, Investigation Report No. 99-0141-CA, Chemical Safety Hazard Investigation Board, Washington, D.C.; summarized in *Loss Prevention Bulletin*, Oct. 2002, 167:4–7.
4. T. Gillard. Dangerous occurrence involving an instrument air compressor. *Loss Prevention Bulletin*, Vol. 142, August 1998, p. 16–18.
5. Anon. (2002). Corrosion dangers from redundant pipework. *Loss Prevention Bulletin*, Feb., 163:22.
6. D. S. Giles and P. N. Lodal. Case histories of pump explosions while running isolated. *Process Safety Progress*, Vol. 20, No. 2, 2001, p. 152–156.
7. J. Bickerton. Near-miss during commissioning. *Loss Prevention Bulletin*, Vol. 162, Dec. 2001, p. 7.
8. Anon. (2000). *Safety Alert*, Mary Kay O'Connor Process Safety Center, College Station, Tex., 12 Sept.
9. R. A. Ogle and R. Carpenter. Lessons learned from fires, flash fires, and explosions involving hot work. *Process Safety Progress*, Vol. 20, No. 2, 2001, p. 75–81.
10. Anon. (2002). *Investigation Report: Steel Manufacturing Incident*, Report No. 2001-02-1-N, Chemical Safety and Hazard Investigation Board, Washington, DC.

Entry into Confined Spaces

A woodcutter who spends most of the day sharpening his saw and only the last hour of the day cutting wood, has earned his day's wage.

—Menachem Mendel of Kotzk (1787–1859)

In the same way, time spent preparing equipment for entry is time well spent.

Many people have been killed inside tanks and other confined spaces. Sometimes they have entered without permission to do so or merely put their head inside an open manway to inspect the inside. Sometimes entry was authorized, but not all of the hazardous material had been removed or it had leaked back in because isolation was poor. Sometimes hazardous material was deliberately introduced in order to carry out tests. Sometimes people have entered a confined space to rescue someone who has collapsed inside and been overcome themselves (see Sections 25.3.1. and 28.1.1).

24.1 INCOMPLETE ISOLATION

A trayed column was prepared for inspection. It was emptied, the remaining vapor removed by sweeping with nitrogen, and the nitrogen replaced by air. All the connecting lines were blinded—or so it was thought—and tests showed that no toxic or flammable vapor was present.

All this preparatory work was done on the night shift, but the signing of the entry permit was left to a day superintendent. As he had been assured that all necessary precautions had been taken, he signed the permit. On the way back to his office he passed near the column, so he stopped to have a look at it. He heard a slight hissing noise and traced this to two instrument connections that had not been blinded. These instruments measured the pressure difference between the column and another column mounted on top of it. This column was still in use. The two instrument connections were insulated along with other lines and had been overlooked. The superintendent canceled the entry permit. Tests showed there was hydrocarbon in the column.

We can learn much from this near miss:

- When preparing a vessel for entry, give the maintenance team a list of all lines to be blinded (or disconnected), identify each one with a numbered tag, and, if there is any doubt about their location, mark them on a sketch. Never ask them to blind *all* lines (*similar* is another word that should never be used; see Sections 31.1 and 31.2).
- Check that all the lines have been blinded (or disconnected).
- The person who signs the entry permit should always carry out his or her own check, regardless of any checks carried out by other people. The person whose signature is on the permit is legally and morally responsible if anything goes wrong [1].

Although not relevant to this incident, note that if any lines are already blinded, the blinds should be removed and checked to make sure that they are not weakened or holed by corrosion.

24.2   HAZARDOUS MATERIALS INTRODUCED

24.2.1

A man was cleaning a small tank ($36\,m^3$, 9,500 gallons) by spraying the inside with cyclohexanone. He was killed by chemical exposure, lack of oxygen, or a combination of both. Two other men were killed while trying to rescue him. As well as being toxic, cyclohexanone is flammable [2]. No one should enter a confined space unless the concentration of flammable gas or vapor is <25% of the lower flammable limit. Air masks should be worn if the concentration of toxic vapor is above the threshold limit value or, for very short exposure, above the appropriate limit.

Entry should not normally be allowed even with air masks into atmospheres that are irrespirable, either because the oxygen content is too low or the concentration of toxic gases could cause death or injury in a short time. If such entry is permitted, two people trained in rescue and resuscitation should be on duty outside the vessel. They should have available all the equipment necessary for rescuing the person inside the vessel and they should always keep him or her in view (see Section 11.5).

24.2.2

Three men were overcome while cleaning a tank with trichloroethane. Never take hazardous liquids into a confined space unless the spillage of the total amount introduced will not cause the vapor concentration to exceed the threshold limit value or 25% of the lower flammable limit. Forced ventilation can be used to reduce the concentration of vapor but if so, the concentration should be monitored.

24.2.3

A 20-year-old contract worker who was cleaning the inside of a paint-mixing tank took a bucketful of methlyethyl ketone, a flammable solvent, into the tank. When the vapor exploded, the worker suffered 70% burns and died in the hospital. The source of ignition was static electricity generated when he repeatedly dipped his scouring pad into the bucket. In court, the company said that it now used remote cleaning methods [3]. Why didn't it do so before? The accident was not hard to foresee in the light of previous experience. There is more on static electricity in Chapter 15 and in Sections 25.2.7, 28.2.5, 30.1, and 32.7.

24.2.4

On other occasions, welding torches have been left inside confined spaces during a meal break or overnight. Welding gas has leaked, resulting in a fire or explosion when the torch was lit. Or argon has leaked and the welder has been asphyxiated on reentering the confined space.

24.3   WEAKNESSES IN PROTECTIVE EQUIPMENT

Compressed air supplies to air masks have failed for various reasons:

- Hoses have been attached to connectors by crimped rings or by the type of fasteners used for water hoses in cars. These are not suitable for industrial use. Bolted connections are better [2].
- Air filters have been blocked by ice in cold weather.
- Emergency supplies of compressed air have failed, either because the emergency cylinders were empty or the change-over mechanism failed to operate. Emergency supplies should be tested each time an air mask is used.
- Nitrogen cylinders have been connected to compressed air lines in error. Different types of connections should be used for nitrogen and air. Many people have been overcome by nitrogen—another example follows—and the odorizing of nitrogen has therefore been suggested [4].

24.4   POOR ANALYSIS OF ATMOSPHERE

A nitrogen receiver, 8 m (26 ft) tall and 2 m (6.5 ft) wide was due for inspection. The inlet line was disconnected and the manway cover removed. The manway was near the bottom of the vessel, and there was no opening at the top. The vessel was purged by natural ventilation supplemented by a compressed air hose, and a test showed that the oxygen content was normal.

An inspector entered the vessel and inspected it from a permanently fitted internal cat ladder. The standby man heard a noise, looked into the vessel, and saw that the inspector had fallen off the ladder. When the standby man tried to enter the vessel to rescue the inspector, he found that his self-contained air mask was too big to go through the manway. Fortunately, the emergency services arrived in a few minutes and were able to rescue the inspector, who had suffered more injury from his fall than from the low oxygen content.

Why did the analysis give a false reading? The test was carried out near the open manway at the bottom of the vessel. There was less oxygen near the top. Tests should always be carried out at various parts of a confined space (unless it is very small). This could have been done by removing a blank at the top of the vessel (this would also have improved the ventilation) or by using a long probe [5].

The need to test in more than one part of a vessel is hardly a new discovery (see Section 23.10.2). It has been known for many years and described in published reports. But it was unknown in the plant where the incident occurred (or if known it had been forgotten). Unfortunately, accident reports rarely tell us what training employees had received or what books, magazines, and safety reports were available for them to read.

Before anyone enters a confined space, we should ask how the person would be rescued, and by whom, if he or she collapsed, for any reason. We should make sure that the standby man is properly trained and equipped. The only good feature in this incident is that the standby man did not try to enter the vessel without an air mask. Many people have done so, to rescue someone overcome inside, and themselves been killed or injured (see also Sections 25.3.1 and 28.1.1).

24.5   WHEN DOES A SPACE BECOME CONFINED?

The inside of a storage tank or pressure vessel is obviously a confined space, and before anyone is allowed to enter it, a systematic procedure should be followed. The tank or vessel should be isolated by disconnecting or blinding of connecting lines; it should then be cleaned, the atmosphere tested, and air mask specified if necessary. However, some confined spaces are less obvious. If a tank is being built or a hole dug in the ground, when do they become confined spaces? A rule of thumb often used is to treat them as confined spaces when the depth is greater than the diameter. Leonardo da Vinci's advice on town planning more than 500 years ago was "Let the street be as wide as the height of the houses."

The following incidents show how easily people can unwittingly fail to recognize confined spaces.

24.5.1

Two men used liquid nitrogen to freeze water lines in a trench, as part of a cut-and-weld job. There was too little ventilation to disperse the nitrogen as it evaporated, and they were overcome. They had not worn safety harnesses or used an oxygen meter.

24.5.2

During a plant shutdown, a piece of equipment was removed from a 1.2-m (48-in.) diameter pipe. No one entered the pipe, but the inside was inspected by shining a light into it. Bright sunshine made it difficult to see anything, so a black plastic sheet was draped over the end of the pipe. There was a strong breeze, so to hold the sheet in place two men sat on one edge of the sheet and two others held it over them. The two sitting men then inspected one of the open ends.

They then tried to do the same at the other open end of the pipe. Unfortunately, there was a flow of nitrogen coming out of the end of the pipe and the two men were overcome. One died and the other recovered after five days in the hospital.

Both the man who died and his co-worker were men of great experience. The day before, one of them had asked for nitrogen to be injected in order to protect the catalyst. The injection point was $\approx$ 50 m (150 ft) and several floors away and he may not have realized that the nitrogen would exit through the 48-in. pipe. He certainly did not realize that a plastic sheet held loosely over the end of the pipe turned it into a confined space [6].

The company's entry procedure did not draw attention to the hazards of temporary enclosures. Obviously it should have, but even if it had, would the men have remembered this fact? Instructions are no substitute for knowledge and understanding, that is, knowledge that confined spaces can easily be formed; knowledge that nitrogen in quite small amounts can reduce the oxygen level to a dangerous extent; and knowledge that what goes in must come out and that whenever we put anything into a plant we should ask where it or something else will exit. The root cause of the accident was the failure of the company to give their employees this understanding of the hazards (see also Sections 29.4, 29.5, 30.12, and 36.5).

24.5.3

To save energy, a company decided to use a flammable and toxic waste gas (known as tail gas) to run a diesel engine and generate power. The gas first had to be cooled and this was carried out in the equipment shown in Figure 24-1. Two pumps were located inside the skirt of the column. One pumped some of the wash water from the base of the column into the venturi and the other circulated the bulk of the water through a cooling tower and back to the top of the column. There were four arched openings in the base of the column so it was not considered a confined space. However, the location was congested, and this reduced ventilation.

An electrician and an engineering student were asked to repair the circulation pump. The procedure they were told to follow, never written down, was to close the knife-gate valve, thus stopping the flow of gas to the column, and then to get rid of the gas already in the column as well as any leaking past the valve—a type that does not give complete isolation—by draining the water seal. There were no valves in the suction lines to the two pumps.

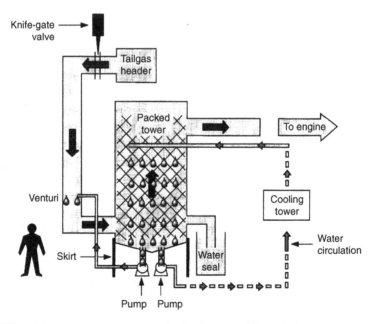

FIGURE 24-1   A man was overcome by leaking gas while replacing a pump in a confined space. From reference 7. (*Reprinted with the permission of the Institution of Chemical Engineers.*)

Unfortunately, the electrician and the student forgot to drain the water seal. Nevertheless, they removed the pump and blanked the open end without incident. They then refilled the tower with gas to prevent air from leaking in. Later that day they replaced the pump, presumably using the same procedure as before. While they were working on their hands and knees, they felt unwell. Before they could get out, the student noticed that the electrician had become unresponsive. Fortunately, the student was able to pull him out and he soon recovered.

One of the causes of the incident was the failure to recognize that the space inside the skirt was a confined space. There was also much else wrong:

- It is not good practice to locate a pump (or any equipment that needs regular maintenance) inside a column skirt (or anywhere else where access is poor). Maintenance will be better as well as safer when access is good.
- Pumps are normally fitted with suction and delivery valves, which are closed before the pumps are removed or repaired. The method followed on the plant: removing a pump (or any other equipment) without isolating it by closed valves and then trying to blank the open ends of the connecting pipes before much of the contents leak out or air leaks in is not good practice.
- All instructions except the simplest should be written down, not just passed on verbally. Many tasks are part of the skill of the craft, and it is not necessary to tell skilled craftspeople how to carry them out; however, that does not apply to detailed and unusual procedures.
- Normally, a member of the operating team prepares equipment for maintenance and then issues a permit-to-work, which the senior member of the maintenance team accepts. The involvement of two people with different functions and the filling-in of a permit provide an opportunity to check that all necessary precautions have been taken. This opportunity is lost when the same person prepares the equipment and then carries out the repairs. In such cases, it is good practice for this person to complete a checklist—in effect, issuing a permit to him- or herself—or for a colleague to do so.

Underlying these detailed causes were managerial failures. The project was being carried out by a special team whose members undertook their own maintenance, independently of the normal operating and mainte-nance organizations. Research and development workers often believe that they can carry out work on plants without being confined to the nor-mal safety procedures. It should be made clear to them that they cannot. Also, to quote from the report [7], "the deadlines were seen as very chal-lenging by those involved," a euphemism suggesting that speed was put before safety.

24.6   MY FIRST ENTRY AND A GASHOLDER EXPLOSION

After I graduated, I spent the next seven years in the research department of Imperial Chemical Industries (ICI) at Billingham, United Kingdom. After two years I was sent down to the factory for about six weeks to see how the company earned its profits. One quiet Saturday afternoon, one of the shift foremen asked me if I would like to go inside a gasholder. It was the dry type in which a movable disc separates the gas in the lower part of a cylinder from the air in the upper part and there is a tar and canvas seal between the disc and the cylinder walls.

At a guess, the volume of the gasholder was several thousand m^3 and its height was $\approx 3\times$ its diameter. To get inside it, we had to go up a staircase onto the roof, through an opening in the center, and then down a Jacob's ladder, a folding ladder, onto the disc. As Figure 24-2 shows, during half the descent we were clinging to the wrong side of the sloping ladder. Fortunately, the gasholder was nearly empty and the angle of the ladder was not too great.

I cannot remember what was in the gasholder. It may have been coke oven gas, water gas ($H_2 + CO$), or producer gas ($N_2 + CO_2$). The

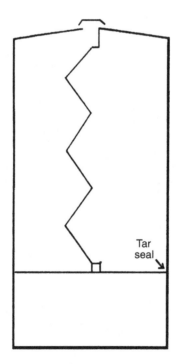

FIGURE 24-2   Diagram of dry gasholder showing Jacob's ladder.

atmosphere above the disc in the gasholder was not tested before we entered and there was no entry permit and no standby man, although the foreman mentioned that strictly speaking there should have been one (this was many years ago). The whole experience was eerie, and I have never forgotten it.

The foreman mentioned that following an explosion in Germany, dry gasholders were out of favor, ICI would not build others, and the long-term plan was to replace them with wet gasholders. I never found out what had happened in Germany, but I found an article on a dry gasholder explosion there, which is worth recounting [8].

The gasholder contained coke oven gas. A section of the bypass pipe was removed for cleaning as it was partially blocked with naphthalene. On the inlet side, the section of pipe was isolated by a closed valve (Figure 24-3) and on the outlet side by a blind slip-plate. When the missing section was replaced, it was found that the pipe coming from the valve had sunk and the two pipes could not be lined up. It was then decided to remove the support at the end of the replaced section so that it would also sink. This involved welding. It ignited gas that had leaked through the closed valve and the resultant explosion tore the outlet main close to the gasholder. The flame from this much larger leak went up the side of the gasholder and five minutes later the gasholder exploded. Either the heat distorted the walls or evaporated the tar. Either way, this would allow gas to bypass the disc and mix with the air in the upper

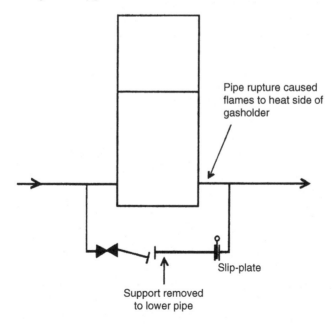

FIGURE 24-3 Gas leaked through a closed valve while burning was in progress nearby.

portion of the gasholder. (See Section 34.3 for an account of another leak through closed valves.)

Perhaps this was the explosion in Germany to which the ICI foreman referred. However, he said that the German explosion occurred because the disc in a gasholder tilted and jammed and gas got past it, and I recall seeing a tilt alarm on the disc in the ICI gasholder.

24.7   FAILURE OF A COMPLEX PROCEDURE

Confined spaces are usually entered through a manway or similar opening that has to be unbolted before anyone can enter. This makes unauthorized entry difficult, though not impossible, because someone has to unbolt the entrance. At one plant there was a room containing hazardous materials and equipment, which was treated as a confined space. Entry was through a door, which was normally locked shut. To make sure that no one could be accidentally locked inside the room, the following rather complex procedure was followed:

- The key to the room was kept in a box fitted with several locks, each operated by a different key. The process foreman kept all the keys. Normally only one lock was closed.
- If someone needed to enter the room, the foreman first established that it was safe to do so. He then issued a permit-to-work and gave the box key to the person who was going to enter. He or she then opened the box, got out the door key, opened the door, *locked the door key back in the box,* and kept the box key.
- If more than one person was entering the room, each of them was given one of the box keys and they each locked the box with the door key inside. The box had a window so that it was possible to see whether or not the door key was in it. It was therefore impossible to lock the door of the room until everyone inside had left and unlocked the box.

Before you read on, please make sure you understand the way the system works, because some of the people at the plant did not.

This system seems good, though complex, but like all systems it could and did degrade. One day a mechanic had to enter the room for a quick job, so he left the door key in the door. (Afterward the foreman admitted that this was often done for quick jobs.) Before the mechanic had finished, two other men arrived to erect scaffolding for a later job. As the door was open, they did not bother to get box keys from the foreman. When the mechanic finished his job, he left the door key in the door so that the scaffolders could lock the door when they finished. He did not remind them to do this; he assumed they knew what to do.

Before the scaffolders had finished, another mechanic arrived to carry out the second job. He went first, as usual, to the foreman's office to get a permit-to-work and a box key. The foreman was not there but the fitter saw the permit for the first job on the table. He added his job to the permit and signed it. He took another box key (and one for his co-worker), and both of them locked the box but they did not notice that the door key was not in the box.

Later someone noticed that the door was open but the key was in the door.

24.7.1   What Went Wrong?

- It is bad practice to allow people to add extra work to an existing permit-to-work—it had become custom and practice at the plant— and they should certainly never do so without the signed agreement of the person who issued the permit.
- If more senior people had kept their eyes open, they might have spotted sooner that procedures were not always followed. Fortunately, this incident drew it to their attention before an accident happened. It shows the value of following up dangerous occurrences.
- Many of the people at the plant did not fully understand the procedure or the reasons for it.
- Could a simpler procedure be devised? For example, each person could padlock the door open with his or her own padlocks when an entry is in force. Simpler procedures reduce the temptation to take shortcuts.
- If simplified procedures are allowed for quick jobs, or become custom and practice, how do you deal with quick jobs that become long ones? (Every do-it-yourself enthusiast knows that five-minute jobs can take all day.)
- Several months earlier, two men had followed the correct procedure before entering the room. But they had to enter again soon afterward, and this time they left the key in the door while they were in the room. Subsequent inquiries showed that they did not understand the reasons for the procedure and were just following it blindly. No wonder they could not be bothered to go through it a second time. If this incident had been followed up more thoroughly, the second incident might not have occurred.

Section 36.6 describes an overly complex instrumented system for controlling entry, which also failed.

24.8   EPIDEMICS OF UNSAFE ENTRIES

I have left the worst entry incidents for the end. They were epidemics rather than isolated occurrences. The first one occurred in an organization where we would not have expected it, the Royal Australian Air Force

(RAAF). The following description is based on a book by Hopkins [9] who was a member of the board of inquiry.

From the late 1970s onward, RAAF maintenance employees worked inside the fuel tanks of F111 bombers repairing the liners, and they suffered prolonged and sometimes intense exposure to the toxic chemicals used for removing old, damaged linings and replacing them with new ones. The RAAF did not realize that it had a serious problem on its hands until 2000, by which time the health of more than 400 people had been ruined.

The protective clothing issued to the workers was grossly inadequate. It was permeable to some of the chemicals used, and many of the respirators contained filters that provided protection only against dust but not against chemicals. The cooling in the buildings where the work was carried out was not in use after 4 p.m. or on weekends, though much overtime was worked and this increased the already high temperature inside the tanks. Realizing that the protective equipment was useless as well as uncomfortable, the workers often failed to use it.

The immediate technical causes of the ill health were thus obvious, but why was nothing done about them for 30 years? Hopkins describes the main factors. It was not a simple case of management not caring about safety. Flight safety had a very high profile and the standard was high, but this attitude was not carried over into the maintenance function, for several reasons:

1. *Undermanning.* During one period, the young officer in charge of the fuel tank repairs was also responsible for six other groups and 170 employees but had no significant management experience. He left the supervision of the work on the fuel tanks to the noncommissioned officers and did not even know that many of the people for whom he was responsible were suffering ill health.
2. *The "can do" attitude.* Employees at all levels had a strong "can do" attitude, a reluctance to admit that any task was beyond them. Such an attitude encourages initiative and self-reliance. After downsizing or an increase in workload, many people try to do the best they can, but managers should ensure that it does not go so far that corners are cut and safety neglected.
3. *The helicopter fallacy.* Many senior managers believe that they do not need to be involved in the details and instead take a helicopter view. They rely on summaries of performance that others have prepared for them. They are like those queen bees who are so busy producing eggs that they have no time to eat and digest and instead rely on predigested food prepared by other bees. It is true that managers should not duplicate their subordinates' jobs, but from a helicopter we see only forests. If we want to know what is really happening, we have to

land the helicopter and look at some of the twigs and leaves. Managers should walk round the areas under their control, keeping their eyes open and talking informally to other employees. This was not the practice at Amberley where the F111s were serviced. As already stated, it is no defense for a manager to say after an accident, "If I had known this was happening, I would have stopped it." It is a manager's job to know.

4. *Reporting*. The RAAF, like many airlines, had a good system for reporting faults, but in practice it applied only to faults in the aircraft. Senior managers ignored fault reports by those who repaired the fuel tanks, as they did not realize the seriousness of the problems and wanted to get the repairs done as soon as possible.

5. *Silos*. There was "a culture of silos" or partitions, in which different groups of people in the same organization pursued their own group objectives, failed to communicate with other groups, and did not see the whole picture. (A silo is "an air or water tight chamber" and thus, by extension, a knowledge-tight chamber.) In the RAAF this produced a very high standard of safety for some of the employees (pilots) but outrageously bad standards for others (some maintenance workers). In many airlines, the risk to maintenance workers is far higher than for passengers.

24.8.1 Similar Experience Elsewhere

The five headings above apply, to varying extents, to many companies and other organizations, though they are rarely allowed to continue unchecked for so long. The RAAF experience shows what can happen if people are overloaded after downsizing, if they have a macho attitude to getting things done, if managers ignore details and do not know what happens at the operating level, and if people work in ignorance of their colleagues. If people are told only what someone thinks they need to know, they will never learn from the experience of their colleagues in other functions or departments.

During the early days of polyvinyl chloride (PVC) production, operators used to enter the batch reactors to clean them between batches. During the 1970s, many of these operators contracted cancer and it was realized that vinyl chloride was carcinogenetic. Methods of cleaning the reactors using high-pressure water, without entering them, were then developed. This early problem has haunted the PVC industry ever since, and opponents of the industry make periodic attempts to prohibit or limit the use of PVC in case traces of monomer are still present. Although it is a hazardous chemical, it is not as hazardous as some critics make out, and there would be less opposition to its use today if a better method of cleaning the reactors had been developed at the start. However, once the chemical industry realized there was a problem, it reacted promptly, unlike the RAAF.

24.8.2   Avoiding the Need

An important lesson that can be drawn from the two "epidemics" is that the most effective way of overcoming the hazards of entry is to avoid the need for it. The following are the common reasons for entry, besides cleaning, and possible ways of avoiding the need:

- Workers often enter spaces to inspect or repair equipment inside the vessel: To avoid the need for entry, withdraw the equipment from the vessel.
- Alternatively, if doctors can inspect the insides of our stomachs, bladders, and bowels from outside (and display the insides on a screen while doing so), engineers should be able to do the same with vessels.
- Workers often enter spaces to operate or maintain valves on vessels in pits. To avoid the need for entry, do not put vessels in pits but, if you have already done so, consider remote operation of valves. If you insist on putting vessels in pits, provide a generous space between the vessel and the walls of the pit.
- Blockages often occur when gravity causes solids to flow out of a vessel. The probability of a choke depends on the shape of the lower part of the vessel, which should be designed to minimize the risk of blockage. If nevertheless a blockage occurs, it should be cleared by a vibratory or mechanical device, not by people entering the vessel [10]. People have been asphyxiated because the solid has collapsed while they were trying to do so.
- Could the internal parts of low-pressure vessels be constructed from the outside? In the construction of some U.K. railway carriages, components are fixed to the floor, roof, and the two walls before these four pieces of steel are bolted together. Fitting equipment to what is going to be the ceiling is much easier when it is in a convenient position [11].

The worst case ever of widespread entry to hazardous confined spaces was the employment of children as young as seven to clean chimneys by climbing up them. In 1850, 800 boys were working as chimney cleaners in London. Once machines for cleaning the chimneys had been invented, a campaign to make "entry" illegal was started in 1803 in the United Kingdom, but it did not achieve its aim until 1875 (see Section 37.6).

References

1. T. Gillard, Entry into vessels—a near miss. *Loss Prevention Bulletin*, Vol. 143, Oct. 1998, p. 18.
2. T. Donaldson, Confined space incidents—a review. *Loss Prevention Bulletin*, Vol. 154, Aug. 2000, p. 3–6.
3. Anon., DuPont in dock over fireball death. *Loss Prevention Bulletin*, Vol. 154, Aug. 2000, p. 27.
4. S. Turner, Odorise your nitrogen. *The Chemical Engineer*, Vol. 661, 9 July 1988, pp. 20–21.
5. Anon., Vessel inspector overcome by nitrogen. *Loss Prevention Bulletin*, Vol. 154, Aug. 2000, p. 20.
6. P. L. Hill, G. P. Poje, A. K. Taylor, and I. Rosenthal, Nitrogen asphyxiation. *Loss Prevention Bulletin*, Vol. 154, Aug. 2000, p. 9.
7. S. R. Porter and P. J. Mullins, Waste gas leads to near fatality. *Loss Prevention Bulletin*, Vol. 154, Aug. 2000, p. 15.
8. Anon., The disaster at the Neunkirchen Iron Works, *The Gas World*, 22 April 1933, p. 397 (Translation from *Das Gas und Wasserfach*, No. 14, Apr. 8, 1933).
9. A. Hopkins, *Safety, Culture and Risk*, CCH Australia, Sydney, NSW, Australia, 2005, Part C.
10. S. Dhodapkar and M. Konanur, Selection of discharge aids for bins and silos, *Chemical Engineering*, Part 1, 112(8): (2005) 27–32; Part 2, 112(10):71–82.
11. J. Abbott, Turbostar comes of age. *Modern Railways*, Vol. 56, No. 615, 1999, p. 891.

Changes to Processes and Plants

Midas, a legendary king of Phrygia, asked the gods to make everything he touched turn into gold. His request was granted, but as his food turned into gold the moment he touched it, he had to ask the gods to take back their favor.

Unfortunately, the gods are less obliging today and will not reverse the results of ill-considered modifications.

It is now many years since the explosion at Flixborough in 1974 (see Section 2.4) brought home to the process industries the need to look systematically for possible consequences before making any change to plants or processes. Many publications [1–3] have described accidents that occurred because no one foresaw the results of such changes and suggested procedures for preventing such accidents in the future. Nevertheless, as the following examples show, unforeseen consequences still occur. Sometimes there is no systematic procedure, sometimes the procedure is not thorough or is not followed, and sometimes the change is so simple that a formal review seems unnecessary. There is also a reluctance in many companies to look in the literature for reports of similar situations. According to an experienced process safety engineer:

> People make very little preparation for a management of change or process hazards analysis (PHA) by looking at the literature or making a search for events in similar facilities. We can sometimes prompt them to look at events within their own facility ... but getting them to spend any reasonable time reviewing other events is tantamount to pulling teeth. ... I would estimate that less than one in twenty PHA practitioners expend more than a very small effort in such preparations [4].

Chapter 2 described mainly changes to equipment—following Flixborough that seemed the main problem—so to restore balance, this chapter describes more changes to processes. The next chapter describes changes to organization.

doi:10.1016/B978-1-85617-531-9.00025-1

25.1   CHANGES TO PROCESSES

25.1.1   Scale-Up Is a Modification

In scaling up a process from the laboratory to production-scale, a company changed it from semibatch operation, in which the second reactant is added gradually, to batch operation, in which the entire quantities of both reactants are added at the outset. Twenty percent of the batches showed temperature excursions, but the operators were able to bring them under control by manual operation of cooling water and steam valves. The company then increased the reactor size from 5 to 10 m^3 (1,350 to 2,700 gal) and increased the quantities of reactants by 9%. The proportion of batches showing temperature excursions rose to half, and ultimately the operators failed to keep a batch under control. The manhole cover was blown off the reactor, and the ejected material caught fire. Nine people were injured.

The company failed to use its management-of-change procedure and also failed to respond to the rising number of temperature excursions [5].

Failures to understand scale-up go back a long way. Canned food was introduced in 1812. In 1845 it became part of regular British Royal Navy rations. Some time later there was an outbreak of food poisoning. Larger cans had been used, and the heat penetration became insufficient to kill the bacteria in the middle [6].

25.1.2   Unrecognized Scale-Up

In his biography, *Homage to Gaia* [7], James Lovelock describes an incident that occurred when he was working for a firm of consultant chemists. There had been a sudden deterioration in the quality of the gelatin used for photographic film, and he and another chemist were sent to visit the manufacturers. They asked the foreman if anything had changed. He replied that nothing had changed; everything was exactly as before. Lovelock's colleague noticed a rusty bucket next to one of the vessels and asked what it was for. The foreman said that a bucketful of hydrogen peroxide was added to each batch of gelatin, but as the bucket was rusty he had bought a new one the previous week. "We soon solved the firm's problem when we found that the new bucket was twice the volume of the old one." Its linear dimensions were only 25% greater, but the foreman had not realized that this doubled the volume.

25.1.3   Ignorance of a Reaction

Lovelock also described a modification that nearly took place but was prevented in time. The United Kingdom Gas Board, at the time the monopoly supplier of natural gas, decided to label the gas in one of their major high-pressure gas pipes with sulfur hexafluoride to detect leaks along the pipeline. The technique would have worked well. Unfortunately, those involved did not know that a mixture of methane and sulfur hexafluoride will explode almost as violently as a mixture of methane and oxygen. Fortunately, they found out in time and abandoned their plan [8].

25.1.4 Changes Made to Handle Abnormal Situations

A coker is a large vessel, typically $\approx$ 12 m (40 ft) tall, in which hot tar-like oil, after being heated in a furnace, is converted to lighter oils, such as gasoline and fuel oil, leaving a tarry mass in the vessel. On cooling, usually with steam and then water, this forms coke, which is dug out. A power failure occurred when a coker was 7% full and the plant was without steam for 10 hours. The inlet pipe became plugged with solid tar, and the operators were unable to inject steam.

There were no instructions for dealing with this problem, although a somewhat similar one had occurred two years earlier. On that occasion it had been possible to inject water to cool the contents; nevertheless, when the bottom cover was removed from the coker, a torrent of water, oil, and coke had spewed out. When the second incident occurred, the supervisor therefore decided to let the coker cool naturally before opening it. Two days later, the temperature of the outside of the bottom flange of the coker had fallen from its usual value of 425°C (800°F) to 120°C (250°F), so the supervisor decided to go ahead. The operators injected some steam—presumably through a different route than the normal one—to remove volatile products, and then started to open the coker. The top cover was removed without incident. The bottom cover was unbolted while supported as usual by a hydraulic jack. When the jack was lowered, hot vapor and oil gushed out and immediately ignited. It was probably above its auto-ignition temperature. Six people, including the supervisor, were killed.

The immediate cause was a failure to realize that the temperature of the middle of the vessel was far higher than that of the walls, high enough to continue to convert the tar to gasoline. Afterward, calculations showed that it would take two weeks, not two days, for the temperature to fall to a level at which it would be safe to open the coker. (Sections 30.7 and 35.7 describe the results of other failures to calculate effects.)

The controls for the hydraulic jack should have been located farther away from the coker, and so many people should not have been allowed so near.

One underlying cause was the failure to plan in advance for a loss of power. Plans should have been made for this foreseeable event but never were, even though an event had occurred two years before and caused a serious spillage.

Another underlying cause was the lack of technical support. The supervisor seems not to have been a professional engineer or recognized the need to consult one. The report [9] does not say whether or not there had been any downsizing or reduction in support, but the incident is rather similar to that at Longford (see Section 26.2) where the operating team was also unaware of a well-known fact, in this case that metal becomes brittle when cold.

We have all been given, at some time, a food such as pasta or rice pudding, straight from the oven in the dish in which it was cooked. If it is too hot to eat, experience tells us that the outside bits are cooler and we eat them first. We know the outside cools faster than the inside. Unfortunately, we find it difficult to apply in one situation the lessons we have learned in another; they are kept in different parts of our minds.

25.1.5   An Abnormal Situation Produced by a Process Change

Powdered aluminum chloride, a catalyst, was added to a reaction mixture. A change was made: aluminum powder was used instead as it was expected to form aluminum chloride by reacting with the hydrogen chloride already there. Unfortunately, the reactor became choked with sludge. The aluminum was much denser than the aluminum chloride, and the agitator was unable to prevent it from settling. If there was a management of change procedure—the report [9] does not say—no one considered the results of more or less mixing, an obvious question to ask when a hazard and operability study (Hazop) is carried out on a vessel in which mixing takes place.

The problem now was how to get the sludge out of the reactor. A chemist examined a sample. It reacted with water, producing heat, so he recommended that a large amount of water, eight times the weight of the sludge, should be put into the reactor as rapidly as possible. Someone suggested that a short burst of steam should first be put into the reactor to break up the sludge. The day supervisor agreed and gave his instructions by telephone to the afternoon shift supervisor who told the night shift supervisor who told an operator. By this time the instruction had become distorted and steam was added continuously for several minutes. The reactor exploded. Fortunately, no one was seriously injured.

An immediate cause was the sloppy method of passing on instructions despite the fact that the addition of water was known to generate heat. The instructions should have been precise and in writing and should have specified the duration of the steam burst. *Short*, like *all* and *similar* (see Section 31.1), is an imprecise word and should never be used in plant instructions.

Another immediate cause was the failure to calculate what temperature would be reached by the addition of steam and then water and the amount of gas that would be driven off. In the laboratory test, it dispersed easily. At the plant, the size of the vent was inadequate.

Underlying these causes was the assumption, as in the last item, that the supervisor can improvise changes in procedure to cope with an abnormal situation. In emergencies, he or she may have to do so but when possible these situations should be foreseen and planned for in advance. In the case just described, a delay of a few hours, or even a day, while the proposed change was discussed by a group of people (including professional staff) and approved at an appropriate level, would have mattered little. Blowing up the reactor caused more delay and cost $13 million.

25.2   CHANGES TO PLANT EQUIPMENT

25.2.1   Changes in the Direction of Flow

Figure 25-1a shows the original design of a reactor. Hot feed gas (PF [preheated feed]) was passed upward through catalyst tubes 3 m (10 ft) long. The heat of reaction was removed by circulating a molten salt (HTS [heat transfer salt]) through the shell. Note that the flow of this liquid was also upward.

The hot exit gases were cooled in a waste heat boiler. During a review of the model, the contractors pointed out that the boiler required an expensive structure to support it. They could avoid this cost, they said, by reversing both flows through the reactor, as shown in Figure 25-1b, and putting the boiler at ground level. This was agreed.

Soon after startup, some temperatures were erratic, violent vibrations occurred, and then the shell ruptured at a point opposite the liquid inlet line.

The investigation [10] showed that gas bubbles had been trapped under the top tubesheet. The tubes near it overheated and the nitrate coolant reacted with the iron shell in a thermite-type reaction, the iron replacing the positive ions in the salt. The investigation also showed that gas bubbles can collect under a horizontal surface when the flow is downward but not when it is upward. The plant was rebuilt to the original design.

It might have been possible to retain the downward flow and to vent any gas bubbles back to the suction tank in the salt system. However, preventing the formation of bubbles is better than letting them form and then getting rid of them.

The change in design was made without following the company's normal procedure for control of change. Reversing the flow seemed

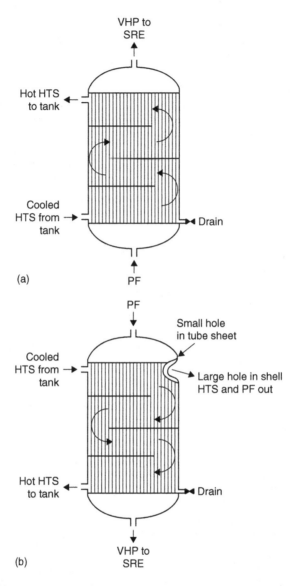

FIGURE 25-1 (a) Original design of reactor. (b) Reversing the flow allowed a gas bubble to be trapped below the top tubesheet and led to overheating of the tops of the tubes. PF = preheated feed; HTS = heat transfer salt; VHP = very hot product gases; and SRE = steam raising exchanger. From reference 10. *(Reprinted with the permission of the Institution of Chemical Engineers.)*

such a minor change (and one that would save so much money) that it received no systematic appraisal.

Another incident was the result of an even more minor change in flow [11]. Four 16-m³ (4,200-gal) tanks that held an odorizing liquid had to be taken out of service and cleaned. Three tanks were cleaned without incident. The procedure was to empty each tank, wash it with methanol, and then remove the last traces of the odorizing liquid by washing with a sodium hypochlorite solution. The tanks were normally blanketed with natural gas, and this was left in use during the cleaning operation, the natural gas flowing through the ullage space of the tanks and then to the flare system.

When the fourth tank was cleaned, the arrangement of the pipework made it impossible for the natural gas to flow through the tank so the natural gas was just connected to it in order to maintain the pressure in the tank. During the hypochlorite wash, there was an explosion in the tank and flames were discharged through the relief valve. Tests then showed that there was 80% oxygen in the ullage space of the tank. The oxygen was probably formed by decomposition of the hypochlorite, catalyzed by the nickel in the stainless steel of the tank. While the first three tanks were cleaned, the continuous flow of natural gas swept out the oxygen as fast as it was formed.

Natural gas or other fuel gases are often used for blanketing when nitrogen is not available. They are just as effective as nitrogen in maintaining a pressure in the equipment and preventing air leaking in; however, if air does leak in, a fire or explosion is more likely to occur.

25.2.2 Two Changes in Firefighting

Two buildings were 23 m (75 ft) apart. The same fixed firefighting unit served both buildings, as they were too far apart for a fire in one to spread to the other. Eight years later, a new building was built between the two original ones, and one of the originals was demolished. The gap between the two buildings was now only 4 m (12.5 ft). When a fire occurred in one building, it spread to the other and the fixed equipment was too small to control both fires [12].

The heating system in a building had to be shut down for repair over a weekend. There were fears that the water in the sprinkler system might freeze, so it was replaced by ethanol. You can guess what happened.

25.2.3 Adding Insulation Is a Modification

To save energy, a company decided to insulate a valve, shown in Figure 25-2, which operated at 310°C (600°F). The three long bolts expanded and a leak occurred and ignited. The flames were 12 m (40 ft) long. The valve body, in direct contact with the hot liquid, would hardly have been affected by the insulation but the long bolts rose in temperature. A rise by 250 degrees C (450 degrees F) would increase their length by 1 mm (0.04 in.) [13].

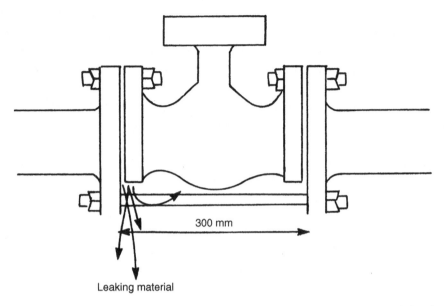

Leaking material

FIGURE 25-2 When this valve was insulated to save heat, the long bolts expanded and the flanges leaked.

When the company managers decided to use valves with long bolts, they should have considered this as a modification and looked for possible consequences. These valves are not suitable for equipment that undergoes changes in temperature.

Another company was more successful. The company went through the modification procedures when it reviewed a proposal to fit acoustic insulation to some pipework. Management then realized that the acoustic insulation would also act as thermal insulation and prevent the cold gas in the pipes from picking up heat from the atmosphere. The insulation was still fitted, but other changes were made to handle the change in temperature.

25.2.4   Two Unauthorized Changes

Figure 25-3 shows a three-way cock in which the top of the central bolt has been marked to show the position of the cock. This was presumably done because the marks on the plug itself are hard to see. They are just faintly visible in the photograph.

Originally the marks on the bolt corresponded to those on the plug. At some time, two washers were placed underneath the bolt. It could no longer be screwed right in and the marks no longer corresponded.

FIGURE 25-3 The central bolt was marked to show the position of the cock. When two washers were inserted under the bolt, it could not be screwed down as far as before and the marks no longer corresponded with the position of the cock.

Most of the operators on the unit set the cock according to the marks on the bolt. Ultimately, this led to misdirection of a process stream, formation of an explosive byproduct, and an explosion.

25.2.5 A Very Simple Change

A company decided to display hot-work permits on the job. They were fixed to any convenient item of equipment. On one unit, they were pushed into the open end of a 1.5-in pipe. The employee who did this probably thought it was a scaffold pole or a disused pipe. The pipe actually supplied a controlled air bleed into a vacuum system to control or break the vacuum. The hot-work permits were sucked into the pipe and blocked the motor valve in the pipe. Product was sucked into a condenser, and the unit had to be shut down for cleaning for two days. Several permits were removed from the valve.

25.2.6 A Temporary Change

A drum-filling machine was causing problems during a Friday night shift. The shift foreman decided to change over to manual filling until the maintenance team returned to work on the following Monday. Until then the valve on the filling machine had to be opened with an adjustable wrench. In addition, the filling head could not be lowered into the drum, and the drum had to be carefully positioned under the filling head.

The maintenance team was too busy to attend to the filling machine, and a week later the temporary system was still in use when the inevitable happened. A drum was not positioned accurately and the liquid hit the top of the drum, splashing the operator's face.

The report blamed poor communication [14]. The shift foreman's note in his log and in the job list did not draw attention to the fact that the temporary work method was hazardous, so the job got the priority given to an inconvenience, not a hazard. However, this is not convincing. The unit manager, the other shift foremen, the fillers, and the safety representative, if there was one, should have spoken to the maintenance team and drawn attention to the hazard. In a well-run organization, written messages are for confirmation, precision, and recording; things get done by talking to the people who will have to do the work, asking them, persuading them, sweethearting them—call it what you will [15].

It is sometimes necessary to make changes on short notice in order to keep a plant running. In such cases, the normal procedures for the control of modifications should be carried out as soon as possible and no later than the next working day.

25.2.7   Another Trivial Change

Filters removed dust from a ventilation system. The dust fell into a 55-gal drum. From time to time, the drum was removed by a forklift truck, emptied by vacuum, and replaced. At some point in the life of the plant, the operators found that it was easier to replace the drum and position it correctly if they kept it on a wheeled trolley. They did not realize that, as the wheels had rubber tires, the drum was now an ungrounded conductor— and could accumulate a charge of static electricity, either during the vacuuming operation, during transport, or as a result of dust falling into it. The trolley was in use for a considerable time before conditions were just right for an explosion. While the trolley was being replaced, a charge passed from the drum to grounded metal nearby, igniting a small cloud of dust that fell into the drum from the filters at just that moment. As so often happens, the small initial explosion disturbed dust that had settled and was followed by a larger and more damaging explosion.

Static discharges may have occurred before but they happened at times when no dust cloud was present. On the day of the explosion, the atmospheric temperature was very high ($\approx 38°C$ [100°F]), and this would have lowered the ignition energy of the dust and made an explosion more likely [16].

This incident shows the limitations of instructions and the need to give operators an understanding of the hazards of the materials and equipment used. However many instructions we write, we can never cover every possibility. If we try to do so, our instructions get longer and more complex and fewer people read them. It is better to educate people so that they understand the hazards.

Auditors should look at plant instructions. Sometimes they are out of date or cannot be found. More often they are spotlessly clean, like poetry books in public libraries, showing that they are rarely consulted. (There is more on static electricity in Sections 24.2.3, 28.2.5, 30.1, and 32.7.)

25.2.8 Unintended Changes

These occur when suppliers supply the wrong process materials (see Section 31.3) or construction materials. For example, a chlorine vaporizer was shut down for repair and inspection. Soon after it was started up, a spiral-wound gasket, changed at the shutdown, blew out. The metal winding in the gasket could not be found. Trace metal analysis showed that it had been made from titanium, which reacts rapidly with chlorine, instead of nickel.

If use of the wrong process material or material of construction can have serious effects on safety, then all incoming materials should be tested before acceptance or use. This became commonplace in the 1970s, after a number of serious incidents, but many companies abandoned their checking programs when their suppliers obtained quality certification. How many more incidents do we need before they are reintroduced?

25.2.9 A Change to the Type of Valve

In carbon dioxide absorption plants, the gas is absorbed in potassium hydroxide, which becomes potassium carbonate. Control valves let down the potassium carbonate solution from high to low pressure. One plant used a motorized ball valve instead. When the jet from this type of valve impinges on a surface, it produces a ring-shaped corrosion groove. A disc of metal was blown out of a bend downstream of the valve.

Sensing the loss of pressure, the automatic controller opened the ball valve fully, discharging hot potassium carbonate solution out of the hole. Unfortunately, the pipe was opposite the control room window. The window was broken and all the operators were killed. It was shift change time, and more operators than usual were present [17].

25.2.10 A Change in the Cooling Agent

A reactor was cooled by circulating brine through the jacket. The brine system was shut down for repair, so town water was connected to the jacket. The gauge pressure of the town water (9 bar, or 130 psi) was greater than the design pressure of the jacket's inner wall, which gave way.

The works modification approval form, which had been completed by the supervisor and the maintenance engineer, asked 20 questions, one of which was "Does the proposal introduce or alter any potential cause of over/underpressurizing the system or part of it?" They had answered no [18].

25.2.11 A Failure to Recognize the Need for Consequential Change

When one item in a plant is changed, others may have to be changed to match, but companies do not always recognize this fact. A company filled drums with liquid chlorine. One was overfilled and bulged when the temperature rose so for protection a high-weight alarm was fitted to the filling and weighing machine. It was set at 1,400 kg. A change was made to smaller drums that were completely full at a gross weight of 1,335 to 1,350 kg, but no one remembered to change the alarm setting. Either there was no procedure for the management of change, or the change was considered so insignificant that the procedure was not followed. About three years later, another drum was overfilled and bulged. The company then decided to check weigh the drums and ordered an additional weighing machine.

Another three years later, this machine had arrived but had not yet been installed and a third drum was overfilled and bulged, this time at a customer's premises. The cause was a minor one-time change in the filling procedure. As the storage space was full, a drum was left connected to the filling machine overnight and the drum-filling valve was leaking or was not fully closed. Check weighing would have prevented this incident [19].

Fortunately, none of the overpressured drums burst or leaked, though they were taken well above their design pressures. The large difference between the design pressure and the rupture pressure is a good example of defense in depth. Most pressure vessels can withstand several times their design pressure before they rupture, but not all equipment is as strong (e.g., low-pressure storage tanks are quite fragile). In contrast, most equipment can withstand only a small percentage increase in absolute temperature. The life of furnace tubes is shortened if they are exposed to an increase of a few percentage points in absolute temperature for a short time.

25.2.12 An Example from the Railways

In the early days of railways, the gaps between the rails caused almost intolerable vibration. To reduce it, some railway companies cut the ends of the rails diagonally so that they overlapped and formed a smoother joint. Unfortunately, the spikes holding down the rails sometimes failed to do so, the end of the rail rose, the wheel went underneath it, and the pointed end of the rail went though the floor of the carriage. The person sitting above it was likely to be speared and impaled against the roof [20].

25.2.13 Another Historic Incident

Malaria and yellow fever, both spread by mosquitoes, hindered the building of the Panama Canal. The cause? To prevent ants climbing up the legs of hospital beds, they were set in pans of water—which unfortunately created an ideal breeding ground for mosquitoes [21].

25.3 GRADUAL CHANGES

If a frog is put into hot water, it jumps out. If it is put into cold water and the temperature is gradually raised, it stays there until it dies. In a similar way, we often fail to notice gradual changes until they have gone so far that an accident occurs. Section 2.9 describes several examples, including a gradual reduction in the flow through a steam main as the result of recession in the industry. The steam traps were barely adequate; this did not matter when the flow was large, but when it became lower, condensate accumulated and water hammer ruptured the pipe.

25.3.1 A Gradual Change in Concentration

Natural gas liquids were dried by passing them through a bed of molecular sieves, which also absorbed some hydrogen sulfide; the sieves were then regenerated by a stream of hot gas. They had to be changed every three or four years. The old ones were wetted with a fire hose in case any pyrophoric materials were present and to keep down dust and then poured down a chute into a high-sided tipper truck for disposal.

The sieves formed a mound shape and had to be spread level in the truck. A man who entered the truck to spread them collapsed. Three other men entered the truck to rescue him. All three collapsed; two of them and the first man died, poisoned by hydrogen sulfide. The sieves had a greater affinity for water than for hydrogen sulfide and released the gas when wetted.

There was much wrong. The high-sided truck was not recognized as a confined space, so the entry procedure (see Chapter 24) was not followed; the men filling the truck had not been warned that toxic gas might be present; many of the operators and staff did not know that it could be released; and no hydrogen sulfide detectors were supplied. But underlying all these shortcomings was the fact that over the years the amount of hydrogen sulfide in the natural gas liquids had gradually increased without anyone realizing that it had reached a level where change in procedures was necessary.

25.3.2   A Gradual Change in Maintenance

Another incident occurred in the U.K. railway system. This heretofore single organization was split, in 1996, into many independent private companies in the hope that this would provide competition and reduce costs. (It did not.) One company owned the track, other companies maintained it, yet more companies owned the trains, and a fourth set of companies maintained them. In an attempt to reduce cost, there was a gradual tendency to reduce maintenance but still maintain specifications. However, "[b]oth sides of the wheel/rail interface may be operating within their respective safety-based standards, but the combined effect of barely acceptable wheel on barely acceptable rails is unacceptable" [22]. This led to rolling contact fatigue of the track (often called gauge corner cracking), a train crash at Hatfield near London in October 2000 that killed four people, and a consequent upheaval while hundreds of miles of rail were replaced.

The engineering principle involved is hardly new. In 1880, Chaplin showed that a chain can fail if its strength is at its lower limit and the load is at its upper limit [23] (see Section 37.5). The Hatfield crash did not occur because engineers had forgotten this but because there were no engineers in the senior management of the company that owned the track. They had all been moved to the maintenance companies. This accident is therefore also an example of the need for the management of organizational change, which is discussed in the next chapter.

25.3.3   Gradual Changes in Procedures

Gradual procedural changes are more frequent than gradual changes in equipment or process conditions. Procedures corrode more rapidly than steel and can disappear once managers lose interest. A procedure is relaxed, perhaps for a good reason. Perhaps there are technical reasons why the normal procedure for isolating equipment for maintenance cannot be followed on a particular piece of equipment. Nothing happens and the simpler procedure is used again just to save time or effort. Before long it has become standard and newcomers are told, "That instruction is out-of-date. We don't do it that way any more." To prevent this sort of gradual change, supervisors and managers should keep their eyes open and also explain why certain procedures are necessary. An effective way of doing this is to describe or, better, discuss accidents that occurred when they were not followed.

25.4 CHANGES MADE BECAUSE THE REASONS FOR EQUIPMENT OR PROCEDURES HAS BEEN FORGOTTEN

This is one of the most common causes of accidents, and I have discussed them at length in my book *Lessons from Disaster—How Organisations Have No Memory and Accidents Recur* [24] and suggested actions that could improve corporate memories (see also Section 38.10).

A recycle stream was found to contain a contaminant that produced a runaway reaction if its concentration was high enough. The stream was therefore routinely analyzed. Several years later, after a change in management, the analysis was stopped. A few months later, an explosion occurred [25].

Responsibility was shared, I suggest, between the original supervisor or supervisors who never documented the reasons for the tests in a readily accessible form (if they documented them at all) and the new supervisor or supervisors who stopped the test without knowing why it had been started. *Never stop or change a procedure unless you know why it was introduced. Never stop using equipment unless you know why it was provided.*

In medieval England there were officials called Remembrancers whose job was to remind the king's courts of matters that they might otherwise forget [26]. (The job still exists, but the duties are now ceremonial.) Every process plant needs such a person.

Sections 30.1, 30.2, 30.5, 30.7, 30.9, 36.7, and 36.8 describe other accidents that would not have occurred if the results of changes had been foreseen. For guidance on the methods used to control change, see Sections 2.12 and 26.8.

References

1. M. S. Mannan (editor), *Lees' Loss Prevention in the Process Industries*, 3rd edition, Elsevier, Boston, MA, 2005, Chapter 21.
2. T. A. Kletz, *What Went Wrong?—Case Histories of Process Plant Disasters*, 4th edition, Gulf, Houston, TX, 1998, Chapter 2.
3. R. E. Sanders, *Chemical Process Safety—Learning from Case Histories*, Butterworth-Heinemann, Boston, MA, 1999.
4. P. J. Palmer, Private communication, 2003.
5. Anon., *Chemical Manufacturing Incident*, Investigation Report No. 1998-06-NJ, U.S. Chemical Safety and Hazard Investigation Board, Washington, DC, 1998.
6. H. Fore, Contributions of chemistry to food consumption, *Milestones in 150 Years of the Chemical Industry*, Royal Society of Chemistry, London, 1990.
7. J. Lovelock, *Homage to Gaia—The Life of an Independent Scientist*, Oxford University Press, Oxford, UK, 2000, p. 41.
8. J. Lovelock, *Homage to Gaia—The Life of an Independent Scientist*, Oxford University Press, Oxford, UK, 2000, p. 176.
9. Anon., *Management of Change*, Safety Bulletin No. 2001-04-SB, Chemical Safety and Hazard Investigation Board, Washington, DC, 2001.

10. G. Lawson-Hall, A costly cost-saving modification. *Loss Prevention Bulletin*, Vol. 160, Aug. 2001, pp. 10–11.

11. Anon., Explosion in an odorant plant. *Loss Prevention Bulletin*, Vol. 168, Dec. 2002, pp. 19–20.

12. Anon., Failure to manage can lead to costly losses. *IRI Sentinel*, 1st Quarter, 1996, pp. 3–4.

13. Anon., Thermal insulation on valve causes expansion of flange bolts. *Petroleum Review*, Oct. 1983, p. 34.

14. Anon., Creeping change. *Loss Prevention Bulletin*, Vol. 161, Oct. 2001, pp. 8–9.

15. T. A. Kletz, *By Accident—A Life Preventing Them in Industry*, PFV Publications, London, 2000, p. 33.

16. R. D. Pickup, Dust explosion case study: "Bad things can still happen to good companies." *Process Safety Progress*, Vol. 20, No. 3, 2001, pp. 169–172.

17. M. Schofield, Corrosion horror stories. *The Chemical Engineer*, Vol. 446, March 1988, p. 39.

18. Anon., Modification malady. *Chemical Safety Summary*, Vol. 56, No. 221, 1985, p. 6.

19. T. Fishwick, Distortion of the end of a full drum of chlorine. *Loss Prevention Bulletin*, Vol. 168, Dec. 2002, pp. 15–16.

20. Anon., *Railway Passenger Travel 1825–1880*, Scribners, New York; Reprinted by Americana Review, New York, 1888.

21. R. Milner, The yellow fever panic of 1905. *Natural History*, Vol. 20, July 1991, p. 52.

22. R. Ford, Gauge corner cracking—privatisation indicted. *Modern Railways*, Vol. 59, No. 640, 2002, pp. 19–20.

23. A. G. Pugsley, *The Safety of Structures*, Arnold, London (quoted by N. R. S. Tait (1987)). *Endeavour*, Vol. 11, No. 4, 1966, pp. 192.

24. T. A. Kletz, *Lessons from Disaster—How Organisations have No Memory and Accidents Recur*, Institution of Chemical Engineers, Rugby, UK, 1993.

25. R. E. Knowlton, Dealing with the process safety "Management gap." *Plant/Operations Progress*, Vol. 9, No. 2, 1990, pp. 108–113.

26. Anon., *Encyclopaedia Britannica*, Entry on Remembrancer, 2001.

Changes in Organization

The best of systems is no substitute for experience, or for seeing and listening and sniffing for yourself.

—Financial News, *Daily Telegraph* (London), February 16, 2002

The explosion at Flixborough, United Kingdom, in 1974 drew the attention of the oil and chemical industries to the need to control changes to plants and processes. Many publications (see Chapter 25) have both described accidents that occurred because no one foresaw the results of such changes and suggested ways of preventing such accidents in the future. Only in recent years, however, have we realized that changes in organization can also have unforeseen effects and should likewise be scrutinized systematically before they are made. In some countries, this is now a legal requirement in high hazard industries.

A common organizational change is to eliminate a job and distribute the jobholder's tasks among other workers. Although the jobholder is asked to list all his or her duties, he or she may miss one or two, especially those carried out by custom and practice and not listed in any job description. For example, someone may have built up a reputation as a "gatekeeper"—someone who knows how to get things done—for example, he or she may know where scarce spare parts may be squirreled away. Another person may be the only mechanic who really understands the peculiarities of a certain machine. Only after such individuals have left are their distinctive contributions really recognized. Such changes and their potential consequences are easily missed.

The following are some examples of the unforeseen effects of changes in organization. Sections 25.3.2 and 37.5 describe another.

doi:10.1016/B978-1-85617-531-9.00026-3

26.1   AN INCIDENT AT AN ETHYLENE PLANT

The plant was starting up after a turnaround. At 2 a.m. on the day of the incident, the shift team started the flow of cold liquid to the demethanizer column. A level should have appeared in the base of the column 2 hours later. It did not, but problems elsewhere distracted the shift team and they did not notice this until 7 a.m. By this time the temperature at the top of the column was −82°C (−115°F) instead of the usual −20°C (−4°F) and the level in the reflux drum rose from zero to full scale in 10 min. This should have told the shift team that the column had flooded, overflowed into the reflux drum, and would now be filling the flare knock-out drum (see Figure 26-1). However, neither of the two high-level indicator/alarms on this drum, set at 8% and 22% of capacity, showed any response.

It was noon before anyone had a thorough look at the plant. The staff then found that the wires leading to the column level indicator were disconnected and that the valves between the knock-out drum and its level indicators were closed. (Section 29.1 describes a similar incident.) Both vessels were shrouded with scaffolding and the state of the wires and valves was not easily seen. Liquid was now entering the flarestack.

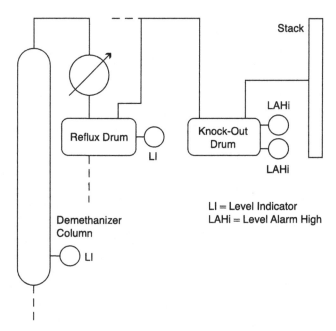

FIGURE 26-1   The level indicator on the column and the level alarms on the knock-out drum were out of order. The column filled with liquid, which overflowed into the drums and then into the stack.

It failed as the result of low-temperature embrittlement but, fortunately, the leaking liquid did not catch fire. No one was injured.

The immediate causes of the incident were the failures to recommission the level instruments before startup and the slowness of the shift teams to realize what was happening. The underlying causes were far deeper and were due to both short-term and long-term changes in organization.

26.1.1 Short-Term Changes

It was the practice at the plant to work 12-hour shifts instead of the usual 8-hour shifts during startups so that there were more people present than during normal operation. On this occasion the operators refused to do so (though they were willing to work overtime if necessary; this would give them more pay than working 12-hour shifts). However, the foremen and shift managers worked 12-hour shifts. They changed shift at 7 a.m. and 7 p.m. while the operators changed at 6 a.m., 2 p.m., and 10 p.m. This pattern of work destroyed the cohesion that had been built up over the years within each shift and lowered the competence of the team as a whole. A report in the local newspaper said that "A major influence over the behavior of the operating teams was their tiredness and frustration." A trade union leader was quoted as saying that the management team members were more tired than the operators as they were working 12-hour shifts.

In addition to the usual shift personnel, two professional engineers were also present on each shift, but their duties were unclear. Were they there to advise the shift manager or, being more senior in rank, could they give him instructions? Should they try to stand back and take an overview or should they get involved in hands-on operations? On the day of the incident, they did the latter and got involved in the details of the problems that distracted everyone from the demethanizer problem.

26.1.2 Long-Term Changes

So far I have followed the published report on the incident [1], but there had also been more fundamental changes. The incident shook the company. It had a high reputation for safety and efficiency and the ethylene plant was considered one of its flagships—one of the least likely places where such a display of incompetence could occur, so what went wrong?

About seven years earlier, there had been a major recession in the industry. As in many other chemical companies, drastic reductions were made in the number of employees, at all levels, and many experienced people left the company or retired early. This had several interconnected results:

- Operating divisions were merged, and senior people from other parts of the company, with little experience of the technology, became responsible for the ultimate control of some production units.
- There was pressure to complete the turnaround and get back on line within three weeks. This pressure came partly from above but also from within the production and maintenance teams, as the members were keen to show what they could do. They should have aborted the shutdown to deal with the problems that had distracted everyone during the night but were reluctant to do so.
- There were fewer old hands who knew the importance, when there were problems, of having a look around and not just relying on the information available in the control room. A look around would have shown ice on the demethanizer column.
- Delayering had produced a large gap in seniority between the manager responsible for the ethylene plant and the person above him. This made it more difficult for the ethylene manager to resist the pressure to get back on line as soon as possible. Previously, an intermediate manager had acted as a buffer between the operating team and other departments, and he prevented commercial people and more senior managers from speaking directly to the startup team. In addition, he would probably have aborted the startup. Senior officers, not foot soldiers, order a retreat.

The company had an outstanding reputation for openness but was reticent about this incident and no report appeared in the open literature—other than the local newspaper—until about 12 years later, after the company had sold the plant.

26.1.3   A Failure to Learn from the Past

After the publication of the first version of the above account [15], Frank Crawley reported [14] that he was present when a similar measurement failure had occurred in the sister plant in the same company and on the same site about 10 years earlier, during the first few days of the initial startup. That incident was handled without any upset. The mass imbalance was recognized, the team was alert to the possibility of carryover, and once carryover was detected feed was stopped until the measurement could be corrected.

This incident should have been in the plant memory bank 10 years later, but it had been forgotten. A year after the first incident there was a change in responsibilities. The shift managers, who were former foremen and lacked technical knowledge, were given greater responsibilities on a complex process, while the professionally qualified plant managers (the equivalent of supervisors in the United States), who previously knew the plant and process in detail, became less involved. The price of this change was the accident on the second plant described previously. Although there was no fire or injury, three months worth of production was lost.

Sections 26.3 and 36.7 describe similar incidents.

26.2 THE LONGFORD EXPLOSION

On September 25, 1998, a heat exchanger in the Esso gas plant in Longford, Victoria, Australia, fractured, releasing hydrocarbon vapors and liquids. Explosions and a fire followed, killing two employees and injuring eight. Supplies of natural gas were interrupted throughout the state of Victoria and were not fully restored until October 14. There was no alternative supply of gas, and many industrial and domestic users were without fuel for all or part of the time the plant was shut down. The accident is described in a detailed official report [2], in a book by Andrew Hopkins that concentrates on the underlying causes [3], and more briefly elsewhere [4, 5].

The purpose of the unit in which the explosion occurred was to remove ethane, propane, butane, and higher hydrocarbons from natural gas by absorbing them in "lean oil." The oil, now containing light hydrocarbons and some methane and known as "rich oil," was then distilled to release these hydrocarbons and the oil, now lean again, was recycled. The heat exchanger that ruptured was the reboiler for the fractionation column. The cold rich oil was in the tubes and was heated by warm lean oil in the shell.

As a result of a plant upset, the lean oil pump stopped. There was now no flow of warm lean oil through the heat exchanger, and its temperature fell to that of the rich oil, $-48°C$ $(-54°F)$. The official report describes in great detail the circumstances that led to the pump stopping. However, all pumps are liable to stop from time to time and the precise reason why this pump stopped on this occasion is of secondary importance. Next time it will likely stop for a different reason. In this case, one of the reasons was the complexity of the plant. It had been designed to recover as much heat as possible and this resulted in complex interactions, difficult to foresee, between different sections.

Ice formed on the outside of the heat exchanger when the flow of warm oil stopped, but no one realized that the low temperature was hazardous. Despite long service, the operators had no idea that the heat exchanger could not withstand low temperatures and thermal shocks and that restarting the flow of warm lean oil could cause brittle failure. More seriously, some of the supervisors and even the site manager, who was away at the time, did not know this. It was not made clear in the operating instructions.

The operators' ignorance does not surprise me. When I worked in production, before I became involved full time in safety, I learned that some operators' understanding of the process was limited. Troubleshooting depended on the chargehands (later called assistant foremen) and foremen, assisted by those operators who were capable of becoming chargehands or foremen in the future. In recent years, I have heard many speakers at conferences describe the demanning and empowerment their companies have carried out and wondered whether the operators of today are really better than those I knew in my youth. At Longford they were not.

Esso claimed that its operators had been properly trained and that there was no excuse for their errors. But the training emphasized the knowledge the operators needed to perform their jobs rather than an understanding needed to deal with unforeseen problems. They were tested after training but only for knowledge, not for understanding. One operator was asked why a certain valve had to be closed when a temperature fell below a certain value. He replied that it was to prevent thermal damage and received a tick for the correct answer. At the inquiry [3], he was asked what he meant by thermal damage and replied that he "had no concept of what that meant." When pressed, he said that it was "some form of pipework deformity" or "ice hitting something and damaging pipework." He had no idea that cold metal becomes brittle and could fracture if suddenly warmed.

Now we come to crucial changes in organization. Two major changes were made during the early 1990s. In the first, all the engineers, except for the plant manager, the senior man on site, were moved to Melbourne. The engineers were responsible for design and optimization projects and for monitoring rather than operations. They did, of course, visit Longford from time to time and were available when required, but someone had to recognize the need to involve them.

In the second change, the operators assumed greater responsibility for plant operations, and the supervisors (the equivalent of foremen) became fewer in number and less involved. Their duties were now largely administrative.

Both of these changes were part of a company-wide initiative by Exxon, the owners of Esso Australia, and were the fashion of the time. There was much talk of empowerment and reduced manning. The report concluded that "The change in supervisor responsibilities . . . may have contributed by leaving operators without properly structured supervision." It added, "Monthly visits to Longford by senior management failed to detect these shortcomings and were therefore no substitute for essential on-site supervision."

On the withdrawal of the engineers, the report said,

> [It] appears to have had a lasting impact on operational practices at the Longford plant. The physical isolation of engineers from the plant deprived operations personnel of engineering expertise and knowledge, which previously they gained through interaction and involvement with engineers on site. Moreover, the engineers themselves no longer gained an intimate knowledge of plant activities. The ability to telephone engineers if necessary, or to speak with them during site visits, did not provide the same opportunities for informal exchanges between the two groups, which are often the means of transfer of vital information.

None of this was recognized beforehand. Chats in the control room and elsewhere allow operators to admit ignorance and discuss problems in an informal way that is not possible when a formal approach has to be made to engineers at the company headquarters. *Empowerment* can become a euphemism for withdrawal of support.

The next example is anecdotal, of course, but supportive of the real-life work problems caused by too much formality. On one occasion when I was a safety adviser with ICI Petrochemicals Division, I was asked to move my small department to a converted house just across the road from the division office block. I objected as I felt that this would make contact with my colleagues a little bit harder because they would be less likely to drop by our offices.

At Longford there were also errors in design. The heat exchanger that failed could have been made from a grade of steel that could withstand low temperatures or a trip could have isolated the flow of cold liquid if the temperatures of the heat exchanged fell too far. These features were less common when the plant was built than they became 30 years later, but they could have been added to the plant. The designs of old plants should be reviewed from time to time. This is particularly important if they have undergone changes that were not individually studied. We cannot bring all old plants up to all modern standards—inconsistency is the price of progress—but we should review old designs and decide how far to go. Esso intended to Hazop the plant, but the study was repeatedly postponed and ultimately forgotten. Another design weakness was the overly complex heat recovery system already mentioned.

Exxon has a high reputation for its commitment to safety and for the ability of its staff. Was Longford a small plant in a distant country that fell below the company's usual standards, or did it indicate a fall in standards in the company as a whole? Perhaps a bit of both. Exxon did not require Esso Australia to follow Exxon standards and the Longford plant fell far below them. Exxon was fully aware of the hazards of brittle failure (see Section 28.3), but its audit of Esso did not discover the ignorance of this hazard at Longford. On the other hand, the removal of the engineers to Melbourne and the reductions in manning and supervision were company-wide changes. It also seems that in the company as a whole, the outstandingly low lost-time accident rate was taken as evidence that safety was under control. Unfortunately, the lost-time accident rate is not a measure of process safety.

Esso was prosecuted for 11 failures "to provide and maintain so far as is practicable for employees a working environment that is safe and without risk to health" and had to pay the largest fine ever imposed by the state of Victoria for such an offense. However, the fine was small compared with the claims for damages caused by the loss of natural gas. In a summary and review [6] of the trial, Hopkins said that it produced no new causal insights. However, it provided an object lesson in how not to handle the defense in such a case. Hopkins concluded that "Esso's decision to plead not guilty, its conduct at the trial and its refusal to accept responsibility led the Judge to conclude that the company had shown no remorse, and the absence of corporate remorse weighed heavily in his decision not to mitigate the penalties in any way."

Although Esso claimed at the inquiry that an operator was responsible for the accident, the company did not claim this at the trial, perhaps because the attempt to blame the operator had produced adverse publicity. It is unusual today for managers to blame the person whose triggering action is the last in a long series of missed opportunities to prevent an accident (see Chapter 38). Perhaps the decision to blame the operator was made by a lawyer who knew nothing about plant operation or human nature.

26.3   THE TEXAS CITY EXPLOSION

The explosion and fire in the BP refinery in Texas City in March 2005 was the most high-profile process industry accident for many years, comparable in the coverage it got with Flixborough and Bhopal. Fifteen people were killed, more than 170 were injured, and damage was extensive.

The accident was somewhat similar to the one described in Section 26.1. Both incidents occurred during startups; in both cases high levels in a column were followed by the overflow of liquid into a stack. In the first case, the wires leading to the level measuring instruments had not been reconnected after the turnaround. At Texas City, the level in a column was left on hand control at a time when it should have been on automatic control. As a result of errors elsewhere there was a higher flow than normal into the column. A high level alarm set at 72% of the level transmitter range operated and was acknowledged. But another alarm set at 78% did not operate. Once the level exceeded the transmitter's range, the level could not be read, though other instruments could have told the operator that the level was out of control. In both incidents there was spillage from the stacks. In the incident described in Section 26.1, the spillage was due to the brittle fracture of the flarestack and by good fortune the spillage did not ignite, but at Texas City the liquid overflowed and caught fire. The source of ignition was probably an idling diesel pickup truck.

If anyone at the Texas City refinery read the report on the Section 26.1 incident, first published in 1999 [1] and, in more detail, in 2003 [15], they would probably have thought that such a display of incompetence could not occur in their refinery. (When I was working in industry, I read many accident reports from other companies and thought that similar errors could not occur where I worked. Months or years later, I was often proved wrong. I fear that many people will say the same about the accidents described in this book.)

At first, the press reported that BP, like Exxon (Section 26.2), blamed the operators for not following procedures, but BP, unlike Exxon, soon made it clear that errors by operators and supervisors were just one of the causal factors.

One of the reasons for the large number of deaths and injuries at Texas City was the presence of temporary buildings for use by maintenance workers close to the site of the explosion.

In both incidents, there were an adequate number of people on the plant at the time, more than during normal operation, but some of them seemed to have lacked necessary knowledge and experience. At Texas City, the senior member of the operating team was absent for part of the time and everything was left to the operators.

The Texas City refinery was acquired by BP when they bought Amoco in 1999. In 2000, BP bought another company, Arco, and acquired seven refineries from the two purchases. All these refineries, except Texas City, were resold.

When a company expands by acquisition it takes time and effort to bring the culture, procedures, and safety standards of the acquired factories into line with those of the new owners. Everyone realizes that it takes time to bring equipment up to a new standard as the equipment has to be designed, built, and installed, but why, some people ask, cannot procedures be changed immediately? We do not live in a society in which that is possible, except perhaps after a major accident. Employees are not putty in the hands of managers. They are more like rubber: their customs and practices restrain and push back. BP had made some progress in changing the culture at Texas City, but in retrospect it was not enough. Many changes were, of course, made after the explosion, in both Texas City and elsewhere.

References 16 and 17 give more information on the Texas explosion. Reference 17 draws attention to what they called the organizational failures at Texas City. This phrase is a euphemism for senior management failures. Organizations have no minds of their own and it is one of the responsibilities of senior managers to recognize and put right weaknesses in organization as well as technology.

As already mentioned, The Texas City explosion received a great deal of publicity, mainly critical, from the press, government, and public. Much of it was directed at BP as a whole rather than just at the Texas City refinery. It was a sign of the times. Press, governments, and the public everywhere are much less tolerant of industrial failures than they used to be. There is sort of positive feedback. The press know that their readers like to read reports on disasters so they report them at length. This reinforces the public's concern and increases the press's willingness to satisfy it.

The principal organizational weaknesses emphasized by the press at Texas city and sometimes more widely, were the following:

- No director was responsible for BP's safety program. BP says this is incorrect.
- The lost-time accident rate was the major measure of safety, though the fact that it does not measure process safety or health problems has been well known for decades.
- Equipment was run until it failed. BP says this is incorrect.
- Safety paperwork was ticked as carried out whether or not it had been.
- Some employees did not report safety problems and therefore, the problems were not investigated. There was no formal procedure for storing and retrieving the lessons learned from past accidents.
- Surveys and audits were carried out, but the results were sometimes ignored.
- Changes to equipment, processes, and organization were not always assessed. (See Chapters 24 and 25.)

26.3.1   Another Industry: Similar Problems

As described in Section 25.3.2, the U.K. railways were reorganized in 1996/1997 and split into many independent companies, with unforeseen and unfortunate results. One of the problems was similar to those described in Sections 26.1 and 26.3:

> The recurring theme is the loss of artisan and supervisor expertise together with contractors, their sub-contractors and agency staff being thrown together in work squads on the day. What we have lost are the "black macs," the warrant officers of the old railway who made things happen through a combination of knowledge and authority based on hard won experience. [T]he recurring theme is the loss of artisan and supervisor expertise.
>
> Training has started again, but it is patchy. The broad railway experience is hard to replicate and, anyway, can only be gained in real time.[18]

Training can be quick or slow, but there are no shortcuts to experience.

26.4   OUTSOURCING

A marketing manager in a company that manufactured ethylene oxide foresaw a market for a derivative. The company operated mainly large continuous plants, whereas the production of the derivative required a batch plant. The derivative was needed quickly, and the company did not want to spend capital on a speculative venture. The manager therefore looked for a contract manufacturer who could make it for the company. He found one able to undertake the task and signed a contract without consulting any of his technical colleagues. The manufacturer was capable but unfortunately was located in a builtup area. When it was realized that ethylene oxide was being handled there, this gave rise to some concern even though the stock on site was small.

A few years later, the buildings in a considerable area around the plant were demolished as part of a slum clearance project. The regulators then refused permission for new ones to be built in their place. Before they could develop the site, the local authority had to pay the contract company to move its plant to a new location.

This incident occurred some years ago, before present-day regulations came into force. It probably could not happen today, but it is a warning that outsourcing of products or services is a change that should be systematically considered before it takes place.

26.5   MULTISKILLING AND DOWNSIZING

Multiskilling presents specific problems, illustrated by the Flixborough explosion. The site was without a mechanical engineer for several months, as the only one—the works engineer—had left and his successor had not arrived. Arrangement had been made for a senior engineer from one of the owner companies to be available when needed but the unqualified engineers who designed and built the temporary pipe that failed did not realize that these tasks were beyond their competence and did not see the need to consult him [7]. Similarly, in many plants there is now no longer an electrical engineer but the control engineer is responsible for electrical matters. An electrical engineer is available for consultation somewhere in the organization, but will the control engineer know when to consult him? *Will the control engineer know what he or she doesn't know?*

The same applies at lower levels. Will the process operator who now carries out simple craft jobs be able to spot faults that would be obvious to a trained craftsperson?

One of the underlying causes of the collapse of a mine tip at Aberfan in South Wales, which killed 144 people, most of them children, was similar. Responsibility for the siting, management, and inspection of tips was given to mechanical rather than civil engineers. The mechanical engineers were unaware that tips on sloping ground above streams can slide and have often done so [7].

On downsizing, according to one report [8],

> No one likes to talk about it, but having less experienced people working in increasingly sophisticated computer-generated manufacturing operations increases the risks of serious and costly mistakes. The investigation into an explosion in one US chemical plant [in 2001] found that the engineer in charge has only been out of college a year, and the operators in the control room at the time of the accident all had less than a year's experience in the unit. Not surprisingly, the explosion was attributed to operator error. . . . And even when errors are not caused by inexperience, diagnosing and fixing them often takes longer when veteran employees are no longer around to help.

Attributing the errors to the operators is, of course, superficial. The underlying cause is either downsizing or employment conditions that failed to retain employees.

26.6   HOW TO LOSE YOUR REPUTATION

The following is an extract from an unofficial report on an explosion that killed one man and caused extensive damage and loss of production. It describes the underlying causes of the explosion, not the technical details. The company had a good safety record, as measured by the lost-time accident rate, and a good reputation for safety, but all that glistens is not gold. When we are under stress we forget our ideals. The gold is found to be fool's gold (iron sulfide).

> The plant was an old one. Some of the units had been Hazoped. The unit where the explosion occurred was scheduled for a Hazop, but it was repeatedly postponed as the staff were having great trouble carrying out problems identified during the Hazops of other units, despite the fact that they knew the legal implications of not actioning things in a timely fashion, that is, they were under resourced.
>
> In addition, it was suspected that the Hazop would show that three large vessels, which had pipe manifolds and pumps right near them, would need to be relocated, at massive cost and process disruption. They knew it was going to be expensive, so they were repeatedly deferring the Hazop.
>
> There had been another incident several months earlier that had reduced the output from the plant. It was behind on production, so there was an almost obsessive focus on making up lost production.
>
> The managers had also been cutting back on maintenance funding, to the extent that people had stopped reporting things as they knew they would not be fixed. This led to a belief on the part of the workforce that the managers were trying to protect themselves, by shifting the blame onto the workforce by documenting everything, even if they knew that operators routinely deviated from documented procedures.

After the explosion the cause of the explosion was said to be failures by operators to follow instructions.

The members of the company engineering group were discouraged from visiting the site too often. It was seen as being more important to spend time in the head office being visible to the managers there. The engineering group looked upon the managers in the head office as their client rather than the operators and the plant.

26.7   ADMINISTRATIVE CONVENIENCE VERSUS GOOD SCIENCE

My final example, from James Lovelock's autobiography, *Homage to Gaia* [9], shows what happened when "administrative convenience ruled and good science and common sense came second," though the results were a decline in effectiveness rather than safety. He was working in a government-funded research center that employed chemists and biologists. It was amalgamated with another similar institution some distance away. To the administrators it seemed sensible to move all chemists to one site and all biologists to the other, as this would avoid the need to duplicate the services each group required. The administrators did not realize, and did not listen to those who did, the numerous research benefits gained from informal day-to-day contact between people from different disciplines. Both institutions declined. As we saw in Section 26.2 on the Longford incident, a similar loss of communication occurred when the professional engineers were moved from a plant to the company's head office.

26.8   THE CONTROL OF MANAGERIAL MODIFICATIONS

As with changes to plants and processes, changes to organization should be subjected to control by a system that covers the following points:

- Approval by competent people. Changes to plants and processes are normally authorized by professionally qualified staff. The level at which management changes are authorized should also be defined.
- A guide sheet or check list. Hazard and operability studies are widely used for examining proposed modifications to plants and processes before they are carried out. For minor modifications, several simpler systems are available [10]. Few similar systems have been described for the examination of modifications to organizations [11]. Some questions that might be asked by those who have to authorize them are suggested here.
- Each modification should be followed up to see if it has achieved the desired end and that there are no unforeseen problems or failures to maintain standards. Look out for near misses and for failures

of operators to respond before trips operate. Many people do not realize that the reliability of trips is fixed on the assumption that most deviations will be spotted by operators before trips operate. We would need more reliable trips if this were not the case.

- Employees at all levels must be convinced that the system is necessary or it will be ignored or carried out in a perfunctory manner. A good way of doing this is to describe or, better, discuss incidents such as those described in the foregoing and that occurred because there was no systematic examination of changes.

26.9   SOME POINTS A GUIDE SHEET SHOULD COVER

Define what is meant by a change. Exclude minor reallocations of tasks between people but do not exclude outsourcing, major reorganizations following mergers or downsizing, or high-level changes such as the transfer of responsibility for safety from the operations or engineering director to the human resources director. Accidents may be triggered by people but are best prevented by better engineering [12].

Nearly half of the companies that replied to a questionnaire on the management of change said that they included organizational change under this rubric [13]. However, they may not include the full range of such changes.

Some questions that should be asked include the following:

- How will we assess the effectiveness of the change over both the short and the long term?
- What will happen if the proposed change does not have the expected effect?
- Will informal contacts be affected (as at Longford)?
- What extra training will be needed, and how will its effectiveness be assessed?
- Following the change, will the number, knowledge, and experience of people be sufficient to handle abnormal situations? Consider past incidents in this way.
- If multiskilling is involved, will people who undertake additional tasks know when experts should be consulted? See Section 26.4 on multiskilling.

Except for minor changes, these questions should be discussed by a group, as in a hazard and operability study, rather than answered by an individual. *None* or *not a problem* should not be accepted as an answer unless backed up by the reasons for the answer. Any proposal for control of changes in an organization should be checked against a number of incidents, such as those described herein, to see if it could have prevented them.

26.10 AFTERTHOUGHTS

If you want to change the industry, you first have to change your company, and if you want to change your company, you first have to change your factory, and if you want to change your factory, you first have to change your plant, and if you want to change your plant, you first have to change your colleagues, and if you want to change your colleagues, you have to change yourself.

—Adapted from a saying by Israel Salanter (1810–1883)

"I have finished the report on risk management and concluded that there is no risk of effective management" [19].

References

1. Anon., A major incident during start-up. *Loss Prevention Bulletin*, Vol. 156, Dec. 1999, p. 3.
2. D. M. Dawson and J. B. Brooks, *The Esso Longford Gas Plant Explosion*, State of Victoria, Australia, 1999.
3. A. Hopkins, *Lessons from Longford*, CCH Australia, Sydney, Australia, 2000.
4. T. A. Kletz, *Learning from Accidents*, 3rd edition, Butterworth-Heinemann, Oxford, UK, and Woburn, MA, 2001, Chapter 24.
5. D. M. Boult, R. M. Pitblado, and G. D. Kenney, "Lessons learned from the explosion and fire at the Esso gas processing plant at Longford, Australia," *Proceedings of the AIChE 35th Annual Loss Prevention Symposium*, Apr. 23–25, 2001.
6. A. Hopkins, Lessons from Longford—The Trial, *Journal of Occupational Health and Safety—Australia and New Zealand*, Vol. 118, No. 6, Dec. 2002, pp. 1–72.
7. T. A. Kletz, *Learning from Accidents*, 3rd edition, Butterworth-Heinemann, Oxford, UK, and Woburn, MA, 2001, Chapters 8 and 13.
8. D. W. De Long, *Research Note: Uncovering the Hidden Costs of Lost Knowledge in Global Chemical Companies*, Accenture Institute for Strategic Change, Cambridge, MA, 2002, p. 2.
9. J. Loveock, *Homage to Gaia*, Oxford University Press, Oxford, UK, 2000, p. 301.
10. M. S. Mannan, *Lees' Loss Prevention in the Process Industries*, 3rd edition, Elsevier, Boston, MA, 2005, Chapter 21.
11. H. Conlin, "Assessing the safety of process operation staffing arrangements," *Hazards XVI—Analysing the Past, Planning the Future*, Symposium Series No. 148, Institution of Chemical Engineers, Rugby, UK, 2002, pp. 421–437.
12. J. Philley, Potential impacts to process safety management from mergers, downsizing, and re-engineering. *Process Safety Progress*, Vol. 21, No. 2, 2002, pp. 151–160.
13. N. Keren, H. H. West, and M. S. Mannan, Benchmarking MOC practices in the process industries. *Process Safety Progress*, Vol. 21, No. 2, 2002, pp. 103–112.
14. F. Crawley, Learning from past accidents—or do we? *Industrial Safety Management*, Vol. 7, No. 1, Jan. 2005, pp. 24–29.
15. T. A. Kletz, (2001). Problems and Opportunities Overlooked, Mary Kay O'Connor Process Safety Center Annual Symposium, College Station, Texas, 30–31 October. Also presented, with some changes, and with the title The Management of Organisational Change, *at Hazards XVII—Process Hazards: Fulfilling our Responsibilities*, Proceedings of a Conference held in Manchester, UK, on 25–27 March 2003, Symposium Series No. 149, Institution of Chemical Engineers, Rugby, UK, 2003, pp. 185–191. It was also published in the first edition of this book (Part B), in 2003.

16. Anon., BP, Fatal Accident Investigation Report: Isomerization Unit Explosion Final Report, Texas City, Texas, USA, Dec. 2005.
17. Anon., *The BP America Refinery Explosion, Texas City*, Chemical Safety Board, Washington, DC, 2007.
18. R. Ford, ORR fines NR £14million for Christmas overruns, *Modern Railways*, 2008, pp. 18–19.
19. Adapted from one of S. Adams's Dilbert cartoons.

This chapter is based in part on a paper presented at the Hazards XVII Conference held in Manchester, United Kingdom, in March 2003, and is included with the permission of the Institute of Chemical Engineers.

Changing Procedures
Instead of Designs

Rigid, repetitive behavior, resistance to change and a lack of imagination are common symptoms.

—Extract from an article on autism, *Daily Telegraph*
(London), April 29, 2002

When we join an organization, especially when we are young, we tend to follow, and are expected to follow, its ways of thinking and acting. It is usually only later, when we have gained experience, that we start to question these default actions. This chapter describes a common, but unfortunate, way many organizations react after an accident.

There are several different actions we can take after we have identified a hazard (as a result of an accident or in some other way), to prevent it from causing another accident or to mitigate the consequences if it does. Our first choice, whenever "reasonably practicable," should be to remove the hazard by inherently safer design. For example, can we use a safer material instead of a toxic or flammable one? Even if we cannot change the existing plant, we should note the change for possible use on the next plant. (*Reasonably practicable* is a U.K. legal phrase that recognizes the impracticability of removing every hazard and implies that the size of a risk should be compared with the cost of removing or reducing it, in money, time, and trouble. When there is a gross disproportion between them, it is not necessary to remove or reduce the risk [1].)

If we cannot remove the hazard, then our second choice should be to keep it under control by adding passive protective equipment—that is, equipment that does not have to be switched on or does not contain moving parts. The third choice is active protective equipment—that is, equipment switched on automatically; unfortunately, the equipment may be neglected and fail to work or it may be disarmed.

The fourth choice is reliance on actions by people, such as switching on protective equipment; unfortunately, the person concerned may fail to act, for a number of reasons, such as forgetfulness, ignorance, distraction, poor instructions, or, after an accident, because he or she has been injured. Changes to procedures instead of designs are often called *work arounds*.

Finally, we can use the techniques of behavioral science to improve the extent to which people follow procedures and accepted good practice. By listing this as the last resort, I do not intend to diminish its value. Safety by design should always be our aim but it is often impossible, and experience shows that behavioral methods can create substantial improvement in the everyday types of accident that make up most of the lost-time and minor accident rates. However, the technique has little effect on process safety. Behavioral methods should not be used as an alternative to the improvement of plant design or methods of working when these are reasonably practicable.

To clarify various ways of preventing incidents, let us consider a simple but common cause of injury and even death, particularly in the home—falls on the stairs.

The inherently safer solution is to avoid the use of stairs by building a single-story building or using ramps instead of stairs.

If that is not reasonably practicable, a passive solution is to install intermediate landings so that people cannot fall very far or to avoid types of stairs, such as spiral staircases, that make falls more likely. An active solution is to install an elevator. Like most active solutions, it is expensive and involves complex equipment that is likely to fail, expensive to maintain, and easy to neglect. This solution is not reasonably practicable for most homes.

The procedural solution is to instruct people always to use the handrails, never to run on the stairs, to keep them free from junk, and so on. This can be backed up by behavioral techniques: specially trained fellow workers (or parents in the home) look out for people who behave unsafely and tactfully draw their attention to the action.

Similarly, if someone has fallen into a hole in the road, as well as asking why it was not fenced or if someone removed the fence or if the lighting should be improved, we should ask if there is a reasonably practicable alternative to digging holes in the road. Could we drill a route for pipes or cables under the road or install culverts for future use when roads are laid out? Must we run pipes and cables under the road instead of overground?

In some companies, the default action after an accident is to start at the wrong end of the list of alternatives and recommend a change in procedures or better observation of procedures, often without asking why the procedures were not followed. Were they, for example, too complex or unclear, or have supervisors and managers turned a blind eye in the past? Changing procedures is, of course, usually quicker, cheaper, and easier than changing the design, but it is less effective. This chapter describes some accidents in which changes in design would have been less expensive but nevertheless only changes in procedures were made. The first two accidents could easily have killed someone; the third is trivial, but they all illustrate the same point. There are other examples in Sections 23.4 and 35.2.

Designers today often consider inherently safer options, but the authors of incident reports do so less often. The very simplicity of the idea seems to make it hard for some people to grasp it. Perhaps they are expecting something more complex or—and this is perhaps more likely—it goes against the widely accepted belief that accidents are someone's fault and the job of the investigation is to find out whose. Having identified the culprit, we are less likely to blame him or her than in the past; we realize that he or she may not have been adequately trained or instructed, and that everyone makes occasional slips, but nevertheless his or her action or inaction caused the incident. In some companies, they blame a piece of equipment rather than the people who design or maintain it. It is hard for some people to accept that the incident is the result of a widespread and generally accepted practice in design and operations.

27.1 MISLEADING VALVE LAYOUTS

27.1.1

The fine adjustment valve A in Figure 27-1 had to be changed. The operator closed the valve below it. To complete the isolation, he intended to close the valve on the other side of the room in the pipe leading to valve A. He overlooked the double bends overhead and closed valve B, the one opposite valve A. Both of the valves that were closed were the third from the ends of their rows. Note that the bends in the overhead pipes are in the horizontal plane. When the topwork of valve A was unbolted, the pressure of the gas in the line caused the topwork to fly off and hit the wall, fortunately missing the mechanic who had unbolted it.

The report on the incident recommended various changes in the instructions to make the duties of people who prepare equipment for maintenance clearer than they were. They were told to trace lines to make sure that the correct isolations have been made.

Color coding of the pipes or valves would have been much more effective but was not considered. The default action of many of the people

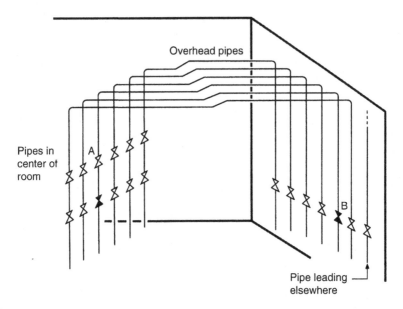

FIGURE 27-1 To change valve A, the operator closed valve B and the valve below valve A. Both closed valves are shown in black. From reference 5. *(Reprinted with the permission of the Institution of Chemical Engineers.)*

in the company was to look first for changes in procedures, to consider changes in design only when changes in procedure were not possible, and to consider ways of removing the hazard rather than controlling it only as the last resort.

The ideal solution, of course, would be to rearrange the pipework so that valves in the same line were opposite each other. To do so in the existing plant would be impracticable but the point should be noted for the future. After a similar incident elsewhere, a design engineer once said to me that it was difficult enough to get all the pipework into the space available without having to worry about such fine points as the relative positions of valves. This may be so, but putting valves in unexpected positions leads to errors.

The changes made after the accident were not even the most effective procedural ones. The incident could have been given widespread publicity, not just immediately afterward but regularly in the future, and made part of the training of people authorized to issue permits-to-work.

27.1.2

Figure 27-2 shows a similar situation. To save cost, three waste heat boilers shared a common steam drum. Each boiler had to be taken off

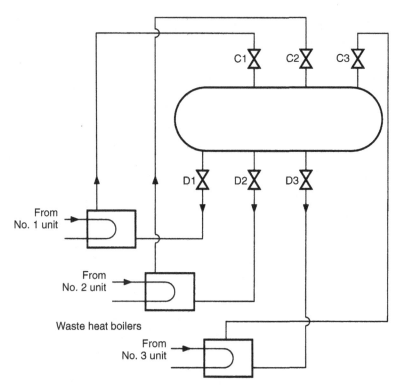

FIGURE 27-2   Note the positions of isolation valves on the common steam drum. *(Reprinted with the permission of the Institution of Chemical Engineers.)*

line from time to time for cleaning. On two occasions, the wrong valve was closed (D3 instead of D2) and an online boiler was starved of water and overheated. The chance of an error was increased by the lack of labeling and the arrangement of the valves—D3 was below C2. On the first occasion the damage was serious. High temperature alarms were then installed on the boilers. On the second occasion they prevented serious damage but some tubes still had to be changed. A series of interlocks were then installed so that a unit had to be shut down before a key could be removed; this key was needed to isolate the corresponding valves on the steam drum.

A better design, used on later plants, is to have a separate steam drum for each waste heat boiler (or group of boilers if several can be taken off line together). There is then no need for valves between the boiler and the steam drum. This is more expensive but simpler and free from opportunities for error. Note that we do not begrudge spending money on complexity but are reluctant to spend it on simplicity.

It is obviously impracticable to change the layout of the existing valves, but perhaps color coding would have been sufficient to prevent further errors. It would have been simpler and cheaper than the mechanical interlocks.

27.2   SIMPLE REDESIGN OVERLOOKED

A bundle of electric cables was supported by cable hangers. The hooks on the ends of the cable hangers were curved over the top of a metal strip (Figure 27-3, *top*). The electric cables had to be lowered to the ground to provide access to whatever lay behind them and then replaced. They were put back as shown in the second part of Figure 27-3. This increased the load on the upper hooks. One failed, thereby increasing the load on the adjacent ones and then they also failed. Altogether, a 60-m (200-ft) length of cables fell down [2].

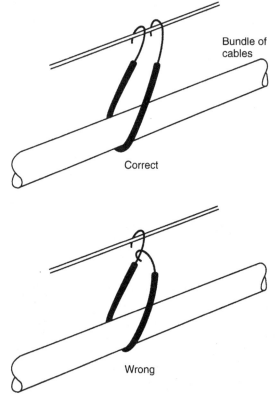

FIGURE 27-3   Different ways of supporting a bundle of cables. *(Reprinted with the permission of the Institution of Chemical Engineers.)*

Many people would fail to see this hazard. Training is impracticable if, as is probably the case, many years will pass before the job has to be done again. The best solution is to use cable hangers strong enough to carry the weight even if they are used incorrectly.

27.3 UNIMAGINATIVE THINKING

Washbasins filled with water were installed at a plant so that anyone splashed with a corrosive chemical could wash it off immediately. The basins were covered to keep the water clean, but people used the covers as tables (Figure 27-4a). Figure 27-4b shows the action taken, and Figure 27-4c shows a better solution.

Perhaps there is something wrong with our educational system or company culture when educated and professionally trained people take the action shown in Figure 27-4b.

(a)

FIGURE 27-4 (a) The cover over the washbasin was used as a table. (b) The solution. (c) A better solution.

(b)

(c)

FIGURE 27-4 Continued

27.4   JUST TELLING PEOPLE TO FOLLOW THE RULES

A tank containing high-level radioactive liquid was fitted with instruments for measuring density and level. They were purged with steam at intervals. Before opening the steam valve, the operator was instructed to check that there was steam in the line by measuring the temperature of a steam trap and checking that it was over 93°C (200°F). However, he merely felt the trap and finding it was hot, he opened the steam valve. Unknown to him, the steam line had been isolated 16 hours beforehand. (Presumably conduction from beyond the isolation valve kept the trap hot.) As the steam cooled, it developed a vacuum and this sucked the radioactive liquid into the steam line. Radioactive alarms sounded, and fortunately no one received a significant dose.

The report [3] drew attention to failures to follow procedures: The people who drained and isolated the steam line did not inform those responsible for purging the instruments, and the operator who was asked to carry out the purging was not adequately trained because he had never done the job before but only watched other people do it.

The report recommended that managers should stress the proper use of procedures, that before carrying out a task operators should stop, think about the task, the expected response and the actions required if it failed to occur, and so on. (Senior managers do not always do this [see Section 26.3 and Chapter 29]. We are all likely to make the same errors.) There was no suggestion that the procedures could be improved, for example, by fitting a warning notice on lines that are out of use, or that the design could be improved. It is surprising that there was no check valve in the steam line. They are not 100% reliable but can greatly reduce the size of any back flow. Check valves with moving parts would be difficult to maintain in a radioactive environment, but fluidic ones would be suitable. Another possibility is the use of a catchpot to catch any liquid that does flow into the steam line.

27.5   DON'T ASSEMBLE IT INCORRECTLY

When an accident occurs because construction or maintenance workers assemble equipment incorrectly, the default action of many managers is to tell them to take more care in the future and to check that it has been assembled correctly. Or perhaps they provide training on the correct method of assembly. They do not realize that equipment should be designed so that it cannot be assembled incorrectly. Even when it is impractical to change the design of existing equipment, we should at least ask the design organization to use a better design in the future.

During rough weather, water entered the engine room of a fishing vessel through the intake of the ventilation fans. It fell onto the switchboard, and a short circuit set it afire; the fire was soon extinguished, but all power was lost. The crew was unable manually to fully close the doors through which the nets were pulled onboard, and water entered through these doors. The ship had to ask for help and was towed back to port.

Why did water enter through the ventilation intake? The louvers in it had been installed incorrectly so that they directed spray and rain into the engine room rather than away from it. (The report said that they had been installed upside down, but the authors must have meant back-to-front.) See Figure 27-5.

The report's first recommendation [4] was that louvers should be checked to make sure they are fitted correctly. It did not suggest that they should be designed so that they could not be fitted incorrectly, so that it was obvious if they were or so that the inside and outside, top and bottom, were clearly labeled. However, the report did recommend

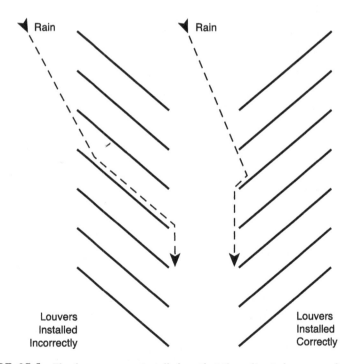

FIGURE 27-5 The louvers were installed so that they directed spray and rainwater through them.

that switchboards should be covered to prevent water entering from above.

27.6   TIGHTEN CORRECTLY OR REMOVE THE NEED

A hose was fastened to its connector with a type of clip used for radiator hoses in cars (known as Jubilee clips in the United Kingdom). The connection leaked. The recommendation in the report on the incident was "Check tightness of Jubilee clips during maintenance." These clips are not robust enough for industrial use, and a better recommendation would have been to replace them with bolted clips.

Similarly, a steel plate fell from a clamp while being lifted because the bolt holding it in position was not tightened sufficiently. The incident was classified as a human failing, and the operator was told to be more careful in the future. It would have been better to use a type of clamp that is not dependent for correct operation on someone tightening it to the full extent [5].

27.7   SHOULD IMPROVEMENTS TO PROCEDURES EVER BE THE FIRST CHOICE?

Improving procedures is often the only possible choice, but are there times when it is more effective than changing designs? This may be the case with road accidents. Up to the late 1970s, the United States had the lowest fatal accident rate per thousand vehicles in the world. The figure has continued to fall and is now about half the rate it was then, a considerable achievement. But other countries have done even better, and the United States is now 13th in the road safety league, behind the United Kingdom, Sweden, the Netherlands, Germany, Canada, Australia, Japan, and several other countries, but not France. The better performance of these countries is not due to better vehicle design, as they all use similar vehicles and the United States tends to use heavier and, therefore, safer cars. Nor is there a significant difference in the design of roads. Leonard Evans [6] suggests that significant differences in the countries with lower accident rates are a more restrictive alcohol policy (which is enforced more rigorously), stricter enforcement of seat belt laws, and prohibition of the sale and use of radar detectors. If so, further improvement in the United States depends on better enforcement of procedures.

If you work in the process industries, the most dangerous task your employer asks you to perform may be to drive between sites.

References

1. Health and Safety Executive, *Reducing Risks, Protecting People*, HSE Books, Sudbury, UK, 2001.
2. Anon, "Worker injured by falling power and data cables." *Operating Experience Weekly Summary, No. 99-08*, Office of Nuclear and Facility Safety, U.S. Dept. of Energy, Washington, DC, 1991. p. 1.
3. Anon, "Radioactive tank contents contaminate steam line." *Operating Experience Weekly Summary, No. 99-34*, Office of Nuclear and Facility Safety, U.S. Dept. of Energy, Washington, DC, 1999. pp. 11–13.
4. Anon, "Upside down louvres." *Safety Digest—Lessons from Marine Accident Reports, No. 2/2001*, Marine Accident Investigation Branch of the UK Department of Transport, Local Government and the Regions, London, 2001. p. 40.
5. T. A. Kletz, *An Engineer's View of Human Error*, 4th edition, Institution of Chemical Engineers, Rugby, UK, 2001, p. 43.
6. L. Evans, Traffic crashes. *American Scientist*, Vol. 90, No. 3, 2002, pp. 244–253.

28

Materials of Construction (Including Insulation)

Lay not up for yourself treasure upon earth: where the rust and moth doth corrupt.

—Book of Common Prayer

At school we all knew people who were often present when there was any trouble but they were rarely identified as the major culprit. In preparing the index for the 4th edition of my book, *What Went Wrong—Case Histories of Process Plant Disasters* (now revised as Part A of this book), I was surprised to find that certain words appeared, often as secondary or incidental causes, much more often than I expected. I expected to find (and did find) frequent references to fires, explosion, pumps, tanks, modifications, and maintenance, but I was surprised how many references there were to rust, insulation, and brittle failure. This chapter describes these incidents and some others. There are more details in Part A (and elsewhere when another reference is quoted).

28.1 RUST

28.1.1 Rust Formation Uses up Oxygen

A tank was boxed up with some water inside. Rust formation used up oxygen, and three men who entered the tank were overcome; one of them died. No tests were carried out before the men were allowed to enter the tank as it had contained *only water*. In a similar incident, three men were sent to inspect the ballast tanks on a barge at an isolated wharf. The first man to enter collapsed, and a second man who tried to rescue him also collapsed; one of them died. In a third incident, rust formation caused a tank to be sucked in. Rusting is usually slow but can be rapid under certain conditions; it increases rapidly when the humidity is high. (Chapters 11 and 24 describe other accidents in confined spaces.)

28.1.2   Rust-Jacking

To avoid welding in a plant that handled flammable liquids, an extension was bolted onto a pipebridge. The old and new parts were painted, but water penetrated the gap between the bolted surfaces and rusting occurred. Rust has seven times the volume of the iron from which it is formed, and it forced the two surfaces apart. Some bolts failed and a steam main fractured. Fortunately, the pipes carrying flammable gases did not fail [1]. Similarly, corrosion of reinforcing bars in concrete can cause the concrete to crack and break away.

28.1.3   Liquid Can Be Trapped Behind Rust

The roof on an old gasoline tank had to be repaired by welding. The tank was emptied and steamed, and tests showed that no flammable vapor was present. However, the tank had been made by welding overlapping plates together along the outside edge only, a method no longer used. Some gasoline was trapped by rust in the space between the overlapping plates. Welding vaporized it and ignited the vapor; it blew out the molten weld and singed the welder's hair.

A similar accident had worse results. Heavy oil trapped between overlapping plates was vaporized and exploded. The roof of the tank was lifted. One man was killed and another badly burned.

28.1.4   Rust as Catalyst

Rust can initiate the polymerization of ethylene oxide at ambient temperature. Once the temperature reaches 100°C (212°F), the reaction becomes self-sustaining and may lead to explosive decomposition. An explosion in an ethylene oxide distillation column may have been started by rust, which had accumulated in a dead-end space. Rust on the inside surface of a tank can promote the polymerization of other substances. Even if the liquid has been treated with an inhibitor, this will not prevent polymerization of vapor, which can condense on the walls or roof [2].

A vigorous reaction between chlorine and a steel vaporizer, described as burning, led to a leak and the loss of a ton of chlorine. The steam was supplied to the vaporizer at a gauge pressure of ≈1 bar (15 psi) and a temperature of ≈100°C (212°F), so it was not nearly hot enough to ignite the chlorine-iron reaction, which starts at ≈200 to 250 degrees C (390 to 480 degrees F). The bottom of the vaporizer was found to be packed with scale containing 80% iron oxide. It is possible that this material had catalyzed the reaction. However, it is more likely that the culprit was traces of methanol, which had been used to clean the vaporizer and had not been thoroughly purged afterward. Methanol, like many other organic compounds, can react with chlorine and generate enough heat to start the chlorine-iron reaction [3].

28.1.5   Rust Jams a Valve

A chlorine cylinder was left standing, connected to a regulator, for eight months. The valve rusted and seemed to be fully closed, though it was not. When someone disconnected the regulator, gas spurted into his face. Four people were hospitalized.

28.1.6   Thermite Reactions

If rusty steel is covered by aluminum paint (or smeared with aluminum in any other way) and then hit by a hard object, such as a hammer, a thermite reaction can occur: the iron oxide reacts with the aluminum to form aluminum oxide and iron. A temperature of 3,000°C (5,400°F) can be reached, and this can ignite any flammable gas, vapor, or dust that is present.

A thermite reaction can also occur between rust and any other metal that has a greater affinity for oxygen. A fractionation column was packed with bundles of 0.1-mm-thick corrugated titanium sheets that had become coated with a layer of rust only 25 microns (μm) thick. During a shutdown, it was decided to check that the correct construction materials had been used. This was done by passing a grinding wheel lightly and quickly across the surface of components and noting the characteristics of the sparks. They ignited the titanium and set off a thermite reaction. The fire spread rapidly, causing extensive damage [4] (see also Section 32.1).

28.1.7   Rust Formation Weakens Metal

Some handheld fire extinguishers are fitted with rubber or plastic feet to protect the bases. If water enters the space between the foot and the extinguisher, it can cause rusting. In at least two cases, extinguishers have ruptured while in use and killed the person who was holding them. Manufacturers advise users to remove the plastic feet and check for corrosion yearly, but it would be better not to use extinguishers of this type [5].

The 7-in. diameter exit pipe from the superheater of a steam boiler was threaded and screwed into a flange. The gauge pressure was 17 bar (250 psi). The joint leaked, causing substantial damage to both pipework and the building roof. The investigation showed that seepage of steam along the threads of the screwed joint had caused corrosion and that all the gaps in the threads had contained rust. Many of the grooves were full of it. There was evidence that the two threads were never tightly engaged and that there were gaps between them from the start. The report [6] recommended that screwed joints should not be used on large pipes, say, those over a 2 in. diameter. However, many companies do not allow them at all except for low-pressure cold water lines and for small bore instrument lines after the first isolation valve and then only for nonhazardous materials. The report also suggested that existing screwed joints should be opened for inspection every few years. (See also Section 34.5.)

28.1.8   Old Plants and Modern Standards

The boiler described in the previous incident was a very old one and raises the question, how far should we go in bringing old plants up to modern standards? Some changes are easy, for example, installing gas detectors for the detection of leaks. Some are impossible, such as increasing the spacing between different parts of the unit. In between there are changes that are possible but expensive, such as replacing pipework by grades of steel that are less likely to corrode or can withstand lower temperatures. Some companies carry out a *fitness for purpose* study of such suspect equipment, replacing some, radiographing or stress-relieving some, fitting extra measurements, alarms, or trips on some, or training operators to pay particular attention to the operating conditions [7]. If a *fitness for purpose* study had been carried out on the boiler after the incident, the conclusion would, I think, have been to replace the screwed joints by flanged ones.

28.1.9   Stainless Steel Can Rust

Stainless steel can rust if it is exposed to particularly aggressive conditions, either physically (for example, by cleaning with steel wool or wire brushes) or chemically [8].

In all these cases, rust was not the major culprit. Proper procedures should be followed before vessels are entered, equipment should be designed without pockets in which liquids or rust can collect, coated or stainless steel should be used if rust can affect materials in contact with it, flammable mixtures should not be tolerated except under rigidly defined conditions where the risk of ignition is accepted, cylinder valves should not be left open for months, and people should be made aware of these hazards and of the properties of rust. (See also Sections 23.6.2 and 34.1.)

28.2 INSULATION

Insulation, like rust, is often mentioned in incident reports, though rarely as the major culprit. It has many benefits, but we should be aware of its drawbacks.

28.2.1 Insulation Hides What Is Beneath It

On several occasions, small diameter branches have been covered by insulation, overlooked, and not isolated before maintenance. Short tags on blinds (slip-plates) may not be noticed on insulated lines. In one incident, a check (nonreturn) valve was hidden by insulation and a new branch was installed on the wrong side of it. As a result, a relief valve was bypassed and equipment was overpressured and it ruptured.

Most important of all, insulation can hide corrosion. The commonest cause of corrosion beneath insulation is ingress of water, especially water contaminated with acids or with chlorides; the latter can cause stress-corrosion cracking of stainless steel. Sections of insulation should be removed periodically for inspection of the metal below, making sure that no gaps are left when it is restored. During inspection, special attention should be paid to places where corrosion is likely, such as insulation supports and stiffening rings, which can trap water, gaps in the insulation around nozzles, and insulation around flanges and valves. Nonabsorptive insulation should be used when possible. Make sure insulation is not left lying around where it can get wet before installation. Remember that while warm equipment may dry out wet insulation, the rate of corrosion doubles for every 15 to 20 degrees C (27 to 36 degrees F) rise in temperature. Reference 9 reviews the subject. (See also Section 23.2.)

Corrosion and a leak of propylene took place beneath insulation on equipment that had been in use for 15 years. The corrosion had occurred only on those parts of the unit that operated between about 0 to 5 degrees C (32 to 40 degrees F). These parts were frequently wetted by condensation from the atmosphere. Some of the equipment was replaced with stainless steel and the rest was inspected more frequently [10].

Supports can corrode as well as equipment. The corroded legs of a 2,000 m^3 (530,000 gal) liquefied petroleum gas (LPG) sphere collapsed during a hydrotest, when it was 80% full of water. One man was killed and another seriously injured. It was then found that the following had occurred:

- Water had penetrated the gap between the concrete insulation and the legs, as the cap over the concrete was inadequate.
- There were also vertical cracks in the concrete.
- Repairs to the concrete had not adhered to the old concrete, leaving further gaps.

- The deluge system had been tested with seawater.
- Inspection was inadequate.

Underlying all these problems, according to the report, was a poor maintenance system, poor management, and ignorance of what could occur and what precautions should be taken.

The sand foundation below a fuel oil tank subsided. This was not noticed as the insulation came right down to the ground. Water collected in the space that was left and caused corrosion. The floor of the tank collapsed, and 30,000 tons of hot oil came out. The bottom 0.2 m (8 in.) of the tank walls should have been left free of insulation so that they could be inspected easily.

28.2.2 Wet Insulation Is Inefficient

If insulation is allowed to get wet, it not only encourages corrosion but also loses much of its efficiency: 4% moisture by volume can reduce the thermal efficiency by 70% as water has a thermal conductivity of up to $20\times$ greater than most insulation materials [11].

28.2.3 Spillages on Insulation Can Degrade and Ignite

When organic liquids are spilled on hot insulation, they can degrade and their auto-ignition temperatures can fall by 100 to 200 degrees C (180 to 360 degrees F). In one incident, ethylene oxide leaked through a hairline crack in a weld on a fractionation column onto insulation and reacted with moisture to form polyethylene glycols. When the metal covering on the insulation was removed in order to gain access to an instrument, air leaked in and the polyethylene glycols ignited. The fire heated a pipe containing ethylene oxide. It decomposed explosively, and the explosion traveled into the fractionation column, which was destroyed. A leak of ethylene oxide from a flange may have caused a similar incident on another column.

On another unit, a spillage onto insulation was the result of filling the heat transfer section with oil until it overflowed. A month later, the oil caught fire; the flames caused a leak of gas, which exploded, causing further damage. Solution: contaminated insulation should be removed promptly.

28.2.4 Some Insulation Is Flammable

Large tanks are sometimes insulated with plastic foam, which is lighter and cheaper than nonflammable insulating materials. However, the foams can, and do, catch fire.

28.2.5 Metal Coatings over Insulation Should Be Grounded

When a glass distillation column cracked, water was sprayed onto it to disperse the leak of flammable vapor. The water droplets were charged and the charge collected on the metal insulation cover, which was not grounded. A spark was seen to jump from the insulation cover to the water line, but fortunately it did not ignite the leak.

28.2.6 Insulation Can Fall Off

If 10% of thermal insulation falls off (or is removed for maintenance or inspection and not replaced), then we lose 10% of its effect. However, if 10% of fire insulation is missing, we lose all the effect as the bare metal will overheat and fail. Missing insulation should be replaced promptly.

28.3   BRITTLE FAILURE

This is a third subject often mentioned in accident reports. A famous case is discussed in Section 26.2.

28.3.1 Temperature Too Low as a Result of Adiabatic Cooling

Most materials become brittle if they are cooled below the brittle-ductile transition temperature. This is the commonest cause of brittle failure and often occurs when the pressure on a liquefied gas is reduced. Vessels, heat exchangers, and road tankers containing liquefied petroleum gas have been cooled below their transition temperatures by deliberate or accidental venting and have then failed when subjected to a sudden shock. We should use construction materials that can withstand foreseeable reductions in temperature outside normal operating conditions.

28.3.2 Temperature Too Low as a Result of Adding Cold Fluids

A vessel broke into 20 pieces when it was filled with cold nitrogen gas from a liquid nitrogen vaporizer. Vehicle tires have exploded in contact with liquid nitrogen. A large pressure vessel failed during a pressure test at the manufacturers as the water used was too cold. A piece weighing 2 tons went through the workshop wall and traveled 15 m (50 ft).

28.3.3   Manufacturing Flaws

As a result of a flaw during manufacture 40 years earlier, a tank containing 15,000 tons of diesel oil opened up like a zipper. For most of those 40 years, the tank had been used to store warm fuel oil and the high temperature prevented brittle failure. The flaw could have been spotted if the tank had been adequately radiographed. A liquid carbon dioxide vessel failed catastrophically as result of poor quality welding. The triggering event was the failure of a heater that was intended to prevent evaporative cooling.

28.3.4   Use of Unsuitable Materials

Cast iron is brittle and cannot withstand sudden shocks. A 6-in. cast iron steam valve failed spectacularly when subjected to water hammer.

A 2,000 m^3 (530,000 gal) propane tank opened up like a zipper, as it was not made from a crack-arresting material. The designers had assumed incorrectly that cracks could be prevented and, unfortunately, when one occurred it spread rapidly. The cause of the crack may have been attack of a weld by bacteria in the seawater used for pressure testing followed by a poor repair. It is difficult to be certain that a crack will *never* occur. It is good practice to prevent the spreading of any that do occur by using grades of steel that can withstand the temperatures reached during normal and abnormal operation.

28.4   WRONG MATERIALS OF CONSTRUCTION

28.4.1   Wrong Materials of Construction and Contaminants

A 316L grade of stainless steel was specified for a fractionation column and its connecting pipework. Seven years later, the bottom section was replaced with a taller one so that the column could be used for a different purpose. A few months later, leaks occurred in some of the old connecting pipework. It was then found that it had been made from 304L steel instead of 316L. This did not matter in its previous use, but an acidic byproduct formed and attacked the 304L steel as a result of the newer use.

During the 1970s, many incidents occurred because the wrong grade of steel was supplied (see Section 16.1). Many companies introduced materials identification programs: every piece of steel entering the site—pipes, flanges, and welding rod as well as complete items of equipment—was tested to check that it was made of the material specified. Many of these programs were abandoned when suppliers were able to show that the quality of their procedures had met approved standards. I question if this is wise, as many suppliers do not seem to understand that supplying a "similar" grade of steel instead of the one specified can have serious results (see also Section 31.7).

In addition, in the 304L/316L stainless steel incident just described, the new shell also leaked and some tray supports gave way. This was traced to chlorine in the steam used for cleaning the column between runs. It is well known that chlorine can cause stress corrosion cracking of stainless steel, but we do not expect to find chlorine in steam. It had picked up the chlorine from deposits in the base of the column and carried it into the column. There were, unknown to everyone, traces of chlorine ions in the feedstock [12].

Sometimes, of course, the wrong construction material is specified. When some vessels made from 316L stainless steel corroded in contact with strong acids, they were replaced in 317L, a grade that is usually more resistant. However, at the same time the temperature was raised from 90 to 110 degrees C (195 to 230 degrees F). Corrosion still occurred. The plant chemist noticed that the amount of iron, nickel, and chromium in the product had increased, but this was not recognized as evidence of corrosion [12].

Unforeseen corrosion in heat exchangers in a sugar refinery was traced to a combination of a less-than-ideal material of construction, presence of chlorine in an additive, and copper carried over from an earlier part of the process. Metallic contamination in the food industry is a well-recognized problem and must be closely monitored. The corrosion took place when the exchangers were being cleaned with acids to remove deposits, and although the plant had been Hazoped it was not clear that the cleaning process had been included in the study. It is, in effect, a different process carried out using the same equipment and should have been the subject of a separate study [13].

28.4.2 A Hasty Reaction When the Plant Leaked

Nitric acid leaked through a plug on a ring main, which normally operated at a gauge pressure of 7 bar (100 psi). It caused some corrosion of the equipment on which it dripped. The plug was one of several fitted by the design and build contractor in case it was found necessary to install additional instruments, though this is not certain as no one who worked on the plant was involved during design. It seems that the company was never consulted about the need for screwed plugs. None of them had been unscrewed during the eight years that had elapsed since the plant was built. The plug that leaked was made from mild steel though the pipework was stainless steel, and the plug was incorrectly seated. A polytetrafluoroethylene (PTFE) wrapping around the threads prevented an earlier leak.

It is sometimes necessary to install temporary plugs to aid pressure testing, to assist draining, or, as in this case, to make it easier to install additional instruments. Their positions should be registered, and if it becomes clear that they are not needed, they should be welded up. However, do not seal weld over an ordinary screwed plug as if the thread corrodes the full pressure inside the equipment is applied to the seal. Use a specially designed plug with a full strength weld. Also, do not use sealing compounds in joints that are going to be welded, as the welding will vaporize the sealing compound and make the weld porous. If sealing compound has been used, the joint should be cleaned before welding. (For other incidents, see Section 34.5 and Sections 7.1 through 7.5.)

When the leak was discovered, the process supervisor immediately decided to drain the ring main via several different drain points so that it was emptied as soon as possible. He did not open the vent at the highest point in the ring main. The opening of several drain points produced complex pressure and vacuum transients in the ring main and unpredictable movements of slugs of liquid. As a result, a column of liquid 3 m (10 ft) high was discharged from a vent point. The plug was never found, so it may have been sucked into the ring main by a transient vacuum. In total, 100 liters (25 gal) of nitric acid were spilled.

The supervisor's action was understandable, but once the pressure in the ring main was reduced, a short delay would not have mattered. Twenty minutes spent discussing possible methods would have been time well spent.

There may also have been some air pockets in the ring main, and a revised filling procedure was adopted when the ring main was refilled. Nevertheless, another pressure discharge occurred from an open vent during refilling. This shows how difficult it is to estimate the pressures developed when complex pipework—there were many changes of elevation—is being filled or emptied. It is equally difficult to avoid such features during design. There was no standing instruction on how to drain the ring main. Complex systems have complex problems, and their causes are much more difficult to understand than outsiders realize.

The investigation disclosed that although nitric acid was used infrequently, nevertheless the ring main was kept up to pressure at all times. It need not have been. Was a ring main really needed?

According to the company report, the most important lesson was not to rush into action, but I think a more important one is the need to plan ahead for jobs that will have to be done sooner or later and not leave the people on the job to improvise when the time comes.

Another lesson is that we should question the need for every plug and look out for plugs that design or construction staff members insert for their own convenience. Those added by construction staff are usually not shown on any drawing.

28.5 CORROSION SENDS A COLUMN INTO ORBIT

Corrosion occurred in an absorber tower, 18 m (60 ft) tall and 2.6 m (8.5 ft) in diameter, in which liquid monoethanolamine removed hydrogen sulfide from gaseous propane and butane. After four years of service, the base of the column was replaced without any postwelding heat treatment. Two years later, a Monel liner was fitted to reduce corrosion but it did not cover the repair weld. After another eight years, a circumferential crack formed. In places it extended nine-tenths of the way through the 1-in. thick wall. Once it broke through, it grew rapidly and the upper part of the column landed over a kilometer away (see Figure 28-1). The escaping gas was ignited, perhaps by a welder's torch, and exploded. Gasoline tanks were damaged and the contents ignited; the flames impinged on a liquefied petroleum gas tank, which ruptured, producing a boiling liquid expanding vapor explosion (BLEVE). Seventeen people were killed and damage was extensive.

The investigation showed that the welding of the new bottom section, without any postwelding heat treatment, had produced a hard

FIGURE 28-1 The result of an absorber failure. From reference 18. *(Reprinted with the permission of Gulf Professional Publishing.)*

microstructure that was susceptible to hydrogen attack and brittle failure [14–16].

Unfortunately, as so often happens, the published reports give no indication of the underlying reasons for the managerial failings. Did the company have any material scientists on its staff? Did it hire an inexperienced contractor and leave it to him or her? Did the senior managers believe that every welder is capable of welding everything? The incident is a warning to companies who think that knowledge and experience are inessential luxuries, that it is okay to be a naïve client and leave everything to a contractor. Elsewhere [17] I have described many accidents that occurred, from the nineteenth century to the present day, because companies placed too much trust in contractors.

Stress corrosion cracking is common in amine gas absorption columns. Reference 18 recommends polymer coating of construction materials.

28.6   UNEXPECTED CORROSION

Corrosion of a pipe led to a leak of >2 tons of a mixture of gaseous chlorine, hydrochloric acid, and hydrogen fluoride. As soon as the leak was detected, the affected section of the plant was isolated from the rest by remotely operated emergency isolation valves, but there was no valve of any sort between the leaking pipe and a vessel. The leak was stopped after two hours when a fitter, wearing full protective clothing and an air mask and standing on a ladder, succeeded in clamping a rubber sheet over the leak.

The plant was seven years old. The process materials were known to be corrosive, the most suitable materials of construction were used, and a life of five years was expected. The vessels were inspected regularly and some had been replaced, but the pipe that failed had never been inspected or renewed. It seems odd to inspect vessels regularly but never inspect the pipes connected to them. What do you do?

The acid gases caused considerable damage to the electronic control equipment. The cost of replacing them and the affected pipework was too great and the plant was demolished.

The protective clothing used during the emergency was rarely used, and much of it was found to be in poor condition and unusable. All emergency equipment should be scheduled for regular inspection.

28.7   ANOTHER FAILURE TO INSPECT PIPEWORK

Many companies that inspect pipes carrying hazardous materials do not inspect those that carry nonhazardous ones, but that does not mean

FIGURE 28-2   This low-pressure steam main was not scheduled for regular inspection and the damage was undetected for months, perhaps years.

that they never fail. Figure 28-2 shows an anchor on a low-pressure steam main with axial expansion joints (bellows) on both sides. The damage was due to operation at a higher temperature than design and probably occurred months or even years before it was noticed. All expansion joints should be registered for regular inspection.

28.8   HOW NOT TO WRITE AN ACCIDENT REPORT

An operator noticed a small leak of hot nitric acid vapor from a pipe. It seemed to be coming from a small hole on a weld. Radiography showed that there was significant corrosion of the weld and of a condenser just below it. A temporary patch was fitted to the leak, and plans were made to replace the condenser and pipe at a turnaround scheduled to take place a few months later.

Full marks to the company for writing a report on the incident and circulating it widely within the company—but the report left many questions unanswered:

- How long had the equipment been in use?
- Had it been radiographed previously?

- Was the original welding to the standard specified?
- Was a positive materials identification program in force when the plant was built, and were the pipe, welding rods, and condenser checked to make sure that they were made from the correct grade of steel? Was a suitable grade specified?

References

1. M. Schofield, "Corrosion horror stories," *The Chemical Engineer*, Vol. 446, Mar. 1988, pp. 39–40.
2. M. S. Mannan (editor), *Lees' Loss Prevention in the Process Industries*, 3rd edition, Elsevier, Boston, Mass, 2005, Sections 11.18.7 and 22.20.7.
3. T. Fishwick, "A chlorine leak from a vaporizer," *Loss Prevention Bulletin*, Vol. 139, Feb. 1998, pp. 12–14.
4. Anon, "A fire in titanium stucture packing," *Loss Prevention Bulletin*, Vol. 166, Aug. 2002, pp. 6–8.
5. Anon, *Safety Alert on a Fire Extinguishing Fatality*, Mary Kay O'Connor Process Safety Center, College Station, Tex., 2 Oct. 2000.
6. T. Fishwick, "Failure of a superheater outlet joint resulting in equipment damage," *Loss Prevention Bulletin*, Vol. 146, Apr. 1999, pp. 5–7.
7. T. A. Kletz, *Learning from Accidents*, 3rd edition, Elsevier, Boston, Mass, 2001, Section 30.7.
8. P. Elliott, "Overcome the challenge of corrosion," *Chemical Engineering Progress*, Vol. 94, No. 5, 1998, pp. 33–42.
9. K. Posteraro, "Thwart corrosion under industrial insulation," *Chemical Engineering Progress*, Vol. 95, No. 10, 1999, pp. 43–47.
10. J. Lindley, "In brief—corrosion," *Loss Prevention Bulletin*, Vol. 163, Feb. 2002, pp. 5–6.
11. K. Collier, "Insulation," *Chemical Engineering Progress*, Vol. 98, No. 10, 2002, p. 47.
12. D. Lyon, "Suspected weld failures in process equipment," *Loss Prevention Bulletin*, Vol. 163, Feb. 2002, pp. 7–9.
13. B. Fabiano, "Corrosion in a sugar refinery," *Loss Prevention Bulletin*, Vol. 163, Feb. 2002, pp. 10–12.
14. Anon, "Union Oil amine absorber fire," *Loss Prevention Bulletin*, Vol. 163, Feb. 2002, pp. 20–21.
15. R. Ornberg, "On the job—Illinois," *Firehouse*, Vol. 176, Oct. 1984, pp. 74–76.
16. C. H. Vervalin, "Explosion at Union Oil," *Hydrocarbon Processing*, Vol. 66, No. 1, 1987, pp. 83–84.
17. T. A. Kletz, *Learning from Accidents*, 3rd edition, Elsevier, Boston, Mass, 2001, Chapters 5, 13.7, 16.5, and 22.
18. M. G. Mogul, "Reduce corrosion in amine gas absorption columns," *Hydrocarbon Processing*, Vol. 78, No. 10, 1999, pp. 47–56.

Operating Methods

Human nature will instinctively modify what should be done into what can be done especially if this makes the job easier or keeps the job moving in some way.

—Loss Prevention Bulletin, October 2000

29.1   THE ALARM MUST BE FALSE

We all know of occasions when operators have said, "The alarm must be false" and sent for the instrument technician. For example, the high-level alarm on a storage tank operates. The operator knows the tank is empty and ignores the alarm. By the time the technician arrives, the tank is overflowing. Someone has left a valve open and the liquid intended for another tank has flowed into the first one.

Here is an incident where the operator had a good reason for thinking that the alarm was false. The three reactors on a plant were being brought back on line after a turnaround. Number 1 had been stabilized to normal operating conditions, but numbers 2 and 3 were still at the early stages of startup. The temperature on number 2 started to rise, and the high-temperature alarm sounded. It seemed impossible that any reaction could have occurred at so early a stage and all other readings were normal, so the operator decided that the instrument must be faulty and sent for a technician. A little while afterward, a pipe on number 1 reactor was found to be growing red hot. During the shutdown, work had been done on the temperature-measuring instruments on the three reactors, and the leads from numbers 1 and 2 were accidentally interchanged. (Section 26.1 describes a similar error.)

It is a good practice to test all trips, interlocks, and alarms after a shut-down, or at least those that have been maintained. The incident also shows the value of a walk around the plant when anything is out of the ordinary. We may not know what we are looking for, but we never know what we may find.

Someone had a similar experience after collecting her car following the repair of some minor accident damage. On the way home she had to make several turns, and on each occasion other cars hooted her. When she got home, she found that the rear direction indicators had been con-nected up the wrong way round so that when she signaled a left turn, the right indicators flashed. When she telephoned the repair company, the company at first insisted that its technicians always checked direction indicators to make sure that they were wired correctly. In fact, only the front ones had been checked.

29.2   A FAMILIAR ACCIDENT—BUT NOT AS SIMPLE AS IT SEEMED

Moving liquid into the wrong vessel is one of the most common acci-dents in the chemical industry and is usually, and often unfairly, blamed on an error by the operator. An unusually frank and detailed report shows the superficiality of such a view.

Some liquid had to be transferred from one vessel to another. Such movements were common, though transfers between the two vessels involved on this occasion were unusual. The foreman asked an experi-enced operator to carry out the operation, but before this operator could do so, he was called to a problem elsewhere on the unit and the job was left to a new and inexperienced operator, with another experienced oper-ator keeping an eye on him from time to time.

The trainee went to the transfer pumps where there was a diagram of the pipework. At one time all the vessels and valves were numbered. Unfortunately, painters had painted over many of the labels, which had then been removed as illegible and never replaced. The trainee opened a wrong valve. As a result, the liquid went into a vessel that was out of use and ready for refurbishment. Some of the liquid leaked out of a faulty thermocouple pocket. About 50 liters (13 gal) of a corrosive liquid were spilled inside a building, and some of it dripped down to the floors below.

The trainee checked that the level in the suction vessel was falling, but he could not check that the liquid was arriving in the intended delivery vessel as other streams were entering at the same time. In addition, the level indicator and alarm on the vessel into which the liquid was actually being pumped had been disconnected as the vessel was out of use.

29.2.1   What Can We Learn?

- If a job had to be left to an inexperienced operator, was manning adequate or had downsizing gone too far? A few years later, following a more serious incident, manning was increased.
- In a piece of unfortunately common *management-speak*, the report blamed the operators for not reporting the missing labels. Of course, they should have reported them, but the supervisors and managers (and the auditors) should also have seen and reported them. If the operators had reported the missing labels, would they have been replaced? This is the sort of minor, complex job that maintenance teams often never get around to doing. If operators report this sort of fault and nothing is done, they do not report such faults again. Following the incident just mentioned, many missing labels were found on other units.
- The pipeline leading to the vessel that was out of use should have been isolated by a blind or at least by a locked valve. The valve handle had been removed, but on this plant that merely indicated that the valve was used infrequently, not that it should be kept shut.
- If equipment is not positively isolated, by blinding or disconnection, then its level instrumentation should be kept in operation. The levels in tanks that were supposed to be out of use have often changed.
- If toxic, flammable, or corrosive liquids are liable to leak inside buildings, the floors should be liquid-tight.

The recommendations were followed up on the unit where they occurred, but because the spillage was small, it had little impact elsewhere in the plant and company. This is a common failing. After the tires on a company vehicle were inflated to such a high pressure that they burst, the recommended inflation pressures were painted above the wheels of all the site vehicles, but only in the factory where the burst occurred, not anywhere else.

29.2.2   Another Similar Accident

The official report [6] on this accident described, more thoroughly than usual, the managerial causes as well as the immediate technical ones. It occurred on a ship, but similar ones have occurred in all, or almost all, chemical and oil plants.

The 3,100-ton ship carried two tanks with a combined volume of 3,219 m^3 (850,000 gal) and had carried 2,800 tons of refrigerated vinyl chloride under a moderate pressure from Rotterdam to Runcorn, United Kingdom. On arrival the chief officer, who was in charge of offloading, was asked to provide a sample. He started up the pump on one of the tanks to circulate the contents but did not check that the valves leading to the other tank were shut. Most of the valves on the piping system were duplicated in case one leaked, but most had been left open and so some of the liquid entered the other tank, the pressure in it rose, and the pressure control valve opened at a gauge pressure of 9 bar (130 psi) and discharged vapor through the vent that was located on the mast. As a result, 600 kg (1,320 lb) of vinyl chloride vapor was discharged, and it was estimated that the flammable cloud had a radius of 50 m (165 feet). Fortunately, it did not ignite. No one was in a location where they could have experienced any acute toxic effects.

The report drew attention to the following:

- The boat had no fixed equipment for the detection of flammable gases, though it did carry two portable detectors.
- The boat had a water spray system that could have been used to help disperse the leak. It was not used, though a hose was used to disperse ice that had formed on the deck.
- It was custom and practice to leave as many valves as possible open to save time and effort on arrival in port. The company's operating manual stated that "All valves in the cargo system not required to be open for the operation are [to be] shut." The chief officer had signed a statement that he understood the instructions. He was making only his second trip as chief officer since joining the crew of the boat.
- The previous chief officer had been promoted to master of the ship, though the company's policy stated that newly promoted and relatively inexperienced masters and chief officers should not sail together.
- The chief officer was trying to supervise too many jobs at once. He should have asked for assistance.
- The operating instructions required a breathing apparatus to be worn during sampling, but the chief officer was not wearing it.

The report concluded that "terminal and vessel operators should check and ensure that safety management systems are working in practice and that cargo operations, in particular, are always conducted in accordance with industry guidelines." Clearly, in this case they were not. The report says that the company, which owned the import and export terminal as well as the boat, carried out too few inspections and that it was not thorough, as the inspectors had no experience with marine gas carriers. They did not understand what was happening.

The main lessons for chemical and oil plants from this sorry story are familiar ones: managers should visit the areas under their control and keep their eyes open as they walk around; regular audits should be carried out by people who know what to look for. One of the reasons for the collapse, during construction, of a tunnel at Heathrow Airport in London was that the airport decided to save money by using its own auditors to monitor construction. Unfortunately, the auditors knew a lot about auditing but little or nothing about tunneling [7, 8].

When the cross-channel roll-on/roll-off ferry *Herald of Free Enterprise* sank in 1987 with the loss of 186 passengers and crew, the chairman of the holding company was reported as saying, "Shore-based management could not be blamed for duties not carried out at sea" [9]. However, boats on short journeys are as easy to audit as fixed plants.

29.3   MORE RELUCTANCE TO BELIEVE THE ALARM

Another incident made worse because measurements were not available occurred on a ferryboat but has lessons for the process industries. The exhaust gas from the engines was used to raise steam in two waste heat boilers. One of them developed a steam leak and was shut down. The steam lines were isolated and the boiler drained, but the exhaust gas continued to pass through the boiler as there was no way of bypassing it. The high-temperature alarm on the boiler sounded, but nothing wrong could be found. The inspection—and another carried out when the vessel reached port and the engines had been stopped—could not have been thorough because when the engines were restarted a few hours later, an expansion joint (bellows) was found to be glowing red hot. The passengers were told to leave the ship, and the fire service was called.

The lack of a bypass was a weakness in the design but the ship's crew seems to have been too ready to believe that the alarm was false (as described in Section 29.1). The inspection carried out in the port may not have been thorough because shutting down the engines disabled the alarms and the crew may not have known this (compare Section 29.2.1, penultimate bullet).

Once the hot expansion joint was found, the incident was dealt with correctly and efficiently. In my experience, the same is true in the process industries. Poor design and operation may have led to an incident, but once it occurs, the right action is usually taken. In my time at the plant, when the fire alarm sounded, maintenance workers left the area—rightly as they were not trained to deal with fires—whereas operators ran toward it. Most fires were extinguished even before the site fire service arrived.

The report on this incident comes from a periodic review of marine accident reports [1]. Most of them are of nautical incidents such as ships running aground or colliding with other ships but a surprising number are process incidents, such as the one just described, unsafe entries to tanks, a foamover (see Section 30.12) because hot oil was put into a tank containing a water layer, a fire in a galley because butane from an old aerosol can leaked into a cupboard, choked vents, and many failures of lifting gear. Some other marine incidents are described in Section 37.2.

29.4   THE LIMITATIONS OF INSTRUCTIONS

However many instructions we write, we never think of everything and so people should be given the knowledge and understanding they need to handle situations not covered in the instructions. This is usually illustrated by descriptions of complex accidents such as the nuclear accident at Three Mile Island [2] in Middletown, Pennsylvania, United States, but a simple incident illustrates the same theme.

The plant handled a toxic material. When filter cartridges contaminated with this material were changed, the old ones were placed in sealed plastic bags and taken to another building for cleaning and disposal. If the bags were dropped, they might easily rupture and so instructions stated that the bags must be moved on a trolley. The trolleys were conveyed downstairs in the elevator.

What would you do if you were asked to move a bag and the elevator was out of order (or if you were this operator's foreman)? The person who wrote the instructions never foresaw this problem.

The man asked to move the bag did as many people would have done: he carried the bag downstairs. He could then have put it on a trolley, but having carried it so far he carried it the rest of the way to the foreman's office and put it on the table. The bag slid off and punctured, and the room had to be evacuated and cleaned.

The inquiry brought to light the fact that the operators and the foremen did not fully appreciate the hazards of the material on the filters. People will follow instructions to the letter only when they understand the reasons for doing so. We do not live in a *Charge of the Light Brigade* society in which people will unthinkingly obey every command (see also Sections 24.5.2, 30.12, and 36.5).

The inquiry also revealed that bags containing contaminated filters had been carried downstairs on at least two other occasions when the elevator was out of order, but nothing was said. Perhaps the foreman preferred not to know or, more likely, he never brought together in his mind the two contradictory facts: the elevator was out of use and a bag had got downstairs. (Section 36.6 describes an incident in which a computer "believed" two contradictory facts.)

29.5   THE LIMITATIONS OF INSTRUCTIONS AGAIN

Thirty gallons of sludge were being pumped into a 55-gal drum. To avoid splashing, two operators fitted the lid on the drum. They did not realize that with no vent for the air to escape, the pressure in the drum would rise. After a while, they noticed that the flow had stopped and that the drum was bulging. They then realized what had happened and decided to remove the lid. As one of them was doing so, the lid flew off, injuring the operator and splashing him with a toxic sludge.

The report [3] emphasized the need to prepare better instructions and hazard check lists for all jobs but, as stated in the previous item, we cannot cover every possibility in our instructions and the longer we make them, the less likely they will be read. We can tell people everything they should do, but we cannot tell them everything they should not do. To quote a judgment from the United Kingdom's supreme court, the House of Lords [4], "[A person] is not, of course, bound to anticipate folly in all its forms, but he is not entitled to put out of consideration the teachings of experience as to the form that those follies commonly take." We could replace *folly* by *human error*.

Accidents such as the one just described are best prevented by better training rather than better instructions—that is, by giving people an understanding of basic scientific principles, in this case that if something is put into a vessel either something, usually air, has to get out or the pressure will rise (but if water is put into a vessel containing a soluble gas, such as ammonia, the pressure will fall).

29.6 EMPTY PLANT THAT IS OUT OF USE

An official report [5] drew attention to a hazard that is easily overlooked. A vessel was not used for several months. It had been pumped dry, but unknown to the operators a layer of solid residue had been left behind in the vessel. When the vessel was brought back into use, on the same duty as before, the fresh reactants reacted with the residue, causing a rise in temperature and the emission of gas into the working area.

The report recommended the following:

- Whenever possible, equipment that is going to be left out of use for longer than usual should be emptied completely.
- If that cannot be done (or has not been done), then the material remaining should be tested with the materials that are to be added to see if there is any unforeseen reaction.
- In some cases, it may be possible to prevent deterioration of residues by covering them with a layer of water or other solvent.

29.7 A MINOR JOB FORGOTTEN—UNTIL THERE WAS A LEAK

A solution of a toxic liquid was kept in a storage tank fitted with a cooling coil, which developed a leak. As cooling was not necessary for the liquid now stored in the tank, a blind was fitted to the cooling water outlet line and it was decided to cap the inlet line. This could not be done immediately because of pressure of other work; therefore, for the time being the inlet line was kept up to pressure so that water leaked into the tank rather than the reverse. The capping job was repeatedly postponed and ultimately forgotten.

Five years later, the water pumps were shut down for a short time and the drop in pressure allowed some of the liquid in the tank to enter the water line. A leak occurred from a sample valve on the water inlet line near the tank—the variation in pressure may have caused the sample valve to leak—and some of the toxic solution leaked out. The leak went into a duct underneath the sample valve. From there it should have flowed to a drain. However, instead it flowed down a *temporary* line, not shown on any drawing, that the construction team had installed and never removed, and dripped down the building. Some water was poured down in an attempt to dilute and sweep away the leak, but its effectiveness was doubtful.

29.7.1   What Went Wrong?

It is hardly necessary to say that jobs should not be forgotten. The fact that capping the cooling water outlet line was forgotten suggests the lack of a good safety management system or of the resources necessary to operate and maintain it. This incident was trivial, but several years later the company was in trouble with a regulatory authority for its failure to maintain and operate adequate safety management systems. Coming events cast their shadows before they arrive, and incidents like this one can serve as warnings that all is not well. Of course, the report said all the right things about the need to log outstanding jobs and so on, and things may have improved at the unit involved, but there was no serious attempt to look at and if necessary change methods elsewhere in the plant and the company.

The sample point, and similar ones on other cooling water lines, were rarely used and were removed after the incident. Removing redundant or temporary equipment is one of those jobs that is frequently postponed and then forgotten. Meanwhile the equipment is not maintained and ultimately gives trouble. (See the notes on plugs in Section 28.4.2.) Construction teams should be asked to list any temporary pipes, supports, drains, or plugs that they have installed for their own convenience and have not removed.

29.8   DESIGN ERROR + CONSTRUCTION ERROR + OPERATING ERROR = SPILLAGE

Figure 29-1 shows the layout of the relevant pipework on an experimental unit that was used intermittently. In the original design, the pump

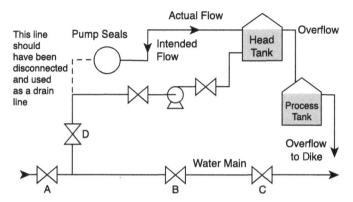

FIGURE 29-1   When the unit was modified, the contractors forgot to disconnect the dotted pipeline. As a result, water flowed in the opposite direction to that intended and into the head tank.

seals were supplied with water via valve D. However, the pressure in the main varied and therefore, to provide a more consistent supply, a small head tank was installed and the old supply line to the pump seals was then used as the drain line. During construction, the need to disconnect it from the water main was overlooked and it remained connected. Perhaps the designers of the modification failed to tell the construction team that the line should have been disconnected, or perhaps they were told but failed to do so. Either way, nobody checked the job thoroughly (if at all) after completion and the company's procedure for the control of modifications was ignored, as the plant was only an experimental one.

Valves A, B, and C were opened to supply water for an hour or so to another part of the unit. Valves B and C were then closed but A was left open. Valve D was already open. Water flowed in the opposite direction to that intended and into the head tank. There was a ballcock valve on the normal inlet but not, of course, on what was intended to be the outlet line so the water filled the head tank and overflowed into the dike. The high-level alarm in the dike sounded but no one heard it. As result of demanning, there was no one present in this part of the plant during the night!

To add to the problem, the head tank did not overflow directly into the dike but into another tank that contained a solution of a process material in water. This tank overflowed, so what could have been a spillage of water turned into a spillage of process liquid. The unusual arrangement of the pipework probably arose because the plant was an experimental one so as few pipes as possible were rerouted when the duties of the tanks were changed.

Figure 29-1 looks simple but it shows only a few pipes. There were many more, and many of them followed circuitous routes as the result of earlier modifications and changes of use. The isometric drawing attached to the full report looks like a plate of spaghetti.

29.8.1   What Went Wrong?

- Complexity in pipework (and everything else) leads to errors. Simplicity is worth extra cost. (It is usually cheaper but not always. See Section 27.1.2.)
- Modification control procedures should not be skipped and should be applied to experimental units as well as production plants and to changes in organization, such as demanning, as well as changes to processes (see Chapters 25 and 26).
- Checking of completed pipework should not be left to construction teams, in-house or contracted. The operating team should check thoroughly. They are the ones who suffer the consequences of errors in construction.

References

1. Anon. According to the book *Safety Digest—Lessons from Marine Accident Reports*, No. 1/2002, Marine Accident Investigation Branch of the UK Department of Transport, Local Government and the Regions, London, 2002. pp. 30–31.
2. T. A. Kletz, *Learning from Accidents*, 3rd edition, Elsevier, Boston, Mass, 2001, Chapter 11.
3. Anon. Propelled drum lid injures research assistant. *Operating Experience Weekly Summary*, No. 99-4, Office of Nuclear and Facility Safety, US Dept of Energy, Washington, DC., 1999. pp. 1–3.
4. M. Wincup, *The Guardian* (London), 7 Feb. 1966
5. Nuclear Installations Inspectorate (1974). *Report by the Chief Inspector of Nuclear Installations on the Incident in Building B204 at the Windscale Works of British Nuclear Fuels Ltd on 26 September 1973*, Her Majesty's Stationery Office, London.
6. Anon. *Report on the Investigation of the Escape of Vinyl Chloride Monomer on Board* Coral Acropora, Runcorn, 10 August 2004, UK Marine Accident Investigation Branch, Southampton, UK, 2004.
7. T. A. Kletz, *Learning from Accidents*, 3rd edition, Elsevier, Boston, Mass, 2001, Section 5.8.
8. Health and Safety Executive, *The Collapse of NATM Tunnels at Heathrow Airport*, HSE Books, Sudbury, UK, 2000.
9. J. McIlroy, *The Daily Telegraph* (London), 10 Oct. 1987.

30

Explosions

At 11 minutes past 11 on the morning of November 27th, 1944, the Midlands was shaken by the biggest explosion this country has ever known. 4,000 tons of bombs stored 90 ft down in the old gypsum mines in the area, blew up, blasting open a crater 400 ft deep and ¾ mile long. Buildings many miles away were damaged. This pub had to be rebuilt and one farm, with all its buildings, wagons, horses, cattle and 6 people completely disappeared.

— Notice outside the Cock Inn, Fauld,
near Burton-on-Trent, England

The immediate cause of most explosions—that is, violent releases of energy—is an exothermic chemical reaction or decomposition that produces a large amount of gas. However, some explosions, such as those described in Sections 30.2, 30.10, and 30.12, have a physical cause. The Fauld explosion is described in reference 1. The ignition source was thought to be the rough handling of a sensitive detonator that was being removed from a bomb.

30.1   AN EXPLOSION IN A GAS-OIL TANK

An explosion followed by a fire occurred in a 15,000 m³ (4 million gal) fixed-roof gas-oil tank while a sample was being taken; the sampler was killed. The explosion surprised everybody as the gas oil normally had a flash point of 66°C (150°F). However, the gas oil had been stripped with hydrogen to remove light materials, instead of the steam originally used, and some of the hydrogen had dissolved in the gas oil and was then released into the vapor space of the storage tank. The change from steam to hydrogen had been made 20 years earlier.

Calculations showed that 90% of the dissolved hydrogen would be released when it was moved to the storage tank and would then only slowly diffuse through the atmospheric vent. Samples were tested for flash point, but the hydrogen in the small amounts taken would have evaporated before the tests could be carried out.

When the change from steam to hydrogen stripping was made, it seems that nobody asked if hydrogen might be carried forward into the storage tank. No management-of-change procedure was in operation at the time.

The source of ignition was probably a discharge of static electricity from the nylon cord used when lowering the sample holder into the tank. Cotton, the recommended material, usually contains enough moisture to conduct electricity, whereas synthetic cords are usually nonconducting. A charge could have built up on the nylon cord as a result of friction between the sampler's glove or cloth and the nylon, while the sample holder was being lowered into the tank, and then discharged to the walls of the tank. No liquid had been moved into the tank during the previous 10 hours, so any charge on the gas oil had ample time to discharge.

During the investigation, a 1995 standard for tank sampling was found. It stated, "In order to reduce the potential for static charge, nylon or polyester rope, cords or clothing should be used." A copy of the accident report [2] was sent to the originators of the standard. They replied apologizing for the omission of the word *not*!

A similar accident, another explosion in a tank containing gas oil contaminated with hydrogen, was reported 14 years earlier [3]. Unfortunately, this incident was not known to anyone in the plant where the second explosion occurred.

30.1.1   Lessons Learned

This explosion, like many others, shows that the only effective way of preventing explosions and fires of gases or vapors is to prevent the formation of flammable mixtures. Sources of ignition are so numerous and the amount of energy needed for ignition is so small (0.2 mJ in this case) that we can never be sure that we have eliminated all sources of ignition. (Energy of 0.2 mJ is the amount released when a one cent coin falls 1 cm. This amount, concentrated into a spark or speck of hot metal, will ignite a mixture of hydrogen and air.)

Nevertheless, we should do what we can to remove sources of ignition. Sample holders lowered into tanks should be held by conducting cords; in addition, as the holder is lowered, the cord should touch the side of the opening in the tank roof so that any charge generated is removed.

Changes in processes, such as replacing steam by hydrogen, should be Hazop studied as well as changes to plant design. The teams should ask, "What will be the result if any materials present at earlier stages of the process are still present?" "They can't be" is rarely, if ever, an adequate answer.

Note that the crucial change, replacing steam by hydrogen, took place long before the explosion. Twenty years went by before the right combination of circumstances for an explosion arose. This is typical of incidents triggered by static electricity. Note that I did not write *caused*. I regard the change from steam to hydrogen as the cause or, more fundamentally, the lack of adequate study of possible consequences before the change. Like most accidents, this one had happened before.

There is more on static electricity in Chapter 15 and Sections 24.2.3, 25.2.7, 28.2.5, and 32.7.

30.2   ANOTHER SORT OF EXPLOSION

A tank with a capacity of $726\,m^3$ (200,000 gal) was used for the storage of methylethyl ketone (MEK). The contents were moved into a ship. The transfer pipeline was not emptied immediately afterward, as it was used frequently for this product, but on the evening of the following day it was decided to empty it by blowing the contents back into the tank with nitrogen at a gauge pressure of 5 bar (75 psi), the usual method.

About five minutes after the level in the tank had stopped rising and before the nitrogen was shut off, there was an explosion in the tank followed by a fire. The roof separated from the walls along half the circumference. As MEK is explosive (flash point 11°C [52°F]), everyone assumed at first that there had been a chemical explosion. However, there was no obvious source of ignition, and static electricity could be ruled out as MEK has a high conductivity. Any static formed will flow to earth through the tank walls in a fraction of a second (as long as the walls are grounded). Someone then asked why the explosion occurred when it did rather than at another time, always a useful question to ask when investigating an accident, especially an explosion. Had anything changed since the last time when the transfer pipeline had been blown with nitrogen?

The answer was yes. A few hours before the explosion, the 2-in. diameter open vent on the tank had been replaced by a filter pot containing alumina, presumably to prevent moist air contaminating the MEK in the tank. The pressure drop through the alumina was sufficient to allow the pressure in the tank, designed for an 8-in. water gauge (2 kPa or 0.3 psi), to be exceeded and to reach the rupture pressure, probably about a 24-in. water gauge (6 kPa or 0.9 psi). The nitrogen flow rate was estimated to be $\approx 11\,m^3$/minute (400 ft^3/minute) so the rupture pressure could be reached in about an hour. Unfortunately, the report does not say how much time elapsed between the start of blowing and the rupture. The fire that followed the explosion could have been the result of sparks produced by the tearing of the roof-to-wall joint.

Once again we see a change made without adequate consideration given to the possible consequences. In this case, the result was inevitable and occurred soon after the change. In the previous case history, the result was probabilistic and did not occur for many years.

30.3   ONE + ONE = MORE THAN TWO

We are familiar with synergy: two (or more) drugs or parts of the body work together to produce a greater effect than the sum of their individual parts. The same is true of hazards, as the following example shows.

The report [4] starts with the words, "It was the best of times; it was the worst of times. The economy was booming; some of the booms were due to plant explosions." One occurred in a power station boiler in a car factory in February 1999. The primary fuel was pulverized coal, but natural gas was also used. There were two gas supply lines, each of which supplied three burners. The boiler was shutting down for overhaul. One of the natural gas lines was isolated and blinded; the valves between the blind and the burners were opened, and the line swept out with nitrogen. The other line had not yet been blinded. In addition, the valves in this line had been opened in error (or perhaps left open). Gas entered the furnace. There were no flame-sensing interlocks to keep the inlet motor valves closed when there was no flame and after 1.5 minutes, an explosion occurred. The ignition source was probably hot ash. The explosion inside the boiler set off a secondary explosion of coal dust in the boiler building and in neighboring buildings. Six employees were killed and many injured. Damage was estimated at $1 billion, making it at the time the most expensive industrial accident in U.S. history.

There were thick accumulations of coal dust in the damaged buildings. Even after the explosions, the dust was an inch thick. On many occasions, a primary explosion has disturbed accumulations of dust and resulted in a far more damaging secondary explosion. The hazards of dust explosions and the need to prevent accumulation of dust are well established. Henry Ford is reputed to have said that history was bunk. Did the car factory personnel still believe Ford's words in 1999?

The same paper also describes another furnace explosion, killing three employees, also in February 1999. There seems to have been a flame-out and, as in the first incident, there were no flame-sensing interlocks to close the fuel gas motor valves when flames went out. In addition, one fuel gas valve was leaking. As in the first incident, the primary explosion disturbed dust, resin this time, and caused a secondary explosion.

30.4 "NEAR ENOUGH IS GOOD ENOUGH"

Anyone who has bought a new house (at least, in the United Kingdom) knows how difficult it can be to get the builders to finish everything and the new owners often move in when a few jobs are still outstanding. It may not matter if the builders have not finished laying the paths or fitting out the guest room, but would you start up plant equipment before it was complete? Here is the story of a company that did, perhaps because it was just a storage tank, not a production plant [5].

Three low-pressure storage tanks were being modified for the storage of crude sulfate turpentine, an impure recovered turpentine with an unpleasant smell and a flash point that can be as low as 24°C (75°F). Several changes were being made:

- A fixed foam firefighting system was being installed, with a pumper connection outside the dike.
- To prevent the smell reaching nearby houses, a carbon bed would absorb any vapors in the vents.
- Flame arrestors in the common vent system would prevent an explosion in one tank igniting the vapor in the others.

Movement of turpentine into the tanks was started six weeks before all the protective equipment could be fitted. Each tank contained ≈ 800 m³ (200,000 gal). The local authority who gave permission for the storage was informed and seems to have raised no objection. Twelve weeks later, the protective equipment was still not complete but the vent absorption system was ready and was brought into use. All three tanks were connected to the vent absorption system and no longer vented directly to the atmosphere.

The manufacturer's instructions said the carbon bed had to be kept wet. It was not and got too hot. During the day, the oxygen content in the tank was too low for ignition but rose in the evening when the tank cooled and air was sucked in. The hot carbon ignited the vapor, and there was an explosion. It spread to the other two tanks through the common vent collection system, as the flame arrestors had not been delivered.

The fixed foam firefighting system on the tanks could not be used, as the piping connection outside the bund had not been installed.

The explosion damaged three other tanks in the same dike. One contained an acidic liquid and another an alkaline one. The acid and alkali reacted and produced hydrogen sulfide. Incompatible liquids should not be stored in the same dike.

The vent absorption system was intended to prevent pollution. Because it was operated incorrectly, because the missing flame arrestors allowed the explosion to spread, and because the firefighting equipment was incomplete, the result was an environmental disaster. Two thousand people were evacuated from their homes for several days and 10 to 15 hectares (25 to 40 acres) of marsh were contaminated.

Note that before the carbon bed was commissioned, an explosion was possible but unlikely. Commissioning it before the rest of the new safety features were ready and not keeping it wet made an explosion inevitable.

What, I wonder, were the qualifications, abilities, knowledge, and experience of the people in charge of the plant involved in this incident? What pressure, I wonder, was put on them to bring the tank into use prematurely? Near enough may not be good enough.

30.5   ANOTHER EXPLOSION IGNITED BY A CARBON BED

A carbon absorption bed was added to the vent system of an ethylbenzene tank to absorb vapor emissions. It was designed to handle only the emissions caused by changes in the temperature of the tank. The much larger emissions produced when the tank was being filled were sent to a flarestack.

One day, when the tank was being filled, the operator forgot to direct the vent gases to the flarestack. When the tank was 25% full, he remembered and promptly corrected the error. This is understandable. When we realize that we have forgotten to carry out a task, we tend to do it at once, without stopping to ask if there might be any adverse result. When the filling was complete, he sent the emissions back to the carbon bed. Within minutes the carbon bed caught fire; damage was severe.

When the carbon bed received far more vapor than it was designed to absorb, it overheated. When the vapor was sent to the flarestack, the carbon bed and the absorbed vapor could not burn, as there was no air (or not enough air) present. When filling was complete and the tank was again connected to the carbon bed, it received a supply of air from the vent and was still hot enough to ignite the absorbed ethylbenzene and then the carbon.

As already stated, the operator's error was understandable (see also Section 35.1). However, during the design of the system, someone should have asked what would occur if the vent stream was wrongly directed. A Hazop would have raised this question. After the fire, various protective devices were considered:

- An interlock to prevent vapor being directed to the carbon bed while the tank was being filled
- A high-temperature alarm on the carbon bed
- A carbon monoxide detector on the carbon bed to detect smoldering
- Nitrogen blanketing of the tank

The report [6] does not say what was actually done, but the last proposal is the best as it will prevent explosions from all sources of ignition.

Many other fires and explosions have occurred in vent collection systems, installed without sufficient thought, for the commendable purpose of improving the environment (see Section 2.11). Two more follow. Under the section, "Green Intention, Red Result," reference 7 describes these and other changes that were made to improve the environment but had adverse effects on safety (see also Section 30.9).

30.6   AN EXPLOSION IN AN ALTERNATIVE TO A CARBON BED

Alternative methods of removing volatile and flammable contaminants from a stream of air are to burn them in a furnace or oxidize them over a catalyst. The concentration of vapor is kept below the lower flammable limit (LEL) to avoid an explosion. The concentration is measured continuously and if it approaches the LEL, the operation is automatically shut down.

A trip on a new oxidizer kept operating. A check with a portable combustible gas detector showed that the plant instrument was reading high. The startup team therefore decided to take the trip system out of use while the reason for its high reading was investigated but to continue with the startup without it.

Many people have taken a chance like this and got away with it. The team on this unit was not so lucky. Within two hours, there was an explosion with flying debris. It is not clear from the report [8] whether or not the plant instrument was really reading high, but it is clear that there were occasional peaks in the vapor concentration.

30.7   ONLY A MINOR CHANGE

A reactor vent discharge containing 100 ppm benzene in nitrogen was sent directly to the atmosphere at a rate of $8.5\,m^3/hour$ ($5\,ft^3/minute$). To meet new emission standards, the company installed an electric flameless destruction system. The vent discharge had to be diluted with air before entering this system and the air rate was set so that the total flow was $170\,m^3/hour$ ($100\,ft^3/minute$). This dilution ensured that the mixture was well below the lower flammable limit of benzene, even during occasional spikes when the benzene concentration rose briefly to 15%.

Shortly after the destruction unit was installed, the vent discharge from a storage tank was also directed into it. The increase in flow rate was only 6.7%. Everyone assumed that this was too small to matter and no one made any calculations. However, the lower flammability limit was exceeded during the spikes in benzene concentrations in the main contributor to the flow. The destruction unit was hot enough to ignite the vapors, and there was an explosion. A high concentration of combustible gas in the gas stream sounded an alarm, but it operated too late to prevent the explosion. Although damage was considerable, the explosion did not travel back to the reactor and tank as both were blanketed with nitrogen.

30.7.1   Lessons Learned

Consider the possible consequences of changes before authorizing them (see Chapter 24). Never dismiss a change in quantity as negligible before calculating its effects. Consider transient and abnormal conditions as well as normal operation. Sections 25.1.2, 25.4, and 25.5 describe other incidents that occurred because no one made simple calculations.

Estimate the response time of every alarm and trip to see if it is adequate. Check it during testing if there is significant delay. Most measuring instruments respond quickly, but analytical instruments are often slow, although it is usually the sampling system rather than the measuring device that causes the delay.

The report [6] says that pollution control equipment should not be treated like a domestic garbage can, something into which anything can be dumped. Every proposed addition should be thoroughly evaluated. On a chemical plant or in a chemical laboratory this applies to all waste collection equipment. Many fires, toxic releases, or rises in pressure have occurred because incompatible chemicals were mixed in the same waste drum (see Section 30.11).

30.8   AN EXPLOSION IN A PIPE

The pipe (C) (Figure 30-1) transferred fractionation residues from a batch distillation vessel (A) to residue storage tank (D) via the reversible pump (B). Distillation residues from other units and condensate from vent headers also went into tank D. When D was full, the contents were moved to A for fractionation and recovery. As the residues were viscous, pipe C was steam-traced.

This part of the plant operated only five days per week. It was left one Friday evening after the contents of D had been moved to A, ready for distillation on Monday. Over the weekend, a discharge reaction ruptured pipe C.

Analysis of the remaining material in other parts of pipe C showed that decomposition and self-heating started at ≈ 140°C (280°F) and that the rate of temperature rise soon exceeded 1,000 degrees C (1,800 degrees F) per minute. This was surprising as the residues reached 140°C in normal

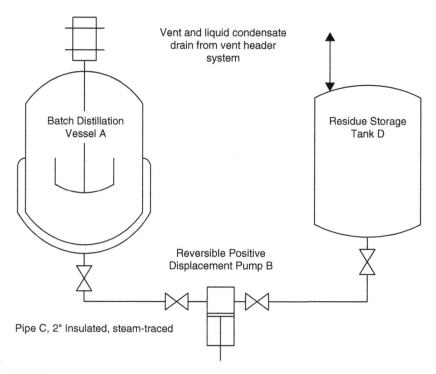

FIGURE 30-1   Some of the residue moved from D to A and left in pipe C decomposed and ruptured the pipe. From reference 9. *(Reprinted with the permission of the American Institute of Chemical Engineers.)*

operation and had never shown signs of decomposition or exothermic activity. Further investigation showed that the instability was due to the presence of 3% water that had entered vessel D with the condensate from the vent headers. Water is a very reactive substance and can form unstable mixtures with many other compounds. The disaster at Bhopal, India, was due to the contamination of methyl isocyanate with water.

As almost always, something else was also wrong. The steam supply to the tracing on line C came from an 8.3 bar gauge (120 psig) supply via a letdown valve, which had failed in the open position. This raised the temperature of the pipe to 170°C (340°F), high enough for decomposition to start.

30.8.1   Lessons Learned

A relief valve was fitted downstream of the steam let-down valve. An alternative and inherently safer solution would have been to use a heating medium that could not rise above 140°C (280°F).

Because water is so reactive and present most everywhere, we should, during Hazop studies, ask, under the heading *Other Than*, if water could be present and, if so, what its effects would be. (A Hazop was carried out elsewhere on an existing plant in which some valves were operated by high-pressure compressed gas. The team was asked if water could be present in the gas and the members all agreed that this was impossible. None of them knew that during shutdowns, when no high-pressure gas was available, the maintenance team occasionally used high-pressure water to operate the valves. See also Section 36.1.7.) We should also always ask if other common contaminants such as rust, lubricating oil, and any material used elsewhere in the process (see Section 30.1.1) could be present.

The vent line drains did not originally go to tank D but were diverted there to reduce waste. Perhaps because this was obviously a good deed, its possible consequences were not thought through. Chapter 25 describes other changes that had unforeseen results. As the report [9] says, "No good deed goes unpunished."

As shown by many of the other incidents described in this book (e.g., Sections 25.2.7 and 30.1), a plant can operate for many years without incident until a slight change in conditions results in an accident.

30.9   A DUST EXPLOSION IN A DUCT

The exhaust stream from a drier that contained volatile organic compounds and some flammable dust was discharged to the atmosphere through a vertical vent stack. To comply with legislation, the vent stack was replaced by an incinerator. There was no room for it near the drier, so it was built 90 m (300 ft) away and connected by a long duct. The dust settled out in the duct and was removed every 6 months, by which time it was 3 to 25 mm ($\frac{1}{8}$ to 1 in.) thick.

The drier was shut down for maintenance, but the incinerator was left on line. When the drier was brought back on line there was an explosion, which killed one man and caused extensive damage. The probable cause was a pressure pulse from the startup of the drier, which disturbed enough of the dust in the duct to produce a small explosion, which then disturbed and ignited much of the remaining dust in the duct. A layer of dust <1 mm thick can, if disturbed, produce an explosion in a building.

There should have been an explosion detection-and-suppression system or explosion vents in the duct. Or better still, filters to remove the dust before it entered the duct [6].

Once again, we see with this example that a change meant to reduce pollution was made as cheaply as possible and without adequate consideration of the hazards. It seems that when people are faced with an environmental problem, a sort of tunnel vision can set in and all thoughts of side effects are brushed aside (see Sections 30.5 through 30.8).

30.10   OBVIOUS PRECAUTIONS NEGLECTED

An underground concrete tank, 27 m (88 ft) in diameter, 4 m (13 ft) tall, and with a capacity of 720 m³ (190,000 gal), had been out of use for several years. It had a concrete roof supported by 27 internal columns and covered by a meter of soil. It was decided to recondition the tank. Two holes were cut in the roof of the tank for the insertion of new instruments. Before work started, the concentration of flammable vapor in the tank was checked and found to be <1% of the lower flammable limit.

During the weekend, no work was carried out on the tank but several loads of product arrived by barge and were transferred into neighboring tanks. The last was a load of premium gasoline. It was followed by a water flush, directed at first into the gasoline tank and then after 10 minutes into the concrete tank.

On Monday morning, three welders started work again. No flammability tests were carried out. When the first torch was lit, the tank exploded. The three welders were blown off the top of the tank and killed. Soil was thrown almost 100 m (325 ft).

Calculations showed that a gauge pressure of at least 0.43 bar (6.2 psi) would have been needed to lift the roof off its supports. The tank was thus much stronger than the usual atmospheric pressure storage tank, which will rupture when the gauge pressure in it exceeds 0.06 bar (0.9 psi) (see Section 30.2). The explosion of as little as 0.7 m³ (180 gal) of gasoline could have developed sufficient pressure to lift the roof off its supports.

30.10.1   What Went Wrong?

- The inlet line to the tank should have been blinded before welding started in order to prevent anything from leaking through it while the transfer line was in use.
- Even it there had been no movements over the weekend, the atmosphere in the tank should have been checked before work was resumed that Monday. A test on Friday (or earlier) does not prove that equipment is still safe on Monday.
- It was naïve to assume that no gasoline would be left in the transfer pipe after flushing with water for only 10 minutes.
- Did the owners leave testing to the reconditioning contractors? Did they know what had occurred during the weekend? The report [10] does not say.

The design of the tank made it stronger than usual; when it did fail, it failed with greater violence. Stronger does not always mean safer.

30.11   A DRUM EXPLOSION

This was a small explosion and fire compared to those described in the preceding examples, and no one was injured, but it was investigated with commendable thoroughness. The 210-L (55-gal) drum contained a peroxide (*di-t*-butyl peroxide). It was kept in a horizontal position on a cradle and small amounts were withdrawn as needed through a small cock on the lid, weighed, and added to a batch reactor. The weigh station was also used for other materials. The explosion blew the lid off the drum. It landed 15 m (50 ft) away with its outside surface on the ground but with soot on this surface. The location of the soot, and its nature, indicated that the lid had been exposed to fire before the explosion and that the heat from this fire caused explosive decomposition of the peroxide.

The location of the initial fire was either a drip tray underneath the cock or a cardboard box containing flammable materials located under the weigh table. The fire that followed the explosion caused most of the damage. No source of ignition was found, but peroxides are easily ignited. After a fire or explosion, the source of ignition is often never found.

The report [11] recommended that the peroxide should be supplied in 20-L (5-gal) containers in the future, so as to reduce the inventory of this unstable substance ("What you don't have can't decompose"), and that housekeeping should be improved. Although contamination of the peroxide was ruled out on this occasion, it was possible for it to occur. It was decided to use a dedicated weigh station in the future.

Often during the investigation of an accident, several scenarios are considered possible but on the balance of evidence, one is considered more likely than the others. If the others could have occurred, as was the case here, then we should take actions to prevent all possible/likely causes in the future.

Other drums have exploded (or bulged) because they were used for waste materials that reacted with each other. People have been injured when removing the lids from bulging drums as the lid flies off as soon as the closing mechanism is released. Empty drums have exploded because vapor from the previous contents was still inside. New drums may contain traces of solvents used by the manufacturers to clean them. Never use drums as access platforms, especially for hot work.

30.12   FOAM-OVER—THE CINDERELLA OF THE OIL AND CHEMICAL INDUSTRIES

I have often drawn attention to the way the same accidents keep recurring, sometimes in the same company, despite the publicity they get at the time [12]. Unfortunately, after perhaps 10 years, most of the people at a plant have left or moved to another department, taking their memories with them. Their successors do not know the reasons for some of the procedures introduced after an accident and, keen to improve output or efficiency, both very desirable things to do, make a number of changes. The accident then happens again.

One accident that keeps recurring, despite frequent publicity, is a foam-over or slop-over. It occurs when hot oil, over 100°C (212°F), is added to a tank containing a water layer, or oil above a water layer is heated above 100°C. The heat travels down to the water, which then vaporizes with explosive violence, often lifting the roof off the tank and spreading the oil over the surrounding area. The oils involved are usually heavy oils or tars, which have to be heated before they can be pumped, and they cover everything with a thick black coat.

Waste liquids were distilled to remove water and light ends and the residue was used as fuel. It was stored in a vertical cylindrical tank ≈ 12 m (40 ft) tall and 3.6 m (12 ft) diameter, volume ≈ 120 m^3 (30,000 gal). The bottom meter of the tank was conical. As the result of a plant upset, some water got into the tank. When hot oil was being run into the tank, the roof parted company with the walls and about 40 m^3 (10,000 gal) of hot black oil was blown out.

The tank had been filled without incident 10 times since the plant upset. A solid crust probably insulated the water in the conical bottom section of the tank until something caused it to move or crack. Calculations showed that as little as 30 kg (65 lb) of water could have produced enough steam to produce the damage that occurred.

To prevent foam-overs, if heavy oil is being moved into a vessel that may contain water, the temperature of the oil should be kept <100°C (212°F) and a high-temperature alarm should be fitted to the oil line. Alternatively, the following steps should be taken:

- Drain the water from the tank.
- Keep the tank >100°C to evaporate any water that leaks in.
- Circulate or agitate the contents of the tank before starting the movement.
- Start the movement at a low rate.

When heavy oil is moved out of a vessel, drawing it from the bottom prevents small amounts of water from accumulating.

This accident illustrates another feature of many industrial accidents: An operation can be carried out many times before a slight variation in conditions results in an accident. A blind man can walk along the edge of a cliff for some distance without falling, but that does not make it a safe thing to do.

The report [13] on this foam-over does not draw attention to the fact that there have been many similar incidents in the past and many published accounts of them; for example, *Hazards of Water* [14], first published in 1955, contains many accounts of tanks and pressure vessels damaged by the sudden vaporization of water. They are also described in Section 12.2. Why then do they keep occurring? Perhaps people ignore reports of past accidents in the belief that the lessons must surely have been learned and incorporated in instructions and codes of practice. But the reasons for them are often forgotten or ignored. And they can never cover every possibility. They can never prohibit every possible action we should *not* take. The best prophylactic is knowledge of the hazards (see also Sections 24.5.2, 29.4, 29.5, and 36.5).

30.13   EXPLOSIONS OF COLD GASOLINE IN THE OPEN AIR

Most unconfined vapor cloud explosions have followed the leak of a flashing liquid—that is, a liquid under pressure above its normal boiling point, such as liquefied flammable gases at ambient temperature or cyclohexane at 150°C, as at Flixborough (see Section 2.1). They leak at a high rate and then turn to vapor and spray.

Only a few unconfined vapor cloud explosions have followed a leak of gas, not surprisingly, as the mass flow rate through a hole of a given size is much smaller, and the gas may disperse by jet mixing. Few explosions have followed a leak of liquid below its boiling point, as the amount of vapor produced is small. If gasoline or a similar liquid is spilled on the ground, the vapor cloud above it is only a few meters thick, too small for an explosion. But if the cloud is dispersed, an explosion is possible.

Davenport [15] lists 71 unconfined vapor cloud explosions, including some partially confined explosions. The majority were due to leaks of flashing liquids—that is, liquids under pressure above their atmospheric boiling points. Thirteen were due to leaks of gas and three to leaks of naphtha/hydrogen mixtures. Eight were due to leaks of liquids that were below their boiling points, though some may have been close to them. Two of these were the result of rail tankers bursting, so the liquid will have sprayed out with great force; one contained dimethyl ether and one acrolein. One explosion was the result of a foam-over so the liquid presumably boiled. Another was the result of the discharge of hot still bottoms from a storage tank. Three involved cold gasoline. One was a pipeline leak and the other resulted from the overfilling of a tank on a ship; both could have produced a large cloud of vapor.

The sixth explosion was the result of overfilling a floating roof tank and led to the formation of a cloud 450 to 600 m long and 60 to 90 m wide. This incident is interesting as there was no bursting vessel to disperse the liquid and it was not hot. It occurred in Newark, New Jersey, in 1983 and is described in references 16 through 18. Estimates of the amount spilled range from 100 to 400 m^3. There is no doubt that there was an explosion and not just a flash fire, as an empty tank 370 m (1,000 ft) from the leaking tank was flattened. The source of ignition was a drum cleaning plant, also about 370 m away, where there was off site damage.

This was the first detailed report I saw of an unconfined vapor cloud explosion resulting from the spillage of a hydrocarbon at ambient temperature and below its boiling point. Such explosions are rare, but the incidents described below, especially the Newark one, show that they can occur if enough liquid is spilled and dispersed. Formation of a vapor cloud is possible if the liquid is discharged at a high level and spills down the side of a tank or other equipment.

30.13.1 Buncefield

In December 2005, at the Buncefield Oil Storage Depot in Hertfordshire, United Kingdom, a large storage tank containing gasoline overflowed through the vent at the top as the level measuring equipment and high-level alarm on the tank, which at the time was receiving liquid, were out of order.

The liquid splashed down the side of the tank and formed a large cloud. It ignited, and an explosion was followed by a large fire. As a result, 45 people were injured and there was extensive damage not only to the depot but to a neighboring industrial estate, where the ignition probably occurred. The premises of 20 businesses employing 5,000 people were destroyed, and those of 60 businesses employing 3,500 people were damaged and made temporarily unusable. The explosion, said to have been the largest in peacetime Europe, occurred at 6 a.m. on a Saturday. If it had occurred during a working day, the injuries would have been much greater. (The explosion may well have damaged a larger area than earlier ones, but it was not the largest in its results. In 1921, the explosion of a mixture of ammonium nitrate and sulfate at Oppau in Germany killed 500 to 600 people.)

The industrial estate had been sited near the depot and allowed to expand as all those concerned were unaware of the Newark explosion and believed that cold oil could not explode in the open air. The defense of the group of oil companies that owned the depot was that such an event had never been known to have occurred. In fact, as well as the Newark explosion, a number of similar events had occurred and reports on them had been published, for example, at Naples, Italy, in 1995 [19] and at St. Herblain, France, in 1991 [20]. Other examples can be found by searching Google for "Gasoline spills resulting in vapor cloud explosions." Damage at Buncefield was, however, more extensive than at Newark and elsewhere.

Section 5.1 describes other cases of overfilling, and reference 21 describes the various types of level measurement equipment that are available.

The Buncefield Investigation Board has published a report on the disaster [30]. In Appendix 4, page 87, it says, "The violent explosion at Buncefield was deemed to be unprecedented at the time, although a review of the literature revealed that this was not strictly correct." This is an understatement. The published evidence, quoted above, shows that that similar events had happened before, but most of those who ought to have known this never learned it or had forgotten it.

30.14   THE INEVITABILITY OF IGNITION

We are all familiar with the fire triangle: air, fuel, and a source of ignition are necessary for a fire or explosion to occur; remove one of the three, and an explosion is impossible. However, when flammable gases or vapors are handled on an industrial scale, this view, though theoretically true, is misleading. If flammable gases or vapors are mixed with air in flammable concentrations, experience shows that sources of ignition are likely to turn up as the amount of energy required to ignite a flammable mixture can be very small. As already stated, it is as little as 0.2 mJ, the energy produced when a 1 cent coin falls 5 mm, though the energy has to be concentrated into a small time and space as in a spark or speck of hot metal.

As an alternative to the fire triangle, I suggest

$$AIR + FUEL \rightarrow BANG \text{ or } AIR + FUEL \rightarrow FIRE$$

What are these mysterious sources of ignition? Sometimes it is static electricity. A steam or gas leak, if it contains liquid droplets or particles of dust, produces static electricity that can accumulate on an ungrounded conductor—such as a piece of wire netting, a scaffold pole, or a tool—and then discharge to ground. Discharges may even occur from the cloud itself. In other cases, ignition may be due to traces of pyrophoric material, to traces of catalyst on which reactions leading to local high temperatures may occur, to friction, or to the impact of steel on concrete. Reference 22 lists many other possible sources of ignition.

The only safe rule is to assume that mixtures of flammable vapors in air in the explosive range will sooner or later catch fire or explode and should never be deliberately permitted, except under carefully defined circumstances where the risk of ignition is accepted, for example, in a small grounded tank containing a conducting liquid.

This chapter has described several explosions, of varying severity, that occurred because those responsible for design and operations did not realize the inevitability of ignition and assumed that because they had removed the obvious sources of ignition, a fire or explosion was impossible. It is hubris to imagine that we can infallibly prevent a thermodynamically favored event [23].

30.14.1 The Aviation Industry

Whereas most of the chemical and oil industries have now learned that mixtures of flammable vapor and air are likely to ignite, the same is not true of the aviation industry. According to Erdem Ural [24], the vapor spaces of the so-called center wing tanks on large airplanes such as 747s (they are actually located between the wings and above the cabin) are often near heat sources and are flammable for more than one-third of the operating hours. As a result, a number of explosions have occurred. The vapor spaces can become flammable in four ways:

- The flash point of the fuel can be as low as 40°C (105°F) and falls as the air pressure falls.
- The temperature in the tank may rise if the plane is exposed to the sun or if the air conditioning is left running for a long time.
- Cooling and vibration can produce a mist, which will have a much lower flash point than the vapor.
- Oxygen is more soluble in fuel than nitrogen, and therefore the gas released when the pressure falls is enriched in oxygen.

There have been about 18 explosions since 1960, some while airplanes were on the ground but including the explosion on TWA flight 800 in 1996, which killed 296 people. The source of ignition in this case is believed to have been faulty electrical equipment, and similar faults have since been found on other old 747s. Two earlier explosions in the air occurred in 1959 and 1963 [25, 26]. The causes of ignition were lightning strikes, but it seems that at the time many people thought that the lightning alone had destroyed the planes and did not realize that the lightning was just the triggering event that ignited the flammable mixture in one of the fuel tanks, which are vented to atmosphere. I remember reading the newspaper reports at the time and wondering how lightning could harm a plane in midair as distinct from one on the ground.

The aviation industry at first claimed that blanketing the tanks with nitrogen would be so expensive that it would not be "reasonably practicable" (to use the U.K. legal phrase), though Ural claims that it would amount to no more than a few dollars per flight. However, the U.S. Air Force already used inerting systems, and systems for commercial aircraft are being developed. They are designed to reduce the oxygen concentration to 12% instead of the 10% or better usually achieved in ground-based systems, as this reduces the cost by 75%. The minimum oxygen concentration needed for an explosion is said to be just under 12% at ground level but 14.5% at 30,000 ft so the margin of safety will be zero or small [27–29]. However, mixtures that are only just flammable are harder to ignite and develop less pressure. The U.S. Federal Aviation Authority has pressed for the adoption of this system and says it will "close the book" on fuel tank explosions. However, this is not entirely true, as experience in the chemical and oil industries shows that inerting systems are likely to lapse unless a continuous management effort is made to make sure that they are kept in full working order. Frequent or, better, continuous analysis of the oxygen content is necessary.

After the explosion on flight 800, the electrical equipment on all 747s was modified so that it could not ignite a flammable mixture. Some airlines argued that the hazard had been removed and there was now no need for blanketing. This is incorrect as other sources of ignition, including lightning, could ignite a flammable mixture, as described in the first part of Section 30.13.

30.14.2 Conclusions

* Mixtures of flammable gas or vapor and air in the explosive range are likely to ignite and explode even though we try to remove all sources of ignition. The probability of ignition is so high that designers and operators should assume that it is inevitable and design and operate accordingly.
* What is well known in some industries or companies may not be known in others. We should study accidents in other industries as well as our own.
* The knowledge of past incidents is easily forgotten. Companies and industries need a conscious systematic approach to make sure that they learn from experience and do not forget what they have learned (see Chapter 31).

Many of the other incidents in this book support the last two conclusions.

References

1. Anon, The largest explosion in the UK, Fauld 1944. *Loss Prevention Bulletin*, Vol. 103, Feb. 1992, pp. 17–18.
2. Y. Riezel, Explosion and fire in a gas-oil fixed roof storage tank. *Process Safety Progress*, Vol. 21, No. 1, 2002, pp. 67–73.
3. A. H. Searson, Explosion in cone-roof gas-oil tank. *Proceedings of 4th International Conference on Loss Prevention and Safety Promotion in the Process Industries*, Institution

of Chemical Engineers, Symposium Series No. 80, Institution of Chemical Engineers, Rugby, UK, 1983.

4. R. Zalosh, A tale of two explosions. *Proceedings of the AIChE Annual Loss Prevention Symposium*, Mar. 2000.

5. D. Chung, Explosions and fire at Powell Duffryn Terminals, Savannah, Georgia. *Proceedings of the AIChE Annual Loss Prevention Symposium*, March 2000.

6. T. J. Myers, H. K. Kytömaa, and R.J. Martin, Fires and explosions in vapor control systems. *Proceedings of the AIChE Annual Loss Prevention Symposium*, March 2002.

7. T. A. Kletz, *Learning from Accidents*, 3rd edition, Elsevier, Boston, MA, 2001, Chapter 26.

8. R. Baker and A. Ness, Designing and operating thermal oxidisers, *Loss Prevention Bulletin*, Vol. 146, 1999, p. 8.

9. D. C. Hendershot, A. G. Keiter, J. Kacmar, J. W. Magee, P. C. Morton, and W. Duncan, Connections: How a pipe failure resulted in resizing vessel emergency relief systems, *Process Safety Progress*, Vol. 22, No. 1, 2003, pp. 48–56.

10. Q. A. Baker, D. E. Ketchum, and K. H. Turnbull, Storage tank explosion investigation. *Proceedings of the AIChE Annual Loss Prevention Symposium*, March 2000.

11. R. F. Antrim, et al, Peroxide drum explosion and fire, *Process Safety Progress*, Vol. 17, No. 3, 1998, pp. 225–231.

12. T. A. Kletz, *Lessons from Disaster—How Organisations Have No Memory and Accidents Recur*, Institution of Chemical Engineers, Rugby, UK, 1993.

13. R. A. Ogle, Investigation of a steam explosion in a petroleum product storage tank, *Process Safety Progress*, Vol. 17, No. 3, 1998, pp. 171–175.

14. Anon, *Hazards of Water*, Institution of Chemical Engineers, Rugby, UK, 2004.

15. J. A. Davenport, *4th International Symposium on Loss Prevention and Safety Promotion in the Process Industries, Vol. 1, Safety in Operations and Process*, Institution of Chemical Engineers, 2003, p. C1.

16. Anon, Report on the incident at the Texaco Company's Newark storage facility, 7 January 1983, *Loss Prevention Bulletin*, No. 057, June 1984, pp. 11–15. Reprinted in *Loss Prevention Bulletin*, No. 188, Apr. 2006, pp. 10–13.

17. M. F. Henry, NFPA's consensus standards at work, *Chemical Engineering Progress*, Vol. 81, No. 8, 1985, pp. 20–24.

18. T. A. Kletz, Can cold petrol explode in the open air? *The Chemical Engineer*, June 1986, p. 63, Reprinted in *Loss Prevention Bulletin*, No. 188, April 2006, p. 9.

19. G. Russo, M. Maremonti, E. Salzano, V. Tufano, and S. Ditali, Vapour cloud explosion in a fuel storage area; a case study, *Process Safety and Environmental Protection*, Vol. 77, No. B6, 1999, pp. 310–365.

20. J. F. Lechaudet and Y. Mouilleau, Assessment of an Accidental Vapour Cloud Explosion, *Loss Prevention and Safety Promotion in the Process Industries*, Vol. 314, 1995, pp. 377–378.

21. K. Schmidt, *Chemical Engineering*, Vol. 115, No. 7, 2004, p. 34.

22. J. Bond, *Sources of Ignition*, Butterworth-Heinemann, Oxford, UK, 1991.

23. P. G. Urben, Book review: Learning from accidents in industry, *Journal of Loss Prevention in the Process Industry*, Vol. 2, No. 1, 1989, p. 55.

24. E. A. Ural, Airplane fuel tank explosions, *Proceedings of the AIChE Annual Loss Prevention Symposium*, March 2003, pp. 463–481.

25. P. B. Friedrich, *The Explosion of TWA Flight 800*, Chelsea House, Philadelphia, PA, 2002.

26. C. Negroni, *Deadly Departure*, Cliff Street Books, New York, 2000.

27. J. Croft, FAA "breakthrough": Onboard inerting, *Aviation Week and Space Technology*, 6 Jan. 2003, pp. 37–39.

28. F. Fiorino, Reducing risks, *Aviation Week and Space Technology*, 4 Aug. 2003, pp. 36–37.

29. F. Fiorino, Fuel tank rule costly, *Aviation Week and Space Technology*, 3 Feb. 2004, p. 44.

30. T. Newton, (Chairman), *The Buncefield*:11 December 2005. The Final Report of the Major Incident Investigation Board, Vol. 1, Buncefield Major Incident Board, London, 2008.

Poor Communication

I've been round the block a few times and have made mistakes. I'd like to think I can pass on my experiences so we don't make the same mistakes twice. One way I do this is to sit amongst the staff. I don't have a separate office. We mix up the staff so that junior executives sit with senior people. This helps us retain the feeling of being a cohesive team. People who lock themselves away in an office have less of an understanding about what's happening in a business.

—Stephen Morris, quoted by J. Oliver,
in *Daily Telegraph* (United Kingdom)

The whole of this book is about poor communication. When an accident occurs in the process industries, outsiders might think that it happened because no one knew how to prevent it. Although the people at the plant at the time, or the designers, may not have known, the information is almost always available somewhere. Few accidents occur because no one knew that there was a hazard. Sometimes the behavior of a compound or reaction takes everyone by surprise, but in most cases this is the result of inadequate testing, the need for which is well known. In this chapter, we look at communication in a narrower sense.

31.1 WHAT IS MEANT BY *SIMILAR?*

Some changes had to be made to a length of low-pressure ventilation ductwork that was <0.6 m (24 in.) in diameter. To keep the rest of it in operation during the modification, a bypass of almost the same diameter was made around the affected section. To isolate this section,

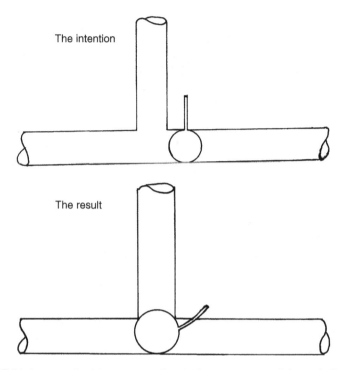

FIGURE 31-1 An inflatable stopper with a rigid stem was wanted, but a *similar* one was fitted instead.

the contractor was told to drill a hole in the main duct and push an inflatable rubber balloon through it. This is a standard item of equipment that had been used successfully on previous occasions. The drawing specified "[manufacturer's name] inflatable pipeline stopper or similar." This manufacturer's stopper is fitted with a metal inflation tube that ensures that the balloon remains in position beneath the insertion hole. The contractor instead used a balloon fitted with a flexible tube. The inflated balloon moved a little way down the duct and blocked the bypass line (see Figure 31-1). Ventilation flow stopped. The operators in the control room had been warned that changing over to the bypass line might cause the low flow alarm to operate, and therefore they ignored it. Some time elapsed before they realized what had happened.

The immediate cause of the incident was therefore the use of the word *similar*. What seems similar to one person seems dissimilar to another. (To some people, bats are similar to birds; to others, a bat—*fledermaus* in German—is more like a flying mouse.) The word *similar* should never be used in specifications or instructions.

Another word that should not be used is *all*. If someone is asked to remove all the slip-plates from a tank or to lubricate all the machines in a unit, he or she does not know whether there are two, three, four, or many.

Each one should be specified by name or number (see Section 24.1). Other words that should not be used are adjectives such as *large*, *small*, *long*, or *short* (see Section 25.1.5).

During an incident investigation, it is often useful to ask why the incident happened when it did and not at some other time. Asking this question disclosed another cause in the incident under discussion.

The process foreman would normally have taken a close interest in the job. He might have spotted that the wrong sort of balloon was being used and he would certainly have been better able than the operators to handle the plant upset that occurred when the ventilation flow stopped. However, he was busy with other changes being made elsewhere at the plant. Why were two major changes being made at the same time? They had both been requested by the regulators who had, as usual, agreed on a timescale with the company, a necessary requirement as otherwise nothing would ever get done. However, design and procurement had taken longer than expected, plant problems had caused delays and the agreed-upon date was only a few days away. If approached, the regulators would probably have agreed to a delay, but the people involved may not have realized this or been reluctant to admit that they could not achieve what they had agreed to achieve. People are sometimes accused of taking chances to maintain production; on this occasion, they took a chance in order to satisfy the regulators and to complete a safety job on time. The result was less safety and lost production.

The story got around, reaching the local press, which reported that the plant upset occurred because the operators had ignored an alarm. This was in a sense true but misleading. They had ignored the alarm because they had been told to do so. Do not believe rumors—or newspapers.

31.2   MORE *SIMILAR* ERRORS

A company had a thorough system of vessel inspection. Most of the vessels were inspected every few years, but if a group of similar vessels were used on a similar duty, just one vessel in the group was inspected every 2 years. If no corrosion was found, the other vessels were not inspected until their turn came around. The maximum period between inspections was 12 years.

However, what is a *similar duty*? After an absorption tower on a nitric acid plant had leaked, it was realized that it operated at 100 to 125 degrees C (212 to 257 degrees F), whereas the other towers in the inspection group operated at 90°C (194°F). The higher temperature increased the rate of corrosion.

Similarly, a change was made to the transmission of a two-rotor helicopter. The manufacturer decided to test the new design on the aft transmission, as in the past it had developed slightly more problems than the forward transmission. The new design passed the tests but failed in service on the forward transmission. The helicopter crashed, killing 45 people [1].

31.3   WRONG MATERIAL DELIVERED

On many occasions, the wrong material has been delivered. Here is an example [2].

A U.K. chemical company ordered a load of epichlorohydrin, a toxic and flammable chemical, from a supply agent and not from the manufacturer. A transport company collected the chemical from the manufacturer and changed the delivery note to one bearing the name of the agent. The man who did so made a slip and entered the number of the wrong tank container, one containing sodium chlorite. This container was therefore delivered to the company and off-loaded into the epichlorohydrin tank. A violent reaction and explosion occurred, several people were injured, and large amounts of fumes and smoke led to the closure of main roads and a major river crossing. Fines and costs amounting to $150,000 were imposed on the company for not testing the material before off-loading it and on the transport company for delivering the wrong material.

There have been many similar incidents. Before accepting any process materials, companies should sample and analyze them to confirm that they are the material ordered. Some companies that used to do this stopped doing so when their suppliers were able to show that their procedures met quality standards. However, there are too many opportunities for error in the course of filling, labeling, and transporting to justify this action, or rather inaction. There are further examples in Chapter 4 and Section 25.2.8 of this book, and Section 28.4.1 shows that engineering materials should also be checked.

In the case described, changing the paperwork en route introduced an avoidable opportunity for error.

31.4   PACKAGED DEALS

When companies buy equipment such as boilers or refrigeration units that are sold already fitted with instruments and relief devices, they do not always check to see that the equipment complies with their usual safety standards, or even with acceptable standards, or that relief valve sizes have been estimated correctly. Here are some incidents that have occurred as a result:

- A contractor supplied nitrogen cylinders complete with a frame for holding them, a reducing valve, and a hose for connecting them to the plant. The end came off the hose and injured the operator, fortunately not seriously. If anyone in the company had tried to order the hose as a separate item, it would never have made it through the purchasing procedure. An engineer would have examined the drawing and found that it did not conform to the company's standards. Furthermore, any hose acquired in this way would have been registered for a regular pressure test. But as part of a *package deal*, it slipped through.

- A reciprocating compressor was started up in error with the delivery valve closed. The relief valve was too small and the packing around the cylinder rod was blown out. The compressor had been in use for 10 years, but the users did not know that the relief valve was merely a *sentinel* valve to warn the operator and that it was incapable of passing the full output of the compressor.

- A reciprocating pump was ordered that was capable of delivering $2 \text{ m}^3/\text{hour}$. The manufacturer supplied his nearest standard size, which was capable of delivering $3 \text{ m}^3/\text{hour}$, but sized the relief valve for $2 \text{ m}^3/\text{hour}$. When the pump was operated against a restricted delivery, the coupling rod was bent. Fortunately, it was the weakest part of the system.

- A specialist contractor was making an underpressure connection to a pipeline when a ¼-in branch was knocked off by a scaffolding plank. The company did not allow ¼-in connections on process lines—all branches up to the first isolation valve were 1 in. minimum—but they did not check the contractor's equipment.

- The support legs on a tank trailer, used to support the tank when it was not connected to a tractor, were designed in such a way that they could not be lubricated adequately. Several failures occurred.

- A relief valve, supplied with a compressor, was of an unsuitable type and was mounted horizontally and vibrated so much that the springs dented the casing. Relief valves should be mounted vertically so that any condensation or dirt that collects in them has the maximum chance of falling out.

- Packaged equipment may not use the same threads as the main plant. This is probably a bigger problem in the United Kingdom than in the United States, as several different types have been in use in the United Kingdom during the lifetimes of old plants.

31.5 "DRAFTSMEN'S DELUSIONS"

Elliott [3] uses the term "draftsmen's delusions" to describe problems that occur because the beliefs of the drawing office differ from the reality of the plant. Others call them misconceptions [4].

For example, a small solvent drying unit was designed to operate at a pressure of 2 bar gauge (30 psig). The drying chambers had to be emptied frequently for regeneration, so a nitrogen connection was needed. The designer looked up the plant specifications and found that the nitrogen supply operated at a pressure of 5.5 bar gauge (80 psig). This was far above the unit's operating pressure, so the designer assumed there was no danger of the solvent entering the nitrogen main by reverse flow and supplied a permanent connection. (He supplied a check valve in the line, but these are not 100% effective. They would be more effective if they were regularly maintained but rarely are; we cannot expect equipment containing moving parts to work forever without maintenance.)

If the designer had asked the operating staff, they would have told him that the unit was to be located near the end of the nitrogen supply line and that its pressure fell to <2 bar gauge when other units were using a lot of nitrogen. If the designer had ever worked at a plant, he would have known that it is by no means uncommon for nitrogen supply pressures to fall, especially when large units are being shut down for maintenance or are being swept out ready for startup.

On the drying unit some solvent, which was flammable, entered the nitrogen main by reverse flow and then entered another vessel where it exploded [5].

If the designer had known that the nitrogen supply was unreliable, he would have fitted a low-pressure alarm to the supply and a more positive isolation on the connection to the plant (such as double-block-and-bleed valves or a hose that can be disconnected when not in use).

There were two other design failings: The vessel in which the explosion occurred was made from thin iron sheets about 1.6 mm ($\frac{1}{16}$ in.) thick. It is impossible to make such a structure gas-tight, and so air leaked in and made the mixture of nitrogen and solvent explosive. The designer was not aware that this could occur. Neither was the construction worker who installed it nor the members of the startup team (or if they did think it looked unsuitable, they may have assumed that the designer knew what he was doing).

The final design failure was the location of the vessel. It was in a zone 2 area—that is, an area in which a flammable mixture of vapor and air is unlikely to occur in normal operation and if it does occur will be present for only a short time. The vessel, which contained electrical equipment, was therefore blanketed with nitrogen. However, if the vessel had been moved a few meters, it would have been in an unclassified (that is, safe) area. It is unusual to find three communication failings communicating to the same incident. All three were necessary for the explosion to occur.

I am not blaming the designers. It was not their fault that they had never worked on the plant or that it was not customary in the company at the time to involve the operations staff in the details of design. A hazard and operability study might have brought the design weaknesses to light. Such a study had been carried out on the main plant, but the little drying unit seemed so simple that such a study seemed unnecessary. Also, if by chance the electrical engineer and the person who decided on the location of the unit had been personal friends and had discussed their current tasks, the desirability of moving the unit might well have come to light. There is more about this incident in Section 32.9 and in reference 5.

There are several actions that can prevent, or make less likely, incidents such as those just described. The most effective is to set up, for every project, a team that includes representatives of all the departments involved: research, process design, mechanical engineering (including certain specialists), construction, and operations. Everyone in the team will be able to see how his or her actions affect, for better or worse, the tasks of other departments. This happens in a hazard and operability study (Hazop) when the line diagrams are examined in detail, but it should also happen at other stages of design. At the earlier stages, the team should be looking for alternatives rather than, or as well as, deviations.

It is also desirable to develop a culture in which people are encouraged to take an interest in other people's tasks, the opposite of a culture in which people are told only what they need to know. We don't know what we need to know until we know what there is to know. If the electrical engineer knew how the electrical area classifications were decided and had easy access to the layout drawings, he might have realized that the drying unit was near the edge of the division 2 area and that moving the unit a short distance would take it into a safe area.

To quote from a recent report [7]:

> We need to move quickly to a position where all engineering graduates are comfortable working as part of an integrated system team, rather than as isolated specialists working within a tightly defined box. This does not imply that they will be "dumbed down" to become low level generalists, but that they will be far more able to make a specialist contribution to a complex system by understanding better what the system, as opposed to the component, is intended to deliver.

31.6   SAME PLANT AND PRODUCT, BUT NO COMMUNICATION

A company manufactured mononitrotoluene (MNT) on two units on the same site. One used a batch process and the other a continuous process. In 1996, the company analyzed the hazards of the batch process. Thermal tests showed that to prevent decomposition, the temperature of the MNT should be kept below 188°C (370°F). This was written into the operating instructions, and an interlock system was installed to prevent the batch distillation equipment from overheating.

In 2002, the continuous distillation column was shut down with 4.5 m^3 (1,200 gal) of MNT in the column. The steam valves to the MNT column were closed, and in addition the steam supply to the whole MNT unit was isolated. Five weeks later the main steam supply was restored, and the valves leading to the MNT distillation column were kept closed. Unfortunately, these valves were leaking and the MNT in the column gradually got hot and then exploded. The top 10 m (35 ft) of the column were blown off, and a part of it punctured a tank containing 380 m^3 (100,000 gal) of MNT and set it alight. Fortunately, no one was killed or seriously injured, though several men were covered in broken glass.

The investigation showed that the results of the 1996 studies of the MNT hazards had never been passed on to the people working on the continuous plant. They did not know that the temperature of MNT should be kept below 188°C (370°F).

As usual, there were other things wrong as well: inspection and maintenance of the steam valves had been neglected, the pressure control system on the distillation column was inadequate, and there was no high-temperature alarm or interlock [8].

31.7   A FAILURE AT THE DESIGN/ CONSTRUCTION INTERFACE

The Yarra River box girder bridge in Melbourne, Australia, collapsed during construction in October 1970, primarily because the construction team did not realize that this then-new type of bridge required a change in construction practice. It was (and probably still is) normal practice in the construction industry to force together two components that did not quite fit. However, in box girder bridges, the components must fit exactly; if they do not, one or both of them should be modified until they do. The designers failed to get this message across to the construction team [9].

31.8   FAILURE OF COMMUNICATION BETWEEN MARKETING AND TECHNOLOGY

A company foresaw a market for an ethylene oxide derivative. The company did not want to spend capital on a speculative venture, so the marketing department personnel therefore looked for a toll manufacturer. They found one able to undertake the task and signed a contract without consulting any of the technical staff. The toll manufacturer was quite capable but unfortunately was located in a builtup area. When it was realized that ethylene oxide was being handled there, this gave rise to some concern.

A few years later, the houses in the area around the plant were demolished as part of a slum clearance scheme. The regulators then refused permission for new ones to be built in their place. Before they could develop the site, the town council had to pay the toll company to move its plant to a new location.

This incident occurred some years ago, before the present-day regulations came into force. It probably could not happen today, but it is a warning that outsourcing of products or services is a change that should, like all changes, be systematically considered before it takes place.

31.9   TOO MUCH COMMUNICATION

A password had to be entered into a control computer before it was possible to override a software interlock. The monthly test of the interlock showed that unknown to the operators it had been overridden. It was then found that the password had been given, officially, to 42 people! We cannot expect every one of 42 people to keep a secret or not to misuse it.

If an interlock, trip, alarm, or any other protective device has to be overridden or taken out of use, via a computer or in any other way, this should be signaled in a clear and obvious way, for example, by a light on the panel, a note on the screen, or a prominent notice.

31.10   NO ONE TOLD THE DESIGNERS

There have been many failures of equipment because the wrong grade of steel was used (see Section 28.4.1 and Chapter 16), but most of them have been the result of errors by suppliers, construction teams, or maintenance teams. Here is one with a different cause.

The failure of a boiler tube in a power plant caused a steam explosion— that is, the rapid vaporization of water. It wrecked the combustion chamber and surrounding equipment. The tube failed because the grade of steel specified by the designer was unsuitable for the duty. What was worse, the same company built an identical boiler, using the same grade of steel, after the failure. It also failed. The underlying cause was not the failure of the steel but the failure of the company's internal communication system [3].

Thus, we end this chapter as we began. Many companies have no formal or informal procedure for passing on information on the causes of accidents and the action needed to prevent them from happening again. In the United Kingdom, the regulators have instructed at least one major company to set up a formal system.

Commenting on the explosion at Longford, Australia, in 1998 (see Section 26.2), Hopkins writes [6], "The operators were quite willing to report. The problem was that the system at the time did nothing with the reports."

31.11 CONCLUSIONS

The need for more formal and informal discussion between people in different functions was described in Section 31.5.

According to S. Sebastian [10], "The most important part of any business or organisation is the interaction of people. Technical individuals tend to disregard this inescapable fact." It is important to interact with our colleagues, but it is also important to interact with those in other departments, those above us and, most important of all, those below us. To quote from an interview with Mark Cooper, a railway maintenance engineer who has improved the reliability of many U.K. railway vehicles [11]:

> He respects Japanese car manufacturers such as Toyota and Nissan for pursuing this ideal. They have successfully inverted the pyramid, so that the guy fitting the windscreen wiper on the production line is the most important person in the organization: everything in it is geared to making his job easier, better, safer and faster. If he needs a piece of kit it is the management's job to run around and make sure he has it.

Equally important, it is management's job to see that all employees understand what they are doing and why and are willing to query what they do not understand. The following quotation is from a former soldier, Sir Philip Trousdell [12]:

> If you can't communicate I don't think you should be in the leadership business. You need to be able to articulate clearly and unambiguously to the people who are going to implement the decision, so that they have no doubts about what's going on. And then you need to start wearing out the shoe leather, going round all the levels of your organisation, explaining where they fit in, what their bit is in this great scheme that you've dreamt up, so that they not only understand what's required but have the chance to ask you questions.

John Timpson, the chief executive of a chain of UK shoe repair shops, writes, "I spend two days every week visiting our shops; blunt talking from the colleagues I meet is much better than any market research [13]."

References

1. UK Air Accidents Investigation Branch, *Report on the Accident to Boeing 234LR, G-BWFC 2.5 miles east of Sumburgh, Shetland Isles on 6 November 1986*, Her Majesty's Stationery Office, London, 1989, p. 36.
2. Anon., Action following the explosion at Avonmouth on 3 Oct. 1996. *Loss Prevention Bulletin*, Vol. 153, June 2000, pp. 9–10.
3. P. Elliott, Overcome the challenge of corrosion. *Chemical Engineering Progress*, Vol. 94, No. 5, 1998, pp. 33–42.
4. B. P. Das, P. W. H. Chung, J. S. Busby, and R. E. Hibbered, Developing a database to alleviate the presence of mutual misconceptions between designers and operators of process plants, *Hazards XVI: Analysing the Past, Planning the Future, Symposium Series No. 148*, Institution of Chemical Engineers, Rugby, UK, 2001, pp. 643–654.

5. T. A. Kletz, *Learning from Accidents*, 3rd edition, Elsevier, Boston, MA, 2001, Chapter 2.
6. A. Hopkins, Lessons from Longford—The Trial. *Journal of Occupational Health and Safety—Australia and New Zealand*, Vol. 118, No. 6, Dec. 2002, pp. 1–72.
7. Anon., Integrated System Design: A New Teaching Challenge, Note issued by the UK Royal Academy of Engineering, London, introducing a new publication, *Creating Systems That Work—Principles of Engineering Systems for the 21st Century*, 2007.
8. Anon., *Reactive Explosion at First Chemical Corporation*, US Chemical Safety and Hazard Investigation Board, Washington, DC, October 2004.
9. Anon., *Report of Royal Commission into the Failure of the West Gate Bridge*, State of Victoria Government Printer, Melbourne, Australia, 1971.
10. S. Sebastian, Succeed in the workplace. *Chemical Engineering Progress*, Vol. 103, No. 1, 2007, pp. 52–53.
11. J. Abbott, Getting Metronet back on track. *Modern Railways*, Vol. 64, No. 700, 2007, pp. 60–63.
12. M. Southon, *Daily Telegraph* (London), 28 August, 2007, p. B4.
13. J. Timpson, Listen but, in the end, trust your instincts, *Daily Telegraph* business news (UK), 30 April, 2009, p. B10.

I Did Not Know …

The recipe for perpetual ignorance is: be satisfied with your opinions and content with your beliefs.

—Elbert Hubbard (1859–1915)

This chapter describes incidents that occurred because some of the properties of the materials and equipment used were unknown to those who handled them.

32.1   … THAT METALS CAN BURN

The use of thin metal packing increased during the 1980s, and this change was followed by an increase in metal fires. Many people did not realize that metals burn readily when they are in the form of powders or thin sheets and can produce higher temperatures than oil fires. Aluminum and iron, not normally considered flammable, as well as titanium and zirconium, can burn when in these forms, and the fires are difficult to extinguish. Small amounts of water may be decomposed into hydrogen and oxygen and can worsen the fire. Water should not be used for firefighting unless a large quantity is available to quickly drench a very small fire. Burning can continue in atmospheres of carbon dioxide, nitrogen, and steam, and the burning metal can react vigorously with other materials. Argon can be used for firefighting, and special agents are also available [1, 2]. If any metal oxides are present, a hot metal with a greater affinity for oxygen can react with it (the thermite reaction). For example, hot aluminum or titanium can react with the oxygen in rust and produce enough heat for a self-sustaining reaction.

Bulk metal is, of course, more difficult to ignite, but self-sustaining fires have destroyed several titanium heat exchangers, as well as their tube sheets. A strong ignition source, such as a welding torch, is needed, but it need not impinge directly on the titanium. The hot slag formed by cutting steel contains iron oxide and can start a thermite reaction [3].

A fire occurred in a column, ≈ 75 m (250 ft) tall and 8 m (25 ft) in diameter, packed with carbon steel. It started when the packing and column internals were being removed; the fire was ignited by hot work. The fuel was the steel packing, possibly supplemented by process materials that had not been completely removed even though the column had been steamed. Because of the high surface area of packing, it is always difficult to be sure that it is completely clean. Even new packing may be coated with oil.

Hot work should be avoided, if possible, above or below packed beds. If it cannot be avoided—for example, by removing the packing first— then the possibility of a fire should be considered and a plan for dealing with it prepared [4].

Increasing the thickness of metal packing makes ignition more difficult, but this increases the heat produced if the packing does ignite. Decreasing the spacing between the metal sheets also makes it harder to ignite them, but they are then more likely to become contaminated by process liquids and harder to clean. Trade-offs have to be made between these factors and the weight and efficiency of the packing.

32.1.1   Another Metal Fire

This fire occurred in a column that was 22 m (73 ft) tall and had a diameter of ≈ 1 m (40 in.), which contained titanium packing. The performance of the column showed evidence of plugging, so it was taken out of service and prepared for entry. Small pieces of titanium were observed on the redistribution tray above the middle of the three beds.

A flash fire occurred in the packing—perhaps not all the process material had been removed—and a few minutes later a bright spot of glowing metal was noticed. It grew rapidly in size and destroyed a whole section of the packing. The most likely source of ignition was pyrophoric deposits, and the fire may have started in the small pieces of titanium. Although titanium in bulk self-ignites at 1,120°C (2,050°F), powdered titanium ignites at 330°C (625°F). It is not known whether the titanium fire started before or after the flash fire [5] (see also Section 28.1.6).

32.2   ... THAT ALUMINUM IS DANGEROUS WHEN WET

Several tank trucks and trailers were on the lower deck of a roll-on, roll-off ferry when a smell of ammonia was detected on the lower deck. Several loads of dangerous goods were on board, but none of the vehicle drivers could explain the smell. It seemed to come from a trailer carrying metal products. Its sides were hot, and water was dripping out of the bottom.

The trailer contained aluminum waste and turnings, which can produce hydrogen when wet. This has been known for many years, and Bretherick quotes reports dating back to 1947, including a patent for a propellant explosive made from aluminum and water [6]. It is difficult to see how ammonia could be formed. According to the International Maritime Dangerous Goods Code, the load should have been classified as "dangerous when wet" (that is, it can produce a flammable gas on contact with water and sufficient energy to ignite it) and should be packed and labeled accordingly. Ventilation was increased, and the ship reached its destination without incident [7].

32.3 ... THAT RUBBER AND PLASTICS ARE PERMEABLE

A former colleague of mine has described a New Year Ball that did not go exactly as planned [8]:

> We wished to make the New Year Ball particularly spectacular and had arranged for a couple of hundred brightly colored balloons to be released among the revelers from a net suspended from the ballroom ceiling. During the afternoon before the event, we decided that manual inflation of the balloons was far too exhausting, and I ordered a cylinder of compressed carbon dioxide to be sent up from the analytical lab. The balloons, all two hundred of them, were inflated in no time at all and the clusters were hoisted to the ceiling in the releasable net. Imagine our chagrin and extreme embarrassment when, upon arriving for the opening of the ball a few hours later, we found that every balloon had shrunk to the size of a small orange and on eventual release fell to the floor with sickening thuds. I had learned my lesson—India rubber is permeable to carbon dioxide!

When plastic water pipes are run through oil-soaked ground, the water may become contaminated with oil (see also Section 34.2).

In some combustible gas detectors, the sample is drawn through a plastic tube to the measuring element. The plastics used absorb some flammable vapors. It is better to use detectors in which the element is at the end of a lead and can be located at the test point such as the inside of a vessel.

Plastic containers used to collect samples of gas for analysis may absorb some constituents of the gas and make the analysis results incorrect.

32.4 ... THAT SOME PLASTICS CAN ABSORB PROCESS MATERIALS AND SWELL

In the early days of nitroglycerine (NG) manufacture, there were many explosions. These became less frequent and less damaging as the

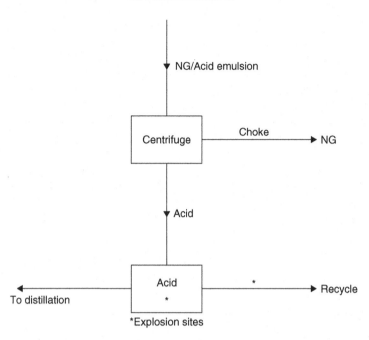

FIGURE 32-1   A choke caused NG to enter the acid line. A Hazop could have foreseen this.

size of reactors was reduced and ultimately the original batch reactors holding approximately a ton of material gave way to continuous reactors holding perhaps a kilogram. Similar reductions in size were made to the washing and separation stages. I have often quoted these changes as an example of intensification and inherently safer design: safer, yes, but not safe [9].

The NG was separated from surplus acid in a centrifuge (Figure 32-1). The NG caused the plastic pipe to swell so that some of the NG passed down the acid line into the acid tank and settled on top of the acid. Two explosions occurred, one in the acid tank and one in the recycle line out of the tank. Vibration probably triggered the first explosion, and the sun's heat probably triggered the second. A Hazop could have prevented the explosions, provided the team realized that "Less of" flow could occur in the NG line.

32.5   ... WHAT LAY UNDERNEATH

Apart from ignorance of the properties of materials, many people are unaware of the way some equipment, particularly old equipment, is constructed. A small tank, capacity $\approx 100\ \mathrm{m}^3$, held 57 tons of a liquid similar

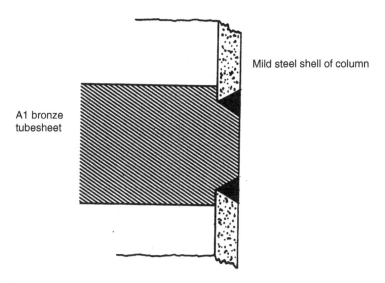

Mild steel shell of column

A1 bronze
tubesheet

FIGURE 32-2 No one at the plant was aware of this unusual method of construction.

to gasoline in its physical properties. All the lines leading to it were disconnected and blanked except for one line in which the two valves were locked off. Nevertheless, in the course of 24 hours the level fell to 50 tons. Dipping confirmed that the fall was real and not an instrument error. No sign of a leak could be seen, even though the tank was sitting on a concrete base in a concrete-lined dike. The tank was emptied and filled with water. Again the level fell.

The drawings were found. To everyone's surprise, they showed that the concrete base was only a concrete ring and that the inside of the ring was filled with sand. Holes were dug round the tank, down to the water table, but no oil was detected. There had been a lot of rain, and the oil had been washed away.

The tank was lifted off its base and the sand replaced by concrete.

32.6   ... THE METHOD OF CONSTRUCTION

An unusual method of construction produced another hidden hazard. A steel fractionation column was fitted with an internal condenser that had an aluminum-bronze tubesheet. It had the same diameter as the vessel but was welded to it in an unusual way, as shown in Figure 32-2. One of the welds cracked in service, and flammable vapor escaped. Fortunately, it did not ignite.

The column had been inspected twice in the five years since it had been built, but nothing unusual was found. It is possible that no special attention was paid to the bimetallic welds, as all the engineers there when the unit was built had left and none of their replacements knew that the construction was unusual.

The underlying cause, of course, was the lack of any system for keeping necessary information extant. Unusual design features and points to watch during inspections should be recorded on vessel registration records and vessel inspection schedules.

Another incident occurred when two laboratory workers, wearing air masks, were attaching a 2-l cylinder of ethylene oxide to some equipment. One of them removed part of the cylinder valve, thinking it was a protective cap. There was an escape of ethylene oxide, which was carried into the ventilation ducting and set off a gas detector alarm. The building was evacuated, and the emergency team removed the cylinder and immersed it in water.

At one time an incident such as this would have been put down to operator error. There were errors, but mainly by other people. According to the report:

- The laboratory workers had not been adequately trained, and their knowledge had not been assessed.
- Some of the instructions were in English rather than in the language of the country where the incident occurred.
- The part that was removed should have been labeled "Do not remove."

Although the alarm system functioned correctly and the correct emergency action was taken, the investigation found that the gas detector was set at maximum sensitivity and often sounded when normal laboratory operations were carried out nearby. This could lead to the alarm being ignored, and its set point was raised [10].

32.7   ... MUCH ABOUT STATIC ELECTRICITY

Static electricity is a common cause of ignition, but many people are not clear about the conditions necessary for it to ignite a flammable mixture. Reports on fires or explosions sometimes quote it as the source of ignition without making it clear exactly how it arose. In the following incident, the people concerned were unsure about the precautions necessary and, in addition, did not realize that the equipment they were using was unsuitable.

A batch plant contained a number of reactors and a number of small storage tanks. Because the spaghetti bowl of fixed piping needed to connect every tank to every reactor would be complex and provide many opportunities for errors and contamination, the plant instead used suction and delivery hoses and a metering pump. There is much to be said for this system, but it introduces different hazards. Hoses are more easily damaged than fixed piping and can be attacked by some process materials. To prevent this, the company specified high-quality hoses, reinforced by metal coils embedded in the plastic, and suitable for all the materials handled.

The metal coils in the hoses were not connected to the end pieces and formed isolated conductors. When a hose became worn, the ends of the spirals protruded into the interior of the hose. The flow of liquid through the hose generated a static charge and an induced charge on the coil; this charge could not flow to ground. Sparks passed between the end of a coil and a metal end piece, which was connected to the plant and therefore grounded. Although most of the time this did not matter, as most of the liquids handled were nonflammable, one process used toluene as a raw material. In this process, a spark could pass through the liquid without igniting it, but once the liquid was displaced by air, a flammable mixture would be formed and an explosion in the vapor space of the inlet vessel was the result. Fortunately, the toluene concentration was near the upper flammable limit and the explosion was not violent [11].

It was unlikely that everyone was ignorant of all the following points, but ignorance was certainly widespread:

- The people who specified the hoses did not ask for the coils to be connected to the end pieces, either because they did not see the need or perhaps because they did not foresee that flammable liquids might be handled.
- The operating staff did not check that the coils were grounded before they used them with a flammable liquid.
- The usual method of checking that a coil is grounded is to measure the conductivity between the two end pieces. However, the hoses contained three separate coils. Even if they were originally connected to the end pieces, this test could not detect a failure of the connections on one coil, as the others would carry the current. Hoses with only a single reinforcing coil should be used (or hoses with external coils so that the each coil could be checked). Three internal coils in a hose may make it mechanically stronger, but there is no easy way of testing their integrity.
- It is not good practice to displace a highly flammable liquid like toluene with air.

Toluene has a low conductivity and any static charge it acquires will drain away only slowly. Conducting liquids lose their charge quickly if their container is grounded. However, this incident could have occurred with any highly flammable liquid, as nongrounded metal acquired an induced charge.

Toluene was moved through the hoses many times before a hose protruded close enough to an end piece for a spark and ignition to occur. This is typical of many accidents. *We have done it this way a hundred times* does not prove an action is safe—unless an accident on the 101st occasion is acceptable.

32.7.1   Another Static Ignition

As already stated, static electricity is often quoted as the source of ignition although it is not clear exactly how it arose. In the following report [12], the investigators went to considerable trouble to establish exactly what probably happened.

Some material was to be added via a hose to an intermediate-size bulk container made from polyethylene and surrounded by a metal cage. There was some water and some highly flammable liquid already in the container, enough to produce a flammable atmosphere. The operator removed the lid from the container and was about to push the hose into the opening when a flash fire burned his face. Nothing had come out of the hose.

Although the operator was grounded, tests showed that contact between his gloves or outer garment and the container could produce an electric charge on the container large enough to ignite the vapor if it discharged as a spark. A wrench and a loose flange were resting on the container, and they could have collected this charge or become charged by induction. The charge from one of them could then have discharged to the metal cage as a spark.

Why did the ignition occur just as the operator was about to insert the hose? (When investigating any fire or explosion, we should always ask why it occurred when it did and not at some other time.) Perhaps the operator, leaning on the container, caused the wrench or the flange to move nearer to one of the bars of the cage, or perhaps the charge passed from the wrench or flange to the operator. Tests showed that such a spark could pass through his clothing.

This incident shows how hard it is to remove all sources of ignition and that the only safe way is to avoid production of a flammable mixture, in this case by inerting the container or perhaps by using a collapsible one so that there is no vapor space.

32.7.2   An Unusual Effect of Static Electricity

A company was filling bags with a powder automatically, using a machine that delivered 50 kg (110 lb) into each bag. Although hand filling

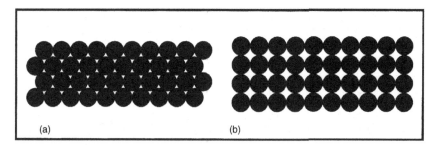

FIGURE 32-3   (a) Close packing of uncharged spheres. (b) Open packing of charged spheres.

showed that this amount of powder would fit into the bags, nevertheless it backed up into the machine and caused it to stall. It had to be stripped down and cleaned before packing could continue.

Experiments showed that rapid filling of the powder lowered the bulk density by 10% as charges of static electricity on the particles pushed them farther apart. Instead of being closely packed as in Figure 32-3a, they were more openly packed as shown in Figure 32-3b. As the output of the plant was large, a machine for eliminating the static charge was designed and installed [13].

There is more about static electricity in Sections 24.2.3, 25.2.7, 28.2.5, and 30.1, and in Chapter 15.

32.8   ... THAT A LITTLE CONTAMINATION CAN HAVE A BIG EFFECT

Some users of X-ray film complained to the manufacturer that there were brown spots on many of the films. The chemist asked to investigate the problem was puzzled. The decisive clue was a comment by another employee that he had not caught any fish in the little local river since a tannery opened 5 to 6 km (3 to 4 miles) upstream. The overalls used in the film factory were washed in this water. Could they be the source of the contamination? Tests showed that the water was contaminated with polyphenols and some of it got on the overalls. A speck of fiber carrying only a few thousand polyphenol molecules and blown onto the films by the ventilation system could produce the spots, though they took two months to develop. The laundry passed the water through an ion exchange purifier before using it, but the ion exchange resin could not remove polyphenols [14].

A tank truck containing a few inches of used motor oil was brought into a workshop for some welding to be carried out underneath the tank.

The welder asked if the contents of the tank were flammable and was told that they were not. When welding started, the tank exploded, causing severe damage but fortunately no injuries. No one realized that used motor oil contains some gasoline, enough to produce a flammable mixture of gasoline vapor and air in the tank, especially on a warm day. This is another example of people not knowing the properties of the material they handled [15].

It was, in any case, bad practice to weld on equipment containing motor oil. All high-boiling-point oils are flammable, though not highly flammable—that is, their flash points are well above ambient temperature. They are not easily ignited, but nevertheless there have been many fires and explosions in equipment containing small amounts of such oils because welding has vaporized the oil and then ignited it (see Section 12.4).

Another incident involving used motor oil occurred when welding was carried out between 1.5 and 3 m (5 and 10 ft) from a small tank containing similar oil. There were several openings in the top of the tank, and welding sparks ignited the gasoline vapor. The tank was mounted on wheels so it could easily have been moved if anyone had realized that it contained an explosive mixture [15].

32.9　... THAT WE CANNOT GET A TIGHT SEAL BETWEEN THIN BOLTED SHEETS

Section 6.1 describes two incidents that occurred because air leaked into ducts made from thin bolted metal sheets. One occurred in a large blowdown system and led to an explosion that was ignited by the flare. The report recommended that joints between nonmachined surfaces should be welded, that there should be a continuous flow of gas to sweep away any leaks that occurred, and that the oxygen content in blowdown systems should be measured regularly.

The second incident occurred in the same plant nine months later because operators at another unit did not carry out the recommendations; perhaps no one told them. A small bolted duct conveyed gland leaks from compressors to a vent stack. Air leaked into the duct, and the mixture of hydrogen, carbon monoxide, and air was ignited by lightning and exploded.

Another incident occurred on a thin metal cabinet containing sparking electrical equipment. As the cabinet was located in a division 2 area, it was purged with nitrogen. The nitrogen supply became contaminated with a flammable liquid by reverse flow from process equipment at a higher pressure (see Section 31.5); later it failed entirely, air leaked in through the bolted joints, and an explosion occurred, injuring one man [16].

32.10   ... THAT UNFORESEEN SOURCES OF IGNITION ARE OFTEN PRESENT

Many incidents have shown that sources of ignition are likely to turn up even though we have tried to remove all those we can foresee. Elimination of ignition should never be accepted as the basis of safety (unless an occasional explosion is acceptable). Nevertheless, explosions still occur because people believe that ignition is impossible.

A vapor-phase oxidation unit consisted of the following:

1. A vaporizer for the raw material
2. A mixing chamber where it was mixed with air
3. A heat exchanger to heat the mixture
4. A flame arrestor
5. A tubular reactor (see Figure 32-4)

The reactor operated in the explosive range but below the auto-ignition temperature. The designers realized that hot spots might form in the reactor and ignite the reaction mixture, so they strengthened the reactor and provided explosion vents. The flame arrestor was installed to prevent the explosion passing back into the heat exchanger. There was no need, they decided, to strengthen or vent the vaporizer, mixer, or heat exchanger as there was no source of ignition in them, or so they thought.

After a two-year operation, an explosion demolished the mixer and damaged the heat exchanger. The probable source of ignition was an

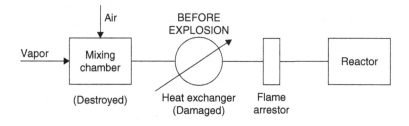

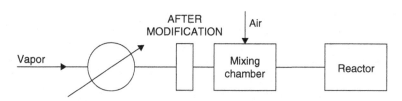

FIGURE 32-4   Layout of equipment in vapor phase oxidation unit before and after explosion in reactor.

unlikely one. The vaporizer had to be cleaned from time to time. Various agents had been used including acids, which had attacked the vaporizer and deposited a mixture of metal and organic residues in the mixer. These oxidized and became hot enough to ignite the flammable mixture of reactant vapor and air in the mixer.

When the plant was repaired, the reactant vapor and air were mixed immediately before entering the reactor. A flammable mixture was then present only in the reactor. This is an inherently safer solution [17]. This could have been done in the original design if someone had realized that flammable mixtures are easily ignited and that we should therefore avoid the need for them when possible and assume they might explode when their presence is essential. In addition to the necessary repairs, the whole plant was strengthened.

Bond [18] summarizes many fires and explosions caused by unsuspected sources of ignition. Other examples of little-known knowledge are presented in Chapter 19.

32.11   ... THAT KEEPING THE LETTER OF THE LAW IS NOT ENOUGH

An explosion in the vapor space of a fixed-roof storage tank caused complete failure of the wall-floor weld and the whole tank, apart from the floor, rose into the air, leaving the contents behind. They caught fire. One man was killed, and eight others were seriously injured. An adjacent tank also lost its contents. Altogether about 4,000 m^3 (1 million gal) of acid was spilled, and some of it contaminated a nearby river.

The tank contained sulfuric acid recovered from an alkylation process and contaminated with a small amount of hydrocarbon, enough to produce a flammable mixture in the vapor space. The tank was supposed to be inerted with carbon dioxide, but its flow rate was too low to prevent air coming in through various openings in the tank, many of which openings had been caused by corrosion. Welding was taking place above the tank, and the probable source of ignition was a spark falling through one of the corrosion holes in the roof or contacting vapor coming out of one of the holes [19].

The following are the main failings that led to the explosion:

- Hot work should not have been allowed so near a tank from which flammable vapor was escaping. The atmosphere was not periodically or continuously monitored.
- The flow of carbon dioxide was too low, either because it was not measured so no one knew what it was or because no one had calculated the flow necessary (see Section 31.5). In this case, both were true. The

carbon dioxide was supplied by a hose pushed through a hole in the roof. Some of it escaped through corrosion holes and some through the overflow pipe, which was shared with tanks vented to the atmosphere.

- The oxygen content of the vapor space was not measured.
- The tank was not provided with a weak seam roof—that is, a wall/roof weld that is weaker than the wall/floor weld so that excessive pressure will cause the wall/roof weld to fail and the liquid will remain in an open cup.
- Thickness measurements and an internal inspection of the tank (and many others) were repeatedly postponed although the company's own inspectors had drawn attention to the need for them and the tank had been emptied several times.
- The dike was big enough to contain the contents of the largest tank within it, but it was not designed to prevent a sudden large release from overflowing. Most dikes are the same [20]. Sudden large releases are rare, but other cases have occurred and there is a case for increasing dike heights if vulnerable sites such as public highways are near them.
- The company claimed that the various regulations on the storage of chemicals did not apply to the contents of the tank. The managers seem to have believed that following the letter of the law in a hair-splitting way was all that was required. Both the law and the managers were at fault. In contrast, in the United Kingdom there is a general requirement to provide a safe plant and system of work and adequate instruction, training, and supervision, so far as is *reasonably practicable* (see the beginning of Chapter 27). The report does not tell us what, if any, training on safety the staff received as students or from their employer.

So many things were below standard at this plant that it is hardly necessary to describe the underlying causes in detail. The senior management of the company seems to have been afflicted by a combination of ignorance and lack of concern. Readers in better-run plants may wonder if there are any lessons for them to learn. However, although it is unusual to find so many faults in one place, each of them has occurred elsewhere on many occasions.

32.12   ... THE POWER OF COMPRESSED AIR

Contractors were removing water from a pipeline, ≈ 1 m (3.3 ft) in diameter, by pushing a foam pig along the line with compressed air at a gauge pressure of 28 bar (400 psig). The water exit line was rather small, ≈ 0.3 m (1 ft) in diameter, so the contractor opened up the end of the pig trap and put a large front end loader in front of it to catch the pig.

The force on the pig was so great, nearly 250 tons, that the pig knocked over the loader and traveled another 150 m (500 ft), destroying a wooden platform on the way. Fortunately, no one was standing in the pig's path at the time.

Many people do not realize the energy in what they call a puff of air or understand the difference between pressure and force. When pressures are measured in pounds per square inch, as they were by the contractor, the full name gets shortened and everyone talks about a pressure of, say, 400 pounds, forgetting or not realizing that this force is exerted on every square inch of the surface. It would be safer to measure pressure in bars or find another name for pounds per square inch.

References

1. M. S. Mannan and H. H. West, Spontaneous combustible substances: a database update, *Proceedings of the Mary Kay O'Connor Process Safety Center Annual Symposium*, College Station, TX, 1999, pp. 267–281.
2. R. Roberts, W. J. Rogers, and M. S. Mannan, Prevention and suppression of metal packing fires, *Proceedings of the Mary Kay O'Connor Process Safety Center Annual Symposium*, College Station, TX, 2002, pp. 123–130.
3. G. E. Mahnken, Watch out for titanium tube-bundle fires. *Chemical Engineering Progress*, Vol. 96, No. 4, 2000, pp. 47–51.
4. M. Kelly, *Safety Alert*, Chevron Phillips, Houston, TX, 18 April 2001.
5. Anon., *Safety Alert*, Mary Kay O'Connor Process Safety Center, College Station, TX, 16 April.
6. P. G. Urben (editor), *Bretherick's Handbook of Reactive Chemical Hazards*, 6th edition, Vol. 1, Butterworth-Heinemann, Oxford, UK and Woburn, MA, 1999, p. 31.
7. Anon., Undeclared dangerous goods problems, *Safety Digest—Lessons from Marine Accident Reports*, No. 1/2001, Marine Accident Investigation Branch of the UK Department of Transport, Local Government and the Regions, London, 2001, p. 29.
8. B. Whitefoot, quoted by D. Claridge, *Memories: Wilton Castle Club*, ICI Chemicals and Plastics, Middlesbrough, UK, 2000, p. 69.
9. N. A. R. Bell, Loss prevention in the manufacture of nitroglycerine, in *Loss Prevention in the Process Industries*, Symposium Series No. 100, Institution of Chemical Engineers, Rugby, UK, 1971, pp. 50–53.
10. T. Fishwick, Ethylene oxide release sets off alarm system. *Loss Prevention Bulletin*, Vol. 167, Oct. 2002, pp. 21–22.
11. T. H. Pratt and J. G. Atherton, Electrostatic ignition in everyday operations: Three case histories, *Process Safety Progress*, Vol. 18, No. 4, 1999, pp. 241–246.
12. G. P. Ackroyd and S. G. Newton, Flash fire during filling. *Loss Prevention Bulletin*, Vol. 165, June 2002, pp. 13–14.
13. A. Pavey, Hidden charges, *Process Safety News*, published by Chilworth Technology, Southampton, UK, No. 4, Autumn/Spring 1997/1998, p. 2.
14. P. Levi, *The Periodic Table*, Michael Joseph, London, 1985, pp. 204–208.
15. R. A. Ogle and R. Carpenter, Lessons learned from fires, flash fires, and explosions involving hot work. *Process Safety Progress*, Vol. 20, No. 2, 2001, pp. 75–81.
16. T. A. Kletz, *Learning from Accidents*, 3rd edition, Elsevier, Boston, MA, 2001, Chapter 2.
17. B. Broeckmann, Explosion protection of mixing unit prior to chemical reactor by pressure resistant design, *Proceedings of the Third World Seminar on the Explosion Phenomenon and on the Application of Explosion Protection Techniques in Practice*, Ghent, Belgium, Feb. 1999.

18. J. Bond, *Sources of Ignition*, Butterworth-Heinemann, Oxford, UK, and Woburn, MA, 1991.
19. Anon., *Investigation Report: Refinery Incident*, Report No. 2001-05-1-DE, US Chemical Safety and Hazard Investigation Board, Washington, D.C., 2000.
20. A. M. Thyer, I. L. Hirst, and S. F. Jagger, Bund overtopping—The consequence of catastrophic tank failure. *Journal of Loss Prevention in the Process Industries*, Vol. 15, No. 5, 2002, pp. 357–363.

33

Control

The main cause of control system failure was inadequate specification.

—*Out of Control* (U.K. Health and Safety Executive)

33.1 INSTRUMENTS THAT CANNOT DO WHAT WE WANT THEM TO DO

33.1.1 Measuring the Wrong Parameter

The pressure of a water supply was normally high enough for it to be used for firefighting. If the supply pressure fell, a low-pressure alarm sounded and an alternative supply of water was then made available. When someone isolated the water supply in error, the trapped pressure in the line prevented the alarm from operating. The instrumentation could do what it was asked to do—detect a low pressure—but not what its designers wanted it to do—that is, detect that the water supply was unavailable (see Figure 33-1).

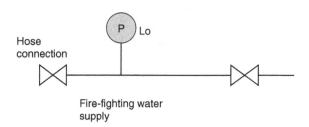

FIGURE 33-1 When the valve on the right was closed, the low-pressure arm became ineffective.

As often happens, something else was wrong as well: the valve in the water line should have been locked open but was not. Valves that are locked open for safety reasons should be listed and checked periodically to make sure that they are still locked. They are part of a protective system.

33.1.2   An Alarm That Immediately Reset Itself

A rotameter was designed to measure a gas flow. If the flow stopped or decreased substantially, the float (bobbin) dropped and interrupted a light beam. This triggered a low-flow alarm.

The design had limitations. If the flow diminished only slightly, the light beam remained broken and the alarm light stayed on after the alarm bell was silenced. However, if the flow fell substantially or stopped completely, the float dropped farther, the light beam was no longer broken, and the alarm light went out (Figure 33-2a). One day when the flow actually failed, the operator canceled the alarm, but with no light to remind him he was distracted by other problems and forgot that the gas flow had stopped. Several hours passed before this was discovered.

Afterward, the design was changed so that the light beam was broken when the flow was normal but fell on the light sensor if the flow changed (see Figure 33-2b). The alarm light then remained lit as long as the low flow continued. As a bonus, it was also activated by a high flow.

Alternatively, the alarm could have been modified so that once it operated, the light stayed on until reset by the operator.

33.1.3   A Trip That Did Not Work under Abnormal Conditions

Carbon dioxide byproduct from an ammonia plant was sent down a long 1,000 m (3,300 ft) pipeline to another unit. The gas normally contained 2% to 3% hydrogen. If the hydrogen content rose >8%, contamination by air could produce an explosive mixture. A trip system was therefore installed to shut down the transfer if this figure was reached. The hydrogen level measurement was based on thermal conductivity.

During shutdowns, the ammonia plant was swept out with nitrogen, which contaminated the carbon dioxide. Nitrogen has twice the thermal conductivity of hydrogen, so the hydrogen measurements were ignored until the nitrogen had been swept out of the pipeline. You have already guessed what happened: air got into the transfer line during this period and an explosion occurred; 850 m (2,800 ft) of the pipeline was destroyed (Figure 33-3).

The source of the air was never identified. Following an earlier incident, different types of connector were used for compressed air and nitrogen hoses, so compressed air could not have been used by mistake

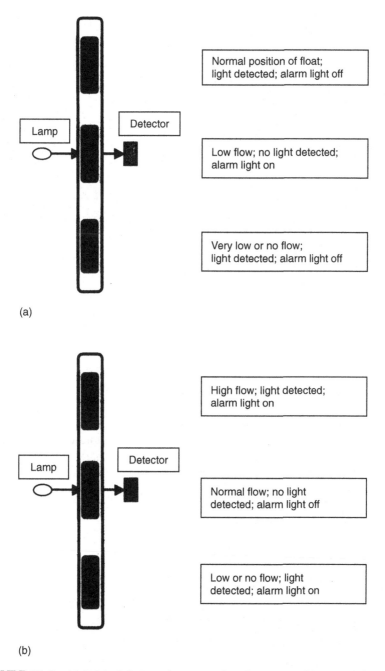

FIGURE 33-2   (a) Original design of rotameter low-flow alarm. (b) Revised design of rotameter low-flow alarm.

FIGURE 33-3 The result of hydrogen and air entering a pipeline. From reference 1. *(Reprinted with the permission of the American Institute of Chemical Engineers.)*

instead of nitrogen for sweeping out the ammonia plant. The source of ignition may have been heat from cutting a bolt.

The report [1] comments, "Looking back it may seem unbelievable.... From management and down there had been a will to make safety a priority. During the previous 10 years, considerable money and resources had been spent. It was a painful surprise. With hindsight anyone can tell how the explosion could easily have been prevented." Afterward, the trip system was modified by making use of a carbon dioxide measurement as well as a hydrogen measurement. And the operators were given a better understanding of the problem.

33.1.4   A Sight-Glass with Limited Range

In this and the following two incidents, the laws of physics prevented the equipment from working in the way the designer intended.

A sight-glass 1.2 m (4 ft) long was connected to vessel branches 0.6 m (2 ft) apart as shown in Figure 33-4. It will indicate the correct level only when the liquid in the vessel is between the two branches. If the liquid level is below the lower branch, the liquid in the sight-glass is isolated and its level cannot fall. If the liquid level is above the upper branch, vapor will be trapped in the upper part of the sight-glass. As the level rises, this vapor will be compressed. If any noncondensable gas is present, the

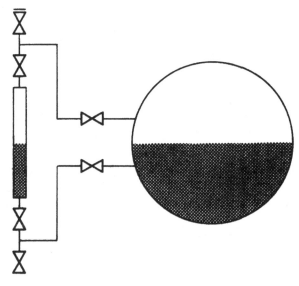

FIGURE 33-4   A level glass with a limited range. From *Chemical Engineering Progress,* July 1980. *(Reprinted with the permission of the American Institute of Chemical Engineers.)*

pressure in the sight-glass will rise and the level in the sight-glass will be depressed below the level in the vessel.

33.1.5   An Explosion in a Nitric Acid Plant

Ammonia was vaporized, mixed with air, and then passed over a catalyst. The ammonia and air flows were measured, and a flow ratio controller was supposed to keep the ammonia concentration below the explosive level (Figure 33-5). The level controller on the vaporizer was out of order, and the level of ammonia was being controlled manually. The level got too high and droplets of ammonia were carried forward. All flow measurements are inaccurate when spray is present, so the flow ratio controller did not detect the increased flow of ammonia and an explosion occurred. The size of the error in the flow measurement depends on the detailed design; if the spray increases the density of the gas by 50%, the flow of vapor and liquid could be 25% higher than the flowmeter reading.

33.1.6   Vapors and Noncondensable Gases Confused

The following has been discovered more than once during hazard and operability studies. A vessel containing a liquefied gas such as liquefied

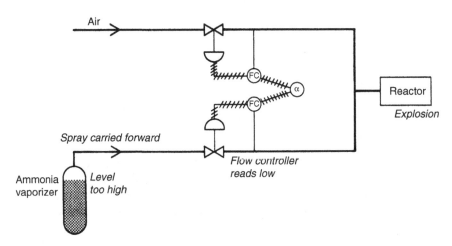

FIGURE 33-5 An increase in the level in the vaporizer led to an explosion in the nitric acid reactor. From *Chemical Engineering Progress*, July 1980. *(Reprinted with the permission of the American Institute of Chemical Engineers.)*

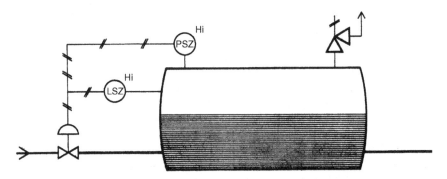

FIGURE 33-6 The designer of this system seemed unaware of the difference between vapors and noncondensable gases. From *Chemical Engineering Progress*, July 1980. *(Reprinted with the permission of the Institution of Chemical Engineers.)*

petroleum gas (LPG) is fitted with a level controller (not shown) and, in addition, a high-level trip to isolate the inlet line if the level gets too high (Figure 33-6). The high-level trip might fail; the relief valve will then lift and discharge liquid to the atmosphere so a high-pressure trip is installed as well.

If the space above the liquid contains some nitrogen or other non-condensable gas, the system will work. As the level rises, the gas will be compressed and the pressure will rise gradually. However, if no noncondensable gas is present and the level rises slowly, the system will not work.

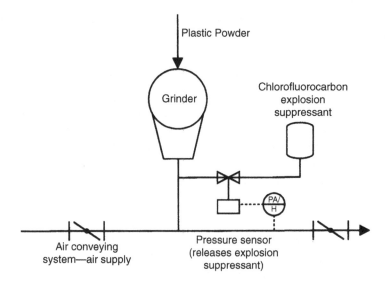

FIGURE 33-7 Accidental operation of the protective equipment—the explosion suppressant—caused an explosion. From reference 2. *(Reprinted with the permission of the American Institute of Chemical Engineers.)*

The vapor will condense and the pressure will not change until the vessel is completely full of liquid. The pressure will then rise too rapidly for the high-pressure trip to operate and the relief valve will lift.

Condensation takes a finite time. If the level rises rapidly, the vapor may not have time to condense and the system will then work.

The designer of the system probably did not understand the difference between a noncondensable gas, such as air or nitrogen, and a vapor.

33.1.7   Protective Equipment Caused an Explosion

A plastics manufacturing plant included a grinder to eliminate oversize particles. A stream of air removed the ground powder. To prevent a dust explosion, there was an explosion suppression system: if a pressure sensor detected a rise in pressure, chlorofluorocarbon (CFC) was released into the grinder and its associated piping to quench the explosion (Figure 33-7).

The system was in use for nearly 20 years but was never called upon to operate. Then the grinder exploded and the cause was the suppression system. An upset in another part of the unit allowed water to get into the grinding system and form a slurry with the powder. Some of the water or slurry worked its way into the branch leading to the pressure detector. This detector was very sensitive—the pressure exerted by only a few inches of water was sufficient to activate it—so the suppression system operated and the CFC was released. The accumulation of slurry prevented the CFC from flowing easily through the system and the door, weighing more than 200 kg (about 500 lb), was blown off the grinder. It hit the wall of the room and bounced back. Fortunately, no one was injured, but operators often stood in front of the grinder to inspect its operation through a window in its front. Many people were surprised that the release of CHCs could blow the door off, but it was held by only four bolts and could withstand an internal pressure of only 1 to 1.4 bar gauge (15 to 20 psig). It was certainly a physical explosion and not a chemical one, as there was no soot or burned material and the powder on the floor was still white [2].

The plant was designed before the days when Hazop was widely used. If a Hazop had been carried out on the design, the possibility of water entering the system could have been recognized (though Section 30.8 refers to an incident in which none of the Hazop team members recognized that water could get into a unit).

Today's explosion prevention systems often measure the rate of pressure rise and other materials are used instead of CFCs (because they affect the ozone layer).

It is good practice when designing any equipment to ask, if it were to become overpressurized, which part(s) will give way, as well as to locate the equipment so that people are unlikely to be in the line of fire. We protect equipment from excessive pressure by relief valves or in other ways but no protective system is 100% infallible.

It is, of course, essential to make sure than no one is in the line of fire from equipment that is designed to discharge—that is, relief valves, rupture discs, and pressure vents. The explosion vents on dust handling equipment can produce far longer flames than most people consider possible. An operator was burned when an explosion occurred in a spray drier and the pressure vents opened [3]. Entrance to the surrounding area was prohibited when the plant was on line, but the operator had gone there to look at a noisy pump. Passive protection—in this case, fitting a duct leading to the outside—is more effective than instructions. If fitting a duct was impossible, then a color photograph of a flame coming out of an explosion vent would have more impact than a written instruction.

In cases like this, it would be a remarkable coincidence if a dust explosion occurred on the one and only occasion that someone defied instructions and went near the vent when the plant was on line. I suspect that the rule had been broken before and that other personnel turned a blind eye.

33.1.8 A Procedure That Cannot Do What We Want It to Do

If tests are being carried out on a vessel, they are often made on the points of a grid. Lines of weakness (such as welds) may then be missed.

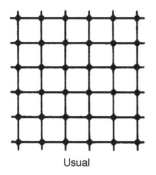

Usual   Better

FIGURE 33-8   If test points are on a grid, it should be tilted so that no lines of weakness are missed.

The grid should be tilted so that the test points are not all above or below each other (Figure 33-8).

Chapter 32 describes mechanical equipment that cannot do what we want it to do.

33.1.9   Preventing Similar Errors

There is no simple way of preventing the errors described in the foregoing. Hazops will help but only if the teams, or at least some of their members, have a good understanding of what is scientifically possible and of the sort of errors that have occurred in the past. The more we discuss our designs with other workers, including those who will have to operate the equipment, the more likely that someone will spot any weaknesses.

33.2   TOO LITTLE INSTRUMENTATION

A tube rupture and fire in a furnace were the result of too little measurement and control. Two furnaces heated the hydrogen supply to a hydrogenation reactor in four parallel streams (Figure 33-9). The check valve in the combined line beyond the heaters was leaking. When the unit was shut down, liquid flowed backward from the reactor to the furnaces and settled in a bend in a low point of one stream. This restricted the flow in that stream and the tube got too hot. It did not rupture immediately but expanded more than usual; this caused a small crack and leak elsewhere in the furnace, in the convection section of the furnace. The leak ignited and the flame impinged on another part of the tube, which ruptured. The resulting fire damaged half the tubes. Replacement took six weeks and cost $1 million, but the consequential loss was much greater.

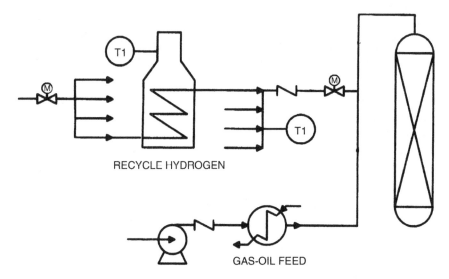

FIGURE 33-9 Restricted flow in one of the parallel streams led to a tube rupture and fire. From reference 4. *(Reprinted with the permission of Elsevier (United Kingdom).)*

When the furnace was rebuilt, changes were made to the design to reduce the stress in the convection section, more temperature measurements were installed, isolation and control valves were installed in each of the four paths, and fire-protected remotely operated emergency isolation valves were fitted so that ruptured tubes could be isolated quickly.

The report [4] does not say whether or not the check valve was inspected regularly. These valves have a reputation for unreliability, but few companies schedule them for regular inspection and we can hardly expect equipment with moving parts to operate for the lifetime of a unit without attention.

The plant was constructed in 1978 and seems to have been constructed without many of the features used elsewhere in similar plants at the time, presumably to save money. Remotely operated emergency isolation valves, for example, were widely used from the 1970s [5] onward.

33.3   DIAGRAMS WERE NOT UP TO DATE

Some changes were made to the alarm system on a boiler, but no one altered the wiring diagram or made any other record of the changes. Later on, when some other changes were made, the old wiring diagram was used. The operators carried out a few checks to make sure that the low-water-level alarm was working. They confirmed that the warning light came on but they did not check that the burners actually went out.

When a low-water level actually occurred during normal operation, the operator was elsewhere. He did not see the warning light, the boiler ran short of water, and it was extensively damaged [6].

There were two major errors in the management system, or rather nonsystem. The first error was the failure to check all trips and interlocks after a turnaround or modification. At the very least, any equipment that has been worked on should be tested thoroughly. On a furnace, for example, the burners should be lit and a check made that they go out when the water level is lowered. The startup takes a little longer, but wrecking a furnace causes rather more expense and delay. This is not a new idea. It is a lesson that the company I worked for learned more than 50 years ago.

The second error was a common one: a failure to keep line and wiring diagrams up to date. Everyone at every company agrees that they ought to do it; many intend to do it, but many more fail to do it. (However, in some countries the law requires it.) Keeping wiring and control cable diagrams up to date is particularly important because we can always trace pipelines to see where they go (unless several lines are insulated together), but it is difficult to trace wires and cables.

33.4 AN AUTOMATIC RESTART FAILS TO RESTART

Following a complete power failure, a rare event expected no more than once every 20 to 30 years, the emergency supply came on line but one safety-critical pump failed to restart. As a result ≈ 2.5 tons of chlorine was discharged into the atmosphere through a vent stack. It was discharged at too high a level to cause any harm, but it could be smelled more than 3 miles away and produced many complaints.

To prevent hunting (i.e., the pump repeatedly switching from one supply to another), the change-over mechanism had to be reset every time it operated. It had been tested several weeks before, by simulating a power failure, confirming that the change-over worked, and then switching the pump back to the normal supply. The tester then forgot to reset the change-over mechanism. This was a foreseeable error but it had not been foreseen. The operators had no way of knowing that the change-over had not been reset. There was an indication in the switch house but none in the control room. It took the operators 15 minutes to puzzle out what had happened as none of them really understood how the change-over worked. Do you have similar equipment at your plant, and do you and your operators know how it works?

Another piece of poor design also misled the operators. A low-flow alarm could have told the operators that the pump had failed. There was such an alarm, but it operated the same alarm window as the high-flow alarm. High-flow rates were common and not safety critical, so the operators did not recognize that the flow was low. They were busy checking that the rest of the plant was okay. Are there any double-duty alarm windows at your plant?

The power failure produced two other learning experiences. Some additional items of equipment needed backup power supplies, and most operators did not know that they could use their radios when the base station was out-of-action but that they had to use them in a different way.

We can sum up this chapter and others, particularly Chapters 25 and 26, by adapting a computer term and rephrasing the quotation at the beginning of this chapter: "What you should have foreseen is what you get" (WYSHFIWYG).

33.5   PROCEDURES: AN ESSENTIAL FEATURE OF CONTROL SYSTEMS

This chapter, except Sections 33.1.9 and 33.3, has described hardware faults and weaknesses in design. Equipment is not the whole of a protective system. In addition, it is necessary to have procedures which describe the correct way to use, maintain, and test it. All protective equipment is likely to fail from time to time, and the failures are often latent (that is, unseen) and can be detected only by testing at regular intervals. To decide on the test frequency, we have first to decide on the acceptable dead time—that is, the fraction of the time that we can tolerate the equipment being out of order. Suppose experience shows that there is a demand on the equipment once per year and that the equipment fails once in two years. If it is tested once per month, then on average it will fail halfway between test intervals and will be out of order on average for two weeks per year—that is, for 4% of the time. On average a demand will occur in this dead period once in 20 years. Is this tolerable? It depends on the seriousness of the result. If it resulted in someone being killed it would not be acceptable and we would need more reliable equipment, or duplication of the equipment, and perhaps more frequent testing [7, 8].

Unfortunately, a lot of people just guess test frequencies and assume that monthly testing will be okay. Or they repeatedly postpone testing because other work is seen as more urgent.

We are all familiar with the bathtub curve (Figure 33-10). Failure rates are high when equipment or living creatures are young, then steady at a low rate, and then rise again when the equipment or living creatures start to wear out. The curve also applies to procedures. The length of the horizontal portion can then be very variable. As soon as supervisors and managers lose interest, the curve starts to rise and the procedure can vanish without a trace. A continuous management effort is needed if use of a procedure is to continue.

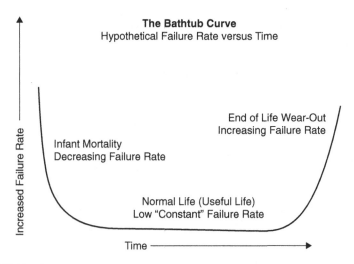

FIGURE 33-10   Bathtub curve.

After an accident has occurred, it is often found that a procedure has not been followed correctly or is ignored entirely and the person responsible may then be blamed for causing the accident (see Sections 26.2 and 26-3 and Chapter 35). But few accidents occur the first time a procedure is not followed. Most procedures can be ignored for a long time before an accident occurs. Managers and supervisors, all those responsible for other people, should keep their eyes open and never turn a blind eye when safety is concerned. A friendly word before an accident is far more effective than punishment afterward.

Sometimes managers and supervisors issue instructions that are difficult or impossible to follow. New procedures (and changes to designs) should be discussed with those who will have to use them. Hazop can help, but only if the team has the right knowledge and experience. Accidents are less likely in those plants where people are willing to tell their supervisor that it is difficult or impossible to follow a procedure. Unfortunately, in some plants, particularly in some cultures, people are reluctant to do so and instead just do the best that they can. Those in authority should be approachable and not isolate themselves in an office with the door closed. Leave it open. (See the quotation at the beginning of Chapter 31.)

Afterthought

Not everything that can be counted counts and not everything that counts can be counted.

—Albert Einstein

References

1. J. O. Pande and J. Tonheim, Ammonia plant: explosion of hydrogen in a pipeline for CO_2 . *Process Safety Progress*, Vol. 20, No. 1, 2001, pp. 37–39.
2. A. M. Dowell, D. C. Hendershot, and G. L. Keeports, Explosion caused by explosion suppression system. *Loss Prevention Bulletin*, Vol. 146, Apr. 1999, pp. 3–4.
3. Anon, Well-known hazard highlighted by an accident. *Loss Prevention Bulletin*, Vol. 95, Oct. 1990, p. 6.
4. B. D. Kelly, Investigation of a hydrogen heater explosion. *Journal of Loss Prevention in the Process Industries*, Vol. 11, No. 4, 1998, pp. 257–259.
5. T. A. Kletz, Emergency isolation valves for chemical plants. *Chemical Engineering Progress*, Vol. 72, No. 9, 1975, pp. 63–71.
6. T. Donaldson, Control failure incidents. *Loss Prevention Bulletin*, Vol. 162, Dec. 2001, pp. 20–22.
7. T. A. Kletz, *Hazop and Hazan; Identifying and Assessing Process Industry Hazards*, 4th edition, Institution of Chemical Engineers, Rugby, UK, 1999, Chapter 3 and Addendum 2.
8. Anon, *Reducing Risk, Protecting People*, HSE Books, Sudbury, UK, 2001.

Leaks

In safety what matters is not who did it, but what was done, and how.

—Roger Ford, *Modern Railways*

Section 9.1 quotes figures showing that most leaks occur from pipes or pipe fittings (such as valves), often because contractors did not follow either instructions or good work practices when details were left to their discretion. The actions suggested to prevent such leaks included specifying designs in detail and carrying out better inspection during and after construction. This is broadly confirmed by a more recent paper [1] that analyzed 270 leaks on offshore oil platforms. The locations of the leaks were as follows:

Small diameter pipes	18%
Other pipes	43%
Valves	12%
Total pipes and pipe fittings	73%
Vessels	8%
Seals	8%
Pumps and compressors	5%
Hoses	5%
Other equipment	2%

The main immediate causes of the leaks were the following:

- Corrosion, erosion, and fatigue 32%.
- Wear and tear, such as loss of flexibility in gaskets and valve packing, and friction between moving parts 26%.
- Poor installation, poor procedures, and failures to follow procedures 39%. Most of these (21%) resulted in an open end; for example, equipment was opened up before the contents were removed.

34.1 LEAKS FROM TANKS

34.1.1 A Leak from a Bad Weld

In 1999, an operator found that a storage tank containing 750 tons of 30% sodium cyanide solution was leaking and that a pool of liquid had formed in the dike. Sixteen tons had leaked, but only 4 tons were recovered as the rest had soaked into the ground. The base of the dike was permeable.

The hazard was primarily an environmental one rather than a safety one as the site is near the River Tees estuary in the United Kingdom. Several decades ago, the river was an open sewer but is now home to salmon and seals. Extensive tests showed no harm to wildlife; nevertheless the company was fined for exceeding its discharge authorizations.

The leak was due to the presence of a piece of welding slag, which had been present since the tank was built 22 years earlier. Water had penetrated between the slag and the weld metal, causing rust to form and creating a leak path.

It is obviously desirable for dikes to have nonpermeable floors, but fitting them to existing dikes is expensive. There are more than 150 tanks on this site alone, of various sizes up to 8,000 m^3 (2 million gal) capacity, and many more throughout the United Kingdom. Making the dike floors impermeable would cost about $150,000 per dike and would exceed the value of the site [2].

Environmental standards have changed since the site was built, and the incident does show the importance, when designing new plants, of asking what changes in safety and environmental standards (and product quality) are likely in the foreseeable future. In some cases, it may be cheaper to meet them now rather than to pay many times more to modify the plant in the future. In other cases, it may be possible to design a plant so that any equipment needed to meet higher standards later can be added on.

The immediate cause of the loss of material was a poor weld. The underlying cause was failure to provide *options for the future*. By 1977, many people realized that environmental standards were going to be tightened.

34.1.2   A Leak from a Plastic Tank

Hydrochloric acid was stored in a polyester tank that was fitted with a drain valve near the base. Drips from this valve corroded the concrete base on which the tank was sitting, despite a coating of tar, and the loss of support caused mechanical failure of the tank. If that was not enough, the dike (which incidentally was too small) leaked through joints in its walls that had been unsuccessfully sealed with tar. The acid then entered the electrical switch house and contaminated a river. It was neutralized with lime 3 miles downstream [3].

Plastic tanks are often used for corrosive materials. We should remember that they are usually not as strong as steel tanks. There are more incidents involving plastic tanks in Section 5.7.

34.1.3   A Leak from a Lined Tank

Caustic soda was stored in a steel tank lined with rubber. Over the years, rainwater seeped into the gap between the base of the tank and the concrete plinth and caused corrosion, accelerated by the high temperature of the liquid (80°C [175°F]). Fortunately, a small leak was noted and the tank was taken out of use and demolished. The outermost foot of the base was badly corroded. The likelihood of corrosion had been noted but nevertheless inspection failed to spot it [3].

34.2   LEAKS FROM LINED PIPES

A leak occurred from a flange on a 1-in. pipe, 6 m (20 ft) long, lined with polytetrafluoroethylene (PTFE). This polymer has a much higher coefficient of expansion than carbon steel, 10× higher averaged over a wide temperature range but up to 75× higher around 20°C (70°F). Unconstrained, a 6-m (2-ft) length will increase by 60 mm (2.4 in.) if the temperature rises from 19 to 30 degrees C (66 to 86 degrees F) but by only 50 mm (2 in.) if it rises from 30 to 100 degrees C (86 to 212 degrees F). The pipe was electrically heated and had been temperature-cycled many times.

When the pipe was heated, the liner tried to expand but could not. However, at the higher temperature, the stress was relieved. When the pipe cooled, it tried to contract and pull itself out of the flange. A similar effect occurred in the transverse direction; as the pipe cooled, it pulled itself away from the walls, making it easier for lengthwise movement to take place. In addition, extra trace heating on the flanges, to compensate for the greater heat loss, resulted in thinning of the liner.

After the incident, the company decided to apply less heat to the flanges and to tighten the flange bolts every year. This had been recommended by the manufacturers [4] but had not been done, perhaps because the reason for it was not explained. It is possible that similar effects apply to other plastic linings.

Another leak from a PTFE-lined pipe caused the loss of 4 kg of fluorochlorohydrocarbons, plus smaller amounts of chlorine, hydrochloric acid, and hydrogen fluoride. Small amounts of these gases can diffuse through PTFE and build up behind the lining, cause it to bulge inward, and restrict the flow. Vent holes were therefore made in the pipe wall to allow the gases to escape. However, combined with the moisture from steam leaks nearby, the gases corroded the pipe beneath the insulation. The lining bubbled out and failed. The vent holes should not have been covered with insulation. For other examples of gases diffusing through plastic, see Section 32.3.

It is well recognized that insulation should be removed from time to time to check for corrosion beneath it (see Section 28.2.1). Places where corrosion is likely should be listed for inspection, and some other places picked at random should also be inspected [3].

34.3   A LEAK THROUGH CLOSED VALVES

Most of the nuclear reactors in the United Kingdom are cooled by carbon dioxide gas. There is a small loss through leaks and purges and when emptying equipment for maintenance; liquid carbon dioxide is therefore stored on site. It is delivered in tank trucks and pumped through a hose into one of a number of refrigerated tanks. A second hose connects the vapor space in the tank to the vapor space in the tank truck so that they are both at the same pressure. When a load, 15 tons, of carbon dioxide is offloaded about ½ ton will flow from the fixed tank back to the tank truck (Figure 34-1).

From the storage tanks, the carbon dioxide is pumped to the reactor cooling system, which operates at a gauge pressure of about 40 bar (580 psi). This gas becomes radioactive. There is also a connection between the reactor and the top of the fixed tanks. It is used to sweep out contaminated gas before maintenance and to remove air afterwards.

A routine test showed a higher level of radioactivity than usual near this gas line. It was then found that three closed valves in this line were all leaking and some gas was flowing backward into the storage tanks. It might therefore have got into the tank trucks during recent deliveries and contaminated the next load, which might have been delivered to a manufacturer of carbonated drinks. Many tests were carried out, no contamination was found, and calculations showed that even if it had occurred, the dose to the public would have been negligible. Anyone drinking a liter of a contaminated carbonated drink would have received

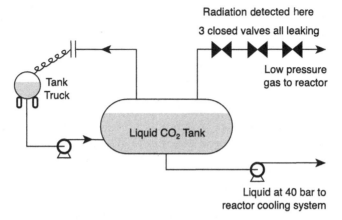

FIGURE 34-1   Contaminated gas leaked through three valves into the fixed tank and from there could have entered the tank truck.

a dose of 1 microsievert (mSv). For comparison, the average background dose in the United Kingdom is 2,200 mSv per year, many times greater in some areas. Nevertheless, the incident aroused considerable concern among the press, politicians, and public.

According to the official report [5], the root cause of the contamination of the tanks was leaking valves, but this was the immediate cause. The root cause was the failure of the designers and operators to be aware of something well known in the chemical and oil industries: a number of valves in series will not provide a positive isolation. For that, a blind, double-block-and-bleed valves, or physical disconnection is necessary. Before entry to a vessel, for example, it is normal practice to isolate all connections by blinds or physical disconnection.

Looking more deeply into the cause, why were the designers and operators unaware that valves are not leak proof? Perhaps it was the insularity of those in the nuclear industry who, like many others, believed that their problems were special and that they could not learn from other industries. Note also that while the immediate cause—leaking valves—is almost certainly correct, the underlying causes are more subjective. They usually are.

After the incident, low-pressure gas for sweeping out the reactor was supplied in a different way, by letdown from the 40 bar supply, and all nuclear power stations were asked to carry out Hazops to see if there were any routes by which their carbon dioxide tanks could be contaminated by radioactive gas.

Note that the incident occurred in a section of the plant devoted to the storage of an inert material. When a plant has serious and obvious hazards, it is a common failing to overlook the hazards in the safer parts of the plant.

The next item, Chapter 23, and Section 35.3 describe other leaks through closed valves.

34.4 A LEAK CAUSED BY SURGE PRESSURE

Surge pressure, particularly water hammer in steam mains, has caused many failures and large leaks of steam and condensate (see Section 9.1.5). Another incident occurred in a 450-mm (17.7-in.) steam pipe operating at a gauge pressure of 13.7 bar (200 psi). The details are complex but the essential features were as follows:

- The steam main went down through a tunnel under a road, rose up on the other side, and was joined by another supply line (see Figure 34-2).
- Following flooding, someone entered the tunnel to inspect the insulation. As the steam trap in the tunnel was blowing, it was isolated before entry was allowed but was not reopened afterward.
- The steam supply valve located before the tunnel was reached, as well as the valve on the other supply line, were both closed and both were passing. The leak in the first valve filled the dip in the main with condensate and the leak in the other valve maintained a steam bubble in the higher part of the main.
- Ultimately, the cold condensate completely filled the dip and overflowed into the horizontal part of the main, causing the steam bubble to condense. The resultant surge pressure ruptured the main at a T-joint, the weakest part.

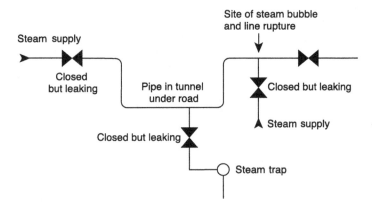

FIGURE 34-2   Condensate filled the dip in the steam main and overflowed into the horizontal section on the right, condensing a steam bubble. The resultant water hammer ruptured the main.

Afterward, the company trained more than 700 employees on the operation of steam systems. Consultants surveyed the steam system, including 3,000 traps. More than 100 were repaired or replaced, and a better system for their inspection and maintenance was set up. However, many other steam mains have also failed due to water hammer. The hazard is well known and has often been described, for example, in the booklet *Hazards of Steam*, first published by Amoco in 1963 and revised by BP in 2004 [9]. A similar incident had occurred in the same power station 25 years earlier. Why did the company not learn from its own and others' experience?

The report [6] does not provide this information, but one can hope that the company improved its procedures for reinstating equipment after isolation for entry or maintenance.

34.5 LEAKS FROM SCREWED FITTINGS

During a pressure test at a gauge pressure of about 350 bar (5,000 psi), a 20-mm (¾-in.) screwed thermowell was blown out at high speed (about 90 mi/h) and seriously injured a man who was looking for possible leaks. The report [7] does not say whether the failure was due to corrosion, damaged threads, failure to fully engage the threads, or incompatibility of the two threads, but all of these have caused other failures of screwed joints (see Section 28.4.2). Many companies do not allow the use of screwed joints except for low-pressure lines handling nonhazardous liquids (hot water is considered hazardous) and for small diameter lines, such as those leading to instruments, and then only after the first isolation valve.

Pressure tests are carried out to confirm that the equipment can withstand the test pressure and, therefore, we should assume that failure is possible and keep everyone out of the way. If we were sure the equipment would not fail, we would not need to test it. Leaks can be detected by testing at the operating pressure.

A screwed nipple and valve blew off an oil line operating at 350°C (660°F). An oil mist 30 m (100 ft) deep covered most of the unit and was sucked into the control room by the ventilation fan. The operators managed to shut down the plant before the oil mist caught fire about 15 minutes later.

Many people do not realize that mists of flammable liquids can burn or explode at temperatures well below the flash point of the vapor. The droplets behave like particles of dust, but there is often some vapor present as well, and thus these explosions may be more powerful than dust ones.

The nipple that failed was installed during construction to aid pressure testing and was not shown on any drawing. If the operating team had known it was there, they would have replaced it with a welded plug. Afterward, they drew up a list of other weak spots in piping systems to be identified and modified if this was practical, and, if not, they inspected the weak points regularly. The following is based on their list [8].

34.6   OTHER WEAK SPOTS IN PIPEWORK

- Vents, drains, and other connections with no obvious function or that are no longer needed for their design function should be replaced with welded plugs as described in Section 28.4.2. If used occasionally, they should be blanked. If used regularly, they should be fitted with double isolations.
- Permanent connections to service lines such as steam, nitrogen, and compressed air should be fitted with check valves and, if the service pressure could be less than the pressure in the process equipment, with double-block-and-bleed valves. A low-pressure alarm should be fitted on the service line (see Section 31.5). If the connection is used only occasionally and the temperature and pressure are moderate, it may be better to use a hose instead of a permanent connection as long as it is certain that the hose can be vented before it is disconnected. However, hoses generally should be used only for temporary jobs.
- Expansion couplings should be avoided on lines carrying hazardous materials. Expansion loops are less likely to fail.
- Small nipples on main pipes should not be less than 1-in. external diameter, although their internal diameter can be less.
- Unusually long runs of small pipe leading to instruments or sample points should be fitted with isolation valves close to the main pipe.
- Inadequately supported small pipes should, of course, be supported.
- Brass valves in process lines are suitable only for low-pressure and low-temperature water lines.
- Unused sample coolers should be removed.
- Equipment (pumps, exchangers, pipes, etc.) no longer needed should be removed.
- Look out for changes in flange ratings in a line. Is the relief setting suitable for the lowest flange rating?
- Screwed joints should be avoided, except for small diameter lines containing nonhazardous materials and then only after the first isolation valve (see the previous section).
- Cast iron fittings, sometimes found on old units, are brittle and can be broken by impact. Replacement should be considered.
- Unnecessarily large liquid draw-off and sample connections should be replaced or, at least, fitted with restriction plates (or a length of

small diameter line, which is less easily removed than a restriction plate).

- Control and electric cables exposed to possible fire damage should be fitted with fire protection as a small fire can cause extensive damage to them that is expensive to repair.
- Hidden connections under insulation should be checked.
- Relief valve tailpipes that discharge into the atmosphere should be checked, especially those that could impinge on other equipment if the discharge is ignited by lightning (see Sections 35.1 and 35.2).
- Are control room air inlets located so that they might draw in contaminated air? This should be assessed.
- Lines that could cause a serious fire or other incident if they leaked should be scheduled for regular inspection.

Some other leaks are described in Chapters 5, 7, and 28.

References

1. Health and Safety Executive, *Offshore Hydrocarbon Release Statistics*, Report No. OTO 200 112, HSE Offshore Division, London, UK; summarized by J. N. Edmondson and D. B. Pratt, Reducing offshore hydrocarbon leaks—A UK regulatory initiative. *Loss Prevention Bulletin*, 168, Dec. 2000, pp. 4–11.
2. A. Whitfield, COMAH and the environment—Lessons learned from major accidents 1999–2000. *Hazards XVI—Analysing the Past, Planning the Future*, Symposium Series No. 148, Institution of Chemical Engineers, Rugby, UK, 2001, pp. 799–899.
3. Anon, Corrosion of support structures. *Loss Prevention Bulletin*, Vol. 163, Feb. 2002, pp. 13–14.
4. R. M. Schisla, S. C. Ernst, and P. N. Lodal, "Case History: PTFE lined pipe failure," *Proceedings of the 35th Annual AIChE Loss Prevention Symposium*, 2001.
5. Anon, *Contamination of the Carbon Dioxide Supply System at Hunterston B Power Station*, Feb. 1997, HSE Books, Sudbury, UK, 1998.
6. C. Galante and S. Pointer, Catastrophic water hammer in steam dead leg. *Loss Prevention Bulletin*, Vol. 167, Oct. 2002, pp. 16–20.
7. Anon, *Process Safety Beacon: Faster Than a Speeding Bullet*, American Institute of Chemical Engineers, New York, Aug. 2002.
8. Anon, Heavy oil fire. *Loss Prevention Bulletin*, Vol. 146, Apr. 1999, p. 20.
9. Anon, *Hazards of Steam*, Institution of Chemical Engineers, Rugby, UK, 2004.

Reactions—Planned and Unplanned

An expert is one who learns through his own experience how painful and deep are the errors one can make even in the most limited field of research.

—Niels Bohr (1885–1962)

35.1 DELAYED MIXING

O-chloronitrobenzene reacts with methanol and caustic soda to produce o-nitroanisole (Figure 35-1).

The reaction is semibatch and operations are normally carried out as follows:

1. The first two reactants are placed in the reactor and mixed.
2. The stirrer is then switched off and the liquid level checked by opening the manway cover.
3. The cover is replaced and the stirrer is switched on.
4. The temperature is raised to 80°C (175°F) by passing hot water through the reactor jacket, and the pressure is raised to 9 bar gauge (130 psig) with nitrogen.
5. A solution of caustic soda in methanol is then added gradually and the temperature kept at 80°C (175°F) by adjusting the flow rate of cold water through the cooling jacket.

One day, after replacing the manway cover (step 3), the operator forgot to switch the stirrer back on. There was no mixing, and the caustic soda plus methanol (added in step 5) formed a separate layer. No reaction occurred,

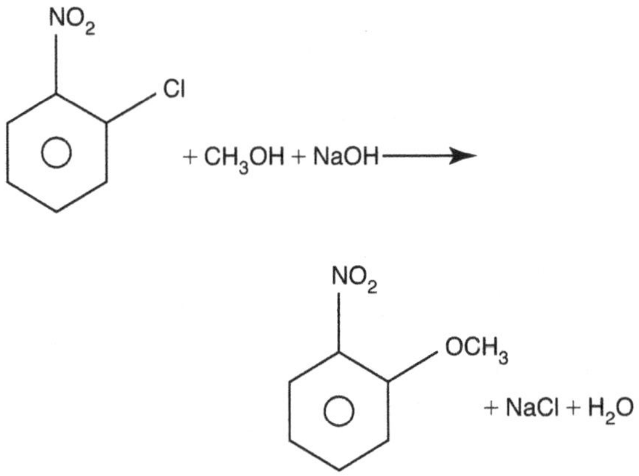

FIGURE 35-1   Late mixing of this reaction led to a leak that covered 75 acres.

and the operator had to apply heat instead of cooling to the jacket to maintain the temperature at 80°C (175°F). When the operator realized that the stirrer was not running, he switched it on. A very rapid reaction occurred, the temperature rose to at least 160°C (320°F), and the pressure rose to 16 bar gauge (230 psig). Most of the contents of the reactor, 10 tons of liquid, were discharged through the relief valve and a yellow deposit was distributed over 300,000 m² (75 acres) of a builtup area, a suburb of Frankfurt, Germany.

At one time, an accident like this would have been blamed on human errors—forgetting to switch the stirrer on after replacing the cover and then switching it on too late. But both errors are easy to make, particularly the second one: when we find we have not done something we should have done, our natural tendency is to do it at once. "Better late than never" is a common saying. However, in this case and many others it proved disastrous (see also Sections 30.5, 35.2, and 35.4). One of the biggest causes of runaways in batch and semibatch reactions is failing to start the stirrer or circulation pump and then starting it late so that large amounts of reactants are suddenly mixed (see Sections 3.2.8 and 22.2). So how can we prevent this from occurring?

There were several weaknesses in the design:

- It should not be necessary to open the reactor to check the level. A level indicator or load cell would have removed the need to switch off the stirrer.
- An agitation detector could have stopped addition of the methanol/ caustic soda mixture if the stirrer was not running. (This would be better than checking the voltage applied to the stirrer motor, as the motor or its coupling can fail.)

- A catchpot after the relief valve would have collected the discharge and prevented it from spreading over the surroundings. This incident occurred in 1993. Nearly 20 years earlier, a discharge at Seveso in Italy had taught us the same lesson (see Section 21.2.5). On that occasion, a large area around the Italian plant had been sprayed with dioxin and 4 km^2 (1,000 acres) made unusable. At the plant involved in the more recent incident, a catchpot had not been fitted because the normal products of the reaction were not all that hazardous. (The caustic soda was well diluted.) However, the designers had overlooked the fact that once a runaway starts, other reactions occur and different and more harmful products are formed.

These design errors could have been avoided if the company had been aware of a similar runaway that occurred in Japan more than 20 years earlier and injured 9 people. The published report on it was brief but it should have been sufficient to alert the German company to the hazard. An underlying cause of the accident was thus a failure to learn from the past, from both Seveso and Japan.

Following the discharge, the adverse reports in the media caused the company to withdraw from the chemical industry. The German authorities imposed further regulations on the industry, affecting all companies [1]. To the public, the chemical industry is a single unified entity and an accident in one company affects others. For this reason, we should share information on the causes of our accidents and the actions we have taken to prevent them from happening again (see Section 38.8).

Another frequent cause of uncontrolled rises in temperature is dissolving caustic soda in water. If the solid caustic soda is added too quickly or with insufficient circulation, some of the solid accumulates at the bottom of the mixing vessel and then slowly dissolves, forming a strong solution. Any sudden mixing causes rapid production of heat, local boiling, and further mixing. It is not necessary to switch on a stirrer to start the mixing process; mechanical shock or vibration may be sufficient [2].

A bucket containing 25% sodium hydroxide solution was used to collect bromine that was dripping from a leak. Unreacted bromine formed a separate layer at the bottom. When the bucket was moved, the two layers mixed and there was a violent eruption [3].

35.2 WAITING UNTIL AFTER THE FOURTH ACCIDENT

A mixture of phenol, formaldehyde, and sulfuric acid—the raw materials for the manufacture of phenol-formaldehyde resin—was discharged onto a roadway four times before the company decided to install a catchpot after the reactor rupture disc.

The first discharge occurred because the operator forgot to add the catalyst—sulfuric acid—at the beginning and then added a larger amount later when a second addition of catalyst was normally made. This was another example of the incorrect belief that it is better to carry out an action late than not carry it out at all (see Sections 30.5, 35.1, and 35.4).

The second discharge occurred because the formaldehyde failed to react, for an unknown reason. When the second addition of catalyst was made, the large excess reacted vigorously.

The third and fourth incidents had similar causes. Part of the heat of reaction was removed by a cooling jacket and part by condensing the vapor given off during reaction. The latter was ineffective, as there was a partial choke in the vapor line where it entered the condenser.

The company's managers did not ignore the first three incidents. They changed the operating procedures. After the fourth incident, they decided that was not enough and they made a change in the design: they installed a catchpot [4].

When a hazard is recognized, by experience or in any other way, the most effective action we can take is to remove it. If that is not possible, we can add on equipment to control it. However, relying on procedures should be our last resort. Moreover, in some companies the default action is to think of procedures first—perhaps because they are cheaper and quicker to install and do not require any design effort—but they are less effective. There are other examples of this in Chapter 27.

35.3   LOWER TEMPERATURE MAY NOT MEAN LESS RISK

In the incident described in Section 35.1, the raw materials did not react because there was no mixing. Raw materials can also fail to react for another reason: because they are too cold.

An aromatic amine was reacted with sulfuric acid and nitrosyl sulfuric acid (NSA) to form a compound, which was then decomposed to form a phenol. A hundred batches had been made every year for several years without incident until a runaway reaction occurred. It produced a large amount of gas, which overpressured and ruptured the 2,270-L (600 gal) reactor. It was driven through the concrete floor while its lid traveled 150 m (500 ft). Fortunately, no one was injured.

The reaction was semibatch. The amine and sulfuric acid were mixed in the reactor and then the NSA was added gradually. When reaction was complete, the mixture was moved to another vessel where it was decomposed to the final product.

The heating and cooling of the reactor were temporarily done manually. Probably for this reason, temperature control was more erratic than usual and at times the reactor was too cold for reaction to occur. About 30% of the NSA failed to react. The temperature then rose above the normal, probably because the valve supplying steam to the reactor was leaking or had not been fully closed. There are so many probables because data recording was rudimentary.

The replacement plant designed and built after the explosion included computer control, data logging, trips and interlocks, and a quench tank filled with water into which the contents of the reactor could be dumped if they got too hot [5].

Note that several hundred batches were made without incident before the runaway occurred. As noted in earlier chapters, a blind man can walk along the edge of a cliff for a long time before he deviates from the correct path far enough to fall over the edge. Section 34.3 describes another accident due to leaking valves.

35.4   FORGETTING TO ADD A REACTANT

The reaction between phenol and formaldehyde to form phenol formaldehyde resins has produced many runaway reactions, most of them the result of the same omission. There are two reaction steps. In the first step, phenol is reacted with formaldehyde in a stirred semibatch reactor, which can be heated or cooled. The phenol is charged to the reactor, a small amount of caustic soda is added as a catalyst, and the formaldehyde is then added gradually. In the second step, much more caustic soda is added gradually, this time as a reactant.

A common error is to forget to add the caustic soda in the first stage. The phenol and formaldehyde do not react, but the operators do not realize this as the automatic temperature control keeps the reactor at the correct temperature. When the addition of caustic soda starts, there is a violent runaway reaction, which may burst the reactor.

To prevent this occurring, the temperature should be measured during the first step and the addition of formaldehyde stopped if it is too low. It should also, of course, be stopped if there is no agitation (see Section 35.1) [1]. Another possibility is to carry out the two stages in two different reactors. In the first reactor, it would be possible to add only small amounts of caustic soda.

For the accidents described so far, the solutions suggested are *add-ons*: more protective devices are proposed, devices that may be neglected or switched off. Could the reactions take place in continuous reactors made from long thin tubes? Has anyone looked for alternative and safer chemical processes? Chemists have been slower than chemical engineers to adopt inherently safer designs.

Experienced process designer Stanley Grossel writes [6]:

> I have been involved in the process design of many chemical processes. Quite often, I have been given a technology transfer package and told to design a suitable plant. When I informed my management that the process was hazardous . . . and that it should be modified to be safer, I was then told that it was too late and that too much time and money had already been expended, and that I should use as many safety measures and as much equipment as necessary to make the process safer.
>
> Based on my often frustrating experiences with a *fait accompli* process, I feel strongly that the concepts of *inherently safer design* should be taught at the undergraduate chemical engineering and chemistry curricula level. It may be even more important for chemists to become aware of this technique, as they are the ones who conceptualize and develop chemical processes. If they were aware of the technique, they might come up with inherently safer processes from the start. . . . This would result in lower initial plant costs and fewer accidents, which then would save replacing of equipment and prevent both business interruption and lawsuits.

35.5   INADEQUATE TESTS

Runaway reactions have occurred because the tests carried out were in some way inadequate [7]. One occurred because the sample was very small—a few milligrams—and was not representative of the reaction mixture. The damage was catastrophic. Afterward, tests were carried out with 5- to 10-gm samples, and incidents still occurred. It is, of course, easy to get a representative sample of a pure compound but not of a mixture. I remember reading this in my university textbook on chemical analysis.

Another incident occurred because the test measured the heat release but did not measure the amount of gas produced. The process worked in the laboratory and in a pilot plant, but when transferred to full-scale operation the vent was too small. The reactor cover lifted and the escaping gas ignited.

A third incident was somewhat similar. The process required two reactants to be mixed and then heated to 85°C (185°F). A third reactant was then added slowly. Tests showed the mixture of the two reactants was stable, so they were premixed in drums ready for charging. Two hours later, the drums started to rupture. The tests had failed to show that gas was slowly produced even at room temperature. Tests should measure the pressure produced as well as the heat produced.

The onset temperature for a runaway is not a fundamental property like the boiling point and can vary by as much as 50°C (90°F), in some cases as much as 100°C (180°F), depending on the method used.

The need for thorough testing is shown by a fourth incident. It occurred in a process for the nitration of an aromatic compound by nitric acid in acetic acid solution using sodium nitrite as catalyst. The solution was dilute, and tests showed only a moderate rise in temperature. However, in further tests, the reaction mixture was allowed to stand at 70°C (160°F) to make sure the reaction was complete. Heat production continued; the temperature reached 180°C (355°F), and the gauge pressure reached 25 bar (360 psi). A literature review showed that the excess nitric acid was reacting with the solvent to produce acetyl nitrite, which decomposes at 70°C (160°F).

The plant was advised to install extra cooling capacity, triggered by a high-temperature measurement, but in the longer term to look for a less reactive solvent. The experience does show that simple screening of the raw materials and the reaction is not enough [8]. Reaction mixtures are always likely to be left, for any number of reasons, when reaction is complete—or sometimes when it is only partly complete. The most notorious example of the latter is the explosion at Seveso, Italy, in 1976, where a partially reacted mixture was left to stand over the course of a weekend [9].

35.6   A HEATING MEDIUM WAS TOO HOT

A product was vaporized and condensed to improve its purity. Vaporization took place in a small jacketed vessel (2.3 m³ [600 gal]). It was kept under vacuum and the contents were heated to 140°C (285°F), the boiling point under the vacuum, by oil at 170°C (340°F). The process had to be shut down for planned maintenance of the steam supply. The vacuum was broken and cooling applied to the jacket; nevertheless the temperature reached 160°C (320°F), the temperature at which the product started to decompose, and then rose rapidly. The glass exit pipe broke, and the escaping liquid caught fire. A high-temperature trip should have automatically switched the jacket from heating to cooling before the temperature reached 160°C, but the sensing device was fouled with tar and read low.

The order of events is not entirely clear but it seems that breaking the vacuum stopped the evaporative cooling and that this took place before the cooling agent replaced the hot oil in the jacket (or possibly before the cooling had time to become effective). One wonders if the operators understood that evaporation provided a cooling effect, which was lost when the vacuum was broken.

The report does not say how often the high-temperature alarm was tested, but after the incident multiple temperature probes were fitted to the vessel. The report [10] does not mention the major weakness in design: using a heating medium hotter than the temperature at which decomposition started. This is inherently unsafe, and the incident shows the inherent weakness of relying on an active protective system (which could and did fail) instead of an inherently safer design. Of course, a cooler heating medium would have needed a larger heating area. A thin film evaporator might have been the best way of achieving this goal.

35.7   AN UNSTABLE SUBSTANCE LEFT STANDING FOR TOO LONG

As described in Section 35.5, runaway reactions have occurred because mixtures of raw materials, intermediates, or products have been left standing for too long. This can also occur with single substances. A peroxide was moved from a weigh tank through a transfer pipe to a reactor and the pipe left empty. One day the pipe was left full of liquid while a leak was repaired. The repair took longer than expected, and the heat from the reactor slowly warmed the liquid in the pipe until it decomposed and ruptured the pipe. The report [11] says, "Luckily, there were no injuries, just a lot of surprised people." I expect they knew the decomposition temperature of the peroxide. What probably surprised them was that the peroxide could get hot enough by conduction along the pipe from the reactor. We all know that metals are good conductors of heat, but most of us have no instinctive grasp of the rate at which heat can flow (or of the rate at which vessels will cool; see Sections 25.1.2, 25.1.4, and 25.1.5, and 30.7).

References

1. J. L. Gustin, How the study of accident case histories can prevent runaway reaction accidents from recurring. *Process Safety and Environmental Protection*, Vol. 80, No. B1, 2002, pp. 16–24.
2. J. Cox, Caustic layering—the forgotten hazard. *The Chemical Engineer*, Vol. 667, 8 Oct. 1998, pp. 25–28.
3. Anon, "Case Histories of Accidents in the Chemical Industry," No. 1636, quoted by Urben, P. G. (editor), *Bretherick's Handbook of Reactive Chemical Hazards*, 6th edition, Butterworth-Heinemann, Oxford, UK, and Woburn, MA, 1999, p. 108.
4. T. Gillard, Loss of reactor contents to atmosphere. *Loss Prevention Bulletin*, Vol. 143, Oct. 1998, pp. 21–22.
5. S. Partington and S. P. Waldram, Runaway reaction during production of an azo dye intermediate. *Process Safety and Environmental Protection*, Vol. 80, No. B1, 2002, pp. 33–39.
6. S. S. Grossel, Safety issues. *Chemical & Engineering News*, Vol. 20, 17 Mar. 2003, p. 4.
7. J. Singh, and C. Sims, "Reactive chemical screening—A widespread weak link," *Proceedings of the Mary Kay O'Connor Process Safety Center Annual Symposium*, College Station, TX, 1999.
8. S. Roe, Are your process materials fully compatible? *Process Safety News*, Vol. 4, Autumn/Spring, 1997/1998, p. 2, published by Chilworth Technology, Southampton, UK.
9. T. A. Kletz, *Learning from Accidents*, 3rd edition, Elsevier, Boston, MA, 2001, Chapter 9.
10. J. Bickerton, Fire in a vacuum still. *Loss Prevention Bulletin*, Vol. 166, Aug. 2002, pp. 12–13.
11. Anon, *Process Safety Beacon: Reactive Chemistry: Not Always When or Where You Want*, American Institute of Chemical Engineers, New York, Mar. 2003.

36

Both Design and Operations Could Have Been Better

You can't solve problems with the same level of knowledge that created them.

—Albert Einstein

Many of the incidents in this book could have been prevented by better design, and many by better operations. Good operations can sometimes compensate for poor design and vice versa, but that is not something on which we should rely.

36.1   WATER IN RELIEF VALVE TAILPIPES

My first example is a simple one. Sections 35.1 and 35.2 describe several incidents in which relief valves discharged process material directly into the atmosphere instead of into a catchpot or other closed system such as a flarestack or scrubber. These were hardly unforeseen incidents. Relief valves are designed to lift, so we should not be surprised when they do and we should design accordingly. Relief valves on steam systems can, of course, safely discharge into the atmosphere, but they are not without hazards. To prevent steam condensing in the tailpipe and filling it with water, a small hole is usually drilled in the tailpipe as a water drain. However, these holes often get blocked with rust and other debris, and then slight leaking of the relief valve or rain causes water to accumulate. This raises the pressure at which the relief valve lifts and when it does a slug of water is blown out. If the relief valve is leaking slightly, this water will be hot.

These drain points are protective systems, and like all protective systems they should be inspected regularly, in this case by rodding to make sure they are clear [1]. I suggest they have at least a 1-in. diameter.

Some companies fit drainpipes to the drain holes, but this can make matters worse. A long narrow tube can choke more readily than a hole. If drain lines are fitted, they should be short, straight, at least 1 in. in internal diameter, and designed so that they can be checked to make sure they are clear [2].

36.2   A JOURNEY IN A TIME MACHINE

This accident occurred in 1998, but it shows such a lack of good practice in both design and operations that I looked at the cover of the report [3] to make sure that it really happened then and not in 1898.

A new unit, alongside an existing one, separated gases from crude oil in three stages. The first-stage separator was designed to withstand a pressure of 95 bar gauge (1,400 psig), and the second a pressure of 35 bar gauge (500 psig). However, operating pressures were lower, 68 bar gauge (990 psig) and 15.5 bar gauge (225 psig). Both separators were fitted with relief valves. The third-stage separator was designed to operate at atmospheric pressure and had neither relief valve nor vent, though a gas exit line and isolation valve were fitted to the top of the vessel. Bypass lines with valves were fitted around all three separators (Figure 36-1). The crude oil was piped from a well about 3 km (2 miles) away.

The new equipment, apart from this pipeline, was freed from air using crude oil from a nearby well. The next job was to sweep the 3-km pipeline free from air using oil from the distant well. The valves were set as shown in Figure 36-1. Note that the two valves in the third-stage bypass should have been open but were shut. No one knew when they were shut or who shut them. The pressure of the crude oil supply pump, designed to pump the oil through a 3-km pipeline, ruptured the separator, killing four people and causing considerable damage to other equipment.

36.2.1   Design Errors

- The major error was the lack of a relief valve on the third separator.
- An isolation valve in the inlet line could have protected the vessel, but this alone would not be considered adequate. It would be an adjunct to a relief valve, not a substitute for it.
- A Hazop or review of the relief system would have disclosed these design errors.

There were no drawings! (Compare the explosion at Flixborough in 1974 where the only drawing for the modification that failed was

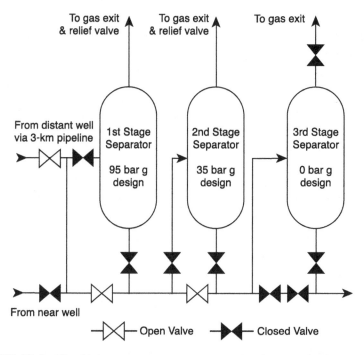

FIGURE 36-1 The third separator was overpressured and ruptured when its bypass valves were closed in error.

a full-size sketch in chalk on the workshop floor.) (See Section 24.4 and reference 4.)

We do not know if the design engineers, who were company employees, were unaware of the codes for vessel design or simply decided (or were told) to do a cheap job. However, the savings in cost were miniscule.

36.2.2   Operating Errors

- There were no written instructions for startup or normal operation. The company policy was to rely on on-the-job training without checking that the messages had been received and understood. Unfortunately, each time procedures or knowledge are passed on, they are likely to be degraded. In verbal communication—and sometimes in written communication—the message sent and the message received are not always the same.
- The valves should have been checked by a responsible person, using a checklist, before purging started.
- The company held monthly safety meetings for all employees, but the report does not say whether these covered process safety or just "hard hat" safety.

- Employees frequently moved between different plants belonging to the same company but received no training on the different designs and procedures.
- Sweeping out pipelines and other equipment with flammable liquids is not good practice. It is safer to first inert the lines with nitrogen. Crude oil has a high conductivity and static electricity is therefore not a hazard but other sources of ignition are possible, though unlikely. *Hazards of Air and Oxygen* [5], first published in 1958, includes pictures of an underground crude oil pipeline in which detonations occurred along a length of 50 km (30 miles). In this case, compressed air was being used to sweep out crude oil. The pictures show soil blown out and projecting pieces of pipe at intervals of some tens of meters.

We do not know if the company managers were amateurs, and thus unaware of the need for good design and operating methods, or trying to do everything on the cheap. It may have been a mixture of both.

36.3   CHOKES IN FLARESTACKS

Many explosions have occurred in flarestacks because a normally continuous flow of gas failed or fell to a very low level and air diffused down the stack, forming an explosive mixture (see Chapter 6). To reduce such diffusion, many companies have installed molecular seals: the gas leaving the stack and any air diffusing down follow a labyrinthine path, as shown in Figure 36-2. However, these seals have disadvantages. Carbon from incompletely burned gas can fall into the bottom of the seal and block the flow. Ideally, material discharged from a relief valve should follow a simple and straightforward path without any equipment that might obstruct the flow. For this reason, some companies have had second thoughts and removed the top of the molecular seal, called the "top hat," as shown by the dotted line in Figure 36-2. This effectively neutralizes the seal but leaves the rest of it in position. On one plant, a steam nozzle was added inside the stack to cool the tip and to reduce smoke formation. Some of the steam condensed and flowed down the drain line into the knock-out drum. In the stack shown in Figure 36-2, the drain line was insulated with the steam line to prevent freezing.

Unfortunately, the drain line had only a 1-in. diameter. It became partially blocked with carbon and a pool of water formed in the bottom of the molecular seal. It filled the outer rings of the seal and overflowed into the inner pipe. Some of it froze when unusually large quantities of cold gas had to be flared during cold weather, partially choking the inner pipe. Finally, there was a rumbling noise and bits of ice and a stream of water were blown out of the stack. The flame was extinguished. Similar

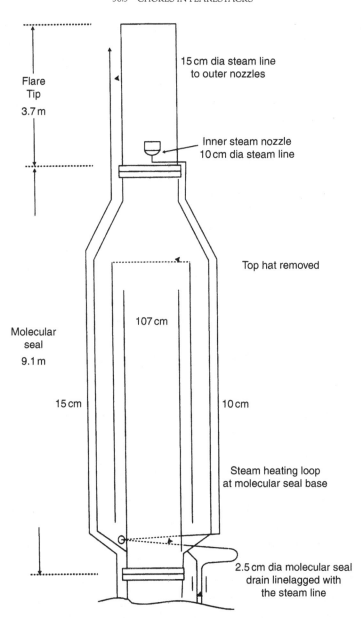

Flare
Tip
3.7 m

15 cm dia steam line
to outer nozzles

Inner steam nozzle
10 cm dia steam line

Top hat removed

107 cm

Molecular
seal
9.1 m

15 cm

10 cm

Steam heating loop
at molecular seal base

2.5 cm dia molecular seal
drain linelagged with
the steam line

FIGURE 36-2 The top of a flarestack fitted with a "top hat." From reference 6. *(Reprinted with the permission of the Institution of Chemical Engineers.)*

incidents have occurred on other plants. On one plant, a narrow stack was completely blocked by ice when steam was injected inside it (see Sections 2.5 and 6.2). It is doubtful if a 107-cm (42-in.) diameter stack could be completely blocked in this way. Nevertheless, the report [6] recommended installation of a much bigger drain line. Removal of the molecular seal—a major job—would not have prevented icing, but all the water formed would have fallen to the bottom of the stack, another example of the unforeseen effects of change.

The major weakness in design was the 1-in. drain line, which was far too small for such an important duty. The operating error was not to consider critically the possible effects of fitting a steam nozzle in the stack.

36.4   OTHER EXPLOSIONS IN FLARESTACKS

For an explosion, we need fuel, air (or oxygen), and a source of ignition. In a flarestack, the fuel is almost always there, as the purpose of the flarestack is to burn it; thus, as the flare is normally there, all we need is the air. Air can leak into the flare lines from the equipment that feeds the lines or through leaks in the lines leading to the stack, or can diffuse down the stack if the upflow stops or becomes very low (see Chapter 6). Another possible cause is development of a vacuum in the flare system so that air is actually sucked into the stack. This is unlikely, but the following are accounts of two explosions that were caused this way [6].

In the first incident, the pressure control valves on two compressor suction drums should have been set to prevent the pressure falling below 2 bar gauge (30 psig) but were set in error at zero. One of them was probably set slightly below zero. This allowed a slight vacuum to form, and it sucked air down the flarestack. Calculations showed that if the control valve was open for only a few minutes, the stack would be completely filled with air and thus it would fill with a flammable mixture of vapor and air in less time. In addition, the nitrogen purge, intended to prevent back flow of air down the stack, was only a third of its normal rate. The resulting explosion deformed the base and blew off parts of the tip. They landed 45 m (150 ft) away.

The second explosion occurred in a plant in which some equipment operated under vacuum. There were rupture discs below the relief valves on a section of the plant that was under a slight pressure, but they had failed and the relief valves were leaking. This leak increased the vacuum elsewhere in the plant and pulled air into the stack, despite the presence of a molecular seal and a flame arrestor in the stack. The initial explosion was followed by two others while the plant was shutting down. The stack was ruptured in three places.

The report [6] suggests that there may have been a split in the top hat of the molecular seal and that the flame arrestor should have been nearer the top of the stack. Flame arrestors are more efficient when they are near the end of a pipe, but they are likely to become dirty and produce a pressure drop. They should not be installed in flarestacks that, as already stated, should provide an uninterrupted path to the atmosphere.

36.5   DESIGN POOR, PROTECTION NEGLECTED

Two large pumps handled a slurry of catalyst and a hydrocarbon similar to gasoline in its physical properties. The pumps, originally one working, one spare, were operated continuously to increase throughput. However, throughput was reduced when one of the pumps had to be maintained. The pump seals were flushed with clean hydrocarbon from one of the plant vessels.

As the result of an upset, the level in this vessel was lost. For the plant as a whole, this was no more than a minor disturbance but the loss of the flush caused the pump glands to leak. The leak was small, so it was decided to repair the seals one at a time. The first repair took eight hours, rather longer than expected. Meanwhile the leak on the other seal gradually worsened and was dispersed with a steam lance. Finally, just as the repair of the first pump was completed, the shift foreman, who had just come on duty, decided that the leak was so bad that the plant should be shut down at once. Unfortunately, the valve actuators on the leaking pump's isolation valves had been removed for repair, there was no easy way of operating the valves manually, and there were no alternative valves that could be closed instead. As a result, the leak could not be stopped immediately but continued until the gauge pressure had fallen to zero. Fortunately, the leaking liquid did not ignite.

36.5.1   What Went Wrong?

- It was a serious design error to depend on an unreliable source for the flushing of the pump seals. There should have been either an alternative supply or a more reliable one. If a Hazop had been carried out, then this weakness would have been discovered, provided that the team had the adequate experience and studied the minor lines as well as the main ones. In some plants, minor lines such as flush lines to pump seals, drain lines, and sample lines are not given a line number and are then overlooked during the Hazop (see Section 36.8).
- There were flowmeters with alarms on the flush lines, but they were out of order. If they had been in working order, the loss of flow might have been noticed sooner and damage would have been less.
- It was a major operating error to remove the valve actuators, even for a short time, with the pump on line and without providing a means of manual isolation. This was not forbidden in the plant operating or

safety instructions. It was not forbidden because the authors never considered that anyone would want to carry out such a foolish act and probably never realized that the valves could not be hand operated. This incident, similar to a number of others, shows the limitations of instructions. They cannot be a substitute for an understanding of the basic principles of safety, scientific knowledge, or plant operation. We cannot forbid folly in all its possible forms (see Sections 24.5.2, 29.4, 29.5, and 30.12). The report does not say at what level the decision to remove the actuators was taken.

- This incident occurred in a company with a commitment to safety and a good record but at a time when the record had worsened as a result of rapid expansion and the construction of many large new plants, larger than those previously operated and operating at higher temperatures and pressures. The need to pay more attention to process safety had been recognized five years earlier and much had been done, but it takes time to change long-established working practices.

36.6   SEVERAL POOR SYSTEMS DO NOT MAKE A GOOD SYSTEM

This incident was the result of too much complexity in design and operations. Unfortunately, this makes the following account rather complex. Please persevere; it contains some messages of wider applicability than you might think at first glance.

Equipment contaminated with radioactivity was cleaned in a shielded room known as a control cell (Figure 36-3). The contaminated equipment

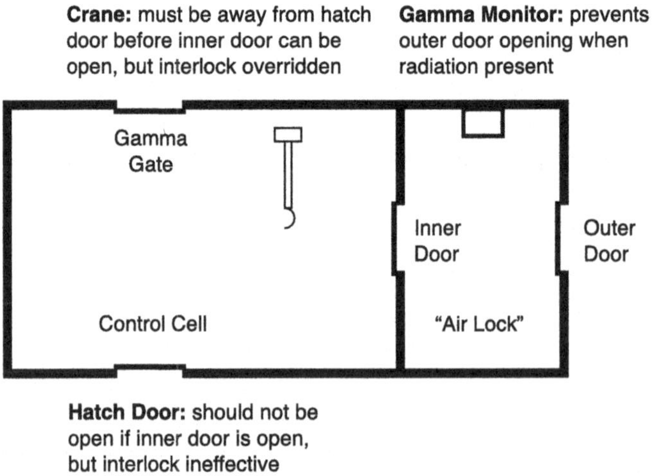

Crane: must be away from hatch door before inner door can be open, but interlock overridden

Gamma Monitor: prevents outer door opening when radiation present

Gamma Gate

Inner Door

Outer Door

Control Cell

"Air Lock"

Hatch Door: should not be open if inner door is open, but interlock ineffective

FIGURE 36-3   An overly complex system of interlocks to prevent both doors being open at the same time, unless maintenance is in progress.

was brought into the cell through a hole in the floor (the hatch door) and removed after cleaning through a hole in the ceiling (the gamma gate). Cleaning and other operations were carried out by remote control. No one was allowed in the cell when radioactive equipment was present but people could enter the cell at other times to maintain the cleaning equipment. They entered through double doors that acted as an air lock. A key exchange system—only one key could be withdrawn at a time—ensured that only one door could be open at a time.

The cleaning equipment needed much more maintenance than expected, and the company decided that both doors should be open during maintenance so that escape, if found to be necessary, could be rapid. To permit this, an override was fitted to the key exchange system, and at about the same time extra layers of protection were added. In addition to the key exchange system, now fitted with an override, there were an additional six layers of protection:

(a) If the inner door was open and radioactive equipment was present, a gamma ray monitor in the space between the doors prevented anyone from opening the outer door. This interlock was hardwired.
(b) A software interlock prevented the opening of the inner door unless the crane was in such a position that it could not open the hatch door.
(c) Another software interlock prevented the opening of the hatch door unless the inner door was closed. Both software interlocks formed part of the programmable electronic system (PES) that operated the other controls. However, an unknown software error made this interlock ineffective. Testing had not detected the fault.
(d) A hardwired interlock prevented the opening of the hatch door when the override was in operation.
(e) A checklist had to be completed before the doors were opened.
(f) A permit-to-work had to be completed before anyone entered the cell.

Despite these precautions, both doors were found open when radioactive equipment was in the cell. Fortunately, no one was in the path of the open doors. If someone had been, he or she could have received a serious dose of radiation.

While preparing the plant for maintenance, the foreman found that the inner door would not open. The foreman did not realize that the crane was in the wrong position and that interlock (b) was keeping the door closed. After the maintenance department had spent most of the day trying to determine why the door would not open, the shift manager decided to override interlock (b) by making the PES "think" that the crane was in the right position. He noted what he had done in his log.

Several days passed before the maintenance in the cell was complete. A different team was then on shift and they had not read all the old logs. By remote control, they opened the hatch door although both cell doors were open and moved active equipment into the cell. Interlock (c) would have prevented this happening if it had not been faulty; interlock (b) would have prevented this happening if it had not been overridden. Interlock (a) was ineffective because there was no radiation present when the outer door was opened. The interlock was not designed to sound an alarm if radiation subsequently appeared. Fortunately, the crane operator noticed on his TV screen that the cell doors were open and he was able to close the inner door.

36.6.1   What Went Wrong?

When the inner door would not open, the foreman assumed it was faulty. He should have checked the state of all the interlocks. Afterwards, a number of interlocks elsewhere on the plant were found to be overridden. Assuming instruments to be faulty is a common failing. For example, when a high-level alarm on a tank sounds, many operators have said, "It can't possibly be full," and sent for the instrument technician. By the time he arrived, the tank was overflowing.

When the maintenance was complete and the plant ready to be brought back into use, no one visited the scene to check that the outer door, at least, was shut. Again, many operators do not realize that a walk around the plant may reveal something that instruments cannot tell us. In contrast, I read some years ago that in London, when a flood warning is received, the first action of the authorities is to send someone down to the river to check the level.

There was no self-checking (logic checking) in the software controlling the crane. It *believed* the crane was in a position where it could not open the hatch door while it was actually opening it! This is artificial stupidity, not artificial intelligence, greater stupidity than any human would have. No human operator would tell someone who could see him carrying out a hazardous task, "Don't worry, I am now somewhere else." However, in one plant (see Section 29.4), certain packages were supposed to be moved downstairs only in the elevator. When the elevator was out of use, no one noticed anything anomalous when a package arrived downstairs.

If it is essential to override a protective system, this should be signaled in a prominent way, for example, by a large red sign, not just by a message in a logbook.

After the incident, the regulators asked the company to replace the software interlocks by hardwired ones. If safety interlocks are based on software, they should at least be on a separate PES from the normal control and operating system.

It is not possible to test software to confirm that every possible fault condition is covered, but it does seem that interlock (c) was not adequately tested.

Because there were so many layers of protection, they were not given the highest safety rating and changes were not studied as thoroughly as they should have been. This was a common mode failure affecting all layers of protection. One strong wall around a castle is better than several weak walls. Several strong walls are better still.

In complex systems, it is difficult for people to understand all possible ramifications. Checklists can help them avoid errors but they are a poor substitute for simplicity.

Underlying these failures were weaknesses in management, in the training of the foreman and shift manager, in the control of modifications and permits-to-work (were they audited?), in the design of control systems and software, and, above all, in the belief that several poor systems make a good system. Replacing a weak system by an entirely new one is better than adding complexity.

The official report [7] on which this description is based aroused little interest outside the nuclear industry. Its title suggested that it was of local interest only. Titles, keywords, and abstracts of accident reports often ignore the lessons of major interest. They tell us about the materials and equipment involved and the result, such as a leak, fire, or explosion, but they do not tell us what we most need to know: the actions needed to prevent a recurrence, such as better control of modifications or better preparation for maintenance.

Some years later, when similar equipment was being designed elsewhere, someone asked if the design team had read the report on this incident. They had but the first reply was that the design standards for this type of equipment had been changed to incorporate the lessons learned. However, reading a standard does not have the same impact on the reader as reading an accident report. Designers should always look for reports on equipment or plants similar to those they are designing. Operating staff moving to a new process should read reports on past accidents on that and similar processes. Needless to say, such reports should be readily available. Section 24.7 describes an overcomplex manual system for a similar situation; it also failed.

36.7 "FAILURES IN MANAGEMENT, EQUIPMENT, AND CONTROL SYSTEMS"

The heading is taken from the official report [8]. So many things were wrong that this account could go in almost any chapter of this book. The incident occurred on the distillation section of a catalytic cracker but the messages are general. Please read on even if you have never seen a catalytic cracker or any oily equipment, and especially if you have ever done any of the following:

- Used computers for process control but not provided overview pages
- Made modifications to plants or processes but did not systematically consider possible consequences or did not provide training and instruction on how to operate after the change
- Installed more alarms than operators can possibly cope with
- Carried out insufficient inspection for corrosion or inspected in the wrong place
- Not learned, reviewed, and remembered the lessons of past experience on plants similar to your own
- Reduced operator manning

If you have never done any of the foregoing, perhaps your halo is obscuring your view.

The details of the incident are complex but were briefly as follows: A lightning strike caused a small fire on a catalytic cracker. During the resulting plant upset, the flow to a fractionation column was lost and the take-off valve on the bottom of the column closed automatically to maintain a level in the column. When the flow to the column was restarted, a light on the panel told the operators that this valve had reopened but it had not. The liquid level in the column rose and the relief valve lifted and discharged some of the contents to a knock-out drum. The operators tried to reduce the level in the column by diverting some of the contents elsewhere, but they reached the knock-out drum by another route. The drum filled and could not be emptied as its contents were automatically pumped back into the fractionation section for recovery. About 20 tons of liquids entered the pipe leading from the knock-out drum to the flarestack. The pipe was weakened by corrosion—its thickness was down to 0.3 mm—and it failed. The drifting cloud of vapor was ignited by the flarestack, 110 m (360 ft) away, producing an explosion, followed by a fire that was allowed to burn for two days as this was the safest way of disposing of the oil.

Damage was extensive, but fortunately no one was killed or seriously injured. By good fortune, it was a Sunday afternoon so there were few people on site. A van carrying contractors was about to enter the area covered by the fireball, but it was still a short distance away and a group of men had just left a building that was demolished.

36.7.1   Better Management Could Have Prevented the Incident

More than four hours elapsed between the lightning strike and the explosion, thus some managers and day supervisors had come into the plant. However, instead of standing back, assessing the situation, and trying to diagnose what was happening, they got involved in hands-on operations (as in Section 26.1). It is not clear from the report whether or not they did so because there were too few operators to cope with the emergency.

The maintenance team had noticed that the flare line was corroding and thus had increased the frequency of inspections but not in the area where the pipe was thinnest, that is, near welds, especially longitudinal welds, as access was difficult at these places. This is rather like the story of the man who was seen after dark on his knees under a lamppost looking for something he had dropped. A passerby joined him and after a while asked if he was sure he had dropped it near the lamppost. "Probably not," said the man, "but the light is better here."

There was no system for reviewing, storing, or recalling incident information from similar plants. As shown in the following, the systems for the control of modifications, for the maintenance of instruments, and for monitoring corrosion were all flawed.

The company was prosecuted and had to pay $500,000 in fines and costs. The damage to the plant amounted to $75 million (at 1994 prices).

36.7.2 Better Control of Modifications Could Have Prevented the Incident

The pump-out system on the knock-out drum was modified a few years before the explosion so that the liquid in it was pumped back to the plant for reprocessing instead of going to slops. This meant that the liquid in it was pumped back to the system from which it had come so no reduction in the amount of liquid in the unit was achieved. It seems that when the modification was designed, no one foresaw that there might be a time when it would be necessary to reduce the total amount of liquid in the unit. They foresaw only a need to reduce the amount of liquid in individual vessels. A Hazop might have disclosed the unforeseen result.

It was possible by valve operation to revert to the original design, but this procedure had fallen into disuse from lack of practice and the absence of any written instructions.

36.7.3 Better Process Control Could Have Prevented the Incident

The failure of the column take-off valve was not an unfortunate and isolated incident. Thirty-nine control loops were examined afterward, and 24 were found to be faulty. Many of the faults were known, but repairs had been left to the next turnaround.

There were five product streams, with the flow rates spread across several display units. This made it difficult for the operators to assess the total output. There was no overview page.

The operators did not understand that some readings, such as temperatures and pressure, are based on direct measurements but others such as valve positions are based on indirect measurements. Thus, the light that indicated that the take-off valve was open really showed that a signal had been sent to the valve telling it to open. It did not inform the operators whether or not the valve had actually opened. (A similar misunderstanding occurred at Three Mile Island [9].) Similarly, many pump running indicators are based on the voltage supplied to the pump motor. Whenever practical, we should measure directly what we want to know, not something from which it can be inferred.

There were far too many alarms for the operators to cope with, 755 out of 1,365 measurements had an alarm fitted, whereas 431 had two alarms. At times, alarms were sounding every few seconds and operators were acknowledging them without realizing what they meant; 275 sounded in the last 11 minutes before the explosion. Records showed that the high-level alarm on the knock-out drum sounded 25 minutes before the explosion but it and the other critical alarms were acknowledged and overlooked. Safety critical alarms were not distinguished from others.

There was, of course, a case for each alarm if it was considered in isolation, but we should never consider each problem on its own without also considering the total effect of our individual decisions.

Sections 26.1 and 26.3 describe similar incidents.

36.8   CHANGES TO DESIGN AND OPERATIONS

Getting rid of waste products produces as many problems in plant design and operation as in the human body. As mentioned in Section 36.5.1, drain lines often have no line number and are overlooked in Hazop studies, whereas sewers are often someone else's problem.

A shallow pit, 0.5 m (20 in.) deep, at a paper mill collected spillages from a tank truck off-loading area. On the day of the incident, sodium hydrosulfide (NaSH) was being offloaded. As contractors were working nearby on a construction project, an operator drained the pit into the sewer. Unfortunately, the flow through the sewer was lower than usual and sulfuric acid was being added to the sewer to control its pH. The acid reacted with the NaSH to form hydrogen sulfide, and within five minutes of the draining, the hydrogen sulfide had escaped through a leaking manway seal. Two contractors were killed and eight injured.

No management of change study was carried out when the pit was connected to the sewer—a change in design; nor when the addition of acid was started—a process change.

No one seems to have realized that mixing NaSH and acid would produce hydrogen sulfide, although the NaSH supplier's material safety data sheet stated that it formed hydrogen sulfide in contact with acid. As a result, there were no monitors, alarms, warning signs, or training on the action to take if it was formed.

The seal on the manway cover was known to be leaking, and repairs had been requested but never made. The leak had not been investigated though it was the company's policy to investigate near misses.

Underlying these findings was the lack of an effective safety policy. There may have been a policy but if so, it was ignored in practice. The incident also shows the importance of giving as much detailed attention to drains and sewers—in design, modification, and operations—as to any other items of equipment [10].

36.9   THE IRRELEVANCE OF BLAME

The report on the last incident illustrates the truth of the following extract from an official U.K. report [11]:

> The fact is—and we believe this to be widely recognized—the traditional concepts of the criminal law are not readily applicable to the majority of infringements that arise under this type of legislation. Relatively few offenses are clear cut, few arise from reckless in difference to the possibility of causing injury, and few can be laid without qualification at the door of a single individual. The typical infringement or combination of infringements arises rather through carelessness, oversight, lack of knowledge or means, inadequate supervision, or sheer inefficiency. In such circumstances prosecution and punishment by the criminal courts is largely irrelevant. The real need is for a constructive means of ensuring that practical improvements are made and preventative measures adopted.

This report was written in 1972 and led to major changes in U.K. legislation. Unfortunately, the report has been forgotten and there is an increasing tendency to look for culprits. The operator who closed the wrong valve is now less likely to be made the culprit; instead we now look higher up the management tree. There is still, however, a simplistic belief that someone must be to blame and a failure to realize that many people had an opportunity to prevent every accident. I have a lot of sympathy with the manager who wrote:

> It is becoming increasingly hard to strike the right balance between the search for total safety and keeping [the plant running] ... what really winds me up is the suggestion that people like me would put "profit before safety." ... As well as an insult to my integrity, I personally find it very offensive. I feel that I carry a weighty responsibility for the lives and livelihoods of the people who entrust themselves to our [operations]. [12]

References

1. F. K. Crawley, Beware of Steam Condensing Downstream of Relief Valves, *Loss Prevention Bulletin*, Vol. 143, Oct. 1998, pp. 19–20.
2. T. Gillard, Seizure of Boiler Relief Valves, *Loss Prevention Bulletin*, Vol. 141, June 1998, pp. 14–15.
3. Anon, *Investigation Report: Catastrophic Vessel Overpressure*. Report No. 1998-002-I-LA, US Chemical Safety and Hazard Investigation Board, Washington, DC, 1998.
4. T. A. Kletz, *Learning from Accidents*, 3rd edition, Butterworth-Heinemann, Oxford, UK, and Woburn, MA, 2001, Chapter 8.
5. Anon, *Hazards of Air and Oxygen*, Institution of Chemical Engineers, Rugby, UK, 2004, Figures 34 and 35.
6. T. Fishwick, Three Flare Stack Incidents, *Loss Prevention Bulletin*, Vol. 142, Aug. 1998, pp. 7–11.
7. HM Nuclear Installations Inspectorate, *Windscale Vitrification Plant Shield Door Incident*, Her Majesty's Stationery Office, London, 1992.
8. Health and Safety Executive, *The Explosion and Fire at the Texaco Refinery, Milford Haven, 24 July 1994*, HSE Books, Sudbury, UK, 1997.
9. T. A. Kletz, *Learning from Accidents*, 3rd edition, Elsevier, Boston, MA, 2001, Chapter 11.
10. Anon, *Investigation Report—Hydrogen Sulfide Poisoning*. Report No. 2002-01-1-AL, US, Chemical Safety and Hazard Investigation Board, Jan. 2003.
11. Anon, *Safety and Health at Work—Report of the Committee (the Robens Report)*, Her Majesty's Stationery Office, London, paragraph 26.1, 1972.
12. M. Holden, *Modern Railways*, Vol. 58, No. 628, Jan. 2001, p. 37.

37

Accidents in Other Industries

During an Alpine hike in 1948, Swiss mountaineer George de Mestral became frustrated by the burs that clung annoyingly to his pants and socks. While picking them off, he realized that it might be possible to produce a fastener based on the burs to compete with, if not obsolete, the zipper.

—Charles Panati, *Extraordinary Origins of Everyday Things*

As this quotation shows, new ideas often result when we transfer an idea from one field of knowledge to another. In the same way, we can learn from accidents in other industries. Their immediate technical causes are not always of interest, but the underlying causes can supplement and reinforce our own experience. Because we are not involved in the technical details, we often see the underlying causes more clearly and reading about them is more recreation than it is work. And it may also be comforting to learn that people in other industries make as many errors as we do.

37.1   AN EXPLOSION IN A COAL MINE

An explosion in a Canadian coal mine in 1992 killed 27 people and led to the bankruptcy of the parent company. There was an explosion of methane, which set off a coal dust explosion. The source of ignition was probably sparks formed by mining machinery striking rock. This was a triggering event rather than a cause as there were so many ongoing faults that an explosion in the end was almost inevitable:

- Inadequate ventilation allowed explosive mixtures of methane and air to form.
- The methods for detecting methane were also inadequate and mining was allowed to continue when such methods were inoperable.
- Too much coal dust was allowed to accumulate.
- It is normal practice to dilute coal dust with stone dust to prevent explosions but not enough stone dust was used as stocks were too low.
- Many of the miners were inexperienced and inadequately trained; 12-hour shifts made them tired; fear of reprisals discouraged the reporting of hazards and those that were reported were not followed up.
- Output was put before safety.

There were other hazards not directly connected with the explosion. Thus falls of roof were common as intersecting fault lines were ignored [1].

These various shortcomings were not, of course, isolated. They all stemmed from a cavalier attitude to safety at all levels but particularly among senior managers as they set the example that others will follow.

This plant was so bad that many readers may feel that it has no message for them. But good plants can deteriorate, perhaps after a change of management (see Chapter 26). Stocks of spares are reduced to save cost, instruments are tested or maintained less frequently, and before long the plant has started going downhill. Every journey, uphill or downhill, starts with one step. Sections 30.4 and 30.9 describe other dust explosions.

37.2 MARINE ACCIDENTS

Many marine accidents are process accidents, similar to many that have occurred, or could occur, in chemical plants. The following are all taken from the periodic reports published by the U.K. Marine Accidents Investigation Branch.

37.2.1 A Misleading Display

A roll-on, roll-off ferry was ready to depart. When one of the ship's officers tried to close the bow door (the type that lifts up like a knight's visor), it refused to move. He sent for the engineers, who decided to close the door manually. They stopped immediately when they realized that the door was buckling. It was then discovered that a bolt that held the door open was still engaged. There was a light on the control panel next to a *Visor open* label, and the operators assumed that this meant that the visor was free to be lowered. The light actually referred to something quite different. The report [2] says:

> The layout of any control panel must be clear and unambiguous. ... If it is capable of being read wrongly, you can be sure it will! Crews come and go, and unless instructions are up to date and clear and easily understood, experience and word-of-mouth explanations get lost.

Visual checks of the locking mechanism were difficult and time consuming. After the incident, this locking mechanism was changed to make inspection easier. As in so many of the incidents described in this book, what looks at first sight to be poor operations could have been prevented, or made less likely, by better design. Sometimes fundamental redesign is needed, but often, as in this case, all we need is more attention to detail.

37.2.2 Stand Clear

When we are carrying out a pressure test, we know that the equipment we are testing might fail and so people should always position themselves to prevent injury if it does. Failures during pressure testing are rare but not unknown. If we were sure the equipment would not fail, we would not need to test it (though testing, as well as proving that the equipment is safe to use at its design pressure, also relieves stress).

In the same way, when moving machinery is started up for the first time after repair, we should remember that it might fail. A centrifugal lubrication oil purifier on a ship had been reassembled after maintenance. It was run empty without trouble, but when oil was admitted the bowl burst. Fortunately, only a minor injury was incurred.

The failure was due to an error in assembly. There was a change of shift during the assembly, and it seems that crew members on the second shift misunderstood exactly what still needed to be done or were not adequately briefed [3]. Equipment should be designed so that it cannot be assembled incorrectly.

A cooling water pump on a passenger ferry failed and one of the engines overheated. The first component to fail was the exhaust gas trunking, which started to melt. This allowed exhaust gases to percolate into the passenger areas. One of the passengers noticed the fumes and reported this to the crew. What, the report wonders, might have happened if he had not raised the alarm at an early stage [4] ?

There was a high-temperature alarm on the engine but it did not operate. Perhaps it was out of order or perhaps the set point was fixed to protect the engine and no one realized that the trunking was more vulnerable. A similar incident occurred in a chemical plant. An electric heater was fitted with a high-temperature trip. The set point was chosen to prevent damage to the heater elements, but no one realized that the body of the heater would be damaged at a lower temperature. In fact, it ruptured.

37.2.3   Wrong Connections

A fishing vessel left port after an overhaul. Soon afterwards the bearings on the engine turbocharger seized. Fortunately, the small fire that followed was soon extinguished and the vessel was towed back to port. It was then found that during reassembly, the bearing oil supply had been connected to the cooling water inlet and the cooling water supply to the bearing oil inlet.

It is easy to say, as the report [5] does, that the ship's crews should always check the work of contractors. Of course they should, but accidents such as this will continue to occur until designers learn to use different types or sizes of connections for different duties. In the meantime, users should paint different connections different colors (or attach colored sticky tape to them).

37.2.4   Preparation for Maintenance

Many accidents have occurred in industry because maintenance was undertaken without adequate consideration of the risk (see Chapter 23). The same is true at sea. For example, a rising engine temperature showed that the seawater inlet was blocked. The mate closed the seacock as far as he could and then removed the cover from the box on the ship's side of the seacock. He removed a plastic bag, which had been sucked into it. As he did so, water started to pour out. He tried to close the seacock further but broke the linkage. The bilge pumps could not cope with the flow, and the ship was abandoned and sank a few hours later [6].

37.2.5   Entry into Confined Spaces

Many people have been killed or overcome because they entered tanks or other confined spaces on ships without authorization or before the atmosphere had been tested. Sometimes the procedures were poor, but often they were ignored (see Chapter 24). The following incident is more unusual.

Frozen fish was being loaded into a refrigerated ship. Once on board, the pallets were stowed by liquefied petroleum gas (LPG)–powered forklift trucks in enclosed decks (the tween decks). The stevedores complained of headaches and nausea, and loading was stopped. The cause was a buildup of carbon monoxide. Most people know that internal combustion engines should not be operated in a confined space, such as a garage. In this case, the low temperature ($-20°C$ [$-4°F$]) reduced the effectiveness of combustion and led to increased production of carbon monoxide, oxides of nitrogen, and unburned fuel.

Electrically powered vehicles should be used in confined spaces. This is stated in the U.K. Code of Safe Working Practice for Merchant Seamen but seems to have been unknown to the ship's officers and the steve-dores' employer [7].

37.2.6 For Want of a Nail, a Ship Was Lost

Two U.K. fishing vessels sank within 10 days of each other. Fortu-nately, the crews were rescued. In both cases, the seawater pipes leaked, probably as the result of corrosion, and flooded the engine rooms. Both ships were fitted with water-level alarms that failed to work, probably because they were not tested regularly, if at all, but possibly because the wiring was not in protective conduits. By the time the flooding was dis-covered, it was too late to close the valves in the seawater lines as they were below the water level. Extended spindles on the valves would have saved the ships [8].

Maintenance and operations on small ships (and small plants?) are often poor but nevertheless we can learn from these incidents. Who has never postponed the testing of alarms and trips because the testers were too busy elsewhere? Who has never overlooked the opportunity to make a cheap change that would add an extra layer of protection? Valves that are normally left open or shut but that might be needed when things go wrong should be operated regularly, say, every week, so that they do not become stiff. Are yours?

37.3 HUMAN ERROR

On ships, as on land, there is a readiness to blame human error—poor maintenance, watch-keepers falling asleep, errors in navigation—instead of looking for underlying causes such as poor training or super-vision, error-prone designs, lack of protective features, overlong hours of work, and so on. In the following extract from a report [9] on marine accidents, I have changed *seamen* to *operators* and made other similar changes:

> There is an abundance of academic literature on human error which quickly lapses into language that leaves the average operator [and engineer] totally bewil-dered, and few will have the foggiest ideas what is meant by "visual/tactile dis-similarity," "cognitive aspects of safety," "rule-based behaviour," "latent conditions and pathogens," or "non-optimised performance related factors." What the operator [and the engineer] needs is a simple explanation about what is meant by human fac-tors so he or she can better understand why it matters and what needs to be done to improve safety and conditions of service.

I have tried to provide such a guide in *An Engineer's Guide to Human Error* [10] (see the introduction and Section 38.3).

The following are two more adopted quotations from a marine report [11], this time without change:

> [When a ship has run aground] giving orders calmly will ensure success. It is not the moment to give the unfortunate helmsman his or her annual appraisal.

> When the draught of your vessel exceeds the depth of water available ... you can always consider the delights of gardening.

Section 26.1 drew attention to the reluctance of some operators to go and look at the plant when it is not operating correctly. The following extract from a letter by a deep-sea pilot [12] describes a view shared by many chemical engineers:

> Modern watch-keepers tend to be wonderful at operating computers and twiddling radars, but abysmal in the basics, such as keeping a visual lookout and correctly applying the collision regulations. Lack of a grounding in mental arithmetic also means that they often cannot roughly estimate their computerized information and realize when it is wrong.

There are other marine accident reports in Sections 27.5 and 29.3.

37.4   TESTS SHOULD BE LIKE REAL LIFE

Section 14.1 describes several tests that did not detect faults because they did not simulate real-life conditions, for example, a high-temperature trip was removed from its case before testing. The tests therefore failed to detect that the pointer was touching the plastic front of the instrument case, and this prevented it from reaching the trip point. The following is an example from another industry.

To the surprise of the manufacturers, a small car failed to pass a rear collision test. It crumpled more than expected. It was then found that the testers had removed the spare wheel before the test, as it seemed unnecessarily wasteful to damage it. However, the spare wheel, correctly inflated, was a necessary part of the energy-absorbing process [13].

37.5   LOAD AND STRENGTH TOO CLOSE

As described in Section 25.3.2, in 2000 a railway accident at Hatfield, United Kingdom, killed four people and drew attention to a literal interface problem. The immediate cause was a cracked rail, but the underlying cause was that British Railways had been privatized and split into a hundred companies. Responsibility for the rails and the wheels now lay with different organizations. To quote the head of the railways' safety organization, "Both sides of the wheel/rail interface may be operating

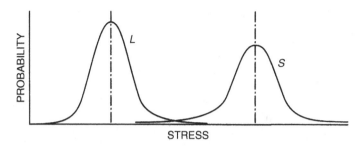

FIGURE 37-1   Overlapping distributions of load (L) and strength (S). *(Reprinted with permission of the Institution of Chemical Engineers.)*

within their respective safety based Standards, but the combined effect of barely acceptable wheel on barely acceptable rails is unacceptable" [14].

Figure 37-1 may make this clearer. In any system, the strength and the load vary to some extent from their design values, and there is inevitably a small overlap between the two asymptotes. Its area is a measure of the probability that the load will exceed the strength and the system will fail, not necessarily immediately but in the long run. Normally, this probability is negligible. In the case of the railways, the wear on the wheel increases the load and cracks in the rail decrease the strength. Both were just within specification, and the overlap was too large. This led to rolling contact fatigue of the track (also called gauge corner cracking), the train crash, and the consequent upheaval while hundreds of miles of faulty rail were replaced.

The engineering principle involved is hardly new. In 1880, Chaplin showed that a chain can fail if its strength is at its lower limit and the load is at its upper limit [15]. The Hatfield crash did not occur because engineers had forgotten this but because there were no engineers in the senior management of the company (Railtrack) that owned the track. They had all been moved to the maintenance companies (or elsewhere) and Railtrack had lost the knowledge that it needed to make it an intelligent customer of the maintenance companies. The change had a further disadvantage:

> As one senior railwayman put it: In an integrated railway you could gain a lot more skills; you could work through managing train crews, signalling, running stations—you really got a feel for everything in the business, you would learn about every activity so that you knew how the railways operated.
>
> Now with so many employees following paths through a single company, their experience is so much more limited in that the broad base of knowledge has disappeared. The steady flow of skilled operators and skilled engineers ready to take up senior management positions has created a dire shortage.
>
> As an example, in the years prior to 1994, the railways took on an average of 40 engineering graduates annually, and then between 1994 and 2002, the total was almost zero [16].

37.6   THE NINETEENTH CENTURY

In a short but interesting book, *What the Victorians Got Wrong* [17], S. and T. Yorke describe the results and causes of many accidents that occurred in the nineteenth century. Many of the causes are familiar, but others were due to ignorance of the properties of the materials handled, such as cast and wrought iron, clay (in dams), coal and coal dust, methane (in mines), town gas, and drinking water. In addition, "Safety was not seen as something that could be influenced; you simply learned where the dangers lay, and tried to recognize when you were at risk. [Many Victorians] were so occupied by the novelty of their inventions that they were unable to think through the 'what if' scenarios." An underlying cause of accidents was that "Victorian society was run by the wealthy who, for generations, had viewed the poor as expendable. However, after long struggles, many improvements were made" (see Section on 24.8.2, last paragraph).

References

1. A. M. Amyotte and A. M. Oehmne, Application of a Loss Causation Model to the Westray Mine Explosion, *Process Safety and Environmental Protection*, Vol. 80, No. B1, 2002, p. 55.
2. Anon, Light reading! *Safety Digest—Lessons from Marine Accident Reports*, No. 1/2001, Marine Accident Investigation Branch of the UK Department of Transport, Local Government and the Regions, London, 2001, p. 13.
3. Anon, The explosive force of high speed rotation! *Safety Digest—Lessons from Marine Accident Reports*, No. 3/2001, Marine Accident Investigation Branch of the UK Department of Transport, Local Government and the Regions, London, 2001, p. 20.
4. Anon, Melting moments, *Safety Digest—Lessons from Marine Accident Reports*, No. 2/2002, Marine Accident Investigation Branch of the UK Department of Transport, London, 2002, p. 13.
5. Anon, The thinking man, *Safety Digest—Lessons from Marine Accident Reports*, No. 2/2002, Marine Accident Investigation Branch of the UK Department of Transport, London, 2002, pp. 36–37.
6. Anon, Ideas for tackling flooding, *Safety Digest—Lessons from Marine Accident Reports*, No. 2/2002, Marine Accident Investigation Branch of the UK Department of Transport, London, 2002, pp. 43–44.
7. Anon, Cold comfort, *Safety Digest—Lessons from Marine Accident Reports*, No. 3/2001, Marine Accident Investigation Branch of the UK Department of Transport, Local Government and the Regions, London, 2001, pp. 18–19.
8. Anon, Defective bilge alarms lead to the loss of two large vessels, *Safety Digest—Lessons from Marine Accident Reports*, No. 2/2001, Marine Accident Investigation Branch of the UK Department of Transport, Local Government and the Regions, London, 2001, pp. 46–47.
9. Anon, A pause for thought, *Safety Digest—Lessons from Marine Accident Reports*, No. 3/2001, Marine Accident Investigation Branch of the UK Department of Transport, Local Government and the Regions, London, 2001, pp. 50–52.
10. T. A. Kletz, *An Engineer's View of Human Error*, 3rd edition, Institution of Chemical Engineers, Rugby, UK, 2001, and Taylor and Francis, Philadelphia, PA.
11. Anon, When the draught of your vessel exceeds the depth of water … , *Safety Digest—Lessons from Marine Accident Reports*, No. 2/2001, Marine Accident Investigation Branch

of the UK Department of Transport, Local Government and the Regions, London, 2001, pp. 60–64.

12. R. Francis, Declining Standards on the Watch, *Daily Telegraph (London)*, 18 Dec. 2002, p. 21.

13. R. Ford, Crashworthiness: An Ethical Issue, *Modern Railways*, Vol. 55, No. 600, Sept. 1998, pp. 576–577.

14. R. Ford, Gauge Corner Cracking—Privatisation Indicted, *Modern Railways*, Vol. 59, No. 640, Jan. 2002, pp. 19–20.

15. A. G. Pugsley, The Safety of Structures, Arnold, London, quoted by N. R. S. Tait (1987). *Endeavour*, Vol. 11, No. 4, 1966, p. 192.

16. T. Miles, Where Are the Aspiring Managers? *Modern Railways*, Vol. 60, No. 653, Feb. 2003, p. 65.

17. S. Yorke and T. Yorke, *What the Victorians Got Wrong*, Counryside Books, Newbury, UK, 2008, pp. 5, 36, 92.

Accident Investigation— Missed Opportunities

If the origin of the human mind is to be understood, it is important to be able to identify signals of distinctly non-human behaviour. Lack of innovation is one of them.

—Roger Lewis, *The Origin of Modern Humans*

Almost all the accidents described in this book need not have occurred. Similar ones have happened before, and accounts of them have been published. Someone knew how to prevent them even if the people on the job at the time did not. This suggests that there is something seriously wrong with our accident investigations, safety training, and the availability of information.

Having paid the price of an accident, minor or serious (or narrowly missed), we should use the opportunity to learn from it. Failures should be seen as educational experiences. The 10 major opportunities summarized in what follows, are frequently missed, the first 7 during the preparation of a report and the other 3 afterward. Having paid the *tuition fee*, we should learn the lessons. The evidence is usually collected adequately; the weakness lies in its interpretation.

38.1 ACCIDENT INVESTIGATIONS OFTEN FIND ONLY A SINGLE CAUSE

Often, accident reports identify only a single cause, though many people, from the chemical and engineering designers down to the last link in the chain, the mechanic who broke the wrong joint or the operator who closed the wrong valve, had an opportunity to prevent the accident. The single cause identified is usually this last link in the chain of events that led to the accident. Just as we are blind to all but one of the octaves in the electromagnetic spectrum, we are blind to many of the opportunities that we have to prevent an accident. But just as we have found ways of making the rest of the spectrum visible, we need to make all the ways of preventing an accident visible.

38.2   ACCIDENT INVESTIGATIONS ARE OFTEN SUPERFICIAL

Even when we find more than one cause, we often find only the immediate causes. We should look beyond them for ways of avoiding the hazards, such as inherently safer design (Chapter 21). For example, could less hazardous raw materials have been used? Also, we should look for weaknesses in the management system. For example, could more safety features have been included in the design? Were the operators adequately trained and instructed? If a mechanic opened up the wrong piece of equipment, could there have been a better system for identifying it? Were previous incidents overlooked because the results were, by good fortune, only trivial? The emphasis should shift from blaming the operator to removing opportunities for error or identifying weaknesses in the design and management systems.

Many of the chapter headings in this book are examples of root causes and, as mentioned in the preface, this has made the allocation of incidents to chapters somewhat arbitrary as most of them have more than one root cause.

When investigators are asked to look for underlying or root causes, some of them simply call the causes they have found root causes (see Section 34.3 for an example). One report quoted corrosion as the root cause of equipment failure, but it is an immediate cause. To find the true root causes, we need to ask if corrosion was foreseen during design and if not, why not? Were operating conditions the same as those given to the designer and if not, why not? Was regular examination for corrosion requested, and if so, had it been carried out and were the results acted upon? Senior managers should not accept accident reports that deal only with immediate causes.

The causes listed in accident reports sometimes tell us more about the investigators' beliefs and background than about the accidents. One company had recognized that failure to learn from past experience was a major cause of accidents and was making strenuous efforts to improve its learning from experience. However, none of their accident reports or the annual summary of them mentioned this as a cause. The members of the investigating panels did not know that similar accidents had happened before.

38.3   ACCIDENT INVESTIGATIONS LIST HUMAN ERROR AS A CAUSE

As mentioned in the introduction, human error is far too vague a term to be useful. We should ask, "What sort of error?" because different sorts of error require different actions if we are going to prevent the errors from happening again [1].

- Was the error a mistake—that is, one due to poor training or instructions, so that the intention was wrong. If so, we need to improve the training and instructions and, if possible, simplify the task. Whereas instructions tell us what to do, training gives us the understanding that allows us to handle unforeseen situations. However many instructions we write, we will never foresee everything that might go wrong. (For examples see Sections 24.5.2, 29.4, 29.5, 30.12, and 36.5.)
- Was the error due to a violation or noncompliance—that is, a deliberate decision not to follow instructions or recognized good practice? If so, we need to explain the reasons for them as we do not live in a society in which people will simply do what they are told. We should, if possible, simplify the task—if an incorrect method is easier than the correct one, it is difficult to persuade everyone to use the correct method—and we should check from time to time to see that instructions are being followed.
- Was the task beyond the ability of the person asked to do it, perhaps beyond anyone's ability? If so, we need to redesign the task.
- Was it a slip or lapse of attention (like many of those described in Chapter 27)? In contrast to mistakes, the intention may have been correct but it was not fulfilled. It is no use telling people to be more careful as no one is deliberately careless. We should remove opportunities for error by changing the design or method of working.

Designers, supervisors, and managers make errors of all these types though slips and lapses of attention by designers and managers are rare as they usually have time to check their work. Errors by designers produce traps into which operators fall—that is, they produce situations in which slips or lapses of attention, inevitable from time to time, result in accidents. Errors by managers are signposts pointing in the wrong directions.

38.4   ACCIDENT REPORTS LOOK FOR PEOPLE TO BLAME

In every walk of life, when things go wrong the default action of many people is to ask who is to blame? The banner headline in my newspaper after a railway accident was "Who is to blame this time?" However, blaming human error for an accident diverts attention from what can be done by better design or methods of operation. To quote James Reason, "We cannot change the human condition but we can change the conditions in which humans work." Even when people ask, "What did we do wrong?" they often find the wrong answer. They find that the instructions were perhaps not clear enough, rewrite them in greater detail and at greater length, and thus reduce the probability that anyone will read them. They should consider the alternative actions listed in Section 38.6 in what follows.

To paraphrase G. K. Chesterton, the horrible thing about all the people who work at plants, even the best, is not that they are wicked, not that they are stupid; it is simply that they have got used to it. They do not see the hazards; all they see is the usual people carrying out the usual tasks in the usual place. They do not see the risks; they see only their own place of work.

One method of jerking people out of their familiarity is to show them slides of the hazards they pass every day without noticing them. On one occasion, I led a discussion of a leak that had occurred from a substandard drain point. Immediately afterward, someone who had been present went into a compressor building that he visited every day. As he walked through the door, he saw a substandard drain point.

38.5   ACCIDENT REPORTS LIST CAUSES THAT ARE DIFFICULT OR IMPOSSIBLE TO REMOVE

For example, a source of ignition is often listed as the cause of a fire or explosion. But it is impossible on the industrial scale to eliminate all sources of ignition with 100% certainty. Although we try to remove as many as possible, it is more important to prevent the formation of flammable mixtures.

Which is the more dangerous action on a plant that handles flammable liquids: to bring in a box of matches or to bring in a bucket? Many people would say that it is more dangerous to bring in the matches, but nobody would knowingly strike them in the presence of a leak and in a well-run plant leaks are small and infrequent. If a bucket is allowed in, however, it may be used for collecting drips or taking samples. A flammable mixture will be present above the surface of the liquid and may be ignited by a stray source of ignition. Of the two *causes* of the subsequent fire, the bucket is the easier to avoid.

I am not, of course, suggesting that we allowed unrestricted use of matches on our plants, but I do suggest that we keep out open containers as thoroughly as we keep out matches. Instead of listing causes, we should list the actions needed to prevent a recurrence. This forces people to ask if and how each so-called cause can be prevented in the future.

38.6 WE CHANGE PROCEDURES RATHER THAN DESIGNS

As discussed in Chapter 27, when making recommendations to prevent an accident, our first choice should be to see if we can remove the hazard—the inherently safer approach. For example, could we use a nonflammable solvent instead of a flammable one? Even if it is impossible at the existing plant, we should note it for the future.

The second best choice is to control the hazard with protective equipment, preferably passive equipment, as it does not have to be switched on. As a last (but frequent) resort, we may have to depend on procedures. Thus, as a protection against fire, if we cannot use nonflammable materials, insulation (passive) is usually better than water spray turned on automatically (active), but that is usually better than water spray turned on by people (procedural). In some companies, however, the default action is to consider a change in procedures first, sometimes because it is cheaper but more often because it has become a custom and practice carried on unthinkingly.

Operators provide the last line of defense against errors by designers and managers. It is a bad strategy to rely on the last line of defense and to neglect the outer ones. Good loss prevention starts far from the top event, in the early stages of design. Blaming users is a camouflage for poor design.

38.7 WE MAY GO TOO FAR

Sometimes after an accident, people go too far and spend time and money on making sure that nothing similar could possibly happen again even though the probability is extremely unlikely. If the accident was a serious one, it may be necessary to do this to reassure employees and the public, but otherwise we should remember that if we goldplate one unit there are fewer resources available to silverplate the others.

As mentioned in Chapter 27, in the United Kingdom the law does not require companies to do everything possible to prevent an accident, only what is *reasonably practicable*. This legal phrase means that the size of a risk should be compared with the cost of removing it, in money, time, and trouble, and if there is a gross disproportion between them, it is not necessary to remove the risk. In recent years, the regulator, the Health and Safety Executive, has provided detailed advice on the risks that are tolerable and the costs that are considered disproportionate [2]. In most other countries, the law is more rigid and, in theory, expects companies to remove all risks. This, of course, is impossible, but it makes companies reluctant to admit that there is a limit to what they, and society, can afford to spend even to save a life. (If this sounds cold blooded, remember that we are discussing very low probabilities of death where further expenditure will make the probability even lower but is very unlikely to actually prevent any death or even injury.)

38.8   WE DO NOT LET OTHERS LEARN FROM OUR EXPERIENCE

Many companies restrict the circulation of incident reports, as they do not want everyone, even everyone in the company, to know that they have blundered. However, this will not prevent the incident from happening again. We should circulate the essential messages widely, in the company and elsewhere, so that others can learn from them, for several reasons as follows:

- Moral: If we have information that might prevent another accident, we have a duty to pass it on.
- Pragmatic: If we tell other organizations about our accidents, they may tell us about theirs.
- Economic: We would like our competitors to spend as much as we do on safety.
- The industry is one: Every accident affects its reputation. To misquote the well-known words of John Donne:

> No plant is an Island, entire of itself; every plant is a piece of the Continent, a part of the main. Any plant's loss diminishes us, because we are involved in the Industry: and therefore never send to know for whom the Inquiry sitteth; it sitteth for thee.

When information is published, people do not always learn from it. A belief that *our problems are different* is a common failing (see Section 34.3).

38.9   WE READ OR RECEIVE ONLY OVERVIEWS

This opportunity is one that many senior managers miss. Lacking the time to read accident reports in detail, they consume predigested summaries of them, full of generalizations such as *there has been an increase in accidents due to inadequate training*. However, as already mentioned, the identification of underlying causes can be subjective and is influenced by people's experience, interests, blind spots, and prejudices. Senior managers should read a number of accident reports regularly and, if necessary, discuss them with their authors to see if they agree with the assignment of underlying causes. In any field of study, reliance on secondary sources instead of primary ones can perpetuate errors.

Senior managers should be aware that mission or policy statements, though legally required in some countries, have little, if any, effect on safety. People do not change their behavior as a result of reading a mission statement. They may change as a result of reading an accident report or, better still, taking part in a discussion of an accident (see Section 38.10.1). Senior managers should also do what they can to stop the spread of the popular view that the consequences of accidents are proportional to the degree of negligence. (Compare Sections 26.1 and 26.3.) Similarly, safety is not proportional to the money spent (see Chapter 21).

38.10   WE FORGET THE LESSONS LEARNED AND ALLOW THE ACCIDENT TO HAPPEN AGAIN

Even when we prepare a good report and circulate it widely, all too often it is read, filed, and forgotten. Every chapter shows that organizations have no memory [3]. Only people have memories and after a few years they move on, taking their memories with them. Procedures introduced after an accident are allowed to lapse, and some years later the accident happens again, even on the plant where it happened before. If by good fortune the results of an accident are not serious, the lessons are forgotten even more quickly (see Section 25.4).

The following are some actions that can prevent the same accidents from recurring so often:

- Include in every instruction, code, and standard a note on the reasons for it and accounts of accidents that would not have occurred if the instruction, procedure, and so on had existed at the time and had been followed. Once we forget the origins of our practices, they become *cut flowers*; severed from their roots they wither and die.
- Never remove equipment before we know why it was installed. Never abandon a procedure before we know why it was adopted.
- Describe old accidents as well as recent ones, other companies' accidents as well as our own, in safety bulletins and discuss them at safety meetings.
- Follow up at regular intervals to see that the recommendations made after accidents are being followed, in design as well as operations.

- Remember that the first step down the road to an accident occurs when someone turns a blind eye to a missing blind.
- Include important accidents of the past in the training of under-graduates and company employees.
- Keep a folder of old accident reports in every control room. It should be compulsory reading for recruits and others should look through it from time to time.
- Read more books, which tell us what is old, as well as magazines, which tell us what is new.
- We cannot stop downsizing, but we should make sure that the remaining employees at all levels have adequate knowledge and experience.
- Devise better retrieval systems so that we can find details of past accidents in our own and other companies more easily than at present, and the recommendations made afterward. We need systems in which the computer will automatically draw our attention to information that is relevant to what we are typing or reading (see Section 38.10.2).

Everyone forgets the past. A historian of football found that fans would condense the first hundred years of their team's history into two sentences and then describe the last few seasons in painstaking detail. (But engineers' poor memories have more serious results.)

38.10.1   Weaknesses in Safety Training

There is something seriously wrong with our safety education when so many accidents repeat themselves so often. (Speaking at a conference on the lessons of Three Mile Island, Norman Rasmussen said that "we do a lot of teaching, it's just that we don't get much learning done in some of these schools" [4].) The first weakness in our safety training is that *it is often too theoretical*. It starts with principles, codes, and standards. It tells us what we should do and why we should do it and warns us that we may have accidents if we do not follow the advice. If anyone is still read-ing or listening, it may then go on to describe some of the accidents.

We should start by describing accidents and draw the lessons from them for two reasons. First, accidents grab our attention and make us read on, or sit up and listen. Suppose an article describes a management system for the control of plant and process modifications. We probably glance at it and put it aside to read later, and you know what that means. If it is a talk, we may yawn and think, *another management system designed by the safety department that the people at the plant will not follow once the novelty wears off*. In contrast, if someone describes accidents caused by modifications made without sufficient thought, we are more likely to read on or listen and consider how we might prevent them in the plants under our control. We remember stories about accidents far better than we remember disconnected advice. Whatever the subject, we should build generalities from individual cases; otherwise they have no foundations.

The second reason why we should start with accident reports is that the accident tells us what actually happened. You may not agree with my recommendations, but I hope you will not ignore the events I have described. If they could happen at your plant, I hope you will take steps to prevent them, though not necessarily the steps that I have suggested.

A second weakness with our safety training is that it usually consists of talking to people rather than discussing safety training with them. Instead of describing an accident and the recommendations made afterward, outline the story and let the audience question you to find out the rest of the facts, those that they think are important and that they want to know. Then let them say what *they think* ought to be done to prevent it happening again. More will be remembered and the audience will be more committed than if they were merely told what to do.

Once someone has blown up a plant, they rarely do so again, at least not in the same way. But when he or she leaves, the successor lacks the experience. Discussing accidents is not as effective a learning experience as letting them happen, but it is the best simulation available and it is a lot better than reading a report or listening to a talk.

We should choose for discussion accidents that bring out important messages such as the need to look for underlying causes, the need to control modifications, the need to avoid hazards rather than to control them, the need for inherently safer design, and so on. You can discuss the accidents described in this book, but it would be better to discuss those that occurred in your own plant. The audience cannot then think, *we would not do anything as stupid as the people at that plant.*

Undergraduate training should include discussion of some accidents, chosen because they illustrate important safety principles. If universities do not provide this sort of training, industry should provide it. In any case, new recruits need training on the specific hazards of the industry.

38.10.2 Databases

Many papers and reports have emphasized the need to learn from experience and to make the information derived from accident investigations and in other ways widely available. There are many databases that try to do this, but they are little used, in part because the information they contain is limited. The information derived, at great expense in suffering and cost, from accidents and research has no value if it is not used.

The following paragraphs describe a program that could lead to much greater use of the available data. Three improvements are desirable:

1. There are so many databases that no one has time to consult more than a few. We need a program similar to Google that can search the whole of the published literature on process safety (or industrial safety): books, reports, papers, articles, and the Internet as well as existing databases. Google searches everything on the Internet, including many items that have now been removed from it, and has plans to copy and search 30 million books. What I propose should not, therefore, be difficult.

2. Searching is hit or miss; we either get a "hit" or we don't. A "fuzzy" search engine will offer us reports on compounds, equipment, operations, results, and recommendations similar to those we are searching for. This is done by arranging the keywords in a sort of family tree. If there are no reports on the keyword, the system will offer reports on its parents or siblings. Filters could prevent it repeatedly referring to the same hazard [5]. Work at Loughborough University in the United Kingdom has demonstrated the feasibility of fuzzy searching [6, 7].

3. In conventional searching, the computer is passive and the user is active. The user has to ask the database if there is any information on, say, accidents involving particular substances, operations, or equipment. The user has to suspect that there may be a hazard or he or she will not look. We need a system in which the user is passive and the computer is active. With such a system, if someone types "X," the computer will signal that the database contains information on this substance, subject, or equipment. A click of the mouse will then display the data. As we type in Microsoft Word, the spellcheck and grammar check programs are running in the background of our computers and drawing attention to our spelling and grammar errors. In a similar way, a safety program in the background could draw attention to any subject on which it has data. Software similar to that already used by Microsoft is needed. A program of this type has been developed for medical use. Without the doctor taking any action, the program reviews the information on symptoms, treatment, diagnosis, and the like already entered for other purposes and suggests treatments that the doctor may have overlooked or not be aware of.

How could the system I have described be funded? Ultimately in the same way as Google is now, by advertising. For example, if the search term was (or included) "check valves," advertisements for suppliers of check valves would appear on the screen and each advertiser would pay a small sum every time someone clicked the advertisement. However, development costs would have to come from institutions or companies willing to support the proposal.

38.10.3 Cultural and Psychological Blocks

There are cultural and psychological blocks that encourage us to forget the lessons of the past. Most of the quotations on memory or forgetfulness in books of quotations say that forgetfulness is advantageous. To quote Paul Tillich [8]:

> Life could not continue without throwing the past into the past, liberating the present from its burden. Without this power life would be without a future; it would be enslaved by the past. ... The earlier stages in the development of a living being are left behind in order to provide space for the future, for a new life.

So if we wish to learn from experience, the technical fixes I have listed here are not enough. We also have to understand the built-in needs of people to come to terms with their failures. Anonymity helps. Whenever possible, accident reports should not say in which plant it occurred and should avoid criticism. We should treat failures as learning experiences. Rather than blame people who made errors, we should tell them that they are now wiser. Looking for people to blame should not be the objective of accident investigations. The reports in this book show that many people had opportunities to prevent almost every accident.

According to research by Brendan Depue of the University of Colorado, people are able "to exert some control over their emotional memories. ... By essentially shutting down specific portions of the brain, they were able to stop the retrieval process of particular memories" [9].

Another psychological block is that we find it difficult to change old beliefs and ways of thinking. This is particularly true of people who have spent all their careers doing the same job in the same department. Sir John Harvey-Jones, a former chairman of ICI, started his career in the navy. He writes that the early recruitment of future naval officers, when they were still schoolboys, resulted in unswerving loyalty but also intolerance of cynicism, experimentation, and novelty [10].

A sociological block is that we live in a society that values the new more than the old, probably the first society to do so. Old used to imply enduring value, whether applied to an article, a practice, or knowledge. Anything old had to be good to have lasted so long. Now it suggests obsolete or at least obsolescent. After a talk in which I had described an accident that had occurred 5 years beforehand, a member of the audience wrote on the comment sheet that he had expected more up-to-date information. A similar accident had occurred 50 years earlier.

The first step toward overcoming these blocks is to realize that they exist and that engineering requires a different approach. We should teach people that "It is the success of engineering which holds back the growth of engineering knowledge, and its failures which provide the seeds for its future development" [11].

References

1. T. A. Kletz, *An Engineer's View of Human Error*, 3rd edition, Institution of Chemical Engineers, Rugby, UK, and Taylor & Francis, New York, 2001.
2. Health and Safety Executive, *Reducing Risks, Protecting People—HSE's Decision Making Process*, HSE Books, Sudbury, UK, 2001.
3. T. A. Kletz, *Lessons from Disaster—How Organisations Have No Memory and Accidents Recur*, Institution of Chemical Engineers, Rugby, UK, and Gulf, Houston, TX, 1993.
4. N. C. Rasmussen, General Discussion. In: T. H. Moss and D. L. Sills (editors). *The Three Mile Island Nuclear Accident—Lessons and Implications*, New York Academy of Sciences, New York, 1999, p. 50.
5. J. Bond, "Linking an accident database to design and operational software," *Hazards XVII: Process Safety—Fulfilling Our Responsibilities, Symposium Series No. 149*, Institution of Chemical Engineers, Rugby, UK, 2003, pp. 491–500.
6. P. W. H. Chung and M. Jefferson, A fuzzy approach to accessing accident databases. *Applied Intelligence*, Vol. 9, 1998, pp. 129–137.
7. R. E. Iliffe, P. W. H. Chung, and T. A. Kletz, "Hierarchical Indexing, Some Lessons from Indexing Incident Databases," *International Seminar on Accident Databases as a Management Tool*, Antwerp, Belgium, Nov. 1998.
8. P. Tillich, Quoted by J. Burnside, Fact, Fiction, History, Myth, Reality, Truth, Lies, *Daily Telegraph* (UK) *Books Supplement*, 4 Mar. 2006, pp. 1–2.
9. B. Depue, Quoted by R. Highfield, *Daily Telegraph* (UK), 13 Aug. 2008. The original article by B. Depue, T. Curran, and M. T. Banich, was published in *Science*, (13 July 2007), **317** (5835): 215–219.
10. J. H. Harvey-Jones, *Getting It Together*, Heineman, London, 1991, p. 89.
11. D. I. Blockley and J. R. Henderson, *Proceedings of the Institution of Civil Engineers*, Part 1, 68, 1980, 719.

An Accident That May Have Affected the Future of Process Safety

All the text books will tell you, stretching back over two decades, that most acquisitions fail to create value for anyone other than the selling shareholders, and that three years after the event the buying company is suffering remorse: they just wish they had never done it.

—Sir John Banham (former director-general of the
U.K. Confederation of British Industry), *Daily Telegraph*
(London), August 11 2005

This short chapter describes an accident—that is, an event that had unforeseen and unexpected results. It led to the end of an independent company, Imperial Chemical Industries (ICI), which had made major changes in process safety, most of which were widely copied.

ICI was formed in 1926 by the merger of the United Kingdom's four largest chemical companies. The dominant partner was Brunner Mond, which was founded in 1874 by Ludwig Mond and John Brunner to manufacture sodium carbonate in Cheshire. After World War I, Brunner Mond expanded, producing ammonia and fertilizers at Billingham in northeast England and later, as part of ICI, producing petrol by the liquefaction of coal. Ludwig's son, Alfred, First Lord Melchett, became the first chairman of ICI. Brunner Mond attached great importance to safety and what we now call human resources. The company's policies were far ahead of industry as a whole, and they became the policies of ICI.

I joined the company at Billingham in 1944 and retired in 1982. My first 7 years were spent in the research department, the next 16 in production, and the last 14 in process safety. The following is a list of the major innovations in process safety made by the company during my time there. All of them were published and made freely available to the process industry as a whole. I was involved to varying degrees in all of them:

1. Hazard and operability studies (Hazop) (see Chapter 18).
2. Quantitative risk assessment (also known as hazard analysis or Hazan). It started in the nuclear industry, but ICI was the first company to apply the technique in the process industries.
3. Inherently safer design (see Chapter 21).
4. Management of change. This became widespread after the explosion at Flixborough in 1974, but ICI started to use it before the Flixborough incident occurred (see Chapters 2 and 25–27).
5. The causes of accidents, including human factors. As early as the 1960s, long before most other companies, ICI staff realized that it was superficial to blame most accidents on operators or other frontline workers and that supervisors and managers could have taken actions that would have prevented the accidents or made them less likely to occur. Such actions included better design, changes in methods of working, and not turning a blind eye to previous failures to follow instructions (see Chapter 38). In April 1969, there were 30 minor accidents in one factory. More than half were recorded by the immediate supervisors of the injured workers as caused by "human failing." The accidents were discussed individually with the supervisors, and in all but two cases they agreed that there was something they could do to prevent the accident from happening again (see Chapter 3).
6. Improving the preparation of equipment for maintenance. A disproportionate number of accidents were, and still are in many companies, due to poor systems for preparing equipment for maintenance or to failures to follow the systems. ICI improved both, particularly in those divisions that handled large quantities of hazardous materials or handled them at high temperatures and pressures (see Chapters 1 and 23).
7. Systematic attempts to remember the lessons of the past. The actions included widespread circulation of accident reports both inside and outside the company, recycling of information, regular discussion of recent and past accidents, and a computer-based index of accident reports and other safety data. These actions have still not been adopted by many other companies where accidents are investigated, reported, and then forgotten and training is unsystematic. The computer-based index lapsed in the 1980s recession. When, in the late 1990s, I asked if I could get a copy, I was told that it could not be found and, if it could be found, it could not be opened. However, the newsletters are

now available on the Institution of Chemical Engineer's web site and can be downloaded without charge. To download, go to www.icheme. org and follow the links "Safety," "Safety Newsletters," and "More Details."

I am not suggesting that ICI was perfect. We had many accidents but at least we learned from them, not just the immediate technical or human causes but the need for the general changes listed earlier. Not all parts of the company were as good as the best divisions, partly because the divisions had considerable autonomy, the head office controlling only rates of pay and major capital expenditures.

39.1   WHY DID ICI, MORE SO THAN OTHER COMPANIES, MAKE THESE CHANGES?

ICI was not dominated by committees. When I was a safety advisor there was no safety committee to tell me what I should do. Committees are a barrier to innovation. If they are considering a change, someone is likely to express reservations, and the chair will then say, "I think we should give further consideration to Dr Cynic's comments." This is the reason why so many official reports miss major recommendations. The reports on the accidents at Flixborough and Bhopal did not recommend reductions in the amounts of hazardous materials in plants and storage (see Section 21.2.1).

In contrast, ICI's policy, never written down, was to pick who the company's managers thought was the right person for a job and give that individual the freedom to achieve his or her objectives in what he or she thought was the most effective way. In every organization there are actions that employees, at each level, can do on their own authority and actions that they can't. In between there is a gray area where, if you ask for permission, it may be refused or postponed, but if you just go ahead, nothing is said. In ICI the gray area was much wider than in most companies [5].

39.2   WHAT WOULD HAVE HAPPENED IF ICI HAD NOT EXISTED?

I think the seven changes listed earlier would have come about but later in time, perhaps even decades later. I think this was because none of them was a complex one such as calculus, relativity, or quantum theory. In contrast, they all seem almost obvious in retrospect. Many people must have said, "Why did we never think of that?"

39.3   WHY DID ICI COME TO AN END?

In the 1980s, ICI Pharmaceuticals Division was about one-third of the company's capital but provided about two-thirds of the profits. It wanted to raise more capital for expansion. By demerging from ICI and renaming itself as Zeneca, it became much easier to do so, and after demerging it soon raised the extra capital. A few other products besides pharmaceuticals became part of Zeneca and were later sold. Later, Zeneca merged with the Swedish company Astra to form AstraZeneca. The Pharmaceuticals Division was different in many ways from the rest of ICI, so these changes made sense.

ICI's normal practice was to breed its own leaders, but in the 1950s an outsider, a former civil servant, was appointed as executive chairman. It was widely accepted in the company that this appointment was not a success. The man concerned is chiefly remembered for a long and unsuccessful campaign to buy Courtaulds. Perhaps this had been forgotten, as after demerging Zeneca, ICI appointed an outsider as managing director and later as chairman. He was previously head of Unilever's chemicals division.

The rest of ICI, once the Pharmaceuticals Division had gone, was mainly bulk production of commodity chemicals, and its sales and profits were very irregular, up in some years and down in others. For a long time it had been the company's policy to gradually increase its involvement in specialty chemicals, by acquisitions and natural growth, as they were less subject to such changes. When Unilever decided to sell its chemicals division, ICI's new managing director saw an opportunity to increase ICI's involvement in a big way. He and the board decided to borrow the money needed to buy his former "toy" and pay for it by selling most of ICI's bulk chemical plants.

Unfortunately, ICI found it much harder than expected to sell these plants and had to accept lower prices than it had hoped to get, thus landing the company with a large debt. The company had to sell some old ICI plants it had intended to keep and then sell some of those it had bought from Unilever. Finally, the Dutch company AkzoNobel bought the remnant that was left. By this time the number of employees had shrunk from 120,000 in the United Kingdom in the 1970s to a total of 10,000. The name Imperial Chemical Industries still exists as that of one of AkzoNobel's subsidiaries, though some of the former ICI plants are now in other parts of AkzoNobel.

There are a lot of similarities between this account and many of the accidents described in this book. The earlier result of bringing in an outsider for the top jobs had been forgotten or dismissed as no longer relevant. More seriously, no one seems to have asked the obvious questions, "What will happen if we cannot get enough money from the sale of ICI's bulk chemical plants to get us out of debt?" and "What is the probability of this happening?" It seems there was no study similar to a Hazop or a risk assessment, qualitative or quantitative. No competent engineer or scientist would make a major change to plant equipment or operation without asking these questions. Yet in the commercial field, major changes are often called bold or resolute.

39.4   WHAT WILL WE MISS IN THE YEARS TO COME?

We know what ICI did in the 80 or so years of its existence, especially in the last half of the twentieth century. We shall never know what innovations it would be making in the twenty-first century if the directors had not killed the goose that laid the golden eggs (perhaps the suggestions made in Section 38.10.2). The only consolation is that AstraZeneca has inherited some of ICI's culture and practices. If Zeneca rather than the other part of the demerged company had kept the ICI name, we might have agreed that ICI had gone from strength to strength after the demerger.

When, in 1969, I wrote my first internal ICI paper on what we now call quantitative risk assessment. I called it "risk analysis." I was then reminded that ICI had produced a book with this title (Assessing Projects: Book 5, *Risk Analysis*, Methuen, London, 1968, and that it dealt with ways of estimating the commercial risks of a project. In the second, 1970, edition the five short books were combined in a single volume.) I therefore called the new technique—new to the company and the industry—"hazard analysis," soon abbreviated to "Hazan." I expect that ICI's commercial staff and the board have long forgotten the book on commercial risk assessment, if they ever saw it.

ICI is not the first company to come to an end because a forecast turned out to be wrong. However, if the Board had copied all or perhaps just some of the practices of their engineers and scientists, the company would still be with us today. They could have:

- Looked systematically for all that might go wrong, as in a Hazop.
- Estimated the probability of these events occurring, as in a quantitative risk assessment.
- Followed a systematic procedure for the management of proposed changes.
- Looked for less risky and inherently safer ways of achieving their objectives, such as gradual change.
- Learned the lessons of the past.

I have seen no detailed obituary of ICI. All I have seen in the newspapers and technical press are a few reminiscences by former employees.

*Good decisions depend on people with imagination. flair and sound judge-
ment and the value of these qualities is greatly enhanced by a grounding in
modern methods of assessing projects.*

—From the Introduction of the ICI book mentioned on the
previous page, by Sir Peter Allen, Chairman of ICI at the
time of publication, page vi

The books listed in References contain more information on the history
of Brunner Mond and ICI.

References

1. J. M. Cohen, *The Life of Ludwig Mond*, Methuen, London, 1956.
2. J. Goodman, *The Mond Legacy*, Weidenfeld and Nicolson, London, 1982.
3. J. Harvey-Jones (executive chairman of ICI, 1982–1987), *Making It Happen: Reflections on
   Leadership*, Collins, London, 1988.
4. C. Hurworth, *Wilton—The First 50 Years*, Falcon Press, Stockton-on-Tees, 1999.
5. T. A. Kletz, *By Accident—A Life Preventing Them in Industry*, PFV Publications, London,
   2000, now available from icheme.org/shop.
6. V. E. Parke, *Billingham—The First Ten Years*, ICI Ltd., Billingham, 1957.
7. W. J. Reader, *Imperial Chemical Industries: A History*, Vol. 1, *The Forerunners, 1870–1926*,
   Vol. 2, *The First Quarter-Century 1926–1952*, Oxford University Press, London, 1975.
8. J. Roeber, *Social Change at Work: The ICI Weekly Staff Agreement*, Duckworth,
   London, 1975.

Relative Frequencies of Incidents

The following is a summary of a paper [1] that discusses the relative frequencies of many of the incidents described in this book. It is based on an analysis of 1,000 major hazards incidents in the process industries. It shows that 20% of the most frequent problems causing such incidents are involved in 70% of the total. Concentrating on these problems will therefore be a cost-effective means of minimizing such incidents. I have added references to the book when accounts of similar incidents are collected in one place, but when they are scattered, for example, those referring to drains and vents, please consult the index.

- Nearly half the incidents were maintenance related in some way (see Chapters 1 and 23), occurring during shutdowns, startups, maintenance, or abnormal operations.
- Well-conducted hazard and operability studies (see Sections 18.7 to 18.10.3) could have prevented about 40% of the incidents (though for the petroleum industry, this fell to 15% after 1990).
- About 25% of the incidents occurred in storage and blending areas (see Chapter 5). The bulk of these were in the petroleum industry. A frequent cause was the presence of flammable vapors (see Section 5.5.2).
- Liquefied petroleum gas (LPG) (see Chapter 5) and gasoline/naphtha were each involved in 12% of the incidents.
- In 24% of the cases where ignition occurred, the source was unknown; where the source was known, auto-ignition and flames each accounted for 21% of incidents.

Primary Causes

- Approximately 18% of the incidents were due to runaway reactions (see Chapters 22 and 35), caused mainly by mixing incompatible reactants, inadequate temperature control, loss of utilities, and reverse flow (see Chapter 18). These were mainly in the chemical rather than the petroleum industry.
- About 12% of the incidents were due to corrosion or erosion (see Chapters 16 and 28) with an unsuitable material of construction the most common problem.
- Abnormal high temperatures were involved in 11% of incidents. The main problems were lack of effective alarms/trips and inadequate procedures.
- Another 10% of the incidents were caused by modifications to plant or operations (see Chapters 2, 25, and 26). All such modifications must be reviewed for safety and Hazops done on the larger ones (see Sections 2.1 to 2.6).
- Roughly 9% of incidents involved flammable vapors in a confined space. Half of these were in storage tanks.
- Approximately 8% were due to uncontrolled flow through drains or vents.
- Another 8% were due to the use of the wrong material of construction (see Section 16.1), 7% to opening up equipment under pressure (Section 1.1), and 7% to the failure of safety instruments (Chapter 14).

Responsibility

- Around 40% of the incidents could have been prevented by better process design. The main problem was lack of safety features that were already in wide use elsewhere in the process industries.
- A third of the incidents could have been prevented by better operating procedures (including handwritten temporary ones) or by replacing missing ones.
- Another 20% of the incidents were attributed to operator error, including errors due to poor training but also errors due to poor labeling or layout, which can be prevented by better design (see Chapters 3 and 4).
- Despite nearly doubling the number of incidents covered and adding more recent data, the most frequent problems remained much the same as in the previous study [2]. In other words, we keep making the same old mistakes.
- To simplify making use of this data, a safety audit is proposed that covers the top 20% of problems detailed here.

- To validate this approach, a cost/benefit analysis is given, which demonstrates that in the long term, the reduction in losses due to minimizing major hazards accidents greatly outweighs the cost of the safety effort required.

Ian Duguid

References

1. I. M. Duguid, "Analysis of past accidents in the process industries," Paper 87 and hand-outs, *Hazards XX: Process Safety and Environmental Protection: Harnessing Knowledge— Challenging Complacency, Proceedings of a Conference held in Manchester on 14–17 April 2008,* Symposium Series No. 154, Institution of Chemical Engineers, Rugby, UK, 2008, p. 1070.
2. I. M. Duguid, *Loss Prevention Bulletin.* No. 142, Aug. 1998, p. 3; No. 143, Aug. 1998, p. 3; and No. 144, Dec. 1998, p. 26.

2

Why Should We Publish Accident Reports?

Some of the reports in this book have come from my own experience. Others were supplied by other people, either privately or through publications. I hope they will help you prevent similar incidents in your plant.

Almost every reader will, if not now then in the future, experience incidents from which others can learn. In return for what you have learned from this book, I hope you will publish accounts of your incidents so that others can learn from them. There are five reasons why you should do so:

1. The first reason is moral. If we have information that might prevent an accident, then we have a duty to pass on that information to those concerned.
2. The second reason is pragmatic. If we tell other people about our accidents, then in return they may tell us about theirs, and we shall be able to prevent them from happening to us. If we learn from others but do not give information in return, we are "information parasites," a term used by biologists to describe those birds, for example, that rely on other species to give warnings of approaching enemies.
3. The third reason is economic. Many companies spend more on safety measures than some of their competitors and thus pay a sort of self-imposed tax. If we tell our competitors about the action we took after an accident, they may spend as much as we have done on preventing that accident from happening again.

doi:10.1016/B978-1-85617-531-9.00044-5

4. The fourth reason is that if one company has a serious accident, the whole industry suffers in loss of public esteem, whereas new legislation may affect the whole industry. So far as the public and politicians are concerned, we are one. To misquote the well-known words of the poet John Donne:

> No plant is an Island, entire of itself; every plant is a piece of the Continent, a part of the main. Any plant's loss diminishes us, because we are involved in the Industry: and therefore never send to know for whom the inquiry sitteth; it sitteth for thee.

5. The fifth reason is that nothing else has the same impact as an accident report. If we read an article that tells us to check modifications, we agree and forget. If we read the reports in Chapter 2, we are more likely to remember.

If your employers will not let you publish an accident report under your own name, perhaps they will let you send it to a journal that will publish it anonymously—for example, the *Loss Prevention Bulletin* (see Recommended Reading)—or perhaps they will let you publish details of the action you took as a result. This may not have the same impact as the report, but it is a lot better than nothing (see Section 8.1.5).

"It's Not Like That Today"

Some of the accidents in this book occurred during recent years. Others go back several decades, a few even earlier. In every walk of life, if we describe something that happened a number of years ago, someone will say, "Schools/hospitals/offices/factories aren't like that any more." Are the old reports still relevant?

In many ways factories, at least, *are* like they used to be. This is not surprising, as human nature is a common factor. We have better equipment but may be just as likely as in the past to cut corners when we design, construct, operate, test, and maintain it, perhaps more likely as there are fewer of us to keep our eyes open as we go around the plant and to follow up on unusual observations. We have access to more knowledge than our parents and grandparents, but are we any more thorough and reliable?

We have got better at avoiding hazards instead of controlling them, as discussed in Chapter 21, but there is still a long way to go.

> The past is our present to the future.
> —Simon Thurley (Chief Executive, English Heritage).

Some Tips for Accident Investigators

DON'T SET A TARGET FOR DANGEROUS INCIDENTS. If you do people will find reasons why some should not be counted and the target will always be met.

DON'T LOOK FOR CULPRITS TO BLAME. Today everybody says they don't but after an accident many revert to old ways of thinking.

AN INDULGENT ATTITUDE TO NON-COMPLIANCE IS USUALLY A PRICE WORTH PAYING TO FIND OUT WHAT REALLY HAPPENED. Remember that many violations occur because people are trying to help; they think they have found a better way of carrying out a task.

TO FIND OUT WHAT HASN'T BEEN REPORTED, KEEP YOUR EYES AND EARS OPEN AND "LUNCH AROUND", that is, don't lunch with the same people every day. If you are asked to approve claims for damaged clothing or overtime for cleaning up spillages, ask if the incident has been investigated and reported. (See the quotation at the end of Chapter 31.)

ALWAYS VISIT THE SITE OF ACCIDENTS and look where others do not, behind and underneath equipment. Look at neighbouring areas for comparison.

PHOTOGRAPH THE SCENE for inclusion in the report and for future use in safety courses and publications. A photograph may tell us more than a thousand words.

doi:10.1016/B978-1-85617-531-9.00045-7

Recommended Reading

Descriptions of other case histories can be found in the following publications.

1. M. S. Mannan (editor), *Lees' Loss Prevention in the Process Industries,* 3rd edition, Elsevier, Boston, MA, 2005.
2. C. H. Vervalin (editor), *Fire Protection Manual for Hydrocarbon Processing Plants,* Vol. 1, 3rd edition, 1985, and Vol. 2, 1981, Gulf Publishing Co., Houston, Texas.
3. *Safety Training Packages,* Institution of Chemical Engineers, Rugby, UK. The notes are supplemented by PowerPoint slides, and some are supplemented by videos.
4. *Loss Prevention Bulletin,* Published every two months by the Institution of Chemical Engineers, Rugby, UK.
5. *Safety Digest of Lessons Learned,* Vols. 2–5, American Petroleum Institute, New York, 1979–1981.
6. *Hazard of Water, Hazard of Air, Safe Furnace Firing,* and the like, 21 booklets written by BP and published by the Institution of Chemical Engineers, Rugby, UK, 2004–2009.
7. *Case Histories,* Chemical Manufacturers Association, Washington, D.C. No new ones are being published, but bound volumes of old ones are available. They are, however, rather brief.
8. R. E. Sanders, *Chemical Process Safety—Learning from Case Histories,* Butterworth-Heinemann, Boston, MA, 1999.
9. *Operating Experience Weekly Summary,* published by the Office of Nuclear and Safety Facility, US Dept of Energy, Washington, DC. The incidents described occurred in nuclear facilities, but many contain lessons of wider interest.
10. J. Atherton and F. Gil, *Incidents That Define Process Safety,* Wiley-Interscience, Hoboken, NJ, 2008.

Reports about safety originally published by Her Majesty's Stationery Office are now supplied by HSE Books, Sudbury, United Kingdom.

doi:10.1016/B978-1-85617-531-9.00050-0

Afterthoughts

One cannot discharge ones duty by making a monumental paper structure and then not implementing it.

—A counsel during the trial following the Longford explosion
(see Section 26.2)

At every safety conference, speakers describe their safety management systems. I often wonder how well they are implemented. Descriptions of their company's accidents might tell us more.

Human language is a spectacular mechanism for transferring ideas from one mind to another, allowing us to accumulate knowledge over many generations. . .

—Daniel Hillis

It is not the strongest of the species that survive, nor the most intelligent, but the ones most responsive to change.

—Charles Darwin

I remember the first time I rode a public bus. . . . I vividly recall the sensation of seeing familiar sights from a new perspective. My seat on the bus was several feet higher than my usual position in the back seat of the family car. I could see over fences, into yards that been hidden before, over the side of the bridge to the river below. My world had expanded.

—Ann Baldwin, *Biblical Archaeology Review*, May/June 1995

We need to look over fences and see the many opportunities we have to learn from accidents.

Some years ago I went to a conference at which a newly appointed director of safety began his presentation with the assertion that "safety management is not rocket science." And he was right. Rocket science is a trivial pursuit compared to the management of safety. There are only a limited number of fuel types capable of lifting a payload into space; but the variety of ways in which harm can come to people is legion. Writing a procedure to achieve some productive aim is not easy, particularly when the task is complex, but it is always possible. In contrast, there are not enough trees in the rainforest to support all the procedures necessary to guarantee a person's safety while performing that activity.

—James Reason, *Transactions of the Institution of Chemical Engineers*, Vol. 80B, May 2002

Several years ago after reading What Went Wrong, *I realized I could use it to "wake up" my people to the dangers and horror others have experienced. All of the line supervisors and managers were given copies of the book and every month, during our regular meetings, each was to talk about something from the book that could happen here, and what we needed to do to be sure it didn't. Not only was this educational and motivational, but it also was a way to get people to discuss and share feelings of vulnerability (something not easily articulated by many of this breed).*

—Shelley Roth, Operations Manager of a chemical plant

Before Columbus made his discovery the Spanish Royal family believed the Straits of Gibraltar to be the last output of the world. Their coat of arms depicted the Pillars of Hercules, the Straits of Gibraltar, with the motto Nec Plus Ultra *(No More Beyond). After Columbus set sail the Royal family, with great economy, did not change their coat of arms. They merely erased the negative so that their motto now read* Plus Ultra *(More Beyond).*

—Danny Abse, *Goodbye, Twentieth Century*

At this point I bring my work to an end [and leave others to go beyond]. If it is found well written and aptly composed, that is what I myself hoped for; if cheap and mediocre, I could only do my best. For just as it is disagreeable to drink wine alone or water alone, so the mixing of the two gives a pleasant and delightful taste, so too variety of style in a literary work charms the ear of the reader. Let this then be my final word.

—The ending of 2 Maccabees (early 1st century BC)

Index

Printed in the United States
By Bookmasters